THE BOOK ®

Renault Clio
Service and Repair Manual

Matthew Minter

(1853-320-11X1)

Models covered

All Renault Clio models, including 16-valve and special/limited editions

1108 cc ohv, 1171 cc, 1390 cc, 1721 cc & 1794 cc sohc and 1764 dohc petrol engines

Does not cover Diesel engine
Does not fully cover RSi model (1794 sohc with multi-point fuel injection)

© Haynes Publishing 1996

ABCDE
FGHIJ
KLM

A book in the **Haynes Service and Repair Manual Series**

ISBN 1 85960 166 9

British Library Cataloguing in Publication Data
A catalogue record for this book is available from the British Library.

Printed by **J H Haynes & Co Ltd, Sparkford, Nr Yeovil, Somerset BA22 7JJ, England**

Haynes Publishing
Sparkford, Nr Yeovil, Somerset BA22 7JJ, England

Haynes North America, Inc
861 Lawrence Drive, Newbury Park, California 91320, USA

Editions Haynes S.A.
147/149, rue Saint Honoré, 75001 PARIS, France

Haynes Publishing Nordiska AB
Box 1504, 751 45 UPPSALA, Sweden

Contents

LIVING WITH YOUR RENAULT CLIO

MOT Test Checks

Roadside Repairs

Routine Maintenance

Contents

Introduction to the Renault Clio

The Renault Clio was first introduced in France in May 1990, and to the UK in March 1991, to replace the popular Renault 5 range of cars. This manual covers models fitted with petrol engines, but other models in the range are available with Diesel engines.

Six petrol engines are available in the Clio range, although not all of the engines are available in all markets. The engines available are a 1.1 litre overhead valve (ohv) type, 1.2, 1.4, 1.7, and 1.8 litre single overhead camshaft (SOHC) types, and a 1.8 litre double overhead camshaft (DOHC) type, which is fitted to the performance-orientated 16-valve model. Apart from the 1.1 litre unit, all of the engines available in the UK use fuel-injection, although carburettor engines are available in other markets. All the engines are of a well-proven design and, provided regular maintenance is carried out, are unlikely to give trouble.

The Clio is available in 3- and 5-door Hatchback and 3-door Van body styles, with a wide range of fittings and interior trim depending on the model specification.

Fully-independent front suspension is fitted, with the components attached to a subframe assembly, and the rear suspension is semi-independent, with trailing arms and torsion bars.

Four- and five-speed manual gearboxes, and three- and four-speed automatic transmissions are available, although the three-speed automatic transmission is not available in the UK. The four-speed automatic transmission is electronically-controlled.

A wide range of standard and optional equipment is available within the Clio range to suit most tastes, including an anti-lock braking system.

The Clio is conventional in design, and the DIY mechanic should find most servicing work straightforward.

Renault Clio 1.4 RT

Renault Clio 16V

Acknowledgements

Thanks are due to Champion Spark Plug, who supplied the illustrations showing spark plug conditions. Certain other illustrations are the copyright of Renault (UK) Limited, and are used with their permission. Thanks are also due to Sykes-Pickavant Limited, who provided some of the workshop tools, to John and Sally Brooks for the use of their car, and to all those people at Sparkford who helped in the production of this manual.

Technical authors who contributed to this project include Andy Legg, Steve Rendle and Mark Coombs.

Project vehicles

The main project vehicle used in the preparation of this manual, and appearing in many of the photographic sequences, was a 1992 right-hand-drive Renault Clio 1.8 RT. Additional work was carried out and photographed on a 1993 right-hand-drive Renault Clio 16V.

Working on your car can be dangerous. This page shows just some of the potential risks and hazards, with the aim of creating a safety-conscious attitude.

General hazards

Scalding

• Don't remove the radiator or expansion tank cap while the engine is hot.
• Engine oil, automatic transmission fluid or power steering fluid may also be dangerously hot if the engine has recently been running.

Burning

• Beware of burns from the exhaust system and from any part of the engine. Brake discs and drums can also be extremely hot immediately after use.

Crushing

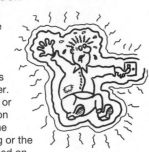

• When working under or near a raised vehicle, always supplement the jack with axle stands, or use drive-on ramps. *Never venture under a car which is only supported by a jack.*
• Take care if loosening or tightening high-torque nuts when the vehicle is on stands. Initial loosening and final tightening should be done with the wheels on the ground.

Fire

• Fuel is highly flammable; fuel vapour is explosive.
• Don't let fuel spill onto a hot engine.
• Do not smoke or allow naked lights (including pilot lights) anywhere near a vehicle being worked on. Also beware of creating sparks (electrically or by use of tools).
• Fuel vapour is heavier than air, so don't work on the fuel system with the vehicle over an inspection pit.
• Another cause of fire is an electrical overload or short-circuit. Take care when repairing or modifying the vehicle wiring.
• Keep a fire extinguisher handy, of a type suitable for use on fuel and electrical fires.

Electric shock

• Ignition HT voltage can be dangerous, especially to people with heart problems or a pacemaker. Don't work on or near the ignition system with the engine running or the ignition switched on.

• Mains voltage is also dangerous. Make sure that any mains-operated equipment is correctly earthed. Mains power points should be protected by a residual current device (RCD) circuit breaker.

Fume or gas intoxication

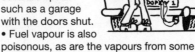

• Exhaust fumes are poisonous; they often contain carbon monoxide, which is rapidly fatal if inhaled. Never run the engine in a confined space such as a garage with the doors shut.
• Fuel vapour is also poisonous, as are the vapours from some cleaning solvents and paint thinners.

Poisonous or irritant substances

• Avoid skin contact with battery acid and with any fuel, fluid or lubricant, especially antifreeze, brake hydraulic fluid and Diesel fuel. Don't syphon them by mouth. If such a substance is swallowed or gets into the eyes, seek medical advice.
• Prolonged contact with used engine oil can cause skin cancer. Wear gloves or use a barrier cream if necessary. Change out of oil-soaked clothes and do not keep oily rags in your pocket.
• Air conditioning refrigerant forms a poisonous gas if exposed to a naked flame (including a cigarette). It can also cause skin burns on contact.

Asbestos

• Asbestos dust can cause cancer if inhaled or swallowed. Asbestos may be found in gaskets and in brake and clutch linings. When dealing with such components it is safest to assume that they contain asbestos.

Special hazards

Hydrofluoric acid

• This extremely corrosive acid is formed when certain types of synthetic rubber, found in some O-rings, oil seals, fuel hoses etc, are exposed to temperatures above $400^{\circ}C$. The rubber changes into a charred or sticky substance containing the acid. *Once formed, the acid remains dangerous for years. If it gets onto the skin, it may be necessary to amputate the limb concerned.*
• When dealing with a vehicle which has suffered a fire, or with components salvaged from such a vehicle, wear protective gloves and discard them after use.

The battery

• Batteries contain sulphuric acid, which attacks clothing, eyes and skin. Take care when topping-up or carrying the battery.
• The hydrogen gas given off by the battery is highly explosive. Never cause a spark or allow a naked light nearby. Be careful when connecting and disconnecting battery chargers or jump leads.

Air bags

• Air bags can cause injury if they go off accidentally. Take care when removing the steering wheel and/or facia. Special storage instructions may apply.

Diesel injection equipment

• Diesel injection pumps supply fuel at very high pressure. Take care when working on the fuel injectors and fuel pipes.

⚠️ *Warning: Never expose the hands, face or any other part of the body to injector spray; the fuel can penetrate the skin with potentially fatal results.*

Remember...

DO

• Do use eye protection when using power tools, and when working under the vehicle.

• Do wear gloves or use barrier cream to protect your hands when necessary.

• Do get someone to check periodically that all is well when working alone on the vehicle.

• Do keep loose clothing and long hair well out of the way of moving mechanical parts.

• Do remove rings, wristwatch etc, before working on the vehicle – especially the electrical system.

• Do ensure that any lifting or jacking equipment has a safe working load rating adequate for the job.

DON'T

• Don't attempt to lift a heavy component which may be beyond your capability – get assistance.

• Don't rush to finish a job, or take unverified short cuts.

• Don't use ill-fitting tools which may slip and cause injury.

• Don't leave tools or parts lying around where someone can trip over them. Mop up oil and fuel spills at once.

• Don't allow children or pets to play in or near a vehicle being worked on.

Dimensions

Note: *Exact dimensions depend on model.*

Overall length . . . . 3709 to 3716 mm
Overall width (excluding door mirrors) . . . . 1616 to 1641 mm
Overall height (unladen) . . . . 1365 to 1395 mm
Wheelbase . . . . 2472 mm
Front track . . . . 1358 to 1372 mm
Rear track . . . . 1324 to 1351 mm
Turning circle (between walls):
   All except 16-valve models . . . . 10.6 metres
   16-valve models . . . . 11.3 metres

Weights

Note: *Exact weights depend on model*

Kerb weight:
   1.1 litre models . . . . 810 to 820 kg
   1.2 litre models . . . . 815 to 825 kg
   1.4 litre models:
      Manual gearbox models . . . . 840 to 905 kg
      Automatic transmission models . . . . 870 to 960 kg
   1.7 litre models . . . . 905 to 955 kg
   1.8 litre models (except 16-valve) . . . . 915 to 975 kg
   16-valve models . . . . 975 to 980 kg
Maximum gross vehicle weight:
   1.1 litre models . . . . 1250 to 1260 kg
   1.2 litre models . . . . 1255 to 1275 kg
   1.4 litre models:
      Manual gearbox models . . . . 1315 to 1330 kg
      Automatic transmission models . . . . 1350 to 1375 kg
   1.7 litre models . . . . 1365 to 1375 kg
   1.8 litre models (except 16-valve) . . . . 1375 to 1385 kg
   16-valve models . . . . 1405 to 1410 kg
Maximum roof rack load (all models) . . . . 70 kg
Maximum towing weight*:
   1.1 and 1.2 litre models:
      Braked trailer . . . . 650 kg
      Unbraked trailer . . . . 405 kg
   1.4 litre models:
      Braked trailer . . . . 750 kg
      Unbraked trailer . . . . 420 kg
   1.7 litre models:
      Braked trailer . . . . 750 kg
      Unbraked trailer . . . . 450 kg
   1.8 litre models (except 16-valve): . . . .
      Braked trailer . . . . 750 kg
      Unbraked trailer . . . . 455 kg
   16-valve models:
      Braked trailer . . . . 800 kg
      Unbraked trailer . . . . 485 kg

Note: Consult a dealer for latest recommendations.

This is a guide to getting your vehicle through the MOT test. Obviously it will not be possible to examine the vehicle to the same standard as the professional MOT tester. However, working through the following checks will enable you to identify any problem areas before submitting the vehicle for the test.

Where a testable component is in borderline condition, the tester has discretion in deciding whether to pass or fail it. The basis of such discretion is whether the tester would be happy for a close relative or friend to use the vehicle with the component in that condition. If the vehicle presented is clean and evidently well cared for, the tester may be more inclined to pass a borderline component than if the vehicle is scruffy and apparently neglected.

It has only been possible to summarise the test requirements here, based on the regulations in force at the time of printing. Test standards are becoming increasingly stringent, although there are some exemptions for older vehicles. For full details obtain a copy of the Haynes publication Pass the MOT! (available from stockists of Haynes manuals).

An assistant will be needed to help carry out some of these checks.

The checks have been sub-divided into four categories, as follows:

1 Checks carried out **FROM THE DRIVER'S SEAT**

2 Checks carried out **WITH THE VEHICLE ON THE GROUND**

3 Checks carried out **WITH THE VEHICLE RAISED AND THE WHEELS FREE TO TURN**

4 Checks carried out on **YOUR VEHICLE'S EXHAUST EMISSION SYSTEM**

---

1 Checks carried out **FROM THE DRIVER'S SEAT**

Handbrake

☐ Test the operation of the handbrake. Excessive travel (too many clicks) indicates incorrect brake or cable adjustment.

☐ Check that the handbrake cannot be released by tapping the lever sideways. Check the security of the lever mountings.

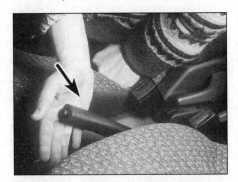

Footbrake

☐ Depress the brake pedal and check that it does not creep down to the floor, indicating a master cylinder fault. Release the pedal, wait a few seconds, then depress it again. If the pedal travels nearly to the floor before firm resistance is felt, brake adjustment or repair is necessary. If the pedal feels spongy, there is air in the hydraulic system which must be removed by bleeding.

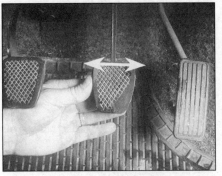

☐ Check that the brake pedal is secure and in good condition. Check also for signs of fluid leaks on the pedal, floor or carpets, which would indicate failed seals in the brake master cylinder.

☐ Check the servo unit (when applicable) by operating the brake pedal several times, then keeping the pedal depressed and starting the engine. As the engine starts, the pedal will move down slightly. If not, the vacuum hose or the servo itself may be faulty.

Steering wheel and column

☐ Examine the steering wheel for fractures or looseness of the hub, spokes or rim.

☐ Move the steering wheel from side to side and then up and down. Check that the steering wheel is not loose on the column, indicating wear or a loose retaining nut. Continue moving the steering wheel as before, but also turn it slightly from left to right.

☐ Check that the steering wheel is not loose on the column, and that there is no abnormal

movement of the steering wheel, indicating wear in the column support bearings or couplings.

Windscreen and mirrors

☐ The windscreen must be free of cracks or other significant damage within the driver's field of view. (Small stone chips are acceptable.) Rear view mirrors must be secure, intact, and capable of being adjusted.

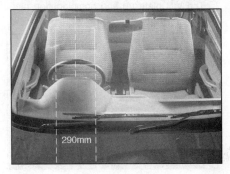

290mm

Seat belts and seats

Note: *The following checks are applicable to all seat belts, front and rear.*

☐ Examine the webbing of all the belts (including rear belts if fitted) for cuts, serious fraying or deterioration. Fasten and unfasten each belt to check the buckles. If applicable, check the retracting mechanism. Check the security of all seat belt mountings accessible from inside the vehicle.

☐ The front seats themselves must be securely attached and the backrests must lock in the upright position.

Doors

☐ Both front doors must be able to be opened and closed from outside and inside, and must latch securely when closed.

2 Checks carried out WITH THE VEHICLE ON THE GROUND

Vehicle identification

☐ Number plates must be in good condition, secure and legible, with letters and numbers correctly spaced – spacing at (A) should be twice that at (B).

☐ The VIN plate (A) and homologation plate (B) must be legible.

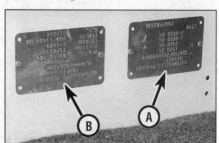

Electrical equipment

☐ Switch on the ignition and check the operation of the horn.

☐ Check the windscreen washers and wipers, examining the wiper blades; renew damaged or perished blades. Also check the operation of the stop-lights.

☐ Check the operation of the sidelights and number plate lights. The lenses and reflectors must be secure, clean and undamaged.

☐ Check the operation and alignment of the headlights. The headlight reflectors must not be tarnished and the lenses must be undamaged.

☐ Switch on the ignition and check the operation of the direction indicators (including the instrument panel tell-tale) and the hazard warning lights. Operation of the sidelights and stop-lights must not affect the indicators - if it does, the cause is usually a bad earth at the rear light cluster.

☐ Check the operation of the rear foglight(s), including the warning light on the instrument panel or in the switch.

Footbrake

☐ Examine the master cylinder, brake pipes and servo unit for leaks, loose mountings, corrosion or other damage.

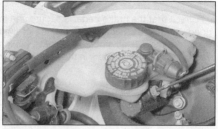

☐ The fluid reservoir must be secure and the fluid level must be between the upper (**A**) and lower (**B**) markings.

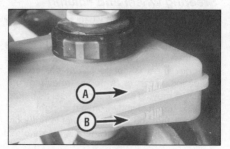

☐ Inspect both front brake flexible hoses for cracks or deterioration of the rubber. Turn the steering from lock to lock, and ensure that the hoses do not contact the wheel, tyre, or any part of the steering or suspension mechanism. With the brake pedal firmly depressed, check the hoses for bulges or leaks under pressure.

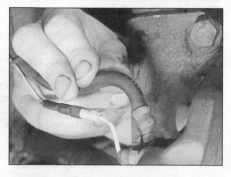

Steering and suspension

☐ Have your assistant turn the steering wheel from side to side slightly, up to the point where the steering gear just begins to transmit this movement to the roadwheels. Check for excessive free play between the steering wheel and the steering gear, indicating wear or insecurity of the steering column joints, the column-to-steering gear coupling, or the steering gear itself.

☐ Have your assistant turn the steering wheel more vigorously in each direction, so that the roadwheels just begin to turn. As this is done, examine all the steering joints, linkages, fittings and attachments. Renew any component that shows signs of wear or damage. On vehicles with power steering, check the security and condition of the steering pump, drivebelt and hoses.

☐ Check that the vehicle is standing level, and at approximately the correct ride height.

Shock absorbers

☐ Depress each corner of the vehicle in turn, then release it. The vehicle should rise and then settle in its normal position. If the vehicle continues to rise and fall, the shock absorber is defective. A shock absorber which has seized will also cause the vehicle to fail.

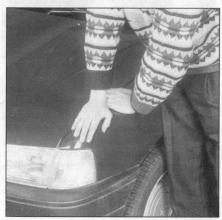

Exhaust system

☐ Start the engine. With your assistant holding a rag over the tailpipe, check the entire system for leaks. Repair or renew leaking sections.

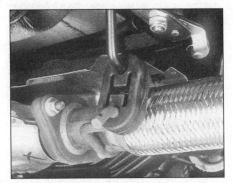

3 Checks carried out **WITH THE VEHICLE RAISED AND THE WHEELS FREE TO TURN**

Jack up the front and rear of the vehicle, and securely support it on axle stands. Position the stands clear of the suspension assemblies. Ensure that the wheels are clear of the ground and that the steering can be turned from lock to lock.

Steering mechanism

☐ Have your assistant turn the steering from lock to lock. Check that the steering turns smoothly, and that no part of the steering mechanism, including a wheel or tyre, fouls any brake hose or pipe or any part of the body structure.

☐ Examine the steering rack rubber gaiters for damage or insecurity of the retaining clips. If power steering is fitted, check for signs of damage or leakage of the fluid hoses, pipes or connections. Also check for excessive stiffness or binding of the steering, a missing split pin or locking device, or severe corrosion of the body structure within 30 cm of any steering component attachment point.

Front and rear suspension and wheel bearings

☐ Starting at the front right-hand side, grasp the roadwheel at the 3 o'clock and 9 o'clock positions and shake it vigorously. Check for free play or insecurity at the wheel bearings, suspension balljoints, or suspension mountings, pivots and attachments.

☐ Now grasp the wheel at the 12 o'clock and 6 o'clock positions and repeat the previous inspection. Spin the wheel, and check for roughness or tightness of the front wheel bearing.

☐ If excess free play is suspected at a component pivot point, this can be confirmed by using a large screwdriver or similar tool and levering between the mounting and the component attachment. This will confirm whether the wear is in the pivot bush, its retaining bolt, or in the mounting itself (the bolt holes can often become elongated).

☐ Carry out all the above checks at the other front wheel, and then at both rear wheels.

Springs and shock absorbers

☐ Examine the suspension struts (when applicable) for serious fluid leakage, corrosion, or damage to the casing. Also check the security of the mounting points.

☐ If coil springs are fitted, check that the spring ends locate in their seats, and that the spring is not corroded, cracked or broken.

☐ If leaf springs are fitted, check that all leaves are intact, that the axle is securely attached to each spring, and that there is no deterioration of the spring eye mountings, bushes, and shackles.

☐ The same general checks apply to vehicles fitted with other suspension types, such as torsion bars, hydraulic displacer units, etc. Ensure that all mountings and attachments are secure, that there are no signs of excessive wear, corrosion or damage, and (on hydraulic types) that there are no fluid leaks or damaged pipes.

☐ Inspect the shock absorbers for signs of serious fluid leakage. Check for wear of the mounting bushes or attachments, or damage to the body of the unit.

Driveshafts (fwd vehicles only)

☐ Rotate each front wheel in turn and inspect the constant velocity joint gaiters for splits or damage. Also check that each driveshaft is straight and undamaged.

Braking system

☐ If possible without dismantling, check brake pad wear and disc condition. Ensure that the friction lining material has not worn excessively, (A) and that the discs are not fractured, pitted, scored or badly worn (B).

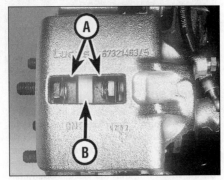

☐ Examine all the rigid brake pipes underneath the vehicle, and the flexible hose(s) at the rear. Look for corrosion, chafing or insecurity of the pipes, and for signs of bulging under pressure, chafing, splits or deterioration of the flexible hoses.

☐ Look for signs of fluid leaks at the brake calipers or on the brake backplates. Repair or renew leaking components.

☐ Slowly spin each wheel, while your assistant depresses and releases the footbrake. Ensure that each brake is operating and does not bind when the pedal is released.

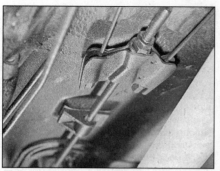

☐ Examine the handbrake mechanism, checking for frayed or broken cables, excessive corrosion, or wear or insecurity of the linkage. Check that the mechanism works on each relevant wheel, and releases fully, without binding.

☐ It is not possible to test brake efficiency without special equipment, but a road test can be carried out later to check that the vehicle pulls up in a straight line.

Fuel and exhaust systems

☐ Inspect the fuel tank (including the filler cap), fuel pipes, hoses and unions. All components must be secure and free from leaks.

☐ Examine the exhaust system over its entire length, checking for any damaged, broken or missing mountings, security of the retaining clamps and rust or corrosion.

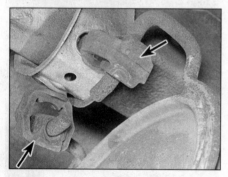

Wheels and tyres

☐ Examine the sidewalls and tread area of each tyre in turn. Check for cuts, tears, lumps, bulges, separation of the tread, and exposure of the ply or cord due to wear or damage. Check that the tyre bead is correctly seated on the wheel rim, that the valve is sound and

properly seated, and that the wheel is not distorted or damaged.

☐ Check that the tyres are of the correct size for the vehicle, that they are of the same size and type on each axle, and that the pressures are correct.

☐ Check the tyre tread depth. The legal minimum at the time of writing is 1.6 mm over at least three-quarters of the tread width. Abnormal tread wear may indicate incorrect front wheel alignment.

Body corrosion

☐ Check the condition of the entire vehicle structure for signs of corrosion in load-bearing areas. (These include chassis box sections, side sills, cross-members, pillars, and all suspension, steering, braking system and seat belt mountings and anchorages.) Any corrosion which has seriously reduced the thickness of a load-bearing area is likely to cause the vehicle to fail. In this case professional repairs are likely to be needed.

☐ Damage or corrosion which causes sharp or otherwise dangerous edges to be exposed will also cause the vehicle to fail.

4 Checks carried out on YOUR VEHICLE'S EXHAUST EMISSION SYSTEM

Petrol models

☐ Have the engine at normal operating temperature, and make sure that it is in good tune (ignition system in good order, air filter element clean, etc).

☐ Before any measurements are carried out, raise the engine speed to around 2500 rpm, and hold it at this speed for 20 seconds. Allow the engine speed to return to idle, and watch

for smoke emissions from the exhaust tailpipe. If the idle speed is obviously much too high, or if dense blue or clearly-visible black smoke comes from the tailpipe for more than 5 seconds, the vehicle will fail. As a rule of thumb, blue smoke signifies oil being burnt (engine wear) while black smoke signifies unburnt fuel (dirty air cleaner element, or other carburettor or fuel system fault).

☐ An exhaust gas analyser capable of measuring carbon monoxide (CO) and hydrocarbons (HC) is now needed. If such an instrument cannot be hired or borrowed, a local garage may agree to perform the check for a small fee.

CO emissions (mixture)

☐ At the time or writing, the maximum CO level at idle is 3.5% for vehicles first used after August 1986 and 4.5% for older vehicles. From January 1996 a much tighter limit (around 0.5%) applies to catalyst-equipped vehicles first used from August 1992. If the CO level cannot be reduced far enough to pass the test (and the fuel and ignition systems are otherwise in good condition) then the carburettor is badly worn, or there is some problem in the fuel injection system or catalytic converter (as applicable).

HC emissions

☐ With the CO emissions within limits, HC emissions must be no more than 1200 ppm (parts per million). If the vehicle fails this test at idle, it can be re-tested at around 2000 rpm; if the HC level is then 1200 ppm or less, this counts as a pass.

☐ Excessive HC emissions can be caused by oil being burnt, but they are more likely to be due to unburnt fuel.

Diesel models

☐ The only emission test applicable to Diesel engines is the measuring of exhaust smoke density. The test involves accelerating the engine several times to its maximum unloaded speed.

Note: *It is of the utmost importance that the engine timing belt is in good condition before the test is carried out.*

☐ Excessive smoke can be caused by a dirty air cleaner element. Otherwise, professional advice may be needed to find the cause.

Jacking, towing and wheel changing

Jacking

The jack supplied with the car should only be used for changing the roadwheels - see 'Wheel changing' later in this Section. When carrying out any other kind of work, raise the car using a hydraulic jack, and always supplement the jack with axle stands positioned under the car jacking points.

When using a hydraulic jack or axle stands, always position the jack head or axle stand head under one of the relevant jacking points (note that the jacking points for use with a hydraulic jack are different to those for use with the car jack). **Do not** jack the car under the sump or any of the steering or suspension components. The jacking points and axle stand positions are shown in the accompanying illustrations.

When raising the front of the car, apply the handbrake and chock the rear wheels. Use a suitable wooden or metal bar positioned under the front subframe as shown. Position the jack head under the centre of the bar. Note that on certain models it will be necessary to cut a notch in the bar to prevent contact with the exhaust system.

When raising the rear of the car, chock the front wheels and engage 1st gear on manual gearbox models, or select position 'P' on models with automatic transmission. Position the jack head under one of the reinforced rear jacking points provided for use with the car jack.

When raising the side of the car, chock the wheels remaining on the ground, apply the handbrake and select a gear, as described previously.

Position a suitable bar under the sill beneath the front door. The bar should have a suitable groove to accommodate the flange on the sill. Ensure that the sill flange is securely engaged with the groove in the bar, then position the jack head under the centre of the bar.

Axle stands should be positioned under the reinforced jacking points provided for use with the car jack.

⚠️ **Warning: Never work under, around, or near a raised car, unless it is adequately supported in at least two places.**

Towing

Towing eyes are fitted to the front and rear of the car for attachment of a tow rope. Always turn the ignition key to position 'M' ('3') when the car is being towed, so that the steering lock is released and the direction indicators and brake lights are operational.

Before being towed, release the handbrake and place the gear lever in neutral on manual gearbox models, or 'N' on automatic transmission models. Note that greater-than-usual pedal pressure will be required to operate the brakes, since the vacuum servo unit is only operational with the engine running. Similarly, on models with power steering, greater-than-usual steering effort will be required.

Note that it is strongly recommended that cars with automatic transmission are transported on a trailer, or towed with the front wheels off the ground. If it is necessary to tow a car with automatic transmission with the front wheels on the ground, the following recommendations must be observed.

Models with three-speed automatic transmission

Towing may be carried out with all four wheels on the ground only if the following conditions are complied with.

(a) Add an extra 2 litres of the specified transmission fluid to the transmission (see 'Recommended lubricants and fluids').

(b) Do not exceed a speed of 18 mph (30 km/h), or a distance of 30 miles (50 km), when towing.

(c) Drain the surplus fluid from the transmission on completion of towing.

Models with four-speed automatic transmission

Towing may be carried out with all four wheels on the ground, or with the rear wheels

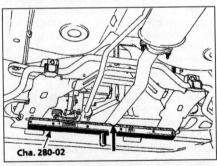

Cha. 280-02

Renault special tool positioned under front subframe to raise front of car with trolley jack. Note notch (arrowed, right) to clear exhaust pipe

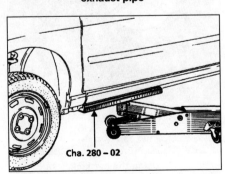

Cha. 280 – 02

Renault special tool positioned under sill to raise side of car with trolley jack

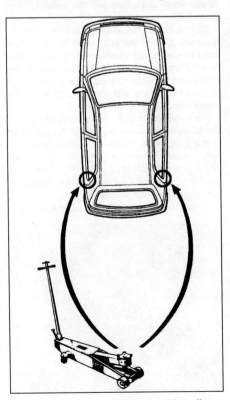

Rear jacking points for use with trolley jack

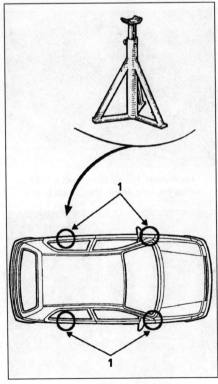

Axle stand locations (1)

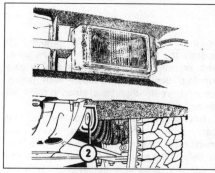

Front towing eye (2) - models with standard front spoiler

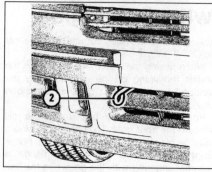

Front towing eye (2) - models with deep front spoiler

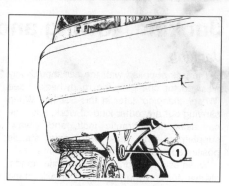

Rear towing eye (1)

raised a maximum of 15 cm off the ground, provided that the speed does not exceed 25 mph (40 km/h), and the distance does not exceed 15 miles (25 km), when towing.

Wheel changing

To change a wheel, first remove the wheel brace, spare wheel and jack.

The wheel brace is located in a bracket on the left-hand side of the luggage compartment. On models with locking wheel covers, the key for the covers is attached to the wheel brace.

The spare wheel is located in a cradle under the rear of the car. To remove the spare wheel, unclip the wheel brace from its location on the left-hand side of the luggage compartment, and use the end of the wheel brace to loosen the spare wheel cradle securing bolt, located

Loosening the spare wheel cradle securing bolt using the wheel brace

Removing the car jack

in the luggage compartment floor (photo). Support the spare wheel cradle, then pull back the safety catch under the rear bumper, and lower the cradle to the ground. Lift the spare wheel from the cradle.

The jack is located under the right-hand access hatch in the windscreen cowl panel. To remove the jack, open the bonnet and lift the access hatch. Loosen the securing nut, then release the clamp and lift the jack from its location (photo).

Apply the handbrake, and place chocks at the front and rear of the wheel diagonally opposite the one to be changed. On manual gearbox models, select first or reverse gear, and on automatic transmission models, place the selector lever in position 'P'. Make sure that the vehicle is located on firm level ground. Where applicable, remove the wheel covers (photo).

Slightly loosen the wheel bolts with the brace provided. Locate the jack head in the jacking point nearest to the wheel to be changed. Note that the lug on the jack head must engage with the cut-out in the jacking point. Engage the end of the wheel brace with the jack, and turn the wheel brace to raise the car.

For additional safety, it is worthwhile sliding the spare wheel under the side of the car, close to the jack, while the car is raised. When the spare wheel is required, put the punctured wheel under the car in its place. This will reduce the risk of personal injury (and of damage to the car), should the car slip off the jack.

When the wheel is clear of the ground,

Using the key to remove a locking type wheel cover

remove the bolts and lift off the wheel. Check that the threads and the wheel-to-hub mating surfaces are clean and undamaged, and that the inside of the spare wheel is clean. The threads may be cleaned with a brass wire brush if rusty. Apply a thin smear of copper-based anti-seize compound to the threads and roadwheel-to-hub mating surfaces, to prevent the formation of corrosion. *Obviously, you are not likely to be able to do anything about this at the roadside, but it is worthwhile attending to it as soon as possible once the car is running again, or changing a wheel may be even more difficult next time!*

Fit the spare wheel, and tighten the bolts moderately by hand. If you put the punctured wheel under the car, remove it now and lower the car to the ground. Tighten the bolts fully using the wheel brace, working in progressive stages and in a diagonal sequence - if a torque wrench is available, tighten the bolts to the specified setting (see Chapter 1 Specifications). Refit the wheel cover where applicable.

Remove the chocks, then stow the jack and punctured wheel. To stow the wheel, place it in the cradle, then lift the cradle sharply upwards until the safety catch engages, and tighten the cradle securing bolt with the wheel brace. Finally, stow the wheel brace in the left-hand side of the luggage compartment.

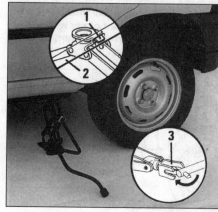

Using the car jack

1 Lug on jack head   2 Jacking point
3 Fork in jack for wheel brace

Booster battery (jump) starting

When jump-starting a car using a booster battery, observe the following precautions:

A) Before connecting the booster battery, make sure that the ignition is switched off.

B) Ensure that all electrical equipment (lights, heater, wipers, etc) is switched off.

C) Make sure that the booster battery is the same voltage as the discharged one in the vehicle.

D) If the battery is being jump-started from the battery in another vehicle, the two vehcles MUST NOT TOUCH each other.

E) Make sure that the transmission is in neutral (or PARK, in the case of automatic transmission).

HAYNES HiNT *Jump starting will get you out of trouble, but you must correct whatever made the battery go flat in the first place. There are three possibilities:*

1 *The battery has been drained by repeated attempts to start, or by leaving the lights on.*

2 *The charging system is not working properly (alternator drivebelt slack or broken, alternator wiring fault or alternator itself faulty).*

3 *The battery itself is at fault (electrolyte low, or battery worn out).*

1 Connect one end of the red jump lead to the positive (+) terminal of the flat battery

2 Connect the other end of the red lead to the positive (+) terminal of the booster battery.

3 Connect one end of the black jump lead to the negative (–) terminal of the booster battery

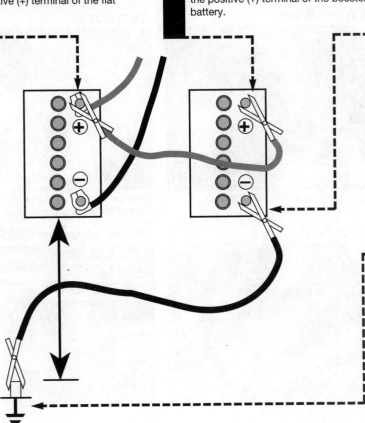

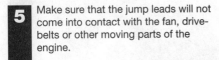

4 Connect the other end of the black jump lead to a bolt or bracket on the engine block, well away from the battery, on the vehicle to be started.

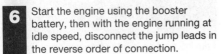

5 Make sure that the jump leads will not come into contact with the fan, drive-belts or other moving parts of the engine.

6 Start the engine using the booster battery, then with the engine running at idle speed, disconnect the jump leads in the reverse order of connection.

Identifying leaks

Puddles on the garage floor or drive, or obvious wetness under the bonnet or underneath the car, suggest a leak that needs investigating. It can sometimes be difficult to decide where the leak is coming from, especially if the engine bay is very dirty already. Leaking oil or fluid can also be blown rearwards by the passage of air under the car, giving a false impression of where the problem lies.

> ⚠ *Warning: Most automotive oils and fluids are poisonous. Wash them off skin, and change out of contaminated clothing, without delay.*

> **HAYNES HiNT** *The smell of a fluid leaking from the car may provide a clue to what's leaking. Some fluids are distictively coloured.*
> *It may help to clean the car carefully and to park it over some clean paper overnight as an aid to locating the source of the leak.*
> *Remember that some leaks may only occur while the engine is running.*

Sump oil

Engine oil may leak from the drain plug...

Oil from filter

...or from the base of the oil filter.

Gearbox oil

Gearbox oil can leak from the seals at the inboard ends of the driveshafts.

Antifreeze

Leaking antifreeze often leaves a crystalline deposit like this.

Brake fluid

A leak occurring at a wheel is almost certainly brake fluid.

Power steering fluid

Power steering fluid may leak from the pipe connectors on the steering rack.

Chapter 1 Routine maintenance and servicing

Contents

Degrees of difficulty

Easy, suitable for novice with little experience	**Fairly easy,** suitable for beginner with some experience	**Fairly difficult,** suitable for competent DIY mechanic	**Difficult,** suitable for experienced DIY mechanic	**Very difficult,** suitable for expert DIY or professional

Specifications

Engine

Oil filter type:

1108 cc (C1E)	Champion type not available
1171 cc (E5F and E7F)	Champion F101
1390 cc (E6J and E7J)	Champion F102
1721 cc (F2N)	Champion F101
1794 cc (F3P)	Champion type not available
1764 cc (F7P)	Champion type not available

Cooling system

Antifreeze mixtures:

	Antifreeze	Water
Protection to -23°C	35%	65%
Protection to -40°C	50%	50%

Fuel system

Air cleaner element type:

1108 cc (C1E) engine	Champion type not available
1171 cc (E5F) engine	Champion W190
1171 cc (E7F) engine	Champion W190
1390 cc (E6J) engine	Champion W190
1390 cc (E7J) engine	Champion W190
1721 cc (F2N) engine	Champion type not available
1794 cc (F3P) engine	Champion type not available
1764 cc (F7P) engine	Champion type not available

Fuel filter element type:

Carburettor engines	Champion L101
Fuel injection engines	Champion L206

Specifications

Fuel system (continued)

Idle speed - carburettor models:
1108 cc (C1E) engine	700 ± 50 rpm

1171 cc (E5F) engine:
With cold enrichment device lead earthed (see text)	870 ± 50 rpm
With cold enrichment device disconnected	800 ± 50 rpm

1390 cc (E6J) engine:
Manual gearbox	800 ± 50 rpm
Automatic transmission (in 'D')	725 ± 25 rpm
1721 cc (F2N) engine	800 ± 50 rpm
Idle speed - fuel injection models	Not adjustable
Idle mixture (CO content) - carburettor models	1.5 ± 0.5 %

Idle mixture (CO content) - fuel injection models:
Models with catalytic converter	0.5 % maximum (not adjustable)
1764 cc engine code F7P-720 (vehicle code C575)	1.5 ± 0.5 %

Ignition system - general

Firing order	1-3-4-2
Location of No 1 cylinder	Flywheel end

Transistor-assisted contact breaker ignition system - 1108 cc (C1E) engine

Contact breaker points gap	0.4 mm
Dwell angle (percentage)	57° ± 3° (63 ± 3%)

Ignition timing*:
Static	10° ± 2° BTDC
At idle (vacuum hose disconnected)	10° ± 2° BTDC

*Observe value stamped on clip fastened to HT lead if this is different

Spark plugs:
Type	Champion N12YCC
Electrode gap	0.80 ± 0.05 mm

Electronic ignition system

Ignition timing	Not adjustable (computer-controlled)

Spark plugs:

Type:
1171 cc (E5F/E7F) and 1390 cc (E6J/E7J) engines	Champion RC9YCC or C9YCX
1721 cc (F2N) engine	Champion RN7YCC or N279YC
1794 cc (F3P) engine	Champion N7YCC
1764 cc (F7P) engine	Champion C7BMC
Electrode gap	0.80 ± 0.05 mm

Braking system

Friction lining minimum thickness:
Front brake pad (including backing plate)	6.0 mm
Rear brake pad (including backing plate)	5.0 mm
Rear brake shoe (including shoe)	2.5 mm

Suspension and steering

Tyre pressures (cold):

	Front*	Rear
145/70 R 13S tyres	2.4 bars (35 lbf/in²)	2.4 bars (35 lbf/in²)
155/70 R 13S and 165/65 R 13T tyres	2.1 bars (30 lbf/in²)	2.3 bars (33 lbf/in²)
145/80 R 13S and 155/80 R 13T tyres	2.1 bars (30 lbf/in²)	2.1 bars (30 lbf/in²)
165/60 R 14H tyres	2.3 bars (33 lbf/in²)	2.3 bars (33 lbf/in²)
185/60 R 14V and 185/55 R 15V tyres	2.2 bars (32 lbf/in²)	2.2 bars (32 lbf/in²)

*Automatic transmission models - add 0.1 bar (1.5 lbf/in²)

Electrical system

Wiper blade type:
Windscreen	Champion X4503
Tailgate	Champion type not available

Torque wrench settings

	Nm	lbf ft
Engine sump drain plug	15 to 25	11 to 18
Spark plugs	24 to 30	18 to 22
Automatic transmission gauze filter	5	4

Automatic transmission sump pan:
	Nm	lbf ft
MB1 transmission	6	4
AD4 transmission	10	7
Roadwheel bolts	90	66

Introduction

This Chapter is designed to help the DIY owner maintain the Clio with the goals of maximum economy, safety, reliability and performance in mind.

On the following pages is a master maintenance schedule, listing the servicing requirements, and the intervals at which they should be carried out, as recommended by the manufacturers. Alongside each operation in the schedule is a reference which directs the user to the Sections in this Chapter covering maintenance procedures or to other Chapters in the manual, where the operations are described and illustrated in greater detail. Specifications for all the maintenance operations are provided at the start of this Chapter. Refer to the accompanying photographs of the engine compartment and the underbody of the vehicle for the locations of the various components.

The Sections in this Chapter covering maintenance procedures contain some additional general checks which are not specified in the maintenance schedule by the manufacturers. These additional checks are recommended in order to detect possible sources of trouble which may develop into more significant problems at a later date. It is up to the individual owner to decide whether or not to carry out these checks.

Servicing your vehicle in accordance with the maintenance schedule and step-by-step procedures will result in a planned maintenance programme that should produce a long and reliable service life. Bear in mind that it is a comprehensive plan, so maintaining some items but not others at the specified intervals will not produce the same results.

No time intervals are specified by the manufacturer, but it is suggested that on vehicles covering less than 12 000 miles (20 000 km) per year, the 6000-mile service be carried out every 6 months, the 12 000-mile service every 12 months, and so on.

Vehicles operating under adverse conditions (extremes of climate, full-time towing, mainly short journeys, etc) may need to be serviced at more frequent intervals. If in doubt, consult a Renault dealer.

The first step in this maintenance programme is to prepare yourself before the actual work begins. Read through all the procedures to be undertaken, then obtain all the parts, lubricants and any additional tools needed.

Every 250 miles (400 km), weekly, or before along journey

- [ ] Check the engine oil level (Section 1)
- [ ] Check the engine coolant level (Section 1)
- [ ] Check the brake fluid level (Section 1)
- [ ] Check the washer fluid level (Section 1)
- [ ] Check the tyres for wear and damage (Section 2)
- [ ] Check the tyre pressures (Section 2)
- [ ] Check the operation of all lights, wipers and washers, instruments, and electrical equipment (Section 3)

Every 6000 miles (10 000 km)

- [ ] Renew the engine oil (Section 4)
- [ ] Renew the oil filter* (Section 4)
- [ ] Check for oil/coolant/fuel leaks (Section 5)
- [ ] Check the operation of the heating and air conditioning systems (Section 6)
- [ ] Check the condition and tension of the auxiliary drivebelts (Section 7)
- [ ] Check the condition and security of the exhaust system (Section 8)
- [ ] Check the engine idle speed and mixture settings (where possible) (Section 9)
- [ ] Check the condition of the spark plugs (Section 10)
- [ ] Check the condition and adjustment of the contact breaker points (where applicable) (Section 11)
- [ ] Check the ignition timing (where possible) (Section 12)
- [ ] Check the manual gearbox oil level or the automatic transmission fluid level (as applicable) (Section 13)
- [ ] Check the condition of the front brake pads (Section 14)
- [ ] Check the braking system for fluid leaks (Section 15)
- [ ] Check the power steering fluid level (where applicable) (Section 16)
- [ ] Check the bodywork and underbody for damage and corrosion (Section 17)
- [ ] Check the headlight beam alignment (Section 18)
- [ ] Check the condition of the battery (Section 19)
- [ ] Perform a road test (Section 20)

*Renew the oil filter after the first 6000 miles (10 000 km), at 12 000 miles (20 000 km), then every 12 000 miles (20 000 km).

Every 12 000 miles (20 000 km)

In addition to all the items listed above, carry out the following:

- [ ] Renew the oil filter (Section 4)
- [ ] Renew the air filter element (Section 21)
- [ ] Renew the spark plugs (Section 10)
- [ ] Check the HT leads, distributor cap and rotor arm (Section 22)

Every 30 000 miles (50 000 km)

In addition to all the items listed above, carry out the following:

- [ ] Check air cleaner temperature control system (Section 23)
- [ ] Renew the fuel filter (Section 24)
- [ ] Check the operation of the clutch/cable - where applicable (Section 25)
- [ ] Renew automatic transmission fluid and filter (Section 26)
- [ ] Check driveshaft rubber gaiters and CV joints (Section 27)
- [ ] Check the condition of the rear brake shoes or pads (as applicable) (Section 28)
- [ ] Check the handbrake adjustment (Section 29)
- [ ] Renew the brake fluid (Section 30)
- [ ] Check the condition of the front and rear suspension and steering components (Section 31)
- [ ] Check the front wheel alignment (Chapter 10)
- [ ] Check the operation and condition of the seat belts (Section 32)
- [ ] Lubricate the hinges and locks (Section 32)

Every 72 000 miles (120 000 km)

- [ ] Renew the engine timing belt (except 1108 cc/C1E engine) (Chapter 2)

⚠️ **Caution: If this operation is neglected, and the timing belt breaks in service, extensive engine damage may result. The author recommends that consideration be given to renewing the timing belt at around 50 000 miles (80 000 km), in order to err on the side of safety.**

Every 2 years, regardless of mileage

- [ ] Renew the coolant (Section 33)

Maintenance Schedule

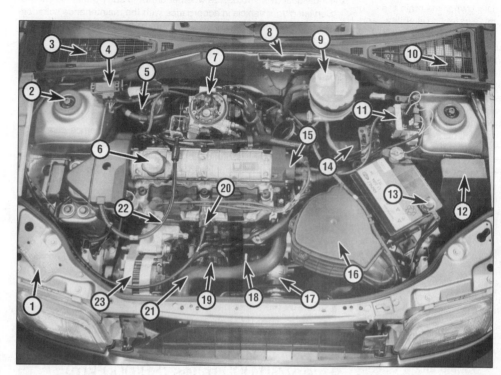

Underbonnet view of 1.8 RT (1794 cc - F3P engine) model

1 VIN plate
2 Suspension strut upper mounting
3 Jack location
4 MAP sensor
5 Brake fluid reservoir
6 Engine oil filler cap
7 Throttle body unit
8 Bonnet lock assembly
9 Coolant expansion tank
10 Washer fluid reservoir location
11 Ignition module
12 Relay box
13 Battery negative terminal
14 ABS modulator assembly
15 Distributor cap
16 Air cleaner assembly
17 Cooling fan motor
18 Engine oil level dipstick
19 Oil filter
20 Knock sensor
21 Radiator top hose
22 Accelerator cable
23 Alternator

Front underbody view of 1.8 RT (1794 cc - F3P engine) model

1 Wheel arch splash shield
2 Sump drain plug
3 Brake caliper
4 Lower suspension arm
5 Track rod balljoint
6 Anti-roll bar
7 Front subframe
8 Gearchange selector rod
9 Catalytic converter
10 Fuel pipes
11 Exhaust downpipe
12 Lower suspension arm balljoint
13 Driveshaft inner constant velocity joint
14 Engine rear mounting link
15 Gearbox cover
16 Engine/gearbox tie-bar

Underbonnet view of 16V (1764 cc - F7P engine) model

1 VIN plate
2 Power steering fluid reservoir
3 Suspension strut upper mounting
4 Jack location
5 MAP sensor
6 Brake fluid reservoir
7 Exhaust heat shield
8 Bonnet lock assembly
9 Coolant expansion tank
10 Washer fluid reservoir location
11 Ignition module
12 Relay box
13 Battery negative terminal
14 Accelerator cable
15 Distributor cap
16 Air cleaner assembly
17 Engine oil filler cap
18 Throttle body unit
19 Idle speed control valve
20 Engine oil level dipstick
21 Engine top cover
22 Vacuum hoses

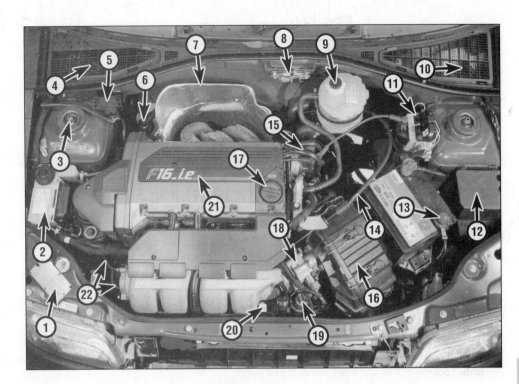

Rear underbody view of 1.8 RT model

1 Fuel filter (fuel injection type)
2 Heat shield
3 Fuel tank
4 Handbrake cables
5 Exhaust pipe
6 Torsion bar
7 Trailing arm
8 Exhaust rear silencer
9 Brake pressure-regulating load-sensitive valve
10 Rear crossmember
11 Anti-roll bar
12 Fuel filler pipe
13 Brake caliper
14 Shock absorber
15 ABS wheel sensor wiring connectors

1

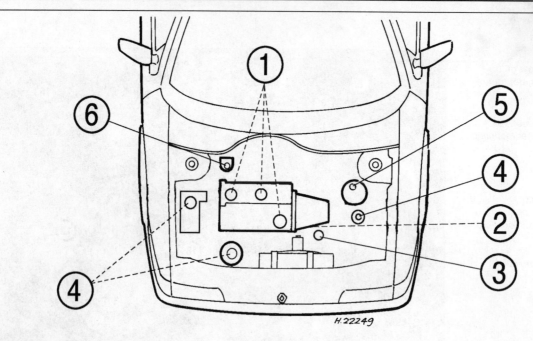

H.22249

Lubricants and fluids

Component or system	Lubricant type/specification
1 Engine Note: *Location of filler cap varies depending on model*	Multigrade engine oil, viscosity 15W/40 to 20W/50, to CCMC G2 or better
2 Manual gearbox	Tranself TRX 80W
3 Automatic transmission (not AD4 final drive)	Elf Renaultmatic D2, Dexron type ATF
3 AD4 automatic transmission final drive	Tranself TRX 80W 140
4 Power steering reservoir Note: *Location of reservoir varies depending on model*	Elf Renaultmatic D2 or Mobil ATF 220
5 Cooling system	Ethylene-glycol based antifreeze (AL Glaceol type C coolant)
6 Brake fluid reservoir	Hydraulic fluid to SAE J1703, DOT 3 or DOT 4

Capacities

Engine oil
Capacity (excluding filter):
1108 cc (C1E) engine	3.0 litres
1171 cc (E5F and E7F) engines	3.5 litres
1390 cc (E6J and E7J) engines	3.5 litres
1721 cc (F2N) and 1794 cc (F3P) engines	4.7 litres
1764 cc (F7P) engine	3.5 litres

Oil filter capacity:
All engines except 1764 cc (F7P)	0.5 litre
1764 cc (F7P) engine	0.2 litre

Difference between MAX and MIN dipstick marks:
1108 cc (C1E) engine	1.0 litre
1171 cc (E5F/E7F) and 1390 cc (E6J/E7J) engines	1.5 litres
1721 cc (F2N) and 1794 cc (F3P) engines	2.0 litres
1764 cc (F7P) engine	0.6 litre

Cooling system
1108 cc (C1E) engine	5.4 litres
1171 cc (E5F/E7F) and 1390 cc (E6J/E7J) engines	5.2 litres
1721 cc (F2N) and 1794 cc (F3P) engines	6.4 litres
1764 cc (F7P) engine	7.0 litres

Fuel tank
All models except 16-valve	43 litres
16-valve models	50 litres

Manual gearbox
Four-speed:
JB0	3.25 litres
JB4	2.75 litres

Five-speed:
JB1 and JB3	3.4 litres
JB5	2.9 litres

Automatic transmission
MB1:
Total capacity	4.5 litres
Drain and refill	2.0 litres

AD4:
Main section total capacity	5.7 litres
Main section drain and refill	3.5 litres
Final drive section	1.0 litre

Power-assisted steering reservoir
Models with mechanical pump	1.1 litres
Models with electric pump	0.7 litre

Weekly checks

1 Fluid level checks

Engine oil

1 The engine oil level is checked with a dipstick that extends through a tube and into the sump at the bottom of the engine. The dipstick is located on the front of the cylinder block (see accompanying illustrations). On models equipped with an oil level gauge, the check can be made by switching on the ignition - the upper and lower limits on the gauge correspond to the upper and lower marks on the dipstick.

2 The oil level should be checked with the vehicle standing on level ground and before it is driven, or at least 5 minutes after the engine has been switched off.

HAYNES HINT *If the oil is checked immediately after driving the vehicle, some of the oil will remain in the upper part of the engine, resulting in an inaccurate reading on the dipstick.*

3 Withdraw the dipstick from the tube, and wipe all the oil from the end with a clean rag or paper towel. Insert the clean dipstick back into the tube as far as it will go, then withdraw it once more. Check that the oil level is between the upper (MAX) and lower (MIN) marks/notches on the dipstick. If the level is towards the lower (MIN) mark/notch, unscrew the oil filler cap from the valve cover, and add fresh oil until the level is on the upper (MAX) mark/notch, then refit the oil filler cap to the valve cover (photos). Note that the difference between the minimum and maximum marks/notches on the dipstick corresponds to 1.0 litre on C-type engines, 1.5 litres on E-type engines, 2.0 litres on 8-valve F-type engines, and 0.6 litre on 16-valve F-type engines.

4 An oil can spout or funnel may help to reduce spillage when adding oil to the engine. Always use the correct grade and type of oil as shown in 'Lubricants, fluids and capacities'.

5 Always maintain the level between the two dipstick marks/notches. If the level is allowed to fall below the lower mark/notch, oil starvation may result which could lead to severe engine damage. It is normal for the engine to consume some oil - up to 1 litre per 600 miles (1000 km) - and to require topping-up between oil changes. If the engine is overfilled by adding too much oil, this may result in oil-fouled spark plugs, oil leaks or oil seal failures.

Coolant

Warning: DO NOT attempt to remove the expansion tank pressure cap when the engine is hot, as there is a very great risk of scalding.

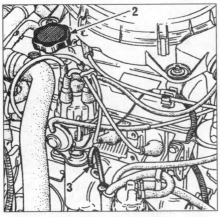

Fig. 1.1 Oil filler cap (2) and engine oil level dipstick (3) on C-type engines (Sec 1)

Fig. 1.2 Oil filler cap (2) and engine oil level dipstick (3) on E-type engines (Sec 1)

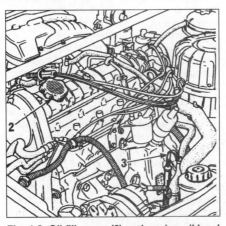

Fig. 1.3 Oil filler cap (2) and engine oil level dipstick (3) on 8-valve F-type engines (Sec 1)

Fig. 1.4 Oil filler cap (2) and engine oil level dipstick (3) on 16-valve F-type engines (Sec 1)

6 All vehicles covered by this manual are equipped with a pressurised cooling system. An expansion tank is located on the left-hand side of the engine compartment. On the C1E (1108 cc) engine, the expansion tank has only one hose, which is connected directly to the radiator. As engine temperature increases, the coolant expands and travels through the hose to the expansion tank. As the engine cools, the coolant is automatically drawn back into the system to maintain the correct level. On all other engines, the expansion tank has a continual flow of coolant, in order to purge air from the cooling system. Hoses are connected to the tank from the top of the cylinder head and from the tank to the water pump inlet.

7 The coolant level in the expansion tank should be checked regularly. The level in the tank varies with the temperature of the engine.

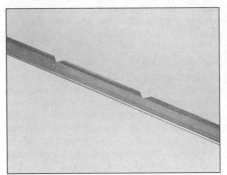

1.3A Engine oil level dipstick MIN and MAX level notches

1.3B Oil is added through the filler on the valve cover

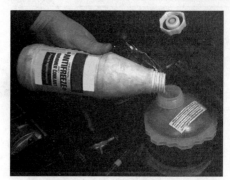

1.9 Use antifreeze, or a mixture of antifreeze and water, for topping-up the cooling system

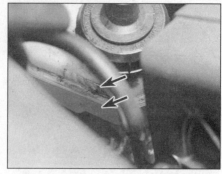

1.11 Check that the brake fluid in the master cylinder reservoir is between the MAX and MIN level marks (arrowed)

1.13 Top-up if necessary using the specified type of brake fluid

When the engine is cold, the coolant level should be between the maximum ('MAX') and minimum ('MIN') level marks on the side of the tank. When the engine is hot, the level may be slightly above the 'MAX' mark.

8 If topping-up is necessary, wait until the engine is cold, then slowly unscrew the pressure cap on the expansion tank. Allow any remaining pressure to escape, then fully unscrew the cap.

9 Add a mixture of water and antifreeze (see below) through the expansion tank filler neck, until the coolant is up to the maximum ('MAX') level mark (photo). Refit and tighten the pressure cap. In an emergency, plain water may be used for topping-up, but remember that this will dilute any antifreeze mixture remaining in the system. Make good the concentration of antifreeze at the first opportunity.

10 With this type of cooling system, the addition of coolant should only be necessary at very infrequent intervals. If frequent topping-up is required, it is likely there is a leak in the system. Check the radiator, all hoses and joint faces for any sign of staining or actual wetness, and rectify as necessary. If no leaks can be found, it is advisable to have the pressure cap and the entire system pressure-tested by a dealer or suitably-equipped garage, as this will often show up a small leak not previously visible.

Brake fluid

11 The brake master cylinder and fluid reservoir assembly is mounted on top of the vacuum servo unit in the engine compartment. The 'MAX' and 'MIN' level marks are indicated on the side of the reservoir, and the fluid level should be maintained between these marks at all times (photo). Note that on some models the 'MIN' mark is formed by the ridge on the reservoir body.

12 The fluid level in the master cylinder reservoir will drop **slowly** as the brake shoes and/or pads (as applicable) wear down during normal operation. Provided the level stays above the 'MIN' mark, there is no need to top-up to compensate for this fall. If the reservoir

requires repeated topping-up to maintain the proper level, or if a sudden fall in level occurs, this is an indication of an hydraulic leak somewhere in the system, which should be investigated immediately.

13 If topping-up is necessary, wipe the area around the filler cap with a clean rag, and disconnect the wiring connector from the fluid level switch. Unscrew the cap and remove it from the reservoir, taking care not to damage the sender unit - also take care not to allow brake fluid to drip onto the vehicle paintwork. When adding fluid, pour it carefully into the reservoir to avoid spilling it on surrounding painted surfaces (photo). Be sure to use only the specified brake hydraulic fluid, since mixing different types of fluid can cause damage to the system. See 'Lubricants, fluids and capacities' at the beginning of this Chapter.

⚠ **Warning: Brake hydraulic fluid can harm your eyes and damage painted surfaces, so use extreme caution when handling and pouring it. Do not use fluid that has been standing open for some time, as it absorbs moisture from the air. Excess moisture can cause a dangerous loss of braking effectiveness.**

14 When adding fluid, it is a good idea to inspect the reservoir for contamination. The system should be drained and refilled if deposits, dirt particles or contamination are seen in the fluid.

15 After filling the reservoir to the proper level, make sure that the cap is refitted securely, to avoid leaks and the entry of foreign matter, and reconnect the fluid level wiring connector.

Washer fluid

16 The washer fluid reservoir is located under the left-hand access hatch in the windscreen cowl panel. For access to the reservoir, open the bonnet and lift the hatch.

17 Check, and if necessary top-up, the washer fluid level in the reservoir. When topping-up the reservoir, a screenwash additive should be added in the recommended quantities.

18 Never use engine antifreeze in the washer fluid, as it can damage the car paintwork.

2 Wheel and tyre maintenance, and tyre pressure check

1 Periodically remove the wheels, and clean any dirt or mud from the inside and outside surfaces. Examine the wheel rims for signs of rusting, corrosion or other damage. Light alloy wheels are easily damaged by 'kerbing' whilst parking, and similarly steel wheels may become dented or buckled. Renewal of the wheel is very often the only course of remedial action possible.

2 To check that the roadwheel bolts are securely fastened, remove the wheel cover (where fitted), then slacken each bolt in turn through one-quarter of a turn and tighten it to the specified torque wrench setting. Refit the trim, where applicable.

3 The tyres originally fitted are equipped with tread wear indicators which will appear flush with the surface of the tread, thus producing the effect of a continuous band of rubber across the width of the tyre, when the tread depth is reduced to approximately 1.6 mm. **At this point, the tyre must be renewed immediately.** Tread wear can be monitored with an inexpensive device known as a tread depth indicator gauge (photo).

4 Note any abnormal tread wear with reference to the illustration. Tread pattern irregularities such as feathering, flat spots and more wear on one side than the other are indications of front wheel alignment and/or

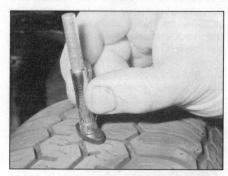

2.3 Checking the tyre tread depth with an indicator gauge

Tyre Tread Wear Patterns

Shoulder Wear

**Underinflation
(wear on both sides)**
Check and adjust pressures

**Incorrect wheel camber
(wear on one side)**
Repair or renew suspension parts

Hard cornering
Reduce speed!

Centre Wear

Overinflation
Check and adjust pressures

If you sometimes have to inflate your car's tyres to the higher pressures specified for maximum load or sustained high speed, don't forget to reduce the pressures to normal afterwards.

Toe Wear

Incorrect toe setting
Adjust front wheel alignment

Note: The feathered edge of the tread which characterises toe wear is best checked by feel.

Uneven Wear

Incorrect camber or castor
Repair or renew suspension parts

Malfunctioning suspension
Repair or renew suspension parts

Unbalanced wheel
Balance tyres

Out-of-round brake disc/drum
Machine or renew

1

balance problems. If any of these conditions are noted, they should be rectified as soon as possible.

5 General tyre wear is influenced to a large degree by driving style - harsh braking and acceleration or fast cornering will all produce more rapid tyre wear. Interchanging of tyres may result in more even wear, but it is worth bearing in mind that if this is completely effective, the added expense is incurred of replacing a complete set of tyres simultaneously, which may prove financially-restrictive for many owners.

6 Front tyres may wear unevenly as a result of wheel misalignment. The front wheels should always be correctly aligned according to the settings specified. Refer to Chapter 10 for further information.

7 Regularly check the tyres for damage in the form of cuts or bulges, especially in the sidewalls. Remove any nails or stones

2.8 Checking the tyre pressures with a tyre pressure gauge

embedded in the tread before they penetrate the tyre to cause deflation. If removal of a nail does reveal that the tyre has been punctured, refit the nail so that its point of penetration is marked, then immediately change the wheel and have the tyre repaired by a tyre dealer. **Do not** drive on a tyre in such a condition. If in any doubt as to the possible consequences of any damage found, consult your local tyre dealer for advice.

8 Ensure that tyre pressures are checked regularly and maintained correctly (photo). Checking should be carried out with the tyres cold, and **not** immediately after the car has been in use. If the pressures are checked with the tyres hot, an apparently-high reading will be obtained, owing to heat expansion. **Under no circumstances** should an attempt be made to reduce the pressures to the quoted cold reading in this instance, or effective under-inflation will result.

9 Under-inflation will cause overheating of the tyre owing to excessive flexing of the casing, and the tread will not sit correctly on the road surface. This will cause a consequent loss of adhesion and excessive wear, not to mention the danger of sudden tyre failure due to heat build-up.

10 Over-inflation will cause rapid wear of the centre part of the tyre tread, coupled with reduced adhesion, harsher ride, and the danger of shock damage occurring in the tyre casing.

11 The balance of each wheel and tyre assembly should be maintained to avoid excessive wear, not only to the tyres but also to the steering and suspension components.

Wheel imbalance is normally indicated by vibration through the car's bodyshell, although in many cases it is particularly noticeable through the steering wheel at certain speeds. Conversely, it should be noted that wear or damage in suspension or steering components may cause excessive tyre wear. Out-of-round or out-of-true tyres, damaged wheels and wheel bearing wear also fall into this category. Balancing will not usually cure vibration caused by such wear.

12 Wheel balancing may be carried out with the wheel either on or off the car. If balanced on the car, ensure that the wheel-to-hub relationship is marked in some way prior to subsequent wheel removal so that it may be refitted in its original position.

13 Legal restrictions apply to many aspects of tyre fitting and usage, and in the UK this information is contained in the Motor Vehicle Construction and Use Regulations. It is suggested that a copy of these regulations is obtained from your local police if in doubt as to current legal requirements with regard to tyre type and condition, minimum tread depth, etc.

3 Electrical system checks

1 Periodically check the operation of all exterior and interior lights, warning lights, and all other electrical equipment, and rectify any faults. Refer to Chapter 12 for further details of individual electrical components.

Every 6000 miles

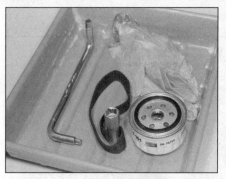

4.1 Tools and materials necessary for the engine oil change and filter renewal

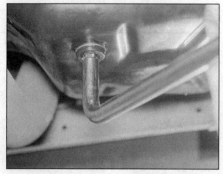

4.2 Using the special drain plug key to unscrew the sump drain plug

4.9 Applying a light coating of clean oil to the sealing ring before fitting the new oil filter

Every 6000 miles

4 Engine oil and filter renewal

HAYNES HiNT *Frequent oil and filter changes are the most important preventative maintenance procedures that can be undertaken by the DIY owner. As engine oil ages, it becomes diluted and contaminated, which leads to premature engine wear.*

1 Before starting this procedure, gather together all the necessary tools and materials (photo). Also make sure that you have plenty of clean rags and newspapers handy to mop up any spills. Ideally, the engine oil should be warm, as it will drain better and more built-up sludge will be removed with it. Take care, however, not to touch the exhaust or any other hot parts of the engine when working under the vehicle. To avoid any possibility of scalding, and to protect yourself from possible skin irritants and other harmful contaminants in used engine oils, it is advisable to wear rubber gloves when carrying out this work. Access to the underside of the vehicle will be greatly improved if it can be raised on a lift, driven onto ramps or jacked up and supported on axle stands. Whichever method is chosen, make sure that the car remains as level as possible, to enable the oil to drain fully.

2 Remove the oil filler cap from the valve cover, then position a suitable container beneath the sump. Clean the drain plug and the area around it, then slacken it half a turn using a special drain plug key (8 mm square) (photo). If possible, try to keep the plug pressed into the sump while unscrewing it by hand the last couple of turns. As the plug releases from the threads, move it away sharply so the stream of oil issuing from the sump runs into the container, not up your sleeve!

3 Allow some time for the old oil to drain, noting that it may be necessary to reposition the container as the oil flow slows to a trickle.

4 After all the oil has drained, wipe off the drain plug with a clean rag and renew its sealing washer. Clean the area around the drain plug opening, then refit and tighten the plug to the specified torque setting.

5 If applicable at this service, move the container into position under the oil filter, located on the front of the cylinder block.

6 On the 16-valve engine there is only limited room between the subframe and the sump, and it is very difficult to reach up to the oil filter. However, it is possible to gain access through this space, and it is not necessary to remove any body or engine components. Note that the oil filter on 16-valve engines is mounted on an oil cooler.

7 Using an oil filter removal tool, slacken the filter initially (on the 16-valve engine, try to unscrew the filter by hand first). Loosely wrap some rags around the oil filter, then unscrew it and immediately position it with its open end uppermost to prevent further spillage of oil. Remove the oil filter from the engine compartment and empty the oil into the container.

8 Use a clean rag to remove all oil, dirt and sludge from the filter sealing area on the engine. Check the old filter to make sure that the rubber sealing ring hasn't stuck to the engine. If it has, carefully remove it.

9 Apply a light coating of clean oil to the sealing ring on the new filter, then screw it into position on the engine (photo). Tighten the filter firmly by hand only - do not use any tools. Wipe clean the exterior of the oil filter.

10 Remove the old oil and all tools from under the car, then (if applicable) lower the car to the ground.

11 Fill the engine with the specified quantity and grade of oil, as described in Section 1. Pour the oil in slowly, otherwise it may

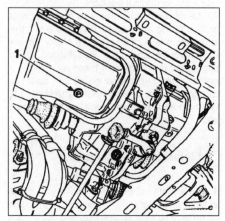

Fig. 1.5 Engine sump drain plug location (1) on 8-valve engines (Sec 4)

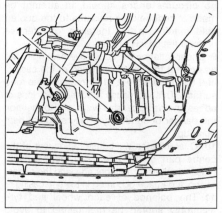

Fig. 1.6 Engine sump drain plug location (1) on 16-valve engines (Sec 4)

overflow from the top of the valve cover. Check that the oil level is up to the maximum mark on the dipstick, then refit and tighten the oil filler cap.

12 Start the engine and run it for a few minutes, checking that there are no leaks around the oil filter seal and the sump drain plug. Note that when the engine is first started, there will be a delay of a few seconds before the oil pressure warning light goes out while the new filter fills with oil. Do not race the engine while the warning light is on.

13 Switch off the engine and wait a few minutes for the oil to settle in the sump once more. With the new oil circulated and the filter now completely full, recheck the level on the dipstick and add more oil if necessary.

14 Dispose of the used engine oil safely with reference to 'General repair procedures' in the reference Sections of this manual.

5 General checks

Engine

1 Visually inspect the engine joint faces, gaskets and seals for any signs of water or oil leaks. Pay particular attention to the areas around the valve cover, cylinder head, oil filter and sump joint faces. Bear in mind that over a period of time some very slight seepage from these areas is to be expected - what you are really looking for is any indication of a serious leak. Should a leak be found, renew the offending gasket or oil seal by referring to the appropriate Chapters in this manual.

2 Also check the security and condition of all the engine-related pipes and hoses. Ensure that all cable ties or securing clips are in place and in good condition. Clips which are broken or missing can lead to chafing of the hoses, pipes or wiring which could cause more serious problems in the future.

Cooling system

3 The engine should be cold for the cooling system checks, so perform the following procedure before driving the vehicle, or after the engine has been switched off for at least three hours.

4 Remove the expansion tank filler cap (see above) and clean it thoroughly inside and out with a rag. Also clean the filler neck on the expansion tank. The presence of rust or corrosion in the filler neck indicates that the coolant should be changed. The coolant inside the expansion tank should be relatively clean and transparent. If it is rust-coloured, drain and flush the system, and refill with a fresh coolant mixture.

5 Carefully check the radiator hoses and heater hoses along their entire length. Renew any hose which is cracked, swollen or deteriorated. Cracks will show up better if the hose is squeezed. Pay close attention to the hose clips that secure the hoses to the

cooling system components. Hose clips can pinch and puncture hoses, resulting in cooling system leaks. If wire-type hose clips are used, it may be a good idea to replace them with worm drive clips.

6 Inspect all the cooling system components (hoses, joint faces, etc) for leaks. A leak in the cooling system will usually show up as white or rust-coloured deposits on the area adjoining the leak. Where any problems of this nature are found on system components, renew the component or gasket with reference to Chapter 3.

7 Clean the front of the radiator with a soft brush to remove all insects, leaves, etc, imbedded in the radiator fins. Be extremely careful not to damage the radiator fins, and be careful not to cut your fingers on them.

Fuel system

⚠️ **Warning: Certain procedures require the removal of fuel lines and connections, which may result in some fuel spillage. Before carrying out any operation on the fuel system, refer to the preCautions given in 'Safety first!' at the beginning of the manual, and follow them implicitly. Petrol is a highly dangerous and volatile liquid, and the precautions necessary when handling it cannot be overstressed.**

8 The fuel system is most easily checked with the vehicle raised on a hoist, or suitably supported on axle stands so the components underneath are readily visible and accessible.

9 If the smell of petrol is noticed while driving or after the vehicle has been parked in the sun, the system should be thoroughly inspected immediately.

10 Remove the petrol tank filler cap, and check that there is no damage or corrosion. There should be an unbroken sealing imprint on the gasket. Renew the cap if necessary.

11 With the vehicle raised, inspect the petrol tank and filler neck for punctures, cracks and other damage. The connection between the filler neck and tank is especially critical (photo). Sometimes a rubber filler neck or connecting hose will leak due to loose retaining clamps or deteriorated rubber.

5.11 Pay particular attention to the condition of the rubber hose (arrowed) which joins the fuel filler neck to the fuel tank

12 Carefully check all rubber hoses and metal fuel lines leading away from the petrol tank. Check for loose connections, deteriorated hoses, crimped lines and other damage. Pay particular attention to vent pipes and hoses which often loop up around the filler neck and can become blocked or crimped. Follow the lines to the front of the vehicle, carefully inspecting them all the way. Renew damaged sections as necessary.

13 From within the engine compartment, check the security of all fuel hose attachments. Inspect fuel and vacuum hoses for kinks, chafing and deterioration.

14 Check the operation of the throttle linkage, and lubricate the linkage components with a few drops of light oil. Similarly check the operation of the manual choke linkage, when applicable.

6 Heating/air conditioning system check

1 Check that the heating system operates correctly. See Chapter 3, Section 12.

2 On models with air conditioning, check that the system operates correctly. Run the engine, and switch the air conditioning on to its maximum setting. After it has been operating for a few minutes, inspect the sight glass on top of the dehydration unit (Fig. 1.7). If a continuous stream of bubbles is visible, the refrigerant level is low, and professional advice should be sought. **Do not** attempt to discharge or recharge the system unless qualified to do so - see Chapter 3.

3 Check the tension and condition of the air conditioning compressor drivebelt. Refer to the auxiliary drivebelt checking procedure in Section 7 of this Chapter, and to Chapter 12, Section 6, for details.

4 During the Winter period, it is advisable to run the air conditioning system occasionally (for 10 minutes or so once a month) in order to ensure correct functioning of the compressor and distribution of the oil within the system.

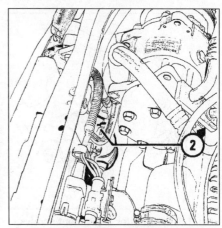

Fig. 1.7 Inspect the air conditioning sight glass (2) with the system in operation (Sec 6)

1

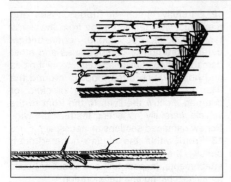

Fig. 1.8 Checking for drivebelt wear - multi-ribbed type shown (Sec 7)

7 Drivebelt(s) condition and tension check

Auxiliary drivebelt

1 The drivebelt(s) is/are driven by the crankshaft pulley located at the right-hand side of the engine compartment.
2 Access to the belt(s) on certain models is difficult. Details can be found in Chapter 12, Section 6.
3 With the engine stopped, inspect the full length of the belt(s) for cracks and separation of the belt plies. It will be necessary to turn the engine in order to inspect the full length of the belt.
4 Twist the belt between the pulleys so that both sides can be viewed.
5 Also check for fraying, and for 'glazing', which gives the belt a shiny appearance.
6 Check the pulleys for nicks, cracks and corrosion.
7 Refer to Chapter 12, Section 6 for details of belt renewal, tension checking and adjustment.

Water pump drivebelt

8 On the 1171 cc and 1390 cc (E-type) engines, the water pump is driven by the timing belt. No specific check is required.

9 On all other engines, the water pump is driven by the auxiliary drivebelt. Refer to the auxiliary drivebelt checking procedure above.

Power steering pump drivebelt

10 The power steering pump drivebelt can be checked as described above. Adjustment and renewal are described in Chapter 12, Section 6.

8 Exhaust system check

1 With the engine cold (at least an hour after the vehicle has been driven), check the complete exhaust system from the engine to the end of the tailpipe. Ideally, the inspection should be carried out with the vehicle on a hoist to permit unrestricted access, but if a hoist is not available, raise and support the vehicle safely on axle stands.
2 Check the exhaust pipes and connections for evidence of leaks, severe corrosion and damage. Make sure that all brackets and mountings are in good condition and tight. Leakage at any of the joints or in other parts of the system will usually show up as a black sooty stain in the vicinity of the leak.
3 Rattles and other noises can often be traced to the exhaust system, especially the brackets and mountings. Try to move the pipes and silencers. If the components can come into contact with the body or suspension parts, secure the system with new mountings or if possible, separate the joints and twist the pipes as necessary to provide additional clearance.
4 Run the engine at idling speed. Have an assistant place a cloth or rag over the rear end of the exhaust pipe, and listen for any escape of exhaust gases that would indicate a leak.
5 On completion, lower the car to the ground.
6 The inside of the exhaust tailpipe can be an indication of the running condition of the engine. The exhaust deposits here are an indication of the engine's state of tune. If the

pipe is black and sooty, the engine is in need of a tune-up, including a thorough fuel system inspection and adjustment.

9 Idle speed and mixture/CO content adjustment

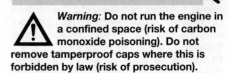

Warning: Do not run the engine in a confined space (risk of carbon monoxide poisoning). Do not remove tamperproof caps where this is forbidden by law (risk of prosecution).

Carburettor models - all engines except 1171 cc (E5F)

1 The procedure for idle speed and CO content adjustment is essentially the same on each of the four carburettor types that may be fitted. Refer to the accompanying illustrations and identify the carburettor type fitted and the adjustment screw locations (Figs. 1.9 to 1.11).
2 Before carrying out the following adjustments, ensure that the spark plugs are in good condition and correctly gapped. On the 1108 cc (C1E) engine, also ensure that the contact breaker points and ignition timing settings are correct.
3 Make sure that all electrical components (and air conditioning, if fitted) are switched off during the following procedure. If the electric cooling fan operates, wait until it has stopped before continuing. On models with power steering, the wheels must be in the straight-ahead position.
4 Connect a tachometer to the engine in accordance with the equipment manufacturer's instructions. The use of an exhaust gas analyser (CO meter) is also recommended to obtain an accurate setting.
5 Remove the tamperproof cap (where fitted) from the mixture adjustment screw by hooking it out with a scriber or small screwdriver.
6 Run the engine at a fast idling speed until it reaches normal operating temperature

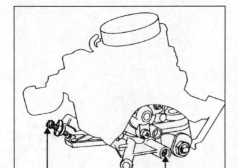

Fig. 1.9 Carburettor adjustment screws - Solex 32 BIS (1108 cc/C1E engine) (Sec 9)

A Idle speed   B Idle mixture/CO

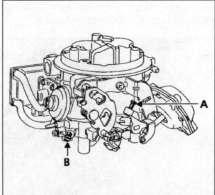

Fig. 1.10 Carburettor adjustment screws - Weber 32 TLDR (1390 cc/E6J engine) (Sec 9)

A Idle speed   B Idle mixture/CO

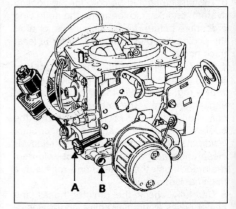

Fig. 1.11 Carburettor adjustment screws - Solex 32-34 Z13 (1721 cc/F2N engine) (Sec 9)

A Idle speed   B Idle mixture/CO

9.7A Adjusting the idle speed - 1390 cc (E6J) engine

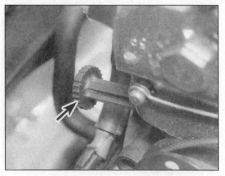

9.7B Idle speed adjustment screw (arrowed) - 1721 cc (F2N) engine

10.2 Remove the engine top cover for access to the spark plugs (16-valve engine only)

(indicated by the cooling fan cutting in and out). Increase the engine speed to 2000 rpm for 30 seconds, and repeat this at three-minute intervals during the adjustment procedure. This will ensure that any excess fuel is cleared from the inlet manifold.

7 With the engine idling, turn the idle speed screw until the engine is idling at the specified speed (photos).

8 Turn the mixture adjustment screw clockwise to weaken the mixture, or anti-clockwise to richen it until the CO reading is as given in the Specifications. If a CO meter is not being used, weaken the mixture as described, then richen it until the maximum engine speed is obtained, consistent with even running.

9 Re-adjust the idling speed if necessary, then check the CO reading again. Repeat until both the idling speed and CO reading are correct.

10 Where required by law, fit a new tamperproof cap to the mixture adjustment screw.

1171 cc (E5F) engine

Note: *The cold enrichment device described here may be found fitted to other carburettor models. If so, the following procedure will apply.*

11 The adjustment procedure on this engine varies as follows:

(a) *Disconnect the wire from the cold enrichment device (Fig. 1.12) and earth it. Adjust as described earlier to obtain the values given in the Specifications.*

(b) *Refit the wire to its holder. The idle speed must drop slightly, and the CO level must fall to around 0.2%.*

Fuel injection models

12 The idle speed on all fuel injection models is electronically controlled, and no adjustment is possible.

13 The CO content is also electronically controlled on all models with a catalytic converter (see Part C of Chapter 4 for details). This means that the only engine on which adjustment is possible is the 1764 cc engine code F7P-720 (vehicle code C575). Adjustment is carried out by turning the screw on the CO adjustment potentiometer (Fig. 1.14).

10 Spark plug check and renewal

⚠️ *Warning: Voltages produced by an electronic ignition system are considerably higher than those produced by conventional systems. Extreme care must be taken when working on the system with the ignition switched on. Persons with surgically-implanted cardiac pacemaker devices should keep well clear of the ignition circuits, components and test equipment.*

1 The correct functioning of the spark plugs is vital for the correct running and efficiency of the engine. It is essential that the plugs fitted are appropriate for the engine, the suitable type being specified at the start of this Chapter. If the correct type of plug is used and the engine is in good condition, the spark plugs should not need attention between scheduled renewal intervals, except for adjustment of their gaps. Spark plug cleaning is rarely necessary, and should not be attempted unless specialised equipment is available, as damage can easily be caused to the firing ends.

2 To remove the plugs, first open the bonnet. On models with the 16-valve engine, remove the engine top cover, which is secured by two screws (photo).

> **HAYNES HiNT** *On all models, check that the HT leads are marked one to four, to correspond to the appropriate cylinder number (number one cylinder is at the flywheel end of the engine). Failure to do this may result in confusion when refitting, if all the HT leads are disconnected at the same time.*

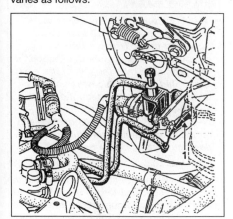

Fig. 1.12 Cold enrichment device wire (1) must be earthed for idle adjustment on 1171 cc (E5F) engine (Sec 9)

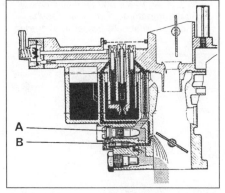

Fig. 1.13 Sectional view of Pierburg 1B1 carburettor (1171 cc/E5F engine) showing adjustment screws (Sec 9)

A Idle speed   B Idle mixture/CO

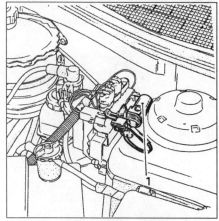

Fig. 1.14 CO adjustment potentiometer (1) - 1764 cc (F7P-720) engine (Sec 9)

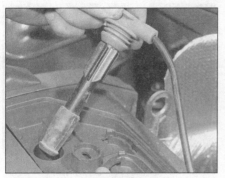

10.3A Disconnecting an HT lead - 16-valve engine

10.3B Disconnecting an HT lead - 1794 cc (F3P) engine

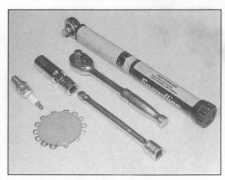

10.5 Tools required for spark plug removal, gap adjustment and refitting

3 Pull the HT leads from the plugs; grip the end fitting, not the lead, otherwise the lead connection may be fractured (photos).

4 It is advisable to remove any dirt from the spark plug recesses using a clean brush, a vacuum cleaner or compressed air before removing the plugs, to prevent the dirt dropping into the cylinders.

5 Unscrew the plugs using a spark plug spanner, suitable box spanner or a deep socket and extension bar (photo). Keep the socket in alignment with the spark plug, otherwise if it is forcibly moved to either side, the ceramic top of the spark plug may be broken off. As each plug is removed, examine it as follows.

6 Examination of the spark plugs will give a

10.11 Measuring the spark plug gap with a feeler blade

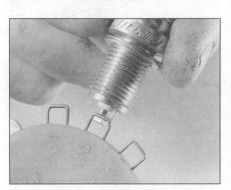

10.12A Measuring the spark plug gap with a wire gauge . . .

good indication of the condition of the engine. If the insulator nose of the spark plug is clean and white, with no deposits, this is indicative of a weak mixture or too hot a plug (a hot plug transfers heat away from the electrode slowly, a cold plug transfers heat away quickly).

7 If the tip and insulator nose are covered with hard black-looking deposits, then this is indicative that the mixture is too rich. Should the plug be black and oily, then it is likely that the engine is fairly worn, as well as the mixture being too rich.

8 If the insulator nose is covered with light tan to greyish-brown deposits, then the mixture is correct and it is likely that the engine is in good condition.

9 If the spark plug has only completed 6000 miles (10 000 km) in accordance with the routine maintenance schedule, it should still be serviceable until the 12 000 mile (20 000 km) service, when it is renewed. However, it is recommended that it be re-gapped in order to maintain peak engine efficiency, and to allow for the slow increase in gap (approximately 0.025 mm per 1000 miles) which occurs in normal operation. If, due to engine condition, the spark plug is not serviceable, it should be renewed.

10 The spark plug gap is of considerable importance as, if it is too large or too small, the size of the spark and its efficiency will be seriously impaired. For best results, the spark plug gap should be set in accordance

10.12B . . . and adjusting the gap using a special adjusting tool

with the Specifications at the start of this Chapter.

11 To set it, measure the gap with a feeler blade, and then bend open, or closed, the outer plug electrode until the correct gap is achieved (photo). The centre electrode should never be bent, as this may crack the insulation and cause plug failure, if nothing worse.

12 Special spark plug electrode gap measuring and adjusting tools are available from most motor accessory shops (photos).

13 Before fitting the spark plugs, check that the threaded connector sleeves are tight, and that the plug exterior surfaces and threads are clean. Apply a little anti-seize compound to the threads.

TOOL TiP

It is very often difficult to insert spark plugs into their holes without cross-threading them. To avoid this possibility, fit a short length of rubber or plastic hose over the end of the spark plug. The flexible hose acts as a universal joint to help align the plug with the plug hole. Should the plug begin to cross-thread, the hose will slip on the spark plug, preventing thread damage to the cylinder head.

14 Remove the hose (if used) and tighten the plug to the specified torque using the spark plug socket and a torque wrench (photo). Refit the remaining spark plugs in the same manner.

15 Wipe clean the HT leads, then reconnect them in their correct order.

10.14 Final tightening of the spark plugs should be carried out using a torque wrench

11.1 Removing the rotor arm from the contact breaker distributor

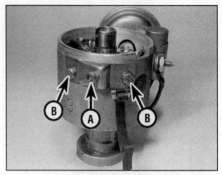

11.4 Contact breaker points adjusting nut (A) and baseplate retaining screws (B)

11 Contact breaker points and condenser check, adjustment and renewal

1 Release the two spring clips and lift off the distributor cap. Pull the rotor arm off the shaft and remove the plastic shield (photo).

2 With the ignition switched off, use a screwdriver to open the contact breaker points, then visually check the points surfaces for pitting, roughness and discoloration. If the points have been arcing, there will be a build-up of metal on the moving contact and a corresponding hole in the fixed contact; if this is the case, the points should be renewed.

3 Another method of checking the contact

breaker points is by using a test meter available from most car accessory shops. Rotate the engine if necessary until the points are fully shut. Connect the meter between the distributor LT wiring terminal and earth, and read off the condition of the points.

4 To remove the points, first unscrew the adjusting nut on the side of the distributor body, and then unscrew the two baseplate retaining screws. Lift off the support bracket (photo).

5 Unhook the end of the adjustment rod from the fixed contact, and slide the rod and spring out of the distributor body (photo).

6 Prise out the small plug and then remove the retaining clip, noting that the hole in the clip is uppermost (photos).

7 Slacken the LT terminal nut and detach the lead (photo).

8 Remove the spring retaining clip from the top of the moving contact pivot post, and take off the fibre insulating washer (photo).

9 Ease the spring blade away from its nylon support, and lift the moving contact upwards and off the pivot post (photo).

10 Unscrew the retaining screw, and remove the fixed contact from the baseplate (photo).

11 The purpose of the condenser (located externally on the side of the distributor body) is to ensure that when the contact breaker points open, there is no sparking across them, which would cause wear of their faces and prevent the rapid collapse of the magnetic field in the coil. Failure of the condenser

1

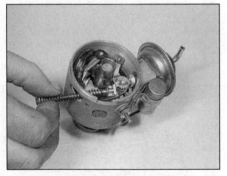

11.5 Removing the adjustment rod and spring

11.6A Prise out the plastic plug . . .

11.6B . . . to gain access to the retaining clip (arrowed) . . .

11.6C . . . which can then be removed with pliers. Note which way up it is fitted

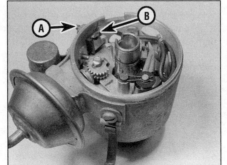

11.7 Slacken the LT terminal nut (A) and detach the lead (B) from the connector

11.8 Remove the retaining clip and the fibre washer from the pivot post

11.9 Withdraw the moving contact from the pivot post

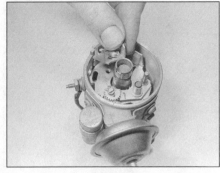

11.10 Removing the fixed contact from the baseplate

11.18 Measuring the points gap using a feeler blade. Adjust if necessary by turning nut (A)

would cause a reduction in coil HT voltage, and ultimately lead to engine misfire.

12 If the engine becomes very difficult to start, or begins to miss after several miles of running, and the contact breaker points show signs of excessive burning, the condition of the condenser must be suspect. A further test can be made by separating the points by hand with the ignition switched on. If this is accompanied by a strong bright flash, it is indicative that the condenser has failed.

13 Without special test equipment, the only sure way to diagnose condenser trouble is to fit a new one, and note if there is any improvement.

14 To remove the condenser, unscrew the nut at the LT terminal post and slip off the lead. Unscrew the condenser retaining screw and remove the unit from the side of the distributor body.

15 Refitting of the condenser is a reversal of the removal procedure.

16 To fit the new contact breaker points, first check if there is any greasy deposit on them, and if necessary clean them using methylated spirit.

17 Fit the points using a reversal of the removal procedure, then adjust them as follows. Turn the engine over using a socket or spanner on the crankshaft pulley bolt, until the heel of the contact breaker arm is on the peak of one of the four cam lobes.

18 With the points fully open, a clean feeler blade equal to the contact breaker points gap, as given in the Specifications, should now just fit between the contact faces (photo).

19 If the gap is too large or too small, turn the adjusting nut on the side of the distributor body using a small spanner until the specified gap is obtained.

20 With the points correctly adjusted, refit the plastic shield, rotor arm and distributor cap. Check the ignition timing as described in Section 12.

21 If a dwell meter is available, a far more accurate method of setting the contact breaker points is by measuring and setting the distributor dwell angle.

22 The dwell angle is the number of degrees of distributor shaft rotation during which the contact breaker points are closed (ie the

period from when the points close after being opened by one cam lobe until they are opened again by the next cam lobe). The advantages of setting the points by this method are that any wear of the distributor shaft or cam lobes is taken into account, and also the inaccuracies of using a feeler gauge are eliminated.

23 To check and adjust the dwell angle, connect one lead of the meter to the ignition coil + terminal and the other lead to the coil - terminal, or in accordance with the meter maker's instructions.

24 Start the engine, allow it to idle and observe the reading on the dwell meter scale. If the dwell angle is not as specified, turn the adjusting nut on the side of the distributor body as necessary to obtain the correct setting. **Note:** *Owing to machining tolerances, or wear in the distributor shaft or bushes, it is not uncommon for a contact breaker points gap correctly set with feeler gauges, to give a dwell angle outside the specified tolerances. If this is the case, the dwell angle should be regarded as the preferred setting.*

25 After completing the adjustment, switch off the engine and disconnect the dwell meter. Check the ignition timing as described in Section 12.

---

12 Ignition timing check and adjustment -
1108 cc (C1E) engine

1 In order that the engine can run efficiently, it is necessary for a spark to occur at the spark plug and ignite the fuel/air mixture at the instant just before the piston on the compression stroke reaches the top of its travel. The precise instant at which the spark occurs is determined by the ignition timing, and this is quoted in degrees before top dead centre (BTDC).

2 The timing may be checked and adjusted in one of two ways, either by using a test bulb to obtain a static setting (with the engine stationary), or by using a stroboscopic timing light to obtain a dynamic setting (with the engine running).

3 Before checking or adjusting the ignition timing, make sure that the contact breaker points are in good condition and correctly adjusted, as described previously.

Static setting

4 Refer to the Specifications at the start of this Chapter, and note the specified setting for static ignition timing. This value will also be found stamped on a clip fastened to one of the HT leads (Fig. 1.15).

5 Pull off the HT lead and remove No 1 spark plug (nearest the flywheel end of the engine).

6 Place a finger over the plug hole and turn the engine in the normal direction of rotation (clockwise from the crankshaft pulley end) until pressure is felt in No 1 cylinder. This indicates that the piston is commencing its compression stroke. The engine can be turned using a socket or spanner on the crankshaft pulley bolt, or by engaging a gear and pushing the car along.

7 Continue turning the engine until the mark

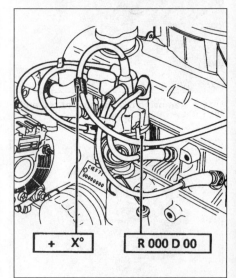

| + X° | | R 000 D 00 |

Fig. 1.15 Ignition timing setting (X° BTDC) is stamped on clip - 1108 cc (C1E) engine only (Sec 12)

R 000 D 00 = Advance curve reference number

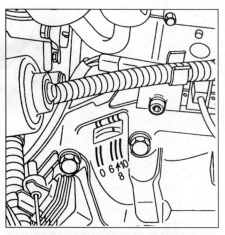

Fig. 1.16 Timing marks (notch on flywheel rim and scale on bellhousing) - 1108 cc (C1E) engine only (Sec 12)

0 = TDC

on the flywheel is aligned with the appropriate notch on the clutch bellhousing (Fig. 1.16).

8 Remove the distributor cap, and check that the rotor arm is pointing towards the No 1 spark plug HT lead segment in the cap.

9 Connect a 12-volt test light and leads between a good earth point and the LT terminal nut on the side of the distributor body.

10 Slacken the distributor clamp retaining nut, and then switch on the ignition.

11 If the test light is on, this shows that the points are open. Turn the distributor slightly clockwise until the light goes out, showing that the points have closed.

12 Now turn the distributor anti-clockwise until the test light just comes on. Hold the distributor in this position and tighten the clamp retaining nut.

13 Test the setting by turning the engine two complete revolutions, observing when the light comes on in relation to the timing marks.

14 Switch off the ignition and remove the test light. Refit No 1 spark plug, the distributor cap and HT lead.

Dynamic setting

15 Refer to the Specifications at the start of this Chapter, and note the setting for ignition

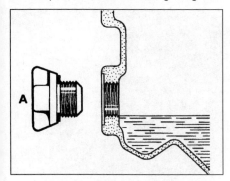

Fig. 1.17 Manual gearbox filler/level plug (A) - correct oil level shown (Sec 13)

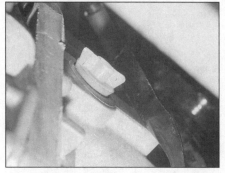

13.2 Filler plug on the front of the gearbox

timing at idle. This initial value will also be found stamped on a clip fastened to one of the HT leads. To make subsequent operations easier, it is advisable to highlight the mark on the flywheel and the appropriate notch on the clutch bellhousing with white paint (typist's correction fluid is ideal) or chalk.

16 Connect the timing light in accordance with the equipment manufacturer's instructions (usually interposed between the end of No 1 spark plug HT lead and No 1 spark plug terminal).

17 Disconnect the vacuum advance pipe from the distributor vacuum unit, and plug its end.

18 Start the engine and leave it idling at the specified speed (refer to the Specifications).

19 Point the timing light at the timing marks. They should appear to be stationary, with the mark on the flywheel aligned with the appropriate notch on the clutch bellhousing.

20 If adjustment is necessary (ie the flywheel mark does not line up with the appropriate notch), slacken the distributor clamp retaining nut and turn the distributor body either anti-clockwise to advance the timing, or clockwise to retard it. Tighten the clamp nut when the setting is correct.

21 Gradually increase the engine speed while still pointing the timing light at the marks. The mark on the flywheel should appear to advance further, indicating that the distributor

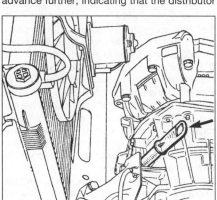

Fig. 1.18 MB1 automatic transmission fluid level dipstick and filler tube location (Sec 13)

13.4 Topping-up the manual gearbox with oil

centrifugal advance mechanism is functioning. If the mark remains stationary or moves in a jerky, erratic fashion, the advance mechanism must be suspect.

22 Reconnect the vacuum pipe to the distributor, and check that the advance alters when the pipe is connected. If not, the vacuum unit on the distributor may be faulty.

23 After completing the checks and adjustments, switch off the engine and disconnect the timing light.

13 Transmission lubricant level check

1

Manual gearbox

1 Position the car over an inspection pit, on car ramps, or jack it up, but make sure that it is level.

2 Unscrew the filler/level plug from the front-facing side of the gearbox (photo).

3 The oil level should be up to the lower edge of the filler hole, as shown in Fig. 1.17. Insert a finger to check the level.

4 Where necessary, top-up the level using the correct grade of oil, then refit and tighten the filler plug (photo).

5 If the gearbox requires frequent topping-up, check it for leakage, especially around the driveshaft oil seals, and repair as necessary.

6 Lower the car to the ground.

Automatic transmission

MB1 (3-speed) automatic transmission

7 Check the fluid level when the car has been standing for some time and the fluid is cold.

8 Position the car on level ground, then apply the handbrake and select 'P' with the selector lever. Start the engine, and allow it to run for a few minutes.

9 With the engine still idling, withdraw the dipstick from the front of the transmission housing, wipe it on a clean cloth, insert it again then withdraw it once more and read off the level. Ideally, the level should be in the centre of the mark on the dipstick. The fluid must never be allowed to fall below the

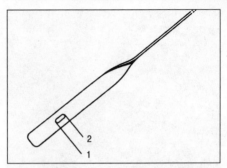

Fig. 1.19 MB1 automatic transmission fluid level dipstick markings (Sec 13)

1 Minimum cold   2 Maximum cold

bottom of the mark, otherwise there is a risk of damaging the transmission. The transmission must never be overfilled so that the level is above the top of the mark, otherwise there is a risk of overheating.

10 If topping-up is necessary, add a quantity of the specified fluid to the transmission through the dipstick tube.

> **HAYNES HiNT** *Use a clean funnel with a fine-mesh screen, to avoid spillage and to ensure that any foreign matter is not introduced into the transmission.*

11 After topping-up, recheck the level again, as described above, refit the dipstick and switch off the engine.

AD4 (4-speed) automatic transmission

Note: *The fluid level checking procedure on the AD4 automatic transmission is particularly complicated, and the home mechanic would be well advised to take the car to a Renault dealer to have the work carried out, as special test equipment is necessary to carry out the check. However, the following procedure is given for those who may have access to this equipment. Note that the final drive oil level is not checked as a routine maintenance task, as it is sealed for life.*

12 First add 0.5 litre (1 pint) of the specified fluid to the transmission by removing the vent from the top of the filler tube and using a clean funnel with a fine-mesh filter. Refit the vent.

13 Position the car on a ramp, or jack it up and support on axle stands.

14 Connect the Renault XR25 test meter to the diagnostic socket, and enter 'DO4' then number '04'. With the selector lever in 'Park', run the engine at idle speed until the temperature reaches 60°C.

15 With the engine still running, unscrew the topping-up plug on the transmission. Allow the excess fluid to run out into a calibrated container for 20 seconds, then refit the plug. The amount of fluid should be more than 0.1 litre (about a fifth of a pint); if it is not, the fluid level in the transmission is incorrect.

16 If fluid is to be added, pour 1 litre (nearly 2

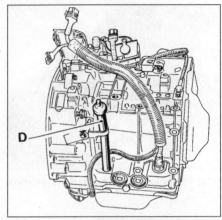

Fig. 1.20 AD4 automatic transmission fluid filler tube (D) and vent (Sec 13)

pints) of the specified fluid through the filler tube after removing the vent. Allow the transmission to cool down, then repeat the checking procedure again as described in the previous paragraphs.

AD4 automatic transmission final drive

17 This is not a routine operation, but it may be considered necessary if there is reason to suspect that the oil level is incorrect - for instance if there has been an oil leak.

18 Position the car on ramps, or jack it up and support on axle stands. Make sure that the car is level.

19 Unscrew the final drive level plug located behind the right-hand driveshaft on the right-hand side of the transmission.

20 Check that the level of the oil is up to the bottom of the plug hole. If not, inject oil of the correct grade into the hole until it overflows.

21 Wipe clean the level plug, then refit and tighten it.

22 Lower the car to the ground.

---

14 Front brake check

1 Firmly apply the handbrake, then jack up the front of the vehicle and support it securely on axle stands. Remove the front roadwheels.

2 For a quick check, the thickness of friction material remaining on each brake pad can be measured through the aperture in the caliper body. If any pad's friction material is worn to the specified thickness or less, all four pads must be renewed as a set. (Pad wear warning contacts are fitted to the inboard pads, but this should not be used as an excuse for omitting a visual check.)

3 For a comprehensive check, the brake pads should be removed and cleaned. This will allow the operation of the caliper to be checked, and the brake disc itself to be fully examined for condition on both sides. Refer to Chapter 9 for further information.

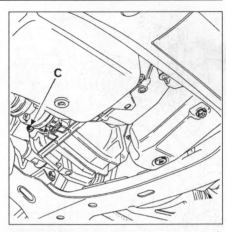

Fig. 1.21 AD4 automatic transmission final drive oil level plug (C) (Sec 13)

---

15 Braking system check

1 Check that the brake pedal pivot is properly greased, and that the pedal moves smoothly and easily through its full travel. When the engine is switched off, the pedal should have a small amount of free play, then firm resistance. If the pedal feels spongy or pedal travel is excessive, the system should be checked further, with reference to Chapter 9. On models with rear drum brakes, if the pedal travel is excessive, first check the handbrake mechanism (as described below) before investigating further.

2 The brake hydraulic system consists of a number of metal hydraulic pipes, which run from the master cylinder around the engine compartment to the ABS modulator (where fitted) and front brakes, then along the underbody to the rear brakes. Flexible hoses are fitted at front and rear to cater for steering and suspension movement.

3 When checking the system, first look for signs of leakage at the pipe or hose unions, then examine the flexible hoses for signs of cracking, chafing or deterioration of the rubber (photo). Bend them between the fingers (but do not actually bend them double, or the casing may be damaged) and check

15.3 Checking the condition of a flexible brake hose

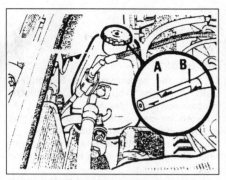

Fig. 1.22 Power steering fluid reservoir and dipstick markings - 1390 cc (E6J and E7J) engined models not equipped with air conditioning (Sec 16)

A - minimum fluid level
B - maximum fluid level

that this does not reveal previously-hidden cracks, cuts or splits. Check that all pipes and hoses are securely fastened in their clips.
4 Carefully work along the length of the metal hydraulic pipes, looking for dents, kinks, damage of any sort or corrosion. Corrosion should be polished off; if the depth of pitting is significant, the pipe must be renewed.

16 Power steering fluid level check

1 The shape and location of the power steering fluid reservoir varies from model to model.
(a) *On 1390 cc (E6J and E7J) engined models not equipped with air conditioning, the reservoir is an integral part of the pump assembly, and is mounted on the front right-hand corner of the cylinder head.*
(b) *On 1721 cc (F2N) and 1794 cc (F3P) engined models equipped with air conditioning, the reservoir is an integral part of the electric pump assembly, and is situated by the side of the battery.*

16.5 Top-up the power steering fluid level to the maximum mark using only the specified type of fluid

16.4 Power steering fluid reservoir MAX and MIN fluid level markings - circular fluid reservoir

(c) *On all other models, the reservoir is either circular in shape and situated at the front right-hand corner of the engine compartment, or is rectangular in shape and is mounted on the right-hand side of the engine compartment, just in front of the suspension turret.*
2 The fluid level should be checked cold, the car should be parked on level ground with the front wheels pointing straight-ahead, and the engine should be stopped. Note that for the check to be accurate, the steering **must not** be operated once the engine has been stopped.
3 On 1390 cc (E6J and E7J) engined models not equipped with air conditioning, wipe the area around the reservoir filler cap, then unscrew the cap and withdraw it from the fluid reservoir. Wipe the dipstick clean, then insert it into the reservoir and withdraw it and check the fluid level. The fluid should be between the marks (A and B, Fig. 1.22) on the dipstick. If not, top-up to the higher mark (B) using the specified type of fluid then refit the reservoir cap, tightening it securely.
4 On models equipped with an electrical power steering pump, or with the circular reservoir which is mounted in the right-hand corner of the engine compartment, the fluid level is visible through the translucent material of the reservoir. The level should be between the 'MAX' and 'MIN' level lines cast on the side of the reservoir (photo). If necessary, wipe the area around the reservoir cap clean, then remove the cap and top-up to the 'MAX' mark using the specified type of fluid.
5 On models equipped with the rectangular fluid reservoir, wipe the area around the cap clean, then unscrew the cap. The fluid level indicator is in the form of a block fixed to the centre of the filler neck filter. The fluid level should be between the upper and lower edges of the block (marks A and B, Fig. 1.23). If not, top-up to the upper edge of the indicator block using the specified type of fluid (photo), then refit the reservoir cap, tightening it securely.
6 On all models, take great care not to allow

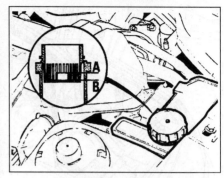

Fig. 1.23 Power steering fluid reservoir and level indicator - rectangular fluid reservoir (Sec 16)

A - maximum fluid level
B - minimum fluid level

any dirt or foreign matter to enter the hydraulic system, and do not overfill the reservoir. When the level is correct, refit the cap. Note that the need for frequent topping-up of the system indicates a leak, which should be investigated immediately.

17 Bodywork damage/corrosion check

1 Once the car has been washed and all tar spots and other surface blemishes have been cleaned off, carefully check all paintwork, looking closely for chips or scratches. Pay particular attention to vulnerable areas such as the front panels (bonnet and spoiler), and around the wheel arches. Any damage to the paintwork must be rectified as soon as possible to comply with the terms of the manufacturer's anti-corrosion warranties; check with a Renault dealer for details.
2 If a chip or light scratch is found which is recent and still free from rust, it can be touched-up using the appropriate touch-up stick which can be obtained from Renault dealers. Any more serious damage, or rusted stone chips, can be repaired as described in Chapter 11, but if damage or corrosion is so severe that a panel must be renewed, seek professional advice as soon as possible.
3 Always check that the door and ventilation opening drain holes and pipes are completely clear, so that water can drain out.
4 The wax-based underbody protective coating should be inspected annually, preferably just prior to Winter, when the underbody should be washed down as thoroughly as possible without disturbing the protective coating (see Chapter 11, Section 1, regarding the use of steam cleaners). Any damage to the coating should be repaired using a suitable wax-based sealer. If any of the body panels are disturbed for repair or renewal, do not forget to replace the coating and to inject wax into door panels, sills and box sections, to maintain the level of protection provided by the vehicle manufacturer.

1

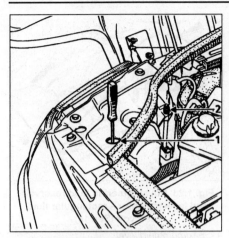

Fig. 1.24 Headlight beam adjustment screws (Sec 18)

1 Vertical adjustment screw
2 Horizontal adjustment screw

18 Headlight beam alignment check

1 Accurate checking and adjustment of the headlight beams is only possible using optical beam setting equipment, and this work should therefore be carried out by a Renault dealer or a garage with the necessary facilities.
2 In an emergency, the headlight beams can be adjusted by turning the screws shown in Fig. 1.24.

19 Electrical system checks

Battery condition

 Warning: Before carrying out any work on the battery, read through the precautions given in 'Safety first!' at the beginning of this manual.

1 The battery is located at the front left-hand corner of the engine compartment.
2 The exterior of the battery should be inspected for damage such as a cracked case or cover.
3 Check the tightness of the battery cable clamp nuts to ensure good electrical connections, and check the entire length of each cable for cracks and frayed conductors.
4 If corrosion (visible as white, fluffy deposits) is evident, remove the cables from the battery terminals, clean them with a small wire brush, then refit them. Corrosion can be kept to a minimum by applying a thin layer of petroleum jelly to the clamps and terminals after they have been reconnected.
5 Make sure that the battery tray is in good condition, and that the retaining clamp is tight.
6 Corrosion on the tray, retaining clamp and the battery itself can be removed with a solution of water and baking soda. Thoroughly rinse all cleaned areas with plain water. Dry the

19.21 Releasing the securing catch to remove a windscreen wiper blade

battery and its surroundings with rags or tissues, which should then be discarded.
7 Any metal parts damaged by corrosion should be covered with a zinc-based primer, then painted.
8 Most models are fitted with a 'maintenance-free' battery which does not require topping-up. If this is the case, the battery will be sealed, and it will not be possible to remove the cell covers.
9 On models fitted with a low-maintenance or a conventional battery, the electrolyte level should be checked periodically as follows.
10 Pull out the cell covers from the top of the battery.
11 Check that the level of electrolyte is approximately 15 mm above the tops of the cell plates.
12 If necessary top-up the level, using only distilled or demineralised water.
13 Refit the cell covers.
14 Further information on the battery, charging and jump-starting can be found in Chapter 12 and in the preliminary Sections of this manual.

Windscreen/tailgate wash/wipe system

15 Periodically check that the fluid reservoir is full, as described previously in Section 1.
16 Check the security of the pump wiring and washer tubing.
17 Check that the washer jets direct the fluid onto the upper part of the windscreen/tailgate, and if necessary adjust the aim of the jets using a pin.
18 Check the operation of the wipers and washers, and rectify any faults immediately. Refer to Chapter 12 for further details.

Wiper blades

19 The wiper blades should be renewed if their cleaning action has deteriorated, or if they are cracked, or if they no longer clean the glass effectively.
20 Lift the wiper arm away from the glass. Note that the tailgate wiper arm can only be lifted a limited distance.
21 To remove a windscreen wiper arm, depress the catch, then turn the blade through 90° and withdraw the blade from the end of the arm (photo).
22 To remove a tailgate wiper blade, lift the catch, then push the blade from the arm.
23 Insert the new blade into the arm, making sure that it locates securely.

20 Road test

Drivetrain

1 Check the performance of the engine, clutch (where applicable), gearbox/transmission and driveshafts.
2 Listen for any unusual noises from the engine, clutch and gearbox/transmission.
3 Make sure that the engine runs smoothly when idling, and that there is no hesitation when accelerating.
4 Check that, where applicable, the clutch action is smooth and progressive, that the drive is taken up smoothly, and that the pedal travel is not excessive. Also listen for any noises when the clutch pedal is depressed.
5 On manual gearbox models, check that all gears can be engaged smoothly without noise, and that the gear lever action is not abnormally vague or 'notchy'.
6 On automatic transmission models, make sure that all gearchanges occur smoothly without snatching, and without an increase in engine speed between changes. Check that all the gear positions can be selected with the car at rest. If any problems are found, they should be referred to a Renault dealer.
7 Listen for a metallic clicking sound from the front of the car as the car is driven slowly in a circle with the steering on full-lock. Carry out this check in both directions. If a clicking noise is heard, this indicates wear in a driveshaft joint, in which case renew the joint if necessary.

Check operation and performance of the braking system

8 Make sure that the car does not pull to one side when braking, and that the wheels do not lock prematurely when braking hard.
9 Check that there is no vibration through the steering when braking.
10 Check that the handbrake operates correctly without excessive movement of the lever, and that it holds the car stationary on a slope.
11 Test the operation of the brake servo unit as follows. Depress the footbrake four or five times to exhaust the vacuum, then start the engine. As the engine starts, there should be a noticeable 'give' in the brake pedal as vacuum builds up. Allow the engine to run for at least two minutes and then switch it off. If the brake pedal is now depressed again, it should be possible to detect a hiss from the servo as the pedal is depressed. After about four or five applications, no further hissing should be heard, and the pedal should feel considerably harder.

Steering and suspension

12 Check for any abnormalities in the steering, suspension, handling or road 'feel'.
13 Drive the car, and check that there are no unusual vibrations or noises.
14 Check that the steering feels positive, with no excessive 'sloppiness' or roughness, and check for any suspension noises when cornering and driving over bumps.

21.3 Removing an air cleaner cover screw - 1794 cc (F3P) engine. Note vacuum hose clipped to cover

21.4 Three screws securing the air cleaner housing - 1764 cc (F7P) engine

21.6A Removing the air cleaner element - 1794 cc (F3P) engine . . .

21.6B . . . and 1764 cc (F7P) engine

Every 12 000 miles

21 Air cleaner element renewal

1108 cc (C1E) engine

1 Release the spring clips and remove the wing nut from the air cleaner cover (Fig. 1.25).

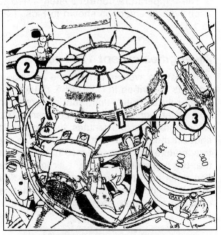

Fig. 1.25 Air cleaner cover removal - 1108 cc (C1E) engine (Sec 21)

2 Wing nut   3 Spring clip

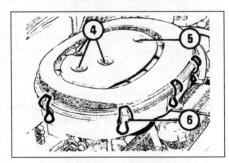

Fig. 1.26 Air cleaner cover removal - E-type engine (Sec 21)

4 Screws   5 Screw   6 Spring clip

1171 and 1390 cc (E-type) engines

2 Remove the four screws from the top of the cover, and release the spring clips (Fig. 1.26).

1721 and 1794 cc (F2N and F3P) engines

3 Remove the five screws from around the edge of the air cleaner cover (photo). Release the vacuum hose from the clips on the cover.

1764 cc (F7P) engine

4 Release the seven spring clips and remove the three screws from the air cleaner housing (photo).

All models

5 Lift off the air cleaner cover, or separate the parts of the housing, as applicable.
6 Remove the element from inside the air cleaner body (photos).
7 Clean the inside of the air cleaner body, being careful not to get dirt into the intake duct. Fit a new filter element.
8 Refit the cover, or reassemble the housing and secure with the screws or wing nut and spring clips.

22 HT leads, distributor cap and rotor arm check and renewal

⚠ *Warning: Voltages produced by an electronic ignition system are considerably higher than those produced by conventional systems. Extreme care must be taken when working on the system with the ignition switched on. Persons with surgically-implanted cardiac pacemaker devices should keep well clear of the ignition circuits, components and test equipment.*

1 The spark plug HT leads should be checked whenever new spark plugs are installed in the engine.
2 Ensure that the leads are numbered before removing them, to avoid confusion when refitting. Pull one HT lead from its plug by gripping the end fitting, not the lead, otherwise the lead connection may be fractured.
3 Check inside the end fitting for signs of corrosion, which will look like a white crusty powder. Scrape out such deposits using a small screwdriver. Push the end fitting back onto the spark plug, ensuring that it is a tight

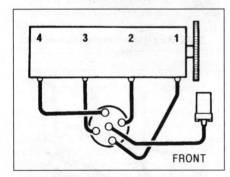

Fig. 1.27 HT lead connection diagram - 1108 cc (C1E) engine only (Sec 22)

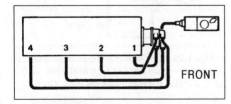

Fig. 1.28 HT lead connection diagram - all other engines (Sec 22)

1

fit on the plug. If not, remove the lead again and use pliers to carefully crimp the metal connector inside the end fitting until it fits securely on the end of the spark plug.

4 Using a clean rag, wipe the entire length of the lead to remove any built-up dirt and grease. Once the lead is clean, check for burns, cracks and other damage. Do not bend the lead excessively or pull the lead lengthwise - the conductor inside might break.

5 Disconnect the other end of the lead from the distributor cap. Again, pull only on the end fitting. Check for corrosion and a tight fit in the same manner as the spark plug end. If an ohmmeter is available, check the continuity of the HT lead by connecting the meter between the spark plug end of the lead and the segment inside the distributor cap. (Resistive leads are commonly fitted, giving an ohmmeter reading of several hundred or thousand ohms per lead.) Refit the lead securely on completion.

6 Check the remaining HT leads one at a time, in the same way.

7 If new HT leads are required, purchase a set suitable for your specific vehicle and engine.

8 Remove the distributor cap, wipe it clean and carefully inspect it inside and out for signs of cracks, carbon tracks (tracking) and worn, burned or loose contacts. Similarly inspect the rotor arm. Renew these components if any defects are found (see Chapter 5). It is common practice to renew the cap and rotor arm whenever new HT leads are fitted. When fitting a new cap, remove the HT leads from the old cap one at a time, and fit them to the new cap in the exact same location - do not simultaneously remove all the leads from the old cap, or firing-order confusion may occur.

9 Even with the ignition system in first-class condition, some engines may still occasionally experience poor starting, attributable to damp ignition components; a spray sealing coat to exclude moisture from the ignition system will help to start a car when only a very poor spark occurs.

Every 30 000 miles

23 Air cleaner temperature control system check

1 In order for the engine to operate efficiently, the temperature of the air entering the inlet system is controlled within certain limits. (This does not apply to the 1764 cc/F7P engine, on which the fuel injection system compensates for changes in air temperature.)

2 The air cleaner has two sources of air, one direct from the outside of the engine compartment (cold air), and the other from a shroud on the exhaust manifold (warm air). A wax capsule thermostat controls a flap inside the air cleaner inlet, either directly or under the influence of a vacuum capsule (photos). When the ambient air temperature is below the predetermined level, the flap directs warm air from the exhaust manifold shroud. As the temperature of the incoming (ambient) air rises, the flap opens to admit colder air from outside the car until eventually it is fully open.

3 The temperature control system fitted to each type of engine is shown in Figs. 1.29 to 1.31.

All engines except 1794 cc (F3P)

4 To test the system, remove the air cleaner body as described in Chapter 4, detach the temperature control unit, and immerse it in cold water (temperature no more than 20ºC). After a maximum of 5 minutes, the flap should shut off the cold air inlet. Now increase the temperature of the water to 40ºC, and check that the flap shuts off the hot air inlet. If the flap does not operate correctly, check first that it is not seized. No adjustment is possible; the unit should be renewed if faulty.

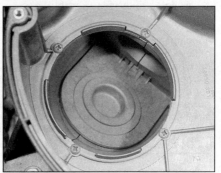

23.2A Air cleaner temperature control flap . . .

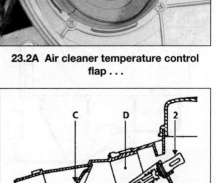

Fig. 1.29 Air cleaner temperature control system - 1108 cc, 1171 cc and 1390 cc (C- and E-type) engines (Sec 23)

A Cold air inlet
B Hot air inlet
C Control flap
D To carburettor
2 Thermostat

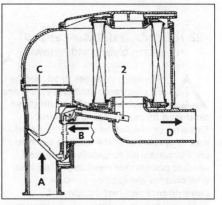

23.2B . . . and wax thermostat. 1721 cc (F2N) engine shown

Fig. 1.30 Air cleaner temperature control system - 1721 cc (F2N) engine (Sec 23)

A Cold air inlet
B Hot air inlet
C Control flap
D To carburettor
2 Thermostat

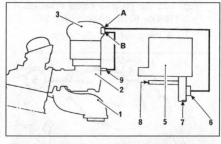

Fig. 1.31 Air cleaner temperature control system - 1794 cc (F3P) engine (Sec 23)

A To vacuum capsule
B From inlet manifold
1 Exhaust manifold
2 Inlet manifold
3 Air intake cover
5 Air cleaner
6 Thermostat/ vacuum capsule
7 Cold air inlet
8 Hot air inlet
9 Manifold vacuum connection

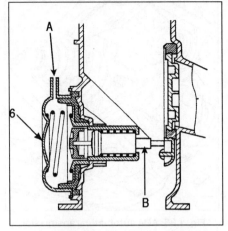

Fig. 1.32 Detail of thermostatic vacuum valve (Sec 23)

A To vacuum capsule   C Ball valve
B From inlet manifold   D Bi-metallic strip

1794 cc (F3P) engine

5 This system is different in that the position of the flap is determined not only by temperature but also by inlet manifold vacuum. The combined thermostat and vacuum capsule in the air cleaner inlet shuts off the cold air inlet at temperatures below - 10°C, and shuts off the hot air inlet at temperatures above +20°C. However, a high manifold vacuum (in excess of 200 mbars) will cause the cold air inlet to be closed again at temperatures up to 35°C. Above 35°C, the vacuum feed to the capsule is cut off by a thermostatic valve located in the air intake cover, and the hot air inlet is closed.

6 Checking of this system for the DIY mechanic is more difficult; begin by making sure that all the vacuum hoses are sound.

7 The combined thermostat and vacuum capsule can be checked as described for the directly-operated system, but place the unit in a freezer to check low temperature operation. Warm the unit gently (for instance with a hairdryer) to check high temperature operation.

8 To check the operation of the thermostatic

valve, disconnect the hose from the manifold side of the valve. Connect a length of plastic or rubber hose to the valve and suck through the hose, or apply vacuum using a hand pump if available. At valve temperatures below 35°C, the application of vacuum should move the flap to close the cold air inlet. At higher temperatures, the application of vacuum should have no effect. The movement of the flap cannot be seen with the air cleaner fitted, but it can be heard.

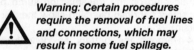

24 Fuel filter renewal

> ⚠ **Warning: Certain procedures require the removal of fuel lines and connections, which may result in some fuel spillage.** *Before carrying out any operation on the fuel system, refer to the preCautions given in 'Safety first!' at the beginning of the manual, and follow them implicitly. Petrol is a highly dangerous and volatile liquid, and the preCautions necessary when handling it cannot be overstressed.*

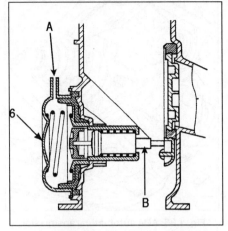

Fig. 1.33 Detail of thermostat/vacuum capsule unit (6) (Sec 23)

A From vacuum valve   E Thermostat

Carburettor models

1 An in-line fuel filter is provided in the fuel pump outlet line. To remove it, release the clips (if fitted) and pull the filter from the hoses (photo).

2 Fit the new filter, making sure that the arrow is pointing in the direction of fuel flow (from the fuel pump towards the carburettor). If crimped-type clips were used to secure the hoses, these should be replaced with worm drive clips on reassembly.

Fuel injection models

3 The fuel filter is located under the car, just in front of the fuel tank. Raise and support the rear of the car.

4 Clamp the hose on the tank side of the filter. Remove the Torx screw from the filter securing clamp, and slacken the hose clips. If crimped-type hose clips are fitted, cut them off and obtain worm drive clips for reassembly. Remove the filter and dispose of it safely.

5 Fit the new filter, making sure that the arrow is pointing in the direction of fuel flow (from the fuel pump towards the engine). Tighten the hose clips and the filter securing clamp (photos).

6 Release the hose clamp and check that there are no leaks, then lower the car to the ground.

25 Clutch check

Note: *The following check is not specified by Renault. but is included by the author.*

1 Check that the clutch pedal moves smoothly and easily through its full travel, and that the clutch itself functions correctly, with no trace of slip or drag. If the movement is uneven or stiff in places, check that the cable is routed correctly, with no sharp turns.

2 Inspect the ends of the clutch inner cable, both at the gearbox end and inside the car, for signs of wear and fraying.

1

24.1 Typical in-line fuel filter fitted to carburettor models

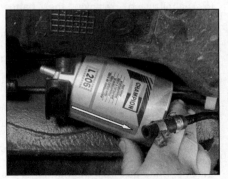

24.5A Fitting a new fuel filter to a fuel injection model. Arrow on filter shows direction of fuel flow

24.5B Tightening the fuel filter securing clamp

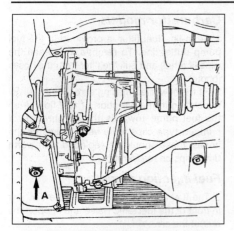

Fig. 1.34 MB1 automatic transmission drain plug (A) (Sec 26)

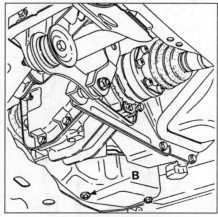

Fig. 1.35 AD4 automatic transmission drain plug (B) (Sec 26)

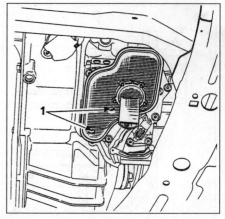

Fig. 1.36 AD4 automatic transmission filter retaining bolts (1) (Sec 26)

26 Automatic transmission fluid and gauze filter renewal

Note: *If working on the AD4 transmission, note that special Renault test equipment is necessary to check the fluid level, and the home mechanic would be well advised to take the car to a Renault dealer to have the work carried out. On all types of automatic transmission, take great care not to introduce dirt into the transmission.*

MB1 (3-speed) automatic transmission

1 The automatic transmission fluid should only be changed when cold.

2 Position the car over an inspection pit, on car ramps, or jack it up, but make sure that it is level. Where applicable, remove the engine splash shield.

3 Place a suitable container beneath the drain plug on the transmission sump pan and at the base of the transmission housing.

4 Remove the dipstick to speed up the draining operation, then unscrew the drain plug and allow the fluid to drain.

5 When all the fluid has drained (this may take quite some time) clean the drain plug, refit it together with a new seal, and tighten securely.

6 Unbolt and remove the sump pan and gasket (it may be necessary to raise the transmission first).

7 Unscrew the filter retaining bolts and remove the filter and sealing ring. Note which way round the filter is fitted.

8 Fit the new filter into position, ensuring it is the correct way round. Tighten the bolts to the specified torque.

9 Clean the sump pan and check that the filter magnets are in position, then refit the sump pan, ensuring that the new gasket is correctly located. Tighten the bolts.

10 Place a funnel with a fine-mesh screen in the dipstick tube, and fill the transmission with the specified type of fluid. Depending on the

extent to which the fluid was allowed to drain, refilling will only require approximately 2 litres (3.5 pints). Add about half this amount, and then check the level on the dipstick. When the level approaches the mark, place the selector lever in 'P', start the engine and allow it to run for approximately 2 minutes. Now check the level and complete the final topping-up as described previously.

AD4 (4-speed) automatic transmission

11 The automatic transmission fluid should only be changed when cold.

12 Position the car over an inspection pit, on car ramps, or jack it up, but make sure that it is level.

13 Place a suitable container beneath the drain plug on the transmission sump pan, unscrew the drain plug and allow the fluid to drain.

14 Remove the battery as described in Chapter 12.

15 Disconnect the wiring harness mounting clips from the automatic transmission and battery mounting.

16 Detach the power-assisted steering hose mounting from the battery mounting.

17 Using a trolley jack and block of wood, raise the automatic transmission far enough to unbolt and remove the battery mounting/transmission mounting bracket.

18 Remove the front left-hand roadwheel, and remove the plastic cover in order to gain access to the transmission pan mounting bolts.

19 Working under the car, unscrew (but do not remove) the rear bolt securing the rear engine mounting link to the underbody. Unscrew and remove the front bolt, and swivel the link downwards.

20 Raise the automatic transmission as far as possible.

21 Unbolt the pan from the bottom of the transmission, and remove the gasket.

22 Unbolt the filter and its seal.

23 Clean the pan and the sediment magnet.

Examine the pan gasket and the drain and topping-up plugs, and renew them if necessary.

24 Fit the new filter together with its seal, and tighten the bolts.

25 Fit the pan together with a new gasket, and tighten the bolts.

26 The remaining procedure is a reversal of removal, but finally refill the transmission with fluid as described in Section 13.

27 Driveshaft rubber gaiter and CV joint check

1 With the car raised and securely supported on axle stands, turn the steering onto full-lock, then slowly rotate the roadwheel. Inspect the condition of the outer constant velocity (CV) joint rubber gaiters while squeezing the gaiters to open out the folds (photo). Check for signs of cracking, splits or deterioration of the rubber, which may allow the grease to escape and lead to the entry of water and grit into the joint. Also check the security and condition of the retaining clips/fasteners. Repeat these checks on the inner CV joints. If any damage or deterioration is found, the gaiters should be renewed as described in Chapter 8.

27.1 Checking the condition of the driveshaft outer constant velocity (CV) joint rubber gaiter

2 At the same time, check the general condition of the CV joints themselves by first holding the driveshaft and attempting to rotate the roadwheel. Repeat this check by holding the inner joint and attempting to rotate the driveshaft. Any appreciable movement indicates wear in the joints, in the driveshaft splines, or a loose driveshaft nut.

28 Rear brake check

Brake shoe, wheel cylinder and drum

1 Chock the front wheels, then jack up the rear of the vehicle and support it on axle stands.
2 Remove the rear brake drums as described in Chapter 9. This will allow the brake shoes and wheel cylinders to be checked, and the brake drum itself to be fully examined for condition. Refer to Chapter 9 for further information.

Brake pad, caliper and disc

3 Chock the front wheels, then jack up the rear of the vehicle and support it on axle stands. Remove the rear roadwheels.
4 For a quick check, the thickness of friction material remaining on each brake pad can be measured through the aperture in the caliper body. If any pad's friction material is worn to the specified thickness or less, all four pads must be renewed as a set.
5 For a comprehensive check, the brake pads should be removed and cleaned. This will allow the operation of the caliper to be checked, and the brake disc itself to be fully examined for condition on both sides. Refer to Chapter 9 for further information.

29 Handbrake check and adjustment

1 The handbrake should be capable of holding the parked vehicle stationary, even on steep slopes, when applied with moderate force. The mechanism should be firm and positive in feel, with no trace of stiffness or sponginess from the cables, and should release immediately the handbrake lever is released. If the mechanism is faulty in any of these respects, it must be checked immediately as follows. **Note:** *On models with rear drum brakes, if the handbrake is not functioning correctly or is incorrectly adjusted, the rear brake self-adjust mechanism will not function. This will lead to the brake pedal travel becoming excessive as the shoe linings wear. Under no circumstances should the handbrake cables be tightened in an attempt to compensate for excessive brake pedal travel.*

2 Chock the front wheels and release the handbrake. Jack up the rear of the vehicle and support it on axle stands. On models with a catalytic converter, undo the heat shield retaining nut(s) and lower the rear of the heat shield to gain access to the handbrake cable adjuster nut.
3 Slacken the locknut, then fully slacken the cable adjuster nut (photo).

Models with rear drum brakes

4 Remove both the rear brake drums as described in Chapter 9.
5 Check that the knurled adjuster wheel on the adjuster strut is free to rotate in both directions. If it is seized, the brake shoes and strut must be removed and overhauled as described in Section 14 of Chapter 9.
6 If all is well, back off the adjuster wheel by five or six teeth so that the diameter of the brake shoes is slightly reduced.
7 Check that the handbrake cables slide freely by pulling on their front ends. Also check that the operating levers on the rear brake trailing shoes return to their correct positions, with their stop-pegs in contact with the edge of the trailing shoe web.
8 With the aid of an assistant, tighten the adjuster nut on the handbrake lever operating rod so that the lever on each rear brake assembly starts to move as the handbrake is moved between the first and second notch (click) of its ratchet mechanism. This is the case when the stop-pegs are still in contact with the shoes when the handbrake is on the first notch of the ratchet, but no longer contact the shoes when the handbrake is on the second notch. Once the adjustment is correct, hold the adjuster nut and securely tighten the locknut. Where necessary, refit the catalytic converter heat shield retaining nuts.
9 Refit the brake drums as described in Chapter 9, then lower the vehicle to the ground.
10 With the vehicle standing on its wheels, repeatedly depress the footbrake to adjust the shoe-to-drum clearance. Whilst depressing the pedal, have an assistant listen to the rear drums to check that the adjuster strut mechanism is functioning; if this is so, a clicking sound will be heard from the adjuster strut as the pedal is depressed.

Models with rear disc brakes

11 Check that the handbrake cables slide freely by pulling on their front ends, and check that the operating levers on the rear brake calipers move smoothly.
12 Move both of the caliper operating levers as far rearwards as possible, then tighten the adjuster nut on the handbrake lever operating rod until all free play is removed from both cables. With the aid of an assistant, from this point adjust the nut so that the operating lever on each rear brake caliper starts to move as

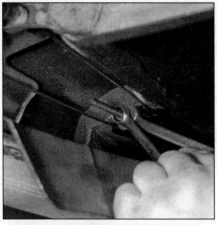

29.3 Adjusting the handbrake cable

the handbrake lever is moved between the first and second notch (click) of its ratchet mechanism. Once the handbrake adjustment is correct, hold the adjuster nut and securely tighten the locknut.
13 Refit the catalytic converter heat shield retaining nuts (where necessary), then lower the vehicle to the ground.

30 Brake fluid renewal

1 The procedure is similar to that for the bleeding of the hydraulic system described in Chapter 9. Before starting, remove as much old brake fluid as possible from the reservoir by syphoning, using a clean poultry baster or similar.
2 Working as described in Chapter 9, open the first bleed nipple in the sequence, and pump the brake pedal gently until nearly all the old fluid has been emptied from the master cylinder reservoir. Top-up to the 'MAX' level with new fluid, and continue pumping until only the new fluid remains in the reservoir and new fluid can be seen emerging from the bleed nipple. Tighten the nipple and top the reservoir level up to the 'MAX' level line.
3 Old hydraulic fluid is invariably much darker in colour than the new, making it easy to distinguish the two.
4 Work through all the remaining nipples in the sequence until new fluid can be seen at all of them. Be careful to keep the master cylinder reservoir topped-up to above the 'MIN' level at all times, or air may enter the system and greatly increase the length of the task.
5 When the operation is complete, check that all nipples are securely tightened and that their dust caps are refitted. Wash off all traces of spilt fluid, and recheck the master cylinder reservoir fluid level.
6 Check the operation of the brakes before taking the vehicle on the road.

1

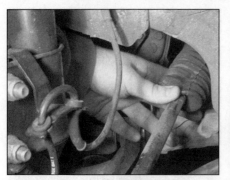

31.2 Checking the condition of a steering gear rubber gaiter

31.4 Checking for wear in the front suspension and hub bearings

31 Suspension and steering check

Front suspension and steering

1 Raise the front of the car and securely support it on axle stands.

2 Visually inspect the balljoint dust covers and the steering gear rubber gaiters for splits, chafing or deterioration (photo). Any wear of these components will cause loss of lubricant, together with dirt and water entry, resulting in rapid deterioration of the balljoints or steering gear.

3 On cars with power steering, check the fluid hoses for chafing or deterioration, and the pipe and hose unions for fluid leakage. Also check for signs of fluid leakage under pressure from the steering gear rubber gaiters, which would indicate failed fluid seals within the steering gear.

4 Grasp the roadwheel at the 12 o'clock and 6 o'clock positions, and try to rock it (photo). Very slight free play may be felt, but if the movement is appreciable, further investigation is necessary to determine the source. Continue rocking the wheel while an assistant depresses the brake pedal. If the movement is now eliminated or significantly reduced, it is likely that the hub bearings are at fault. If the free play is still evident with the brake pedal depressed, then there is wear in the suspension joints or mountings.

5 Now grasp the roadwheel at the 9 o'clock and 3 o'clock positions, and try to rock it as before. Any movement felt now may again be caused by wear in the hub bearings, or in the track rod balljoints. If a balljoint is worn, the visual movement will be obvious. If the inner joint is suspect, it can be felt by placing a hand over the steering gear rubber gaiter and gripping the track rod. If the wheel is now rocked, movement will be felt at the inner joint if wear has taken place.

6 Using a large screwdriver or flat bar, check for wear in the suspension mounting bushes by levering between the relevant suspension component and its attachment point. Some movement is to be expected, as the mountings are made of rubber, but excessive wear should be obvious. Also check the condition of any visible rubber bushes, looking for splits, cracks or contamination of the rubber.

7 With the car standing on its wheels, have an assistant turn the steering wheel back and forth about an eighth of a turn each way. There should be very little, if any, lost movement between the steering wheel and the roadwheels. If this is not the case, closely observe the joints and mountings previously described, but in addition check for wear of the steering column universal joint and the steering gear itself.

Rear suspension

8 Chock the front wheels, then jack up the rear of the car and support it on axle stands.

9 Working as described above for the front suspension, check the rear hub bearings, the suspension bushes and the shock absorber mountings for wear.

32 Bodywork and fittings check

Seat belt check

1 Carefully examine the seat belt webbing for cuts, or any signs of serious fraying or deterioration. If the seat belt is of the retractable type, pull the belt all the way out and examine the full extent of the webbing.

2 The seat belts are designed to lock up during a sudden stop or impact, yet allow free movement during normal driving. Fasten and unfasten the belt, ensuring that the locking mechanism holds securely and releases properly when intended. Check also that the retracting mechanism operates correctly when the belt is released.

Hinge and lock lubrication

3 Lubricate the hinges of the bonnet, doors and tailgate with a light general-purpose oil. Similarly lubricate all latches, locks and lock strikers. At the same time, check the security and operation of all the locks, adjusting them if necessary (see Chapter 11).

4 Lightly lubricate the bonnet release mechanism and cable with a suitable grease.

Every 2 years

33 Coolant renewal

Draining

⚠️ **Warning: Wait until the engine is cold before starting this procedure. Do not allow antifreeze to come in contact with your skin, or with the painted surfaces of the vehicle. Rinse off spills immediately with plenty of water.**

1 If the engine is cold, unscrew and remove the pressure cap from the expansion tank. If it is not possible to wait until the engine is cold, place a cloth over the pressure cap of the expansion tank and **slowly** unscrew the cap. Wait until all pressure has escaped, then remove the cap.

2 Place a suitable container beneath the bottom hose connection to the radiator or water pump.

3 Loosen the clip, then disconnect the bottom hose and allow the coolant to drain into the container. The hose clips fitted as original equipment are released by squeezing the tags together with pliers.

4 Move the container beneath the cylinder block drain plug. On the 1108 cc (C1E) engine, the drain plug is located on the crankshaft pulley end of the engine, below the water pump. On the other engines, it is

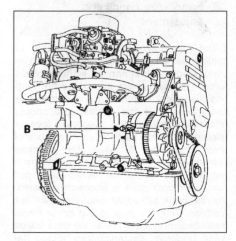

Fig. 1.37 Coolant drain plug (B) on the rear of the cylinder block - E-type engine (Sec 33)

33.4 Coolant drain plug (arrowed) on the rear of the cylinder block - F-type engine (engine removed for clarity)

located on the rear right-hand side of the cylinder block (photo and Fig. 1.37).

5 Unscrew the plug, and drain the coolant into the container.

6 Flush the system if necessary as described in the following paragraphs, then refit the drain plug and secure the bottom hose. Use a new hose clip if necessary. Refill the system as described later in this Section.

System flushing

7 With time, the cooling system may gradually lose its efficiency if the radiator matrix becomes choked with rust and scale deposits. If this is the case, the system must be flushed as follows. First drain the coolant as already described.

8 Loosen the clip and disconnect the top hose from the radiator. Insert a garden hose in the radiator top hose connection stub, and allow the water to circulate through the radiator until it runs clear from the bottom outlet.

9 To flush the engine and the remainder of the system, disconnect the top hose at the water pump on the 1108 cc (C1E) engine, or remove the thermostat as described in Chapter 3 on the other engines. Insert the garden hose, and allow the water to circulate through the engine until it runs clear from the bottom hose.

33.14 Open the bleed screw (arrowed) on the radiator when refilling the cooling system

10 In severe cases of contamination, the radiator should be reverse-flushed. To do this, first remove it from the car, as described in Chapter 3, invert it and insert a hose in the bottom outlet. Continue flushing until clear water runs from the top hose outlet.

11 If, after a reasonable period, the water still does not run clear, the radiator should be flushed with a good proprietary cleaning system.

12 Regular renewal of the antifreeze mixture as specified should prevent severe contamination of the system. Note that as the radiator is made of aluminium, it is important not to use caustic soda or alkaline compounds to clean it. See Chapter 3 for more details.

Filling

13 Refit the cylinder block drain plug, radiator bottom hose and any other hoses removed if the system has just been flushed.

14 Open the coolant bleed screws. On the 1108 cc engine, there are three bleed screws - one each on the heater supply and return hoses, and one in the top hose next to the thermostat. On the other engines, there is only one, situated on one side of the radiator towards the top (photo).

15 Pour the appropriate mixture of water and antifreeze into the expansion tank, and close

each bleed screw in turn as soon as a continuous flow of bubble-free coolant can be seen flowing from it. Continue to fill the expansion tank until the coolant is at the maximum level.

16 Start the engine and run it at 1500 rpm (ie a fast idle speed) for approximately 4 minutes. Keep the expansion tank topped-up to the maximum level during this period.

17 Refit the pressure cap to the expansion tank, and run the engine at 1500 rpm for approximately 10 minutes until the electric cooling fan cuts in. During this period, the coolant will circulate around the engine and any remaining air will be purged to the expansion tank.

18 Switch off the engine and allow it to cool, then check the coolant level as described earlier and top-up if necessary.

Antifreeze mixture

19 The antifreeze should always be renewed at the specified intervals. This is necessary not only to maintain the antifreeze properties, but also to prevent corrosion which would otherwise occur as the corrosion inhibitors become progressively less effective.

20 Always use an ethylene-glycol based antifreeze which is suitable for use in mixed-metal cooling systems. The proportion of antifreeze and levels of protection afforded are indicated in the Specifications.

21 Before adding antifreeze, the cooling system should be completely drained, preferably flushed, and all hoses checked for condition and security.

22 After filling with the correct water/antifreeze mixture, a label should be attached to the radiator or expansion tank stating the type and concentration of antifreeze used and the date installed. Any subsequent topping-up should be made with the same type and concentration of antifreeze.

23 Do not use engine antifreeze in the screen washer system, as it will cause damage to the vehicle paintwork.

Notes

Chapter 2 Engine

Contents

Degrees of difficulty

| Easy, suitable for novice with little experience | Fairly easy, suitable for beginner with some experience | Fairly difficult, suitable for competent DIY mechanic | Difficult, suitable for experienced DIY mechanic | Very difficult, suitable for expert DIY or professional |

Specifications

1108 cc (C1E) engine

General

Type	Four-cylinder, in-line, overhead valve
Designation	C1E
Bore	70.0 mm
Stroke	72.0 mm
Capacity	1108 cc
Firing order	1-3-4-2 (No 1 cylinder at flywheel end)
Direction of crankshaft rotation	Clockwise viewed from pulley end
Compression ratio	8.8 : 1
Maximum power output - kW (bhp)	35 (49) at 5250 rpm
Maximum torque - Nm (lbf ft)	78 (58) at 2500 rpm

Cylinder block

Material	Cast-iron
Liner bore diameter	70.0 mm
Liner height (shoulder to top)	95.005 to 95.035 mm
Cylinder block depth (from top to liner locating shoulder)	94.945 to 94.985 mm
Liner protrusion	0.04 to 0.12 mm
Maximum difference in protrusion between adjacent liners	0.04 mm
Liner base seal thickness:	
Blue	0.08 mm
Red	0.10 mm
Green	0.12 mm

Crankshaft

Number of main bearings	5
Main bearing journal diameter:	
Standard	54.795 mm
1st undersize	54.545 mm ± 0.010 mm
Main bearing running clearance	0.020 to 0.058 mm
Crankpin (big-end) journal diameter:	
Standard	43.980 mm
1st undersize	43.730 mm -0.02 mm
2nd undersize (production repair)	43.480 mm
Big-end bearing running clearance	0.014 to 0.053 mm
Big-end cap side play	0.310 to 0.572 mm (suggested values)
Crankshaft endfloat	0.05 to 0.23 mm

Pistons and piston rings

Piston ring end gaps	Pre-adjusted
Piston ring thickness:	
Top compression ring	1.75 mm
Second compression ring	2.00 mm
Oil control ring	3.5 mm
Piston clearance in liner	0.045 to 0.065 mm (suggested values)

Gudgeon pins

Length	59.0 mm
Outside diameter	18.0 mm
Bore	11.0 mm

Connecting rods

Big-end side play	0.310 to 0.572 mm (suggested values)
Small-end side play	0.310 to 0.572 mm

Cylinder head

Material	Aluminium
Height:	
Nominal	70.15 mm
Repair	69.65 mm
Maximum acceptable gasket face distortion	0.05 mm
Valve seat angle:	
Inlet and exhaust	90°
Valve seat width	1.1 to 1.5 mm

Camshaft

Drive	Chain
Number of bearings	4
Camshaft endfloat	0.05 to 0.12 mm

Valves

Head diameter:	
Inlet	33.5 mm
Exhaust	30.3 mm (1st model) or 29.0 mm (2nd model)
Stem diameter	7.0 mm
Valve spring free length	42.2 mm

Valve clearances

Cold:	
Inlet	0.15 mm
Exhaust	0.20 mm
Hot:	
Inlet	0.18 mm
Exhaust	0.25 mm

Pushrods

Length	172.3 mm (1st model), 176.3 mm (2nd model) or 173.5 mm (3rd model)

Tappets

External diameter:	
Standard	19.0 mm
Oversize	19.2 mm

Lubrication system

System pressure:	
At idle	0.7 bar
At 4000 rpm	3.5 bars
Oil pump type	Two-gear, or bi-rotor
Oil pump clearances:	
Two-gear:	
Gear-to-body (maximum)	0.20 mm
Bi-rotor:	
Inner-to-outer rotors (at centre of peaks)	0.04 to 0.29 mm

Torque wrench settings

	Nm	lbf ft
Rocker shaft	15 to 20	11 to 15
Camshaft sprocket	30	22
Crankshaft sprocket/pulley:		
Bolt 40 mm long	80	59
Bolt 45 mm long	110	81
Connecting rod (big-end) caps	45	33
Main bearing caps	55 to 65	41 to 48
Flywheel (use new bolts and thread locking compound)	50	37
Cylinder head bolts:		
Stage 1	55 to 65	41 to 48
Stage 2	Run the engine for 20 minutes, then allow to cool for 2 1/2 hours	
Stage 3	Loosen bolt 1 fully, then tighten to: 55 to 65	41 to 48
Stage 4	Loosen and tighten remaining bolts to Stage 3 setting	

1171 cc (E5F and E7F) and 1390 cc (E6J and E7J) engines

General

Type	Four-cylinder, in-line, overhead camshaft
Designation:	
Carburettor models	E5F and E6J
Fuel injection models	E7F and E7J
Bore (all engines)	75.8 mm
Stroke:	
E5F and E7F	64.9 mm
E6J and E7J	77.0 mm

2

1171 cc (E5F and E7F) and 1390 cc (E6J and E7J) engines (continued)

General

Capacity:
E5F and E7F	1171 cc
E6J and E7J	1390 cc
Firing order	1-3-4-2 (No 1 cylinder at flywheel/driveplate end)
Direction of crankshaft rotation	Clockwise viewed from pulley end

Compression ratio:
E5F	9.25 : 1
E7F	8.8 : 1 or 9.25 : 1
E6J and E7J	9.5 : 1

Maximum power output (typical) - kW (bhp):
E5F and E7F	43 (60) at 6000 rpm
E6J and E7J	58 (80) at 5750 rpm

Maximum torque (typical) - Nm (lbf ft):
E5F and E7F	85 (63) at 3500 rpm
E6J and E7J	107 (79) at 3500 rpm

Cylinder block

Material	Cast-iron
Liner bore diameter	75.8 mm +0.03 -0.00 mm

Liner height:
Total	130.0 mm
From shoulder to top	91.5 mm +0.035 +0.005 mm
Cylinder block depth (from top to liner locating shoulder)	91.5 mm -0.015 -0.055 mm
Liner protrusion without O-ring	0.02 to 0.09 mm
Maximum difference in protrusion between adjacent liners	0.05 mm

Crankshaft

Number of main bearings	5

Main bearing journal diameter:
Standard	54.795 mm ± 0.01 mm
1st undersize	54.545 mm ± 0.01 mm
Main bearing running clearance	0.020 to 0.058 mm

Crankpin (big-end) journal diameter:
Standard	43.980 mm +0 -0.02 mm
1st undersize	43.730 mm +0 -0.02 mm
Big-end bearing running clearance	0.014 to 0.053 mm
Big-end cap side play	0.310 to 0.572 mm
Crankshaft endfloat	0.045 to 0.852 mm

Pistons and piston rings

Piston ring end gaps	Pre-adjusted

Piston ring thickness:
Top compression ring	1.5 mm
Second compression ring	1.75 mm
Oil control ring	3.0 mm
Piston clearance in liner	0.045 to 0.065 mm (suggested values)

Gudgeon pins

Length	60.0 mm
Outside diameter	19.0 mm
Bore	11.0 mm

Connecting rods

Big-end side play	0.310 to 0.572 mm
Small-end side play	0.310 to 0.572 mm

Cylinder head

Material	Aluminium
Height	113.0 mm ± 0.05 mm
Maximum acceptable gasket face distortion	0.05 mm
Refinishing limit	No refinishing permitted

Valve seat angle:
Inlet . . . . 120°	
Exhaust	90°
Valve seat width	1.7 mm

Camshaft

Drive	Toothed belt
Number of bearings	5
Camshaft endfloat	0.06 to 0.15 mm

Valves

Head diameter:	
Inlet	37.5 mm
Exhaust	33.5 mm
Stem diameter	7.0 mm

Valve clearances (cold)

Inlet	0.10 mm
Exhaust	0.25 mm

Lubrication system

System pressure:	
At idle	1.0 bar
At 4000 rpm	3.0 bars
Oil pump type	Two-gear
Oil pump clearances:	
Gear-to-body (minimum)	0.110 mm
Gear-to-body (maximum)	0.249 mm
Gear endfloat (minimum)	0.020 mm
Gear endfloat (maximum)	0.086 mm

Torque wrench settings

	Nm	lbf ft
Camshaft sprocket bolt	50 to 60	37 to 44
Crankshaft pulley/sprocket bolt	80 to 90	59 to 66
Rocker shaft bolts	21 to 25	16 to 19
Flywheel/driveplate bolts	50 to 55	37 to 41
Exhaust manifold	23 to 28	17 to 21
Inlet manifold	23 to 28	17 to 21
Main bearing cap	60 to 67	44 to 50
Connecting rod (big-end) cap - oiled:		
Stage 1	10	7
Stage 2	45	33
Sump	7 to 9	4 to 7
Oil drain plug	15 to 25	11 to 19
Cylinder head bolts:		
Stage 1	30	22
Stage 2	60	44
Alternative Stage 1	20	15
Alternative Stage 2	Angle-tighten all bolts by 97° ± 2°	
Stage 3	Wait for at least 3 minutes	
Stage 4	Slacken bolts 1 and 2 fully	
Stage 5	Tighten bolts 1 and 2 to:	
	20	15
Stage 6	Angle-tighten bolts 1 and 2 by 97° ± 2°	
Stage 7	Slacken bolts 3, 4, 5 and 6 fully	
Stage 8	Tighten bolts 3, 4, 5 and 6 to:	
	20	15
Stage 9	Angle-tighten bolts 3, 4, 5 and 6 by 97° ± 2°	
Stage 10	Slacken bolts 7, 8, 9 and 10 fully	
Stage 11	Tighten bolts 7, 8, 9 and 10 to:	
	20	15
Stage 12	Angle-tighten bolts 7, 8, 9 and 10 by 97° ± 2°	
Engine/gearbox mountings (see text):		
Rear mounting/lower link bolts	60	44
Right-hand lower bracket to body	55	41
Right-hand upper bracket to cylinder head	45	33
Right-hand mounting centre nut	45	33
Right-hand limiter bolts	55	41
Left-hand mounting bracket to gearbox	55	41
Left-hand mounting bracket/battery mounting to body	20	15
Left-hand mounting pad to bracket	35	26
Left-hand mounting pad centre nut	75	55

2

1721 cc (F2N), 1764 cc (F7P) and 1794 cc (F3P) engines

General

Type	Four-cylinder, in-line, overhead camshaft
Designation	F2N, F7P and F3P
Bore:	
F2N	81.0 mm
F7P	82.0 mm
F3P	82.7 mm
Stroke (all engines)	83.5 mm
Capacity:	
F2N	1721 cc
F7P	1764 cc
F3P	1794 cc
Firing order	1-3-4-2 (No 1 cylinder at flywheel end)
Direction of crankshaft rotation	Clockwise viewed from pulley end
Compression ratio:	
F2N	9.5 : 1
F7P	10.1 : 1
F3P	9.7 : 1
Maximum power output - kW (bhp):	
F2N	66.5 (92) at 5750 rpm
F7P	102 (137) at 6500 rpm
F3P	68.5 (95) at 5750 rpm
Maximum torque - Nm (lbf ft):	
F2N	135 (100) at 3000 rpm
F7P	158 (116) at 4250 rpm
F3P	142 (105) at 2750 rpm

Cylinder block

Material	Cast-iron

Crankshaft

Number of main bearings	5
Main bearing journal diameter:	
Standard	54.795 mm
1st undersize	54.545 mm
Main bearing running clearance	0.020 to 0.058 mm
Crankpin (big-end) journal diameter:	
Standard	48.0 mm
1st undersize	47.75 mm +0.02 +0 mm
Big-end bearing running clearance	0.014 to 0.053 mm
Big-end cap side play	0.22 to 0.40 mm (suggested values)
Crankshaft endfloat	0.07 to 0.23 mm

Pistons and piston rings

Piston ring end gaps	Pre-adjusted
Piston ring thickness:	
Top compression ring	1.75 mm
Second compression ring	2.0 mm
Oil control ring	3.0 mm
Piston clearance in bore:	
F2N engine with de Colmar piston	0.025 to 0.045 mm
F2N engine with SMM normal piston	0.040 to 0.060 mm
F2N engine with SMM reduced-clearance piston	0.035 to 0.055 mm
F3P engine	0.025 to 0.050 mm (suggested values)
F7P engine	0.025 to 0.045 mm

Gudgeon pins

Fit in connecting rod:	
F2N/F3P engine	Interference
F7P engine	Floating

Connecting rods

Big-end side play	0.22 to 0.40 mm (suggested values)
Small-end side play	0.22 to 0.40 mm

Cylinder head

Material	Aluminium
Height:	
F2N and F3P	169.5 mm ± 0.2 mm
F7P	136.5 mm ± 0.05 mm
Maximum acceptable gasket face distortion	0.05 mm
Valve seat angle:	
F2N and F3P:	
Inlet	120°
Exhaust	90°
F7P:	
Inlet	90°
Exhaust	90°
Valve seat width:	
F2N and F3P:	
Inlet and exhaust	1.7 mm ± 0.2 mm
F7P:	
Inlet	1.4 mm
Exhaust	1.7 mm

Camshaft

Drive	Toothed belt
Number of bearings	5
Camshaft endfloat:	
F2N and F3P	0.048 to 0.133 mm
F7P	Not stated

Auxiliary shaft

Endfloat	0.07 to 0.15 mm

Valves

F2N and F3P

Head diameter:	
Inlet	38.1 mm
Exhaust	32.5 mm
Stem diameter	8.0 mm
Valve spring free length	44.9 mm

F7P

Head diameter:	
Inlet	30.0 mm
Exhaust	28.5 mm
Stem diameter	7.0 mm
Valve spring free length	Not stated

Valve clearances (cold)

F2N/F3P engine:	
Inlet	0.20 mm
Exhaust	0.40 mm
F7P engine	Not adjustable

Lubrication system

System pressure:	
At 1000 rpm	2.0 bars
At 3000 rpm	3.5 bars
Oil pump type	Two-gear
Oil pump clearances:	
Gear-to-body (minimum)	0.10 mm
Gear-to-body (maximum)	0.24 mm
Gear endfloat (minimum)	0.02 mm
Gear endfloat (maximum)	0.085 mm

Torque wrench settings

	Nm	lbf ft
Camshaft sprocket	50	37
Camshaft bearing caps:		
8 mm diameter - F2N/F3P engine	20	15
8 mm diameter - F7P engine	24	18
6 mm diameter	10	7

2

1721 cc (F2N), 1764 cc (F7P) and 1794 cc (F3P) engines (continued)

Torque wrench settings (continued)

	Nm	lbf ft
Timing belt idler wheel	20	15
Timing belt tensioner roller nut	50	37
Auxiliary shaft sprocket	50	37
Crankshaft pulley	90 to 100	66 to 74
Oil pump cover:		
6 mm diameter bolts	10	7
8 mm diameter bolts	20 to 25	15 to 18
Connecting rod (big-end) caps	45 to 50	33 to 37
Sump bolts	12 to 15	9 to 11
Sump-to-gearbox stud nuts (F7P engine only)	25	18
Flywheel (use new bolts)	50 to 55	37 to 41
Valve cover	3 to 6	2 to 4
Main bearing caps	60 to 65	44 to 48
Engine/gearbox mountings (see text):		
Rear mounting/lower link bolts	60	44
Right-hand lower bracket to body	55	41
Right-hand upper bracket to cylinder head	45	33
Right-hand mounting centre nut	45	33
Right-hand limiter bolts	55	41
Left-hand mounting bracket to gearbox	55	41
Left-hand mounting bracket/battery mounting to body	20	15
Left-hand mounting pad to bracket	35	26
Left-hand mounting pad centre nut	75	55
Cylinder head bolts:		

F2N and F3P

	Nm	lbf ft
Stage 1	30	22
Stage 2	70	52
Stage 3	Wait for 3 minutes minimum	
Stage 4	Loosen all the bolts completely	
Stage 5	20	15
Stage 6	Angle-tighten by 123° ± 2°	

F7P

	Nm	lbf ft
Stage 1	30	22
Stage 2	50	37
Stage 3	Wait for 3 minutes minimum	
Stage 4	Loosen all the bolts completely	
Stage 5	25	18
Stage 6	Angle-tighten by 107° ± 2°	

Part A: 1108 cc Engine – In-car engine repair procedures

1 General information

This Part of Chapter 2 is devoted to in-car repair procedures for the 1108 cc (C1E) engine. Similar information covering the 1171 cc (E5F and E7F), 1390 cc (E6J and E7J), 1721 cc (F2N), 1764 cc (F7P), and 1794 cc (F3P) engines will be found in Parts B and C. All procedures concerning engine removal and refitting, and engine block/cylinder head overhaul for all engine types, can be found in Part D of this Chapter.

Most of the operations included in this Part are based on the assumption that the engine is still installed in the car. Therefore, if this information is being used during a complete engine overhaul, with the engine already removed, many of the steps included here will not apply.

Engine description

The engine is of four-cylinder, in-line, overhead valve type, mounted transversely in the front of the car.

The cast-iron cylinder block is of the replaceable wet liner type. The crankshaft is supported within the cylinder block on five shell-type main bearings. Thrustwashers are fitted at the centre main bearing, to control crankshaft endfloat.

The connecting rods are attached to the crankshaft by horizontally-split shell-type big-end bearings, and to the pistons by interference-fit gudgeon pins. The aluminium alloy pistons are of the slipper type, and are fitted with three piston rings, comprising two compression rings and a scraper-type oil control ring.

The camshaft is chain-driven from the crankshaft, and operates the rocker arms via pushrods. The inlet and exhaust valves are each closed by a single valve spring, and operate in guides pressed into the cylinder head. The valves are actuated directly by the rocker arms.

A semi-closed crankcase ventilation system is employed, and crankcase gases are drawn from the rocker cover via a hose to the air cleaner and inlet manifold.

Lubrication is provided by a gear-type oil pump, driven from the camshaft and located in the crankcase. Engine oil is fed through an externally-mounted full-flow filter to the engine oil gallery, and then to the crankshaft, camshaft and rocker shaft bearings. A pressure relief valve is incorporated in the oil pump.

Repair operations possible with the engine in the vehicle

The following operations can be carried out without having to remove the engine from the car.

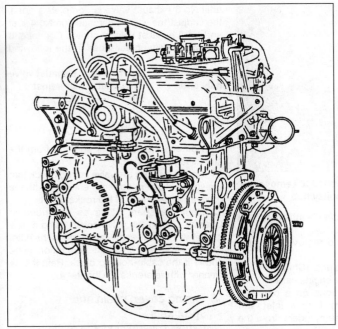

Fig. 2.1 1108 cc (C1E) engine (Sec 1)

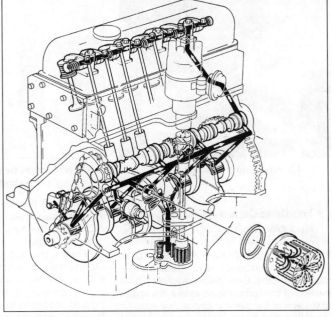

Fig. 2.2 Oil circuit on the 1108 cc (C1E) engine (Sec 1)

(a) *Removal and refitting of the cylinder head.*
(b) *Removal and refitting of the timing cover, chain and gears.*
(c) *Removal and refitting of the sump.*
(d) *Removal and refitting of the connecting rods, pistons and liners*.*
(e) *Removal, overhaul and refitting of the oil pump.*
(f) *Renewal of the crankshaft oil seals.*
(g) *Renewal of the engine mountings.*
(h) *Removal and refitting of the flywheel.*

*Although the pistons and liners can be removed and refitted with the engine in the car, it is better to carry this work out with the engine removed in the interests of cleanliness and improved access. Refer to Part D for details.

2 Compression test - description and interpretation

1 A compression check will tell you what mechanical condition the top end (pistons, rings, valves, head gaskets) of the engine is in. Specifically, it can tell you if the compression is down due to leakage caused by worn piston rings, defective valves and seats or a blown head gasket. **Note:** *The engine must be at normal operating temperature, and the battery must be fully charged, for this check.*
2 Begin by cleaning the area around the spark plugs before you remove them (compressed air should be used, if available, otherwise a small brush or even a bicycle tyre pump will work). The idea is to prevent dirt from getting into the cylinders as the compression check is being done.

3 Remove all of the spark plugs from the engine (see Chapter 1).
4 Disconnect the coil HT lead from the centre of the distributor cap, and earth it on the cylinder block. Use a jumper lead or similar wire to make a good connection.
5 Fit the compression gauge into the No 1 spark plug hole (photo).
6 Have an assistant hold the accelerator pedal fully depressed, while at the same time cranking the engine over several times on the starter motor. Observe the compression gauge - the compression should build up quickly in a healthy engine. Low compression on the first stroke, followed by gradually-increasing pressure on successive strokes, indicates worn piston rings. A low compression reading on the first stroke, which does not build up during successive strokes, indicates leaking valves or a blown head gasket (a cracked head could also be the cause). Deposits on the undersides of the valve heads can also cause low compression. Record the highest gauge reading obtained, then repeat the procedure for the remaining cylinders.
7 Add some engine oil (about three squirts from a plunger-type oil can) to each cylinder, through the spark plug hole, and repeat the test.
8 If the compression increases after the oil is added, the piston rings are definitely worn. If the compression does not increase significantly, the leakage is occurring at the valves or head gasket. Leakage past the valves may be caused by burned valve seats and/or faces, or warped, cracked or bent valves.
9 If two adjacent cylinders have equally low

compression, there is a strong possibility that the head gasket between them is blown. The appearance of coolant in the combustion chambers or the crankcase would verify this condition.
10 If one cylinder is about 20 percent lower than the others, and the engine has a slightly rough idle, a worn lobe on the camshaft could be the cause.
11 If the compression is unusually high, the combustion chambers are probably coated with carbon deposits. If this is the case, the cylinder head should be removed and decarbonised.
12 Actual compression pressures for the engines covered by this manual are not specified by the manufacturer. However, bearing in mind the information given in the preceding paragraphs, the results obtained should give a good indication of engine condition and what course of action, if any, to take.

2.5 Compression gauge in use

2

4.9 Adjusting a valve clearance

5.5 A puller may be needed to remove the crankshaft pulley hub

3 Top Dead Centre (TDC) for No 1 piston - locating

1 Top dead centre (TDC) is the highest point in the cylinder that each piston reaches as the crankshaft turns. Each piston reaches TDC at the end of the compression stroke, and again at the end of the exhaust stroke; however, for the purpose of timing the engine, TDC refers to the position of No 1 piston at the end of its compression stroke. No 1 piston is at the flywheel end of the engine.
2 Disconnect both battery leads.
3 Apply the handbrake, then jack up the front right-hand side of the car and support it on axle stands. Remove the right-hand roadwheel.
4 Remove the plastic cover from within the right-hand wheel arch to give access to the crankshaft pulley bolt.
5 Remove the spark plugs as described in Chapter 1.
6 Place a finger over the No 1 spark plug hole in the cylinder head (nearest the flywheel). Turn the engine in a clockwise direction, using a socket or spanner on the crankshaft pulley bolt, until pressure is felt in the No 1 cylinder. This indicates that No 1 piston is rising on its compression stroke.
7 Look through the aperture at the flywheel end of the engine, and continue turning the crankshaft until the TDC timing mark on the flywheel is aligned with the TDC mark on the gearbox bellhousing. (See Chapter 1, Fig. 1.16.)
8 If the distributor cap is now removed, the rotor arm should be aligned with the No 1 HT lead segment.

4 Valve clearances - adjustment

Note: *This operation is not part of the maintenance schedule. It should be undertaken if noise from the valvegear becomes evident, or if loss of performance gives cause to suspect that the clearances may be incorrect.*

1 Remove the air cleaner as described in Chapter 4.
2 Unclip and remove the cable-holding block from the valve cover. Tie the cables to one side.

3 Disconnect the crankcase ventilation hose from the valve cover.
4 Unscrew the nuts, and remove the valve cover and gasket.
5 Remove the spark plugs (Chapter 1) in order to make turning the engine easier.
6 Draw the valve positions on a piece of paper, numbering them 1 to 4 inlet and exhaust according to their cylinders, from the flywheel (left-hand) end of the engine (ie 1I, 1E, 2I, 2E, 3E, 3I, 4E, 4I). As the valve clearances are adjusted, cross them off.
7 Using a socket or spanner on the crankshaft pulley bolt, turn the engine in a clockwise direction until No 1 exhaust valve is completely open (ie the valve spring is completely compressed).
8 Insert a feeler blade of the correct thickness (see Specifications) between the No 3 cylinder inlet valve stem and the end of the rocker arm. It should be a firm sliding fit. If adjustment is necessary, loosen the locknut on the rocker arm using a ring spanner, and turn the adjustment screw with a screwdriver until the fit is correct. Tighten the locknut and recheck the adjustment, then repeat the adjustment procedure on No 4 cylinder exhaust valve. Note that the clearances for the inlet and exhaust valves are different.
9 Turn the engine in a clockwise direction

until No 3 exhaust valve is completely open, then adjust the valve clearances on No 4 inlet and No 2 exhaust valves (photo). Continue to adjust the valve clearances in the following sequence.

Exhaust valve fully open	Inlet valve to adjust	Exhaust valve to adjust
1E	3I	4E
3E	4I	2E
4E	2I	1E
2E	1I	3E

10 Remove the socket or spanner from the crankshaft pulley bolt.
11 Before refitting the valve cover, check that the spring clips are correctly engaged with the grooves at each end of the rocker shaft.
12 Refit the spark plugs as described in Chapter 1. Refit the valve cover, using a new gasket where necessary, and tighten the nuts. Reconnect the crankshaft ventilation hose and refit the cable-holding block. Refit the air cleaner with reference to Chapter 4.

5 Timing cover, chain and sprockets - removal, inspection and refitting

Removal

1 Remove the alternator drivebelt as described in paragraph 7 of the next Section.
2 Remove the sump as described in Section 7.
3 Refer to Section 3 and bring the engine to TDC, No 1 firing.
4 Lock the starter ring gear to prevent the engine turning, by using a wide-bladed screwdriver inserted between the ring gear teeth and the crankcase. Alternatively, engage 4th gear and have an assistant depress the footbrake pedal. With the engine locked, use a socket or spanner to unscrew the crankshaft pulley retaining bolt - this bolt is very tight.
5 With the bolt removed, lift off the pulley and withdraw the pulley hub. If the hub is tight, carefully lever it off using two screwdrivers, or use a two- or three-legged puller (photo).

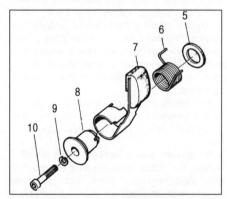

Fig. 2.3 Mechanical timing chain tensioner components (Sec 5)

5 Washer	8 Collar
6 Spring	9 Washer
7 Slipper arm	10 Retaining bolt

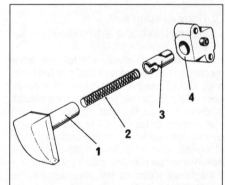

Fig. 2.4 Hydraulic timing chain tensioner components (Sec 5)

1 Piston with tensioner slipper	3 Sleeve
2 Spring	4 Tensioner body

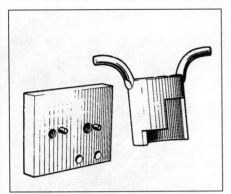

Fig. 2.5 Renault tool for holding the mechanical tensioner together (Sec 5)

5.10 Removing the timing chain and sprockets

5.17 Use a socket or tube to renew the timing cover oil seal

6 Unscrew the nuts and bolts securing the timing cover to the cylinder block. Carefully prise the timing cover off, using a screwdriver to release it. Remove the gasket (where fitted).
7 Observe the components of the timing chain tensioner, noting that one of two types may be fitted. The mechanical tensioner is identified by its coil tensioning spring and single mounting bolt. The automatic hydraulic tensioner is identified by its piston plunger and two mounting bolts.
8 To remove the mechanical-type tensioner, Renault technicians use a special tubular tool which locates over the tensioner and spring, and holds all the components together until the assembly is refitted. The tool is shown in Fig. 2.5, but it is not essential. If the tool is not available, simply unscrew the retaining bolt using an Allen key, hold the tensioner slipper and spring end together, and withdraw the assembly from the cylinder block.
9 If a hydraulic tensioner is fitted, lock the plunger by tying locking wire around the body. Unscrew the mounting bolts, and withdraw the tensioner from the block. Extract the gauze filter from the hole in the cylinder block.
10 Bend back the locktab, then unscrew and remove the camshaft sprocket retaining bolt. Remove the washer. Withdraw the camshaft sprocket and chain from the camshaft, then release the chain from the crankshaft sprocket (photo). Use two screwdrivers or a puller to remove the sprocket from the crankshaft.
11 With the sprockets and chain removed, check that the Woodruff key in the nose of the crankshaft is a tight fit in its slot. If not, remove it now, and store it safely to avoid the risk of it dropping out and getting lost.

Inspection

12 Examine the timing chain. Chain wear can be judged by holding the chain horizontally and observing how far it sags. Renew the chain without question if it has covered upwards of 30 000 miles (50 000 km), or if it has been noisy in operation.
13 Examine all the teeth on the camshaft and crankshaft sprockets. If these are 'hooked' in

appearance, renew the sprockets. It is good practice to renew the sprockets and chain as a set, since the old components will have worn together.
14 If a mechanical-type chain tensioner is fitted, examine the chain contact pad. Renew the tensioner assembly if the pad is heavily scored, or if there is any suspicion that the spring has become weak.
15 If a hydraulic-type chain tensioner is fitted, dismantle it by releasing the slipper piston with a 3 mm Allen key. Examine the piston, spring, sleeve and tensioner body bore for signs of scoring, and renew if evident. Also renew the tensioner if the chain contact pad is heavily scored.
16 If the hydraulic tensioner is serviceable, lubricate the components and reassemble. Lock the sleeve in the slipper piston first by turning it clockwise with an Allen key, then slide this assembly into the tensioner body. Avoid pressing the slipper now, or the sleeve will be released and the whole assembly will fly apart. New hydraulic tensioners are supplied with a 2 mm thick safety spacer inserted between the pad and the tensioner body to prevent the assembly flying apart.
17 Renew the oil seal in the timing cover by driving out the old seal using a suitable drift. Install the new seal using a large socket or block of wood (photo).

Refitting

18 Refit the Woodruff key to the crankshaft slot, and then tap the crankshaft sprocket into position. Ensure that the timing mark on the sprocket is on the side facing away from the engine.
19 Turn the crankshaft if necessary until the timing notch on the flywheel is in line with the TDC mark on the timing scale.
20 Temporarily place the camshaft sprocket in position. Turn the camshaft so that the timing marks on the sprocket faces are aligned as shown in Fig. 2.6. Remove the camshaft sprocket.
21 Fit the timing chain to the camshaft sprocket, position the sprocket in its approximate fitted position, and locate the

chain over the crankshaft sprocket. Position the camshaft sprocket on the camshaft, and check that the marks are still aligned when there is an equal amount of slack on both sides of the chain.
22 Refit the camshaft sprocket retaining bolt using a new locktab, and tighten the bolt to the specified torque. Bend up the locktab to retain the bolt.
23 If a mechanical tensioner is fitted, place it in position and locate the spring ends in the block and over the slipper arm. Refit the retaining bolt and tighten it securely with an Allen key.
24 If a hydraulic-type tensioner is fitted, first insert the gauze filter in the cylinder block. Remove the safety spacer (if fitting a new tensioner) and locate the tensioner on the cylinder block. Insert and tighten the bolts. Push the slipper fully inwards, then release it. The piston should spring out automatically under spring pressure.
25 Oil the chain, sprockets and tensioner. Ensure that the mating faces of the timing cover and cylinder block are clean and dry, with all traces of old sealant removed. A new oil seal must be in place in the timing cover.

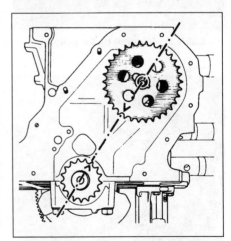

Fig. 2.6 Correct alignment of crankshaft and camshaft sprocket timing marks (Sec 5)

2

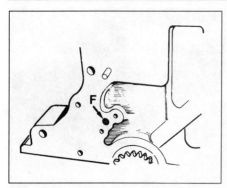

Fig. 2.7 Location (F) of gauze filter in the cylinder block for the hydraulic tensioner (Sec 5)

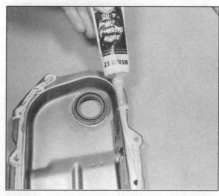

5.27 Apply a bead of sealant to the timing cover face

5.28 Refit the crankshaft pulley hub . . .

26 If there are no locating dowels for the timing cover, a gasket must be fitted. Locate the gasket on the cylinder block, retaining it with a smear of grease. Fit the timing cover, and finger-tighten the nuts and bolts. Oil the pulley hub, and temporarily fit it on the end of the crankshaft so that the timing cover is positioned correctly, then tighten the cover nuts and bolts.

27 Where locating dowels are fitted, apply a bead of CAF 4/60 THIXO paste to the timing cover joint face (photo), then position the cover over the dowels and the two studs. Refit the nuts and retaining bolts, then progressively tighten them in a diagonal sequence.

28 Lubricate the crankshaft pulley hub, and carefully slide it onto the end of the crankshaft (photo).

29 Place the pulley in position, refit the retaining bolt and washer, and tighten the bolt to the specified torque (photos). Hold the crankshaft stationary using one of the methods described in paragraph 4.

30 Refit the sump with reference to Section 7.

31 Refit and tension the alternator drivebelt with reference to Chapter 12, Section 6.

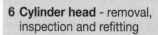

6 Cylinder head - removal, inspection and refitting

Removal

1 Disconnect both battery leads, referring to Chapter 12 if necessary.

2 Drain the cooling system, including the cylinder block, with reference to Chapter 1. The block drain plug is located on the crankshaft pulley end of the engine, below the water pump. It is important to drain the block because, if the wet cylinder liners are disturbed, the coolant will drain into the sump.

3 Drain the engine oil with reference to Chapter 1.

4 Remove the air cleaner with reference to Chapter 4.

5 Disconnect the HT leads at the spark plugs, release the distributor cap retaining clips or screws, and remove the cap and leads.

6 Disconnect the lead at the temperature gauge sender on the water pump (photo).

7 Slacken the alternator mountings and adjustment arm bolt, push the alternator in towards the engine, and slip the drivebelt off the pulleys. Unscrew the bolt securing the alternator adjustment arm to the water pump, and swing the alternator clear of the engine (photos).

5.29A . . . followed by the pulley . . .

5.29B . . . and retaining bolt

6.6 Disconnecting the water temperature gauge sender lead

6.7A Slacken the alternator mountings and slip the drivebelt off the pulleys

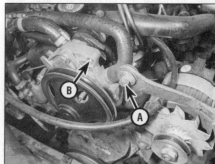

6.7B Unscrew the bolt (A) and remove the alternator adjustment arm from the pump at (B)

6.15 Removing the cable-holding block

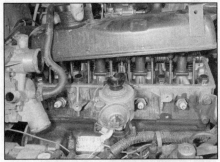

6.17 Unscrew the three nuts and lift off the valve cover

6.18 Unscrew the two nuts and two bolts (arrowed) and withdraw the rocker shaft

8 Release the hose clips and remove the two heater hoses from the water pump. Also remove the support bracket from the rocker cover.

9 Release the hose clip and remove the radiator top hose from the water pump.

10 Disconnect the accelerator cable, choke cable and vacuum pipe from the carburettor with reference to Chapter 4.

11 Disconnect the brake servo hose from the inlet manifold.

12 Unscrew the nut and washer on the inlet manifold and on the cylinder block, and lift off the heat shield (where fitted).

13 Disconnect the fuel inlet pipe and the crankcase ventilation hose at the carburettor.

14 Disconnect the distributor wiring.

15 Unscrew the bolt and release the retaining clip securing the cable-holding block to the side of the cylinder head (photo).

16 Unscrew the two bolts and withdraw the tension springs securing the exhaust front section to the exhaust manifold.

17 Unscrew the nuts and remove the valve cover, complete with gasket, from the cylinder head (photo).

18 Progressively unscrew the two bolts and two nuts securing the rocker shaft pedestals to the cylinder head. Lift the rocker shaft assembly upwards and off the two studs (photo).

19 Lift out each of the pushrods in turn, using a twisting action to release them from their cam followers (photo). Keep them in order of removal by inserting them in a strip of cardboard having eight numbered holes

punched in it. Note that No 1 should be the pushrod nearest the flywheel.

20 Slacken all the cylinder head retaining bolts half a turn at a time in the reverse order to that shown in Fig. 2.8. When the tension has been relieved, remove all the bolts, with the exception of the centre bolt on the distributor side.

21 Using a hide or plastic mallet, tap each end of the cylinder head so as to pivot the head around the remaining locating bolt and unstick the gasket. **Do not** attempt to lift the head until the gasket has been unstuck, otherwise the cylinder liner seal at the base of each liner will be broken, allowing water and foreign matter to enter the sump.

22 After unsticking the cylinder head from the gasket, remove the remaining bolt and lift the head, complete with water pump, manifolds and carburettor, off the engine (photo). **Note:** *The crankshaft must not be rotated with the head removed, otherwise the liners will be displaced*. If it is necessary to turn the crankshaft (eg to clean the piston crowns), use liner clamps or bolts with suitable washers screwed into the top of the block to retain the liners. If the sealing of the liner seals is in any doubt, they must be renewed with reference to Part D of this Chapter.

Inspection

23 The mating faces of the cylinder head and block must be perfectly clean before refitting the head. Use a scraper to remove all traces of gasket and carbon, and also clean the tops

of the pistons. Take particular care with the aluminium cylinder head, as the soft metal is easily damaged. Also, make sure that debris is not allowed to enter the oil and water channels - this is particularly important for the oil circuit, as carbon could block the oil supply to the camshaft, rocker shaft, rocker arms or crankshaft bearings. Using adhesive tape and paper, seal the water, oil and bolt holes in the cylinder block. Clean the piston crowns in the same way.

> **HAYNES HINT** *To prevent carbon entering the gap between the pistons and bores, smear a little grease in the gap. After cleaning the piston, rotate the crankshaft so that the piston moves down the bore, then wipe out the grease and carbon with a cloth rag.*

24 Check the block and head for nicks, deep scratches and other damage. If slight, they may be removed carefully with a file; however, if excessive, machining may be the only alternative.

25 If warpage of the cylinder head is suspected, use a straight-edge to check it for distortion. Also check the protrusion of the cylinder liners. Either of these items can be associated with the head gasket blowing. Refer to Part D of this Chapter if necessary.

26 Clean out all the bolt holes in the block, using pipe cleaners or a rag and a screwdriver. Make sure that all oil is removed, otherwise there is a possibility of the block being cracked by hydraulic pressure when the bolts are tightened.

27 Examine the bolt threads and the threads in the cylinder block for damage. If necessary, use the correct-size tap to chase out the threads in the block, and use a die to clean the threads on the bolts.

Refitting

28 Remove the cylinder liner clamps or washers, and make sure that the faces of the cylinder head and the cylinder block are perfectly clean. Place a new gasket on the cylinder block with the words 'Haut-Top' uppermost. Do not use any kind of jointing compound (photo).

6.19 Take out the pushrods and keep them in order

6.22 Free the cylinder head from the gasket, then lift the head, complete with manifolds and water pump, off the engine

2

6.28 Place the cylinder head gasket in position

6.29A Lower the cylinder head onto the gasket and . . .

6.29B . . . fit the retaining bolts; tighten in the correct sequence to the specified torque

29 Lower the cylinder head into position. Insert the cylinder head bolts and tighten them progressively to the Stage 1 specified torque, following the sequence shown in Fig. 2.8 (photos).

30 Install the pushrods in their original locations.

31 Lower the rocker shaft assembly onto the cylinder head, making sure that the adjusting ball-ends locate in the pushrods (photo). Install the spring washers (convex side uppermost), nuts and bolts, and tighten them progressively to the specified torque.

32 Adjust the valve clearances as described in Section 4.

33 Refit the valve cover using a new gasket, and tighten the nuts.

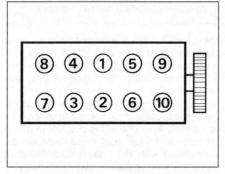

Fig. 2.8 Cylinder head bolt tightening sequence (Sec 6)

34 Connect the exhaust front section to the manifold, then refit the tension springs and tighten the bolts. Refer to Chapter 4 if necessary.

35 Refit the cable-holding block to the side of the cylinder head, and retain it with the clip and bolt.

36 Reconnect the distributor wiring.

37 Reconnect the crankcase ventilation hose and the carburettor fuel inlet pipe.

38 Refit the heat shield (where fitted) and tighten the nut.

39 Reconnect the brake servo hose to the inlet manifold.

40 Reconnect the accelerator cable, choke cable and vacuum pipe to the carburettor with reference to Chapter 4.

41 Reconnect the radiator top hose to the water pump, and tighten the clip.

42 Fit the support bracket to the rocker cover, then reconnect the two heater hoses to the water pump and tighten the clips.

43 Refit the alternator and drivebelt. Adjust the drivebelt with reference to Chapter 12, Section 6.

44 Reconnect the lead to the temperature gauge sender on the water pump.

45 Refit the distributor cap and leads, and connect the HT leads to the spark plugs.

46 Refit the air cleaner with reference to Chapter 4.

47 Refill the engine with oil, with reference to Chapter 1.

48 Refit the cylinder block drain plug. Refill

the cooling system with reference to Chapter 1.

49 Reconnect the battery leads. Run the engine for 20 minutes, then switch off and allow it to cool for at least two and a half hours. Remove the valve cover.

50 Slacken and retighten each cylinder head bolt in turn to the torque wrench setting given in the Specifications.

51 Recheck the valve clearances and adjust if necessary. Finally, refit the valve cover.

7 Sump - removal and refitting

Removal

1 Disconnect both the battery leads.

2 Jack up the front of the car and support it on axle stands.

3 Unscrew the two bolts, noting the position of the spacer, and remove the engine steady rod.

4 Drain the engine oil as described in Chapter 1, then refit and tighten the drain plug using a new washer.

5 Unscrew the bolts and remove the flywheel cover plate from the bellhousing (photo).

6 Where fitted, disconnect the two wires at the oil level sensor on the front face of the sump.

7 Unscrew and remove the bolts securing the sump to the crankcase (photo).

6.31 Refit the rocker shaft assembly

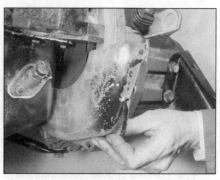

7.5 Remove the flywheel cover plate

7.7 Unscrew the retaining bolts and remove the sump

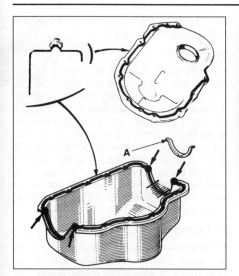

Fig. 2.9 Sealing paste application diagram for the sump and timing cover (Sec 7)

A Alternative rubber seal - if used, do not apply sealing paste to the hatched area shown

8 Tap the sump with a hide or plastic mallet to break the seal between the sump flange, crankcase and timing cover. Lower the sump, and remove the gaskets and rubber seals (where fitted). Note that on some models a gasket is not used, only a sealing compound.

Refitting

9 Check that the mating faces of the sump, timing cover and cylinder block are perfectly clean and dry.

10 Where gaskets are fitted, first locate the rubber seals in the curved sections of the sump. These seals locate on the front timing cover and rear main bearing cap. Apply a little CAF 4/60 THIXO paste to the ends of the side gaskets, then locate them on the sump so that their ends are over the rubber seals.

11 Where gaskets are not used, apply a 3 mm diameter bead of CAF 4/60 THIXO paste on the sump as shown in Fig. 2.9, adding an extra amount at each of the four corners. Where a rubber seal is fitted instead of the previous cork type, do not apply the paste in the hatched area. Do not apply

8.2 Unscrew the retaining bolts and withdraw the oil pump

excessive paste, otherwise it may find its way into the lubrication circuit and cause damage to the engine.

12 If no gaskets are fitted, it is important that the sump is positioned correctly and not moved around after the paste has touched the crankcase. Long bolts or dowel rods may be fitted temporarily to help achieve this.

13 To prevent oil dripping from the oil pump and crankcase, wipe these areas clean before refitting the sump.

14 Lift the sump into position. Insert the bolts and tighten them progressively until they are secure.

15 Reconnect the two wires at the oil level sensor on the front face of the sump, when applicable.

16 Refit the flywheel cover plate to the bellhousing, and tighten the bolts.

17 Fill the engine with oil, with reference to Chapter 1.

18 Refit the engine steady rod and tighten the bolts.

19 Lower the car to the ground.

20 Reconnect the battery.

8 Oil pump - removal, inspection and refitting

Removal

1 Remove the sump as described in Section 7.

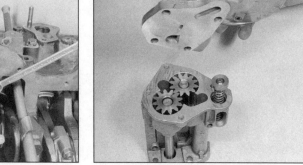

8.4 Lift off the oil pump cover

2 Unscrew the mounting bolts, and withdraw the oil pump from the crankcase and drivegear (photo).

Inspection

3 Clean the exterior of the oil pump.

4 Unscrew the four retaining bolts and lift off the pump cover, taking care not to lose the oil pressure relief valve components, which may be ejected under the action of the spring (photo).

5 Remove the pressure relief valve ball seat, ball, spring and spring seat from the pump body (photos).

6 Lift out the idler gear, the drivegear and shaft (gear-type pump) or the drivegear and rotors (bi-rotor-type pump).

7 Clean the components and carefully examine them for any signs of scoring or wear. Renew the pump if these conditions are apparent. The only spares available are the pressure relief valve components.

8 If the components appear serviceable, use a feeler blade to measure the clearance between the pump body and the gears (photo) or between the inner and outer rotors. If the clearance exceeds the specified amount, the pump must be renewed.

9 If the pump is satisfactory, reassemble the components in the order of removal, fill the pump with oil and refit the cover. If a new pump is being fitted, prime it with oil before fitting.

2

8.5A Remove the pressure relief valve ball seat and ball . . .

8.5B . . . followed by the spring and spring seat

8.8 Check the gear-to-body clearance using feeler blades

8.10 Refitting the oil pump

10.5 Flywheel mounting bolts

Refitting

10 Enter the oil pump shaft into its location in the cylinder block, and engage the shaft with the distributor drivegear (photo).

11 Push the pump up into contact with the block, then refit and tighten the three mounting bolts. Note that a gasket is not used.

12 Refit the sump with reference to Section 7.

9 Crankshaft oil seals - renewal

Right-hand (pulley end) oil seal

1 This procedure is included in the removal and refitting of the timing cover. Refer to Section 5.

2 Before refitting the pulley hub, inspect the seal rubbing surface. If it is grooved or rough, renew the hub.

Left-hand (flywheel end) oil seal

3 Remove the flywheel as described in Section 10.

4 Prise out the old oil seal using a small screwdriver, taking care not to damage the surface of the crankshaft. Alternatively, the oil seal can be removed by drilling two small holes diagonally opposite each other and inserting self-tapping screws in them. A pair of grips can then be used to pull out the oil seal, by pulling on each side in turn.

5 Inspect the seal rubbing surface on the crankshaft. If it is grooved or rough in the area where the old seal was fitted, the new seal should be fitted approximately 2 mm less deep, so that it rubs on an unworn part of the surface.

6 Wipe clean the oil seal seating, then dip the new seal in fresh engine oil and locate it over the crankshaft with its closed side facing outwards. Make sure that the oil seal lip is not damaged as it is located on the crankshaft.

7 Using a metal tube, drive the oil seal squarely into the bore. A block of wood cut to pass over the end of the crankshaft may be used instead.

8 Refit the flywheel with reference to Section 10.

10 Flywheel - removal, inspection and refitting

Removal

1 Remove the gearbox as described in Chapter 7.

2 Remove the clutch as described in Chapter 6.

3 Mark the flywheel in relation to the crankshaft.

4 The flywheel must now be held stationary while the bolts are loosened. To do this, locate a long bolt in one of the transmission mounting bolt holes, and either insert a wide-bladed screwdriver in the starter ring gear, or use a piece of bent metal bar engaged with the ring gear.

5 Unscrew the mounting bolts (photo) and withdraw the flywheel from the crankshaft. Be careful not to drop it - it is heavy. Obtain new bolts for use when refitting.

Inspection

6 Examine the flywheel for scoring of the clutch face, and for wear or chipping of the ring gear teeth. If the clutch face is scored, the flywheel may be machined until flat, but renewal is preferable. If the ring gear is worn or damaged, it may be possible to renew it separately, but this job is best left to a Renault dealer or engineering works. The temperature to which the new ring gear must be heated for installation is critical and, if not done accurately, the hardness of the teeth will be destroyed.

Refitting

7 Clean the flywheel and crankshaft faces, then locate the unit on the crankshaft, making sure that any previously-made marks are aligned.

8 Apply a few drops of locking fluid to the threads of the new bolts. Fit the bolts, and tighten them in a diagonal sequence to the specified torque.

9 Refit the clutch with reference to Chapter 6.

10 Refit the gearbox as described in Chapter 7.

11 Engine/gearbox mountings - inspection and renewal

1 Apply the handbrake, then jack up the front of the car and support it on axle stands.

2 Visually inspect the rubber pads on the engine/gearbox mountings for signs of

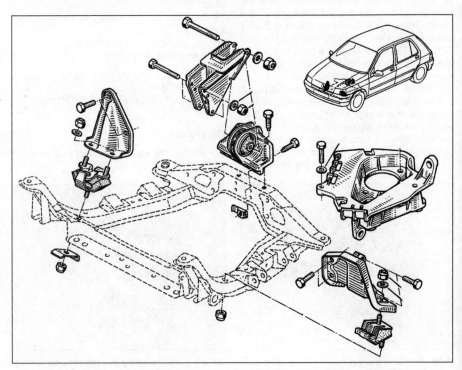

Fig. 2.10 Engine/gearbox mounting components - 1108 cc/C1E engine (Sec 11)

cracking and deterioration. Careful use of a lever will help to determine the condition of the rubber pads. If there is excessive movement in the mounting, or if the rubber has deteriorated, the mounting should be renewed.

Front right-hand mounting

3 Unscrew the nut and remove the plate from under the front subframe.
4 Using a hoist, raise the right-hand side of the engine by 50 to 75 mm (2 to 3 in).
5 Unscrew the nuts and remove the mounting from the bracket.

6 If necessary, unbolt the bracket from the cylinder block.
7 Fit the new mounting using a reversal of the removal procedure.

Front left-hand/gearbox mounting

8 Using a socket through the hole in the subframe, unscrew the lower mounting nut.
9 Take the weight of the gearbox with a trolley jack, then unscrew the upper nut and withdraw the mounting from the engine compartment. If necessary, the mounting bracket can be unbolted from the gearbox.

10 Fit the new mounting using a reversal of the removal procedure.

Rear mounting

11 Support the weight of the engine/gearbox using a hoist or trolley jack.
12 Unscrew and remove the through-bolt.
13 Unbolt the mounting from the subframe, and if necessary unbolt the bracket from the cylinder block.
14 Fit the new mounting using a reversal of the removal procedure.

Part B: 1171 cc and 1390 cc Engines – In-car engine repair procedures

12 General information

This Part of Chapter 2 is devoted to in-car repair procedures for the 1171 cc (E5F and E7F) and 1390 cc (E6J and E7J) engines. Similar information covering the 1108 cc (C1E), 1721 cc (F2N), 1764 cc (F7P) and 1794 cc (F3P) engines will be found in Parts A and C. All procedures concerning engine removal and refitting, and engine block/cylinder head overhaul for all engine types, can be found in Part D of this Chapter.

Most of the operations included in this Part are based on the assumption that the engine is still installed in the car. Therefore, if this information is being used during a complete engine overhaul, with the engine already removed, many of the steps included here will not apply.

Engine description

The engine is of four-cylinder, in-line, overhead camshaft type, mounted transversely in the front of the car.

The cast-iron cylinder block is of the replaceable wet liner type. The crankshaft is supported within the cylinder block on five shell-type main bearings. Thrustwashers are fitted at the centre main bearing to control crankshaft endfloat.

The connecting rods are attached to the crankshaft by horizontally-split shell-type big-end bearings, and to the pistons by interference-fit gudgeon pins. The aluminium alloy pistons are of the slipper type, and are fitted with three piston rings, comprising two compression rings and a scraper-type oil control ring.

The overhead camshaft is mounted directly in the cylinder head, and is driven by the crankshaft via a toothed rubber timing belt which also drives the water pump. The camshaft operates the valves via rocker arms located on a rocker shaft bolted to the top of the cylinder head.

A semi-enclosed crankcase ventilation system is employed.

Lubrication is by pressure feed from a gear-type oil pump, which is chain-driven direct from the crankshaft.

The distributor rotor is driven directly from

Fig. 2.11 Cutaway view of an E-type engine (Sec 12)

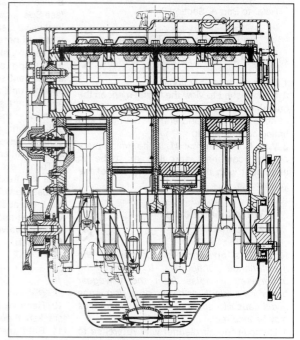

Fig. 2.12 Lubrication circuit of the E-type engine (Sec 12)

the rear end of the camshaft. On carburettor engines, the fuel pump is also operated by the camshaft via an eccentric and lever.

Repair operations possible with the engine in the vehicle

The following operations can be carried out without having to remove the engine from the car.
(a) Removal and refitting of the cylinder head.
(b) Removal and refitting of the timing belt and sprockets.
(c) Renewal of the camshaft oil seal.
(d) Removal and refitting of the camshaft.
(e) Removal and refitting of the sump.
(f) Removal and refitting of the connecting rods, pistons and liners*.
(g) Removal and refitting of the oil pump.
(h) Renewal of the crankshaft oil seals (except the left-hand/driveplate end oil seal on automatic transmission models).
(i) Renewal of the engine mountings.
(j) Removal and refitting of the flywheel (the driveplate on automatic transmission models can only be removed after removing the complete engine/transmission assembly - see Part D).

*Although the operation marked with an asterisk can be carried out with the engine in the car after removal of the sump, it is better for the engine to be removed in the interests of cleanliness and improved access. For this reason, the procedure is described in Part D of this Chapter.

13 Compression test - description and interpretation

Refer to Part A, Section 2.

14 Top Dead Centre (TDC) for No 1 piston - locating

1 Refer to Part A, Section 3.
2 If it is wished to view the timing mark on the camshaft sprocket, refer to Section 16 and remove the right-hand engine mounting and timing cover bracket.
3 With No 1 piston at TDC, the timing mark on the camshaft sprocket must be in alignment with the mark on the end of the valve cover.

15 Valve clearances - adjustment

Note: *This operation is not part of the maintenance schedule. It should be undertaken if noise from the valvegear becomes evident, or if loss of performance gives cause to suspect that the clearances may be incorrect.*

1 Remove the air cleaner with reference to Chapter 4.
2 Disconnect the accelerator cable from the

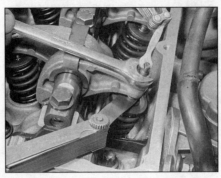
15.8 Adjusting a valve clearance

carburettor or throttle body housing with reference to Chapter 4. On carburettor models, also disconnect the choke cable.
3 Disconnect the crankcase ventilation hoses from the valve cover.
4 Unscrew the bolts, and remove the valve cover and gasket.
5 Remove the spark plugs (Chapter 1) in order to make turning the engine easier.
6 Draw the valve positions on a piece of paper, numbering them 1 to 4 inlet and exhaust according to their cylinders, from the flywheel/driveplate (left-hand) end of the engine (ie 1E, 1I, 2E, 2I and so on). The inlet valves are on the inlet manifold side of the cylinder head, and the exhaust valves are on the exhaust manifold side. As the valve clearances are adjusted, cross them off.
7 Using a socket or spanner on the crankshaft pulley bolt, turn the engine in a clockwise direction until No 1 exhaust valve is completely open (ie the valve spring is completely compressed).
8 Insert a feeler blade of the correct thickness (see Specifications) between the No 3 cylinder inlet valve stem and the end of the rocker arm. It should be a firm sliding fit. If adjustment is necessary, loosen the locknut on the rocker arm using a ring spanner, and turn the adjustment screw with a small open-ended spanner until the fit is correct (photo). Hold the adjustment screw, tighten the locknut and recheck the adjustment. Repeat the adjustment procedure on No 4 cylinder exhaust valve. Note that the valve clearances for the inlet and exhaust valves are different.
9 Turn the engine in a clockwise direction until No 3 exhaust valve is completely open. Adjust the valve clearances on No 4 inlet and No 2 exhaust valves. Continue to adjust the valve clearances in the following sequence.

Exhaust valve fully open	Inlet valve to adjust	Exhaust valve to adjust
1E	3I	4E
3E	4I	2E
4E	2I	1E
2E	1I	3E

10 Remove the socket or spanner from the crankshaft pulley bolt.
11 Refit the spark plugs. Refit the valve cover, using a new gasket where necessary. Reconnect the crankcase ventilation hoses.

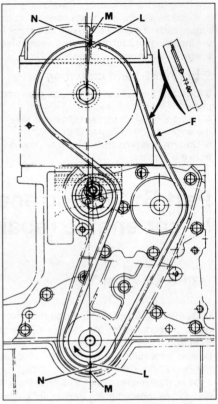
Fig. 2.13 Timing belt and timing marks (Sec 16)

F Belt tension checking point
L Timing marks on the camshaft and crankshaft sprockets
M Fixed timing marks
N Timing bands on the belt

12 Reconnect and adjust the accelerator cable with reference to Chapter 4. On carburettor models, also reconnect and adjust the choke cable.
13 Refit the air cleaner.

16 Timing belt - removal, inspection and refitting

⚠️ **Caution: If the timing belt breaks or slips in service, extensive engine damage may result. Renew the belt at the intervals specified in Chapter 1, or earlier if its condition is at all doubtful.**

Removal

1 Disconnect both the battery leads.
2 Remove the bonnet with reference to Chapter 11.
3 Apply the handbrake, then jack up the front right-hand side of the car and support on axle stands. Remove the roadwheel.
4 Remove the plastic cover from within the right-hand wheel arch, to give access to the crankshaft pulley.
5 Support the weight of the engine with a trolley jack or with an engine hoist. Use a

16.11 Unscrewing the crankshaft pulley bolt

16.13 Timing marks (arrowed) on the crankshaft sprocket and front oil seal cover

16.19A Fitting the timing belt over the camshaft sprocket . . .

piece of wood between the jack and the engine sump, or if using an engine hoist, attach it to the two lifting eyes on the cylinder head.

6 Where fitted, remove the plastic cover from the top of the right-hand engine upper mounting bracket. Mark the position of the bracket in relation to the mounting nut, then unscrew the nut and bolts and remove the bracket from the top of the timing cover.

7 Unbolt the timing cover bracket from the right-hand end of the cylinder head.

8 To give improved access, mark the position of the engine mounting lower bracket, then unbolt it from the body.

9 Remove the alternator drivebelt with reference to Chapter 12, Section 6.

10 Turn the engine in a clockwise direction, using a socket on the crankshaft pulley bolt, until the TDC mark on the camshaft sprocket is aligned with the TDC mark on the end of the valve cover. Looking through the timing aperture in the transmission bellhousing, check that the TDC marks are aligned.

11 Unscrew the crankshaft pulley bolt while holding the crankshaft stationary as follows (photo). Unbolt the ignition timing sensor from the top of the gearbox/automatic transmission, and have an assistant insert a screwdriver in the starter ring gear teeth through the access hole in the top of the gearbox bellhousing.

12 Remove the crankshaft pulley from the

nose of the crankshaft. If it is tight, use a puller.

13 Check that the timing mark on the bottom of the crankshaft sprocket is aligned with the mark on the front oil seal cover (photo).

14 Unbolt and remove the timing belt covers.

15 Loosen the nut securing the timing belt tensioner. Move the tensioner outwards to release the tension from the belt, then retighten the nut.

16 Check if the belt is marked with arrows to indicate its running direction; mark it if necessary. Release the belt from the camshaft sprocket, water pump sprocket and crankshaft sprocket, and remove it from the engine. Be careful not to kink or otherwise damage the belt if it is to be re-used.

17 Clean the sprockets and tensioner, and wipe them dry. Also clean the cylinder head and block behind the timing belt running area.

Inspection

18 Examine the timing belt carefully for any signs of cracking, fraying or general wear, particularly at the roots of the teeth. Renew the belt if there is any sign of deterioration of this nature, or if there is any oil or grease contamination. Renew any leaking oil seals. The belt **must** be renewed if it has completed the maximum mileage given in Chapter 1. Also examine the sprockets and tensioner as described in the next Section.

Refitting

19 Check the directional mark and timing bands on the back of the timing belt. Fit the belt on the crankshaft sprocket so that one of the timing bands is aligned with the marks on the sprocket and cover. The other timing band should be positioned so that it will locate on the camshaft sprocket in alignment with the other timing mark. After engaging the belt with the crankshaft sprocket, pull it taut over the water pump sprocket and onto the camshaft sprocket, then position it over the tensioner wheel (photos).

20 With the belt fully engaged with the sprockets, slacken the tensioner nut and tension the belt, then retighten the nut. The tensioner is not spring-loaded, so it will have to be rotated manually to tension the belt; a screwdriver can be used as a lever between bolts in the holes in the tensioner hub, or use a tool like the one in illustration 2.14. Check that the timing marks are still aligned correctly.

21 The belt deflection must now be checked. To do this, first make a mark on the engine front lifting bracket in line with the timing belt, midway between the camshaft and water pump sprockets. A force of 30 N (7 lbf) must be applied to the timing belt. Belt deflection under this force must be 6.0 ± 0.5 mm (0.236 ± 0.02 in). Renault technicians use a special tool to do this, but an alternative arrangement can be made by using a spring

2

16.19B . . . and tensioner

16.19C Timing band on the belt aligned with the mark on the sprocket (arrowed)

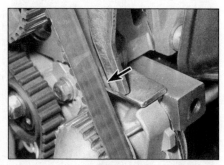

16.21A Timing belt deflection mark made on the front engine lifting bracket (arrowed)

16.21B Checking the timing belt tension using a spring balance and steel rule

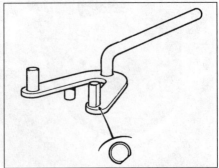

Fig. 2.14 Renault tool for holding the camshaft sprocket stationary (Sec 17)

It is easy to make up a crankshaft sprocket puller using two bolts, a metal bar and the existing crankshaft pulley bolt. By unscrewing the crankshaft pulley bolt, the sprocket is pulled from the end of the crankshaft.

balance and steel rule. Apply the force with the spring balance, and read off the deflection on the steel rule (photos).

22 If the adjustment is incorrect, the tensioner hub will have to be repositioned. The tensioner nut must be tightened securely, since if it were to come loose, considerable engine damage would result.

23 Temporarily refit the crankshaft pulley and its bolt. Using a socket on the pulley bolt, turn the engine through two complete revolutions in its normal direction, then recheck the timing belt tension and make sure that the timing marks are still in alignment. Remove the pulley.

24 Refit the timing belt covers and tighten the bolts.

25 Refit the crankshaft pulley to the nose of the crankshaft, then screw in the bolt. Tighten the bolt to the specified torque, holding the crankshaft stationary using the method described in paragraph 11.

26 Refit the alternator drivebelt with reference to Chapter 12, Section 6.

27 Refit the engine mounting lower bracket in its previously-noted position, and tighten the bolts to the specified torque.

28 Refit the timing cover bracket to the right-hand end of the cylinder head, and tighten the bolts.

29 Refit the engine upper mounting bracket, making sure that it is located in its previously-noted position. (If necessary, check the mounting adjustment with reference to

Section 25). Tighten the bolts and nut to the specified torque. Remove the trolley jack or hoist.

30 Where applicable, refit the plastic cover to the top of the engine mounting bracket, and tighten the screw.

31 Refit the plastic cover inside the right-hand wheel arch.

32 Refit the roadwheel and lower the car to the ground.

33 Refit the bonnet with reference to Chapter 11.

34 Reconnect both the battery leads.

17 Timing belt sprockets and tensioner - removal, inspection and refitting

Removal

1 Remove the timing belt as described in Section 16.

2 To remove the camshaft sprocket, hold the sprocket stationary using a metal bar with two bolts tightened onto it inserted into the holes in the sprocket, then unscrew the bolt. The Renault tool is shown in Fig. 2.14.

3 Remove the sprocket from the end of the camshaft. Note that it has a tab on its inner face which locates in a slot in the end of the camshaft.

4 A puller may be necessary to remove the crankshaft sprocket.

5 Where applicable, remove the Woodruff key from the slot in the crankshaft (on later engines, the key is incorporated in the sprocket).

6 Unscrew the nut and withdraw the tensioner from the stud on the front cover (photo).

Inspection

7 Inspect the teeth of the sprockets for signs of nicks and damage. Also examine the water pump teeth. The teeth are not prone to wear, and should normally last the life of the engine.

8 Spin the tensioner by hand, and check it for any roughness or tightness. Do not attempt to clean it with solvent, as this may enter the bearing. If wear is evident, renew the tensioner.

Refitting

9 Locate the tensioner on the stud on the front cover, then refit the nut and tighten it finger-tight at this stage.

10 Where applicable, fit the Woodruff key in its slot in the crankshaft, making sure that it is level to ensure engagement with the sprocket.

11 Slide the sprocket fully onto the crankshaft; use a metal tube if necessary to tap it into position.

12 Locate the sprocket on the end of the camshaft, making sure that the tab locates in the special slot, then screw in the bolt. Tighten the bolt to the specified torque, holding the sprocket stationary using the method described in paragraph 2.

13 Refit the timing belt as described in Section 16.

18 Camshaft oil seal - renewal

1 Remove the camshaft sprocket as described in Section 17.

2 Note the fitted position of the old oil seal. Using a small screwdriver, prise out the oil seal from the cylinder head.

17.6 Removing the timing belt tensioner

18.4 Fitting a new camshaft oil seal

19.2 Two of the rocker shaft mounting bolts (arrowed)

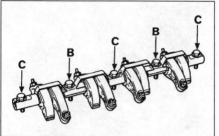

Fig. 2.15 Rocker shaft bolt identification (Sec 19)

B Solid bolts coloured yellow
C Hollow bolts coloured black

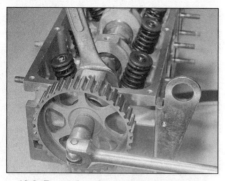

19.3 Removing the camshaft sprocket retaining bolt

3 Wipe clean the seating in the cylinder head.
4 Smear a little oil on the outer perimeter and sealing lip of the new oil seal. Locate the seal squarely in the cylinder head. Drive the seal into position using a metal tube which has an external diameter slightly less than that of the bore in the cylinder head (photo). Make sure that the oil seal is the correct way round, with the lip facing inwards.
5 Refit the camshaft sprocket as described in Section 17.

19 Camshaft - removal, inspection and refitting

Removal

1 Remove the cylinder head as described in Section 20, and place it on the workbench.
2 Progressively unscrew the bolts holding the rocker shaft and retaining plate to the cylinder head, and withdraw the shaft (photo). Note that the bolts are different, and should be identified for position before removal. Numbers 2 and 4 bolts have solid shanks, whereas the other bolts have hollow shanks. Their heads are coloured as shown in Fig. 2.15.
3 Hold the camshaft stationary using a spanner on the special flats provided on the camshaft, then unscrew the bolt and withdraw the sprocket (photo).
4 Using a Torx key, unscrew the two bolts

holding the distributor to the cylinder head, and remove the distributor. There is no need to mark the distributor, as it is not possible to adjust its position. Although there is an elongated slot for one of the bolts, the other bolt locates in a single hole.
5 Using a dial gauge, measure the endfloat of the camshaft, and compare with that given in the Specifications (photo). This will give an indication of the amount of wear in the thrustplate.
6 Unscrew the two bolts and lift the thrustplate out from the slot in the camshaft (photo).
7 Carefully withdraw the camshaft from the sprocket end of the cylinder head, taking care not to damage the bearing surfaces (photo).

Inspection

8 Examine the camshaft bearing surfaces, cam lobes and fuel pump eccentric for wear ridges and scoring. Renew the camshaft if any of these conditions are apparent.
9 Examine the condition of the bearing surfaces both on the camshaft and in the cylinder head. If the head bearing surfaces are worn excessively, the cylinder head will need to be renewed.

Refitting

10 Lubricate the bearing surfaces in the cylinder head and the camshaft journals, then insert the camshaft into the head.
11 Refit the thrustplate, then insert and tighten the bolts.

19.5 Checking the camshaft endfloat with a dial test gauge

12 Measure the endfloat as described in paragraph 5, and make sure that it is within the limits given in the Specifications. Excessive endfloat can only be due to wear of the thrustplate or the camshaft.
13 Refit the distributor and tighten the two bolts using a Torx key.
14 Refit the camshaft sprocket, making sure that the tab engages with the cut-out in the end of the camshaft. Hold the camshaft stationary with a spanner on the special flats, then insert the bolt and tighten it to the specified torque (photo).
15 Refit the rocker shaft and retaining plate, then insert the bolts in their original positions and tighten them to the specified torque.
16 Refit the cylinder head as described in Section 20.

2

19.6 Removing the camshaft thrustplate

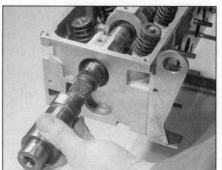

19.7 Removing the camshaft

19.14 Tightening the camshaft retaining bolt

20.11A Removing the valve cover . . .

20.11B . . .and gasket

20.15A Lifting the cylinder head from the block

20 Cylinder head - removal, inspection and refitting

Removal

1 Disconnect both battery leads, referring to Chapter 12 if necessary.
2 Drain the cooling system, including the cylinder block, with reference to Chapter 1. The block drain plug is located on the right-hand rear face of the engine, beneath the inlet manifold. It is important to drain the block because if the wet cylinder liners are disturbed, the coolant will drain into the sump.
3 Drain the engine oil with reference to Chapter 1.
4 Remove the timing belt with reference to Section 16 of this Chapter.
5 Remove the air cleaner with reference to Chapter 4.
6 Disconnect the wires from the following components:
(a) Temperature sender.
(b) Carburettor or throttle body unit.
(c) Ignition module.
7 Disconnect the accelerator cable, and on carburettor models the choke cable, with reference to Chapter 4. Position them to one side.
8 If wished, the inlet and exhaust manifolds can be removed at this stage. Otherwise, disconnect the various fuel, vacuum and coolant hoses and electrical feeds from the

carburettor or throttle body and from the inlet manifold (see Chapter 4). Be prepared for fuel spillage. Also disconnect the exhaust downpipe.
9 Loosen the clips and disconnect the following hoses:
(a) Expansion tank purge hose from the cylinder head outlet.
(b) Radiator top hose from the thermostat housing.
(c) Crankcase ventilation hoses from the valve cover.
10 Disconnect the HT leads from the spark plugs, and pull them carefully from the HT lead holder on the valve cover. Unbolt and remove the holder.
11 Unbolt the valve cover from the cylinder head, and remove the gasket (photos).
12 Remove the thermostat with reference to Chapter 3.
13 Progressively slacken the cylinder head bolts in the reverse order to that shown in Fig. 2.8. Remove all the bolts except the one positioned on the front right-hand corner, which should be unscrewed by only three or four threads.
14 The joint between the cylinder head, gasket and cylinder block must now be broken without disturbing the cylinder liners. To do this, pull the distributor end of the cylinder head forward so as to swivel it around the single bolt still fitted, then move the head back to its original position. If this procedure is not followed, there is a

possibility of the cylinder liners moving and their bottom seals being disturbed, causing leakage after refitting the head.
15 Remove the remaining bolt, and lift the head from the cylinder block, followed by the gasket. Note the location dowel on the front right-hand corner of the block (photos).
16 Note that the crankshaft must **not** be rotated with the cylinder head removed, otherwise the cylinder liners will be displaced. If it is necessary to turn the crankshaft (eg to clean the piston crowns), clamp the liners using bolts and washers, or make up some retaining clamps out of flat metal bar, held in place with bolts screwed into the block (photo).

Inspection

17 The mating faces of the cylinder head and block must be perfectly clean before refitting the head. Use a scraper to remove all traces of gasket and carbon, and also clean the tops of the pistons. Take particular care with the aluminium cylinder head, as the soft metal is easily damaged. Also, make sure that debris is not allowed to enter the oil and water channels - this is particularly important for the oil circuit, as carbon could block the oil supply to the camshaft and rocker arms or crankshaft bearings. Using adhesive tape and paper, seal the water, oil and bolt holes in the cylinder block. Clean the piston crowns in the same way.

20.15B Removing the cylinder head gasket

20.15C Cylinder head locating dowel (arrowed)

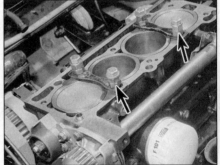

20.16 Clamps (arrowed) holding the liners in place

20.20 Cleaning the cylinder head bolt holes in the cylinder block

20.26A Torque-tightening the cylinder head bolts

20.26B Angle-tightening the cylinder head bolts

> **HAYNES HiNT**
>
> *To prevent carbon entering the gap between the pistons and bores, smear a little grease in the gap. After cleaning the piston, rotate the crankshaft so that the piston moves down the bore, then wipe out the grease and carbon with a cloth rag.*

18 Check the block and head for nicks, deep scratches and other damage. If slight, they may be removed carefully with a file. It may be possible to repair more serious damage by machining, but this is a specialist job.

19 If warpage of the cylinder head is suspected, use a straight-edge to check it for distortion. Also check the protrusion of the cylinder liners. Either of these items can be associated with the head gasket blowing. Refer to Part D of this Chapter for further information.

20 Clean out all the bolt holes in the block using a pipe cleaner, or a rag and screwdriver (photo). Make sure that all oil is removed, otherwise there is a possibility of the block being cracked by hydraulic pressure when the bolts are tightened.

21 Examine the bolt threads and the threads in the cylinder block for damage. If necessary, use the correct-size tap to chase out the threads in the block, and use a die to clean the threads on the bolts.

Refitting

22 Position No 1 piston at TDC, then remove the cylinder liner clamps, and wipe clean the faces of the head and block.

23 Check that the location dowel is in position on the front right-hand corner of the block.

24 Position the new gasket on the block and over the dowel - it can only be fitted one way round.

25 Lower the cylinder head onto the block. Oil the threads and the head undersides of the cylinder head bolts, then insert them and initially screw them in finger-tight.

26 Tighten the cylinder head bolts to the specified torques in the sequence shown in Fig. 2.8, and in the stages given in the Specifications at the beginning of this Chapter. The first three stages pre-compress the gasket, and the remaining stages are the main tightening procedure (photos).

27 If not already done, refit the rocker shaft and retaining plate. Insert the bolts in their original positions, and tighten them progressively to the specified torque.

28 If the cylinder head has been overhauled, it is worthwhile checking the valve clearances at this stage, to prevent any possibility of the valves touching the pistons when the timing belt is being fitted. Turn the crankshaft so that there are no pistons at TDC. Use a socket on the camshaft sprocket to turn the camshaft and check the valve clearances.

29 Refit the thermostat with reference to Chapter 3.

30 Refit the timing belt with reference to Section 16 of this Chapter.

31 If not already done, adjust the valve clearances as described in Section 15 of this Chapter.

32 Refit the valve cover together with a new gasket, and tighten the bolts progressively.

33 Refit the spark plugs if they were removed.

34 Refit the HT lead holder, then connect the HT leads to the spark plugs, and insert the leads in the holder.

35 Reconnect the crankshaft ventilation and coolant hoses.

36 Refit the inlet and exhaust manifolds, or reconnect the various services to them, with reference to Chapter 4. Reconnect the exhaust downpipe.

37 Reconnect the accelerator and choke cables (as applicable), and adjust them with reference to Chapter 4.

38 Reconnect all wires in their correct positions.

39 Refit the air cleaner with reference to Chapter 4.

40 Refill the engine with oil, with reference to Chapter 1.

41 Reconnect both battery leads.

42 Refill and bleed the cooling system with reference to Chapter 1.

21 Sump - removal and refitting

Note: *An engine lifting hoist is required during this procedure.*

Removal

1 Disconnect the battery negative and positive leads.

2 Drain the engine oil as described in Chapter 1, then refit and tighten the drain plug using a new washer.

3 Jack up the front of the car and support on axle stands. Remove the right-hand front roadwheel.

4 Remove the plastic cover from inside the wheel arch for access to the engine.

5 Remove the bonnet as described in Chapter 11.

6 Remove the air cleaner with reference to Chapter 4.

7 Working beneath the car, loosen (but do not remove) the rearmost bolt on the engine rear mounting link. Unscrew and remove the front bolt, and swivel the link downwards.

8 Remove the exhaust downpipe with reference to Chapter 4.

9 Unscrew the bolts securing the engine-to-gearbox tie-bar and cover assembly to the gearbox bellhousing and engine block, and lower it from the sump (photos).

10 Unscrew the bolts securing the sump to the cylinder block, but leave two bolts finger-tight at this stage.

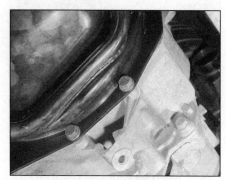

21.9A Unscrew the bolts . . .

21.9B ... and remove the tie-bar and cover assembly

21.12 Lowering the sump from the engine

21.14 Fitting the gasket to the sump

11 The engine must now be raised by approximately 75 to 100 mm (3 to 4 in) before the sump can be removed. To do this, first mark the position of the right-hand engine mounting upper bracket in relation to the outer mounting nut, then unscrew the nut. Attach a hoist to the engine lifting eye next to the timing cover, and raise the engine.

12 Unscrew the two remaining sump bolts, then break the joint by striking the sump with the palm of the hand. Lower the sump over the oil pump and withdraw it (photo).

Refitting

13 Clean all traces of gasket from the cylinder block and sump, and wipe them dry.

14 Locate a new gasket on the sump, making

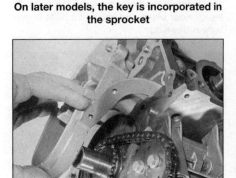

22.3 Removing the Woodruff key (arrowed) from the front of the crankshaft. On later models, the key is incorporated in the sprocket

sure that it is correctly positioned on the curved sections (photo).

15 Offer the sump to the cylinder block, and insert the bolts. Progressively tighten the bolts in diagonal sequence to the specified torque.

16 Lower the engine and refit the right-hand engine mounting upper bracket and nut, tightening the nut to the specified torque. Remove the hoist.

17 Refit the tie-bar and cover assembly to the gearbox bellhousing and engine block.

18 Refit the exhaust downpipe and tighten the bolts. Refer to Chapter 4 if necessary.

19 Refit the engine rear mounting link and tighten the bolts.

20 Refit the air cleaner with reference to Chapter 4.

21 Refit the bonnet as described in Chapter 11.

22 Refit the plastic cover to the inside of the wheel arch.

23 Refit the roadwheel and lower the car to the ground.

24 Refill the engine with oil, with reference to Chapter 1.

25 Reconnect the battery leads.

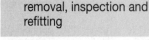

22 Oil pump and sprockets - removal, inspection and refitting

Removal

1 Remove the timing belt and the crankshaft sprocket with reference to Sections 16 and 17.

2 Remove the sump as described in Section 21.

3 Remove the Woodruff key (when fitted) from its slot in the crankshaft (photo).

4 Unbolt the front cover from the cylinder block (photo).

5 Slide off the oil seal spacer.

6 Unscrew the bolts securing the sprocket to the oil pump hub. Use a screwdriver through one of the holes in the sprocket to hold it stationary (photo).

7 Remove the oil pump sprocket, and release the chain from the drive sprocket on the crankshaft.

8 Slide the drive sprocket from the crankshaft.

9 Unscrew the two mounting bolts, and withdraw the oil pump from the crankcase (photo). If the two locating dowels are displaced, refit them in the crankcase.

Inspection

10 Unscrew the four retaining bolts, and lift off the pump cover and pick-up tube.

11 Using a feeler gauge, check the clearance between each of the gears and the oil pump body (photo). Also check the endfloat by measuring the clearance between the gears and the cover joint face. If any clearance is outside the tolerances given in the Specifications, the oil pump must be renewed.

12 Mount the cover in a vice, then depress the relief valve end stop and extract the spring

22.6 Unscrewing the oil pump sprocket bolts

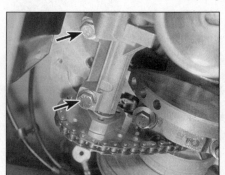

22.9 Oil pump mounting bolts (arrowed)

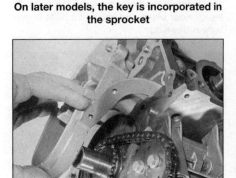

22.4 Removing the front cover from the cylinder block

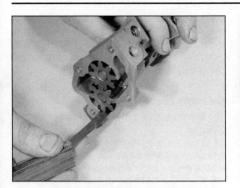

22.11 Checking the clearance between the oil pump gears and body

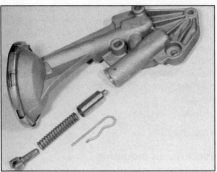

22.12 Oil pump relief valve components

22.21A Removing the oil seal from the front cover

clip. Release the end stop, and remove the spring and piston (photo).

13 Examine the relief valve piston and seat for signs of wear and damage. If evident, renew the oil pump complete.

14 If the components are serviceable, clean them and reassemble in the reverse order to dismantling. Before refitting the cover, fill the pump with fresh engine oil to assist circulation when the engine is first started. A new pump should also be primed with oil.

15 Examine the chain for excessive wear, and renew it if necessary. Similarly check the sprockets.

Refitting

16 Wipe clean the oil pump and crankcase mating surfaces.

17 Check that the two locating dowels are fitted in the crankcase, then position the oil pump on them and insert the two mounting bolts. Tighten the bolts securely.

18 Slide the sprocket onto the crankshaft.

19 Engage the oil pump sprocket with the chain, then engage the chain with the crankshaft drive sprocket, and locate the sprocket on the oil pump hub.

20 Align the holes, then insert the sprocket bolts and tighten them securely while holding the sprocket stationary with a screwdriver.

21 The oil seal in the front cover should be renewed whenever the cover is removed. Note the fitted position of the old seal, then prise it out with a screwdriver and wipe clean

the seating. Smear the outer perimeter of the new seal with fresh engine oil, and locate it squarely on the cover with its closed side facing outwards. Place the cover on a block of wood, then use a socket or metal tube to drive in the oil seal (photos).

22 Clean all traces of sealant from the front cover and block mating faces. Apply a 0.6 to 1.0 mm diameter bead of sealant around the perimeter of the front cover, then refit it to the cylinder block and tighten the bolts securely.

23 Smear the oil seal with a little engine oil, then slide the spacer onto the front of the crankshaft (photo). Turn the spacer slightly as it enters the oil seal, to prevent damage to the seal lip. If the spacer is worn where the old oil seal contacted it, it can be turned around so that the new oil seal contacts the unworn area.

24 Refit the Woodruff key to its slot in the crankshaft, when applicable.

25 Refit the sump with reference to Section 21.

26 Refit the crankshaft sprocket and timing belt with reference to Sections 17 and 16.

23 Crankshaft oil seal(s) - renewal

Note: *On automatic transmission models, it is not possible to remove the transmission alone, so access to the left-hand (driveplate end) oil seal is not possible with the engine in situ.*

Right-hand (pulley end) oil seal

1 This procedure is included in the removal and refitting of the oil pump. Refer to Section 22, paragraphs 1 to 5 and 21 to 26.

Left-hand (flywheel end) oil seal

2 Remove the flywheel as described in Section 24.

3 Prise out the old oil seal using a small screwdriver, taking care not to damage the surface on the crankshaft. Alternatively, the oil seal can be removed by drilling two small holes diagonally opposite each other and inserting self-tapping screws in them. A pair of grips can then be used to pull out the oil seal, by pulling on each side in turn.

4 Inspect the seal rubbing surface on the crankshaft. If it is grooved or rough in the area where the old seal was fitted, the new seal should be fitted slightly less deeply, so that it rubs on an unworn part of the surface.

5 Wipe clean the oil seal seating, then dip the new seal in fresh engine oil, and locate it over the crankshaft with its closed side facing outwards. Make sure that the oil seal lip is not damaged as it is located on the crankshaft.

6 Using a metal tube, drive the oil seal squarely into the bore until flush. A block of wood cut to pass over the end of the crankshaft may be used instead.

7 Refit the flywheel with reference to Section 24.

2

22.21B Driving a new oil seal into the front cover

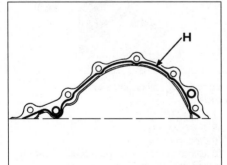

Fig. 2.16 Sealant bead application (H) around the front cover (Sec 22)

22.23 Sliding the spacer onto the front of the crankshaft

24 Flywheel - removal, inspection and refitting

Note: *Refer to Part D, Section 54 for details of removing, inspection and refitting of the driveplate on automatic transmission models.*

1 Refer to Section 10 in Part A of this Chapter, noting the following differences:
(a) *Renewal of the flywheel bolts is not specified.*
(b) *The flywheel locating face on the crankshaft should be coated with Loctite Autoform, or an equivalent compound, when refitting.*

25 Engine/transmission mountings - inspection, renewal and adjustment

Note: *The engine mountings are positioned with a jig at the Renault factory. It is important to mark the exact position of each mounting before removing it, to ensure correct alignment of the engine/gearbox.*

Inspection

1 Apply the handbrake, then jack up the front of the car and support it on axle stands.

2 Visually inspect the rubber pads on the engine/transmission mountings for signs of cracking and deterioration. Careful use of a lever will help to determine the condition of the rubber pads. If there is excessive movement in the mounting, or if the rubber has deteriorated, the mounting should be renewed.

Renewal

Right-hand mounting

3 Unscrew the screw and remove the plastic cover from the top of the mounting. On right-hand drive models, note that the cover holds the accelerator cable in position.

4 Mark the position of the mounting using a scriber or dab of paint on the washer beneath the central nut and around the lower mounting bracket.

5 Unscrew the nut and remove the washer.

6 Support the weight of the engine, using a hoist on the right-hand lifting eye, or a trolley jack and a block of wood beneath the sump.

7 Unbolt the upper mounting bracket from the timing cover bracket on the end of the cylinder head.

8 Unbolt the lower mounting bracket and movement limiter from the body.

9 Fit the new components using a reversal of the removal procedure, but leave the mounting nuts and bolts loose. Finally, carry out the mounting adjustment procedure described later in this Section.

Gearbox mounting

10 Remove the battery as described in Chapter 12.

11 Take the weight of the transmission, using a trolley jack and a block of wood, or by using a hoist on the left-hand engine lifting eye.

12 Mark the position of the mounting pad using a scriber or dab of paint.

13 Unscrew the central nut and the two nuts securing the mounting pad to the upper bracket.

14 Unbolt the upper mounting bracket from the body, and remove the mounting pad.

15 If necessary, unbolt the lower bracket from the gearbox.

16 Refitting is a reversal of the removal procedure, but leave the mounting nuts/bolts loose. Finally, carry out the mounting adjustment procedure described later in this Section.

Rear mounting/lower link

17 Working beneath the car, unscrew the bolts securing the rear mounting link to the brackets on the subframe and cylinder block. Withdraw the link from under the car.

18 If necessary, unbolt and remove the brackets (photo).

19 Refitting is a reversal of the removal procedure, but leave the mounting bolts loose. Finally, carry out the mounting adjustment procedure described later in this Section.

Adjustment

20 If just one mounting is being renewed, fit the new mounting in the same position as the old one, and check that the dimensions are correct. If more than one mounting has been renewed, take the weight of the engine/gearbox assembly on a hoist, loosen all of the mounting bolts and carry out the following procedure.

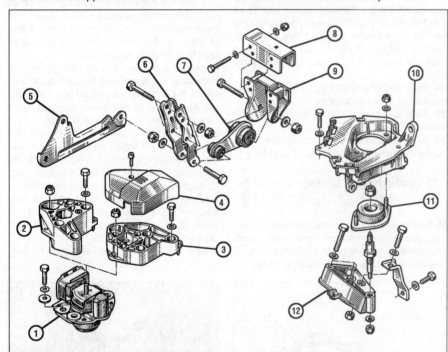

Fig. 2.17 Engine/gearbox mounting components - all engines except C1E (Secs 25 and 43)

1 *Right-hand lower bracket/rubber pad/movement limiter assembly*
2 *Right-hand upper bracket (first type)*
3 *Right-hand upper bracket (second type)*
4 *Cover*
5 *Rear bracket*
6 *Rear bracket*
7 *Lower link*
8 *Rear bracket*
9 *Rear bracket*
10 *Gearbox mounting upper bracket/battery mounting*
11 *Gearbox mounting rubber pad*
12 *Gearbox mounting lower bracket*

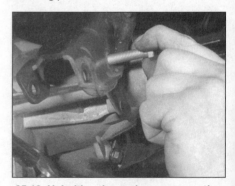

25.18 Unbolting the engine rear mounting bracket

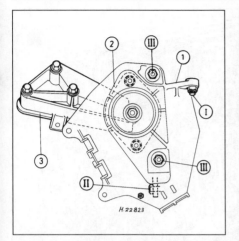

Fig. 2.18 Gearbox mounting components and adjustment. For I, II and III see text (Sec 25)

1 *Upper bracket/battery mounting*
2 *Rubber mounting pad*
3 *Gearbox mounting (lower) bracket*

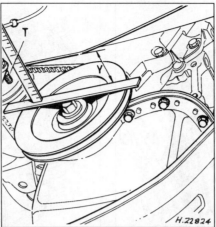

Fig. 2.19 Checking the clearance between the crankshaft pulley and the edge of the body on the E-type engine. For Y see text (Sec 25)

T Strengthening rod

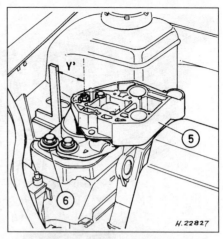

Fig. 2.20 Checking the clearance (Y') between the outer edge of the lower bracket and the centre of its stud (Sec 25)

5 *Upper bracket (unbolted for first measurement)*
6 *Lower bracket*

Gearbox mounting

21 Refer to Fig. 2.18 and pre-tighten the turret nut (I) and lower mounting bolt (II) to 3 Nm (2 lbf ft). Tighten the two upper mounting bolts (III) to 20 Nm (15 lbf ft), then tighten the nut (I) and bolt (II) to the same torque.

22 With the gearbox mounting pad and lower bracket and the right-hand mounting lower bracket in place, check that the mounting bolts are loose.

23 Remove the right-hand front roadwheel, then remove the inner wheel arch plastic panel for access to the crankshaft pulley. Refer to Fig. 2.19, and check that there is a minimum clearance of 48 mm (models without power steering) or 32 mm (models with power steering) between the crankshaft pulley and

the edge of the body next to the strengthening rod. Note that this clearance does not take into consideration the thickness of the straight-edge on the pulley.

24 Maintaining this clearance, centre the gearbox mounting pad in the longitudinal direction in the middle of the upper bracket/battery tray gaps. Tighten the mounting nuts and bolts.

Right-hand mounting

25 Unbolt the upper bracket if it has already been fitted. Measure the dimension Y' (shown in Fig. 2.20) between the outer edge of the lower bracket and the centre of its stud.

26 Refit the upper bracket, and tighten the bolts securing it to the specified torque.

27 With the lower mounting bracket and

movement limiter bolts loose, lower the hoist supporting the engine/transmission, and allow it to rest on its mountings. Measure dimension Y' again. If necessary, adjust the mounting so that the dimension is the same as recorded previously. Tighten the two lower bracket bolts to the specified torque.

28 With the remaining bolts loose, centre the limiter within the hole in the upper bracket so that the clearance is the same on both sides. Tighten the limiter mounting bolts to the specified torque.

29 Working under the car, refit the rear mounting link and tighten the bolts to the specified torque.

30 Recheck dimension Y' on the right-hand mounting, and adjust if necessary.

2

Part C: 1721 cc, 1764 cc and 1794 cc Engines – In-car engine repair procedures

26 General information

This Part of Chapter 2 is devoted to in-car repair procedures for the 1721 cc (F2N), 1764 cc (F7P), and 1794 cc (F3P) engines. Similar information covering the 1108 cc (C1E), 1171 cc (E5F and E7F), and 1390 cc (E6J and E7J) engines will be found in Parts A and B. All procedures concerning engine removal and refitting, and engine block/cylinder head overhaul for all engine types can be found in Part D of this Chapter.

Most of the operations included in this Part are based on the assumption that the engine is still installed in the car. Therefore, if this information is being used during a complete

engine overhaul, with the engine already removed, many of the steps included here will not apply.

Engine description

The engine is of four-cylinder, in-line, single overhead camshaft (8-valve) or double overhead camshaft (16-valve) type, mounted transversely at the front of the car.

The crankshaft is supported in five shell-type main bearings. Thrust washers are fitted to No 2 main bearing to control crankshaft endfloat.

The connecting rods are attached to the crankshaft by horizontally split shell-type big-end bearings and to the pistons by gudgeon pins. The gudgeon pins are a press fit in the connecting rods on all engines except the F7P

(16-valve) engine, and fully floating, retained by circlips, on the F7P engine. The aluminium alloy pistons are of the slipper type and are fitted with three piston rings, comprising two compression rings and a scraper-type oil control ring.

The single overhead camshaft (8-valve engines) or double overhead camshafts (16-valve engine) are mounted directly in the cylinder head, being driven by the crankshaft via a toothed timing belt.

On the 8-valve engines, the camshaft operates the valves via inverted bucket-type tappets which operate in bores machined directly in the cylinder head. Valve clearance adjustment is by shims located externally between the tappet bucket and the cam lobe. The inlet and exhaust valves are mounted

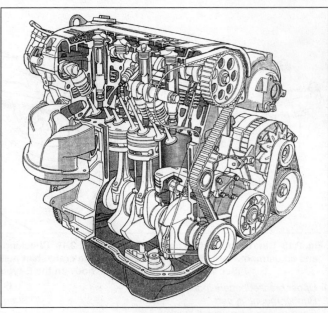

Fig. 2.21 Cutaway view of an 8-valve F-type engine - carburettor version shown (Sec 26)

Fig. 2.22 Cutaway view of the 16-valve F-type engine (Sec 26)

vertically in the cylinder head and are each closed by a single valve spring.

On the 16-valve engine, the camshafts operate the valves via hydraulic tappets which operate in bores machined directly in the cylinder head. Valve clearance is adjusted automatically. The inlet valves are inclined to the front of the engine and are operated by the front (inlet) camshaft. The exhaust valves are inclined to the rear of the engine and are operated by the rear (exhaust) camshaft. Each cylinder has four valves, two inlet and two exhaust, and each pair of valves operates simultaneously.

An auxiliary shaft located alongside the crankshaft is also driven by the timing belt and actuates the oil pump via a skew gear.

A semi-closed crankcase ventilation system

is employed, and crankcase fumes are drawn from an oil separator on the cylinder block and passed via a hose to the inlet manifold.

Engine lubrication is by pressure feed from a gear-type oil pump located beneath the crankshaft. Engine oil is fed through an externally-mounted oil filter to the main oil gallery feeding the crankshaft, auxiliary shaft and camshaft(s). On the 16-valve engine oil jets at the bottom of the cylinders spray cooling oil into the pistons.

The distributor rotor is driven direct from the left-hand end of the camshaft (rear/exhaust camshaft on 16-valve engines). On carburettor models the fuel pump is driven by an eccentric and plunger from the camshaft.

Repair operations possible with the engine in the vehicle

The following operations can be carried out without having to remove the engine from the car:

(a) Removal and refitting of the cylinder head.
(b) Removal and refitting of the timing belt and sprockets.
(c) Renewal of the camshaft oil seal(s).
(d) Removal and refitting of the camshaft(s).
(e) Removal and refitting of the sump.
(f) Removal and refitting of the connecting rods and pistons·.
(g) Removal and refitting of the oil pump.
(h) Renewal of the crankshaft oil seals.
(i) Renewal of the engine mountings.
(j) Removal and refitting of the flywheel.

·Although the operation marked with an asterisk can be carried out with the engine in the car after removal of the sump, it is better for the engine to be removed in the interests of cleanliness and improved access. For this reason, the procedure is described in Part D of this Chapter.

27 Compression test - description and interpretation

Refer to Part A, Section 2.

28 Top Dead Centre (TDC) for No 1 piston - locating

1 Top dead centre (TDC) is the highest point in the cylinder that each piston reaches as the crankshaft turns. Each piston reaches TDC at the end of the compression stroke and again at the end of the exhaust stroke; however, for

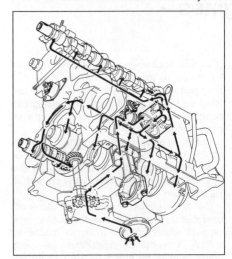

Fig. 2.23 Lubrication circuit on the 8-valve F-type engine (Sec 26)

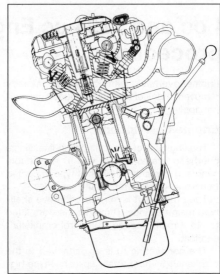

Fig. 2.24 Detail of the lubrication circuit on the 16-valve F-type engine (Sec 26)

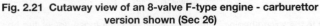

the purpose of timing the engine, TDC refers to the position of No 1 piston at the end of its compression stroke. On all engines in this manual, No 1 piston (and cylinder) is at the flywheel end of the engine.

Basic procedure

2 Disconnect both battery leads.

3 Apply the handbrake, then jack up the front right-hand side of the car and support it on axle stands. Remove the right-hand roadwheel.

4 Remove the plastic cover from within the right-hand wheel arch to give access to the crankshaft pulley bolt. On the 16-valve 1764 cc (F7P) engine, unbolt and remove the lower timing cover, as this covers the pulley completely.

5 Remove the spark plugs with reference to Chapter 1.

6 Place a finger over the No 1 spark plug hole in the cylinder head (nearest the flywheel). Turn the engine in a clockwise direction, using a socket on the crankshaft pulley bolt, until pressure is felt in the No 1 cylinder. This indicates that No 1 piston is rising on its compression stroke. On the 16-valve 1764 cc (F7P) engine, due to the deep recess of the spark plugs in the cylinder head, it will not be possible to place a finger over the spark plug hole; a length of wooden dowel rounded at one end may be used instead, or the spark plug held loosely over the spark plug hole.

7 Remove the air cleaner assembly on the 1764 cc (F7P) and 1794 cc (F3P) engines (Chapter 4).

8 Look through the aperture at the flywheel end of the engine, and continue turning the crankshaft until the TDC timing mark on the flywheel is aligned with the TDC mark on the gearbox bellhousing.

Camshaft sprocket timing marks

9 To view the camshaft sprocket timing marks, proceed as follows. On the 1721 cc (F2N) and 1794 cc (F3P) engines, unscrew the

screw and remove the plastic cover from the top of the right-hand engine mounting.

10 On the 16-valve 1764 cc (F7P) engine, remove the plastic cover from the right-hand engine mounting, then support the engine using a trolley jack and piece of wood beneath the sump; unbolt the engine mounting bracket and upper timing belt cover/engine mounting from the cylinder head.

11 Check that the TDC mark(s) on the camshaft sprocket(s) are aligned with the TDC mark on the rear timing belt cover or end of the valve cover, as applicable.

12 If the distributor cap is now removed, the rotor arm should be in alignment with the No 1 HT lead segment.

TDC setting plug

13 It is possible to lock the crankshaft in the TDC position as follows.

14 Remove the plug on the lower front-facing side of the engine, at the flywheel end, and obtain a metal rod which is a snug fit in the plug hole. Turn the crankshaft slightly if necessary to the TDC position, then push the rod through the hole to locate in the slot in the crankshaft web. Make sure that the crankshaft is exactly at TDC for No 1 piston (flywheel end) by aligning the timing notch on the flywheel with the corresponding mark on the transmission bellhousing. If the crankshaft is not positioned accurately, it is possible to engage the rod with a balance hole in the crankshaft web by mistake, instead of the TDC slot.

29 Valve clearances (8-valve engines only) - adjustment

Note: *This operation is not part of the maintenance schedule. It should be undertaken if noise from the valvegear becomes evident, or if loss of performance gives cause to suspect that the clearances may be incorrect.*

1 On the 1721 cc (F2N) engine, remove the air cleaner as described in Chapter 4. Also unbolt the fuel vapour separator from the front of the engine, but leave the hoses connected.

2 Where applicable, disconnect the crankcase ventilation hose from the valve cover.

3 Unscrew the nuts from the valve cover, and withdraw the cover from the engine (photo). On the 1721 cc (F2N) engine, lift the fuel pipe cluster slightly before removing the cover. Remove the gasket.

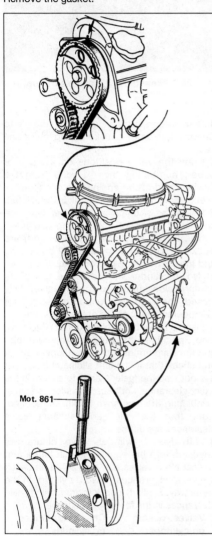

Fig. 2.25 TDC setting on the 8-valve F-type engine (Sec 28)

Mot. 861

2

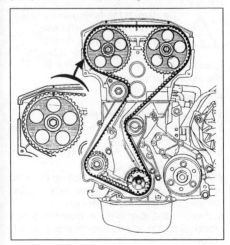

Fig. 2.26 TDC setting on the 16-valve F-type engine (Sec 28)

29.3 Removing the valve cover

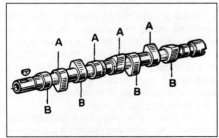

Fig. 2.27 Cam lobe identification on the 8-valve engine (Sec 29)

A Inlet   B Exhaust

29.7 Measuring a valve clearance

29.9 Shim thickness engraved on the underside

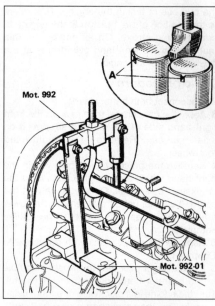

Fig. 2.28 Renault tool for compressing tappet buckets to change tappet shims (Sec 29)

Fit slots (A) at right-angles to the camshaft

4 Remove the spark plugs with reference to Chapter 1, in order to make turning the engine easier.

5 Draw the valve positions on a piece of paper, numbering them 1 to 8 from the flywheel end of the engine. Identify them as inlet or exhaust (ie 1E, 2I, 3E, 4I, 5I, 6E, 7I, 8E).

6 Using a socket or spanner on the crankshaft pulley bolt, turn the engine until the valves of No 1 cylinder (flywheel end) are 'rocking'. The exhaust valve will be closing and the inlet valve will be opening. The piston of No 4 cylinder will be at the top of its compression stroke, with both valves fully closed. The clearances for both valves of No 4 cylinder may be checked at the same time.

7 Insert a feeler blade of the correct thickness (see Specifications) between the cam lobe and the shim on the top of the tappet bucket, and check that it is a firm sliding fit (photo). If it is not, use the feeler blades to ascertain the exact clearance, and record this for use when calculating the new shim thickness required. Note that the inlet and exhaust valve clearances are different.

8 With No 4 cylinder valve clearances checked, turn the engine through half a turn so that No 3 valves are 'rocking', then check the valve clearances of No 2 cylinder in the same way. Similarly check the remaining valve clearances in the following sequence.

Valves rocking on cylinder	Check clearances on cylinder
1	4
3	2
4	1
2	3

9 Where a valve clearance differs from the specified value, then the shim for that valve must be replaced with a thinner or thicker shim accordingly. The shim size is stamped on the bottom face of the shim (photo), but it is prudent to use a micrometer to measure the true thickness of any shim removed, as it may have been reduced by wear.

10 The size of shim required is calculated as follows. If the measured clearance is less than specified, subtract the measured clearance from the specified clearance, and deduct the result from the thickness of the existing shim. For example:

Sample calculation - clearance too small
Clearance measured (A) = 0.15 mm
Desired clearance (B) = 0.20 mm
Difference (B - A) = 0.05 mm
Shim thickness fitted = 3.70 mm
Shim thickness required:
3.70 - 0.05 = 3.65 mm

11 If the measured clearance is greater than specified, subtract the specified clearance from the measured clearance, and add the result to the thickness of the existing shim. For example:

Sample calculation - clearance too big
Clearance measured (A) = 0.50 mm
Desired clearance (B) = 0.40 mm
Difference (A - B) = 0.10 mm
Shim thickness fitted = 3.45 mm
Shim thickness required:
3.45 + 0.10 = 3.55 mm

12 The shims can be removed from their locations on top of the tappet buckets without removing the camshaft if the Renault tool shown in Fig. 2.28 can be borrowed, or a suitable alternative fabricated. On carburettor models, the fuel pump must also be removed if the tool is being used.

13 To remove the shim, the tappet bucket has to be pressed down against valve spring pressure just far enough to allow the shim to be slid out. Theoretically, this could be done by levering against the camshaft between the cam lobes with a suitable pad to push the bucket down, but this is not recommended by the manufacturers.

14 An arrangement similar to the Renault tool can be made by bolting a bar to the camshaft bearing studs and levering down against this with a stout screwdriver. The contact pad should be a triangular-shaped metal block with a lip filed along each side to contact the edge of the buckets. Levering down against this will open the valve and allow the shim to be withdrawn.

15 Make sure that the cam lobe peaks are uppermost when depressing a tappet, and rotate the buckets so that the notches are at right-angles to the camshaft centre-line. When refitting the shims, ensure that the size markings face the tappet buckets (ie face downwards).

16 If the Renault tool cannot be borrowed or a suitable alternative made up, then it will be necessary to remove the camshaft to gain access to the shims, as described in Section 34.

17 Remove the socket or spanner from the crankshaft pulley bolt.

18 Refit the spark plugs with reference to Chapter 1, then refit the valve cover, together with a new gasket where necessary. Reconnect the crankcase ventilation hose and fuel pipe cluster (where fitted). Refit the fuel vapour separator and air cleaner as applicable.

30 Timing belt (8-valve engines) - removal, inspection and refitting

 Caution: If the timing belt breaks or slips in service, extensive engine damage may result. Renew the belt at the intervals specified in Chapter 1, or earlier if its condition is at all doubtful.

Removal

1 Disconnect both battery leads.

2 For extra working room, remove the bonnet with reference to Chapter 11.

3 Remove the air cleaner assembly with reference to Chapter 4.

4 Apply the handbrake, then jack up the front of the car and support it on axle stands. Remove the right-hand roadwheel.

5 Remove the plastic cover from inside the wheel arch for access to the crankshaft pulley.

6 Remove the alternator drivebelt with reference to Chapter 12, Section 6.

30.10A Right-hand engine mounting

30.10B Removing the engine mounting upper bracket

30.11 Removing the engine mounting lower bracket

7 Slacken the crankshaft pulley bolt while holding the crankshaft stationary. To do this, have an assistant insert a screwdriver in the starter ring gear teeth, through the access hole in the top of the gearbox bellhousing. Take care not to damage the ignition timing sensor.

8 Support the weight of the engine on a trolley jack, with a piece of wood inserted between the jack head and sump.

9 Remove the single screw and take off the

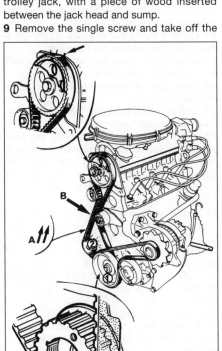

Fig. 2.29 Timing belt run on the 8-valve engine, showing alignment of lines on belt with camshaft and crankshaft sprocket backplate marks (Sec 30)

1 Crankshaft pulley
2 Auxiliary shaft sprocket
3 Camshaft sprocket
A Arrows showing belt running direction
B Point to check timing belt tension

cover from the right-hand engine mounting. Note that the cover forms part of the upper timing cover. On right-hand drive models, the accelerator cable runs through it.

10 Unscrew the bolts and nut, and remove the engine mounting upper bracket (photos).

11 Unbolt the engine mounting lower bracket from the cylinder head (photo).

12 Unbolt and remove the timing belt lower cover.

13 Turn the engine in a clockwise direction, using a socket on the crankshaft pulley bolt, until the TDC mark on the camshaft sprocket is uppermost and in line with the corresponding mark or notch on the metal plate behind the sprocket.

14 Lock the engine in the TDC position by inserting a rod through the TDC setting plug hole. See Section 28, paragraphs 13 and 14.

15 Fully unscrew the crankshaft pulley bolt, and remove the crankshaft pulley from the nose of the crankshaft (photo). If it is tight, use a puller.

16 Double-check that the camshaft sprocket timing mark is aligned with the corresponding mark on the valve cover.

17 Check if the belt is marked with arrows to indicate its running direction; mark it if necessary. Note how the timing belt is located over the sprockets and tensioner (photos).

18 Loosen the nut, turn the timing belt tensioner clockwise to release the tension from the belt, then retighten the nut.

30.17A Timing belt, viewed with the engine removed from the car

19 Release the belt from the camshaft sprocket, idler/tensioner wheels, auxiliary shaft sprocket and crankshaft sprocket, and remove it from the engine. Be careful not to kink or otherwise damage the belt if it is to be re-used.

20 Clean the sprockets and idler/tensioner wheels, and wipe them dry. Do not apply excessive amounts of solvent to the idler/tensioner wheels, otherwise the bearing lubricant may be contaminated. Also clean the cylinder head and block behind the belt running area.

Inspection

21 Examine the timing belt carefully for any signs of cracking, fraying or general wear, particularly at the roots of the teeth. Renew the belt if there is any sign of deterioration of this nature, or if there is any oil or grease

2

30.15 Removing the crankshaft pulley

30.17B Arrows on timing belt show running direction

contamination. Renew any leaking oil seals. The belt **must** be renewed if it has completed the maximum mileage given in Chapter 1. Also examine the sprockets and tensioner as described in Section 32.

Refitting

22 Check that the crankshaft is still at the TDC position for No 1 cylinder, and locked in this position using the metal rod through the hole in the crankcase.

23 Check that the timing mark on the camshaft sprocket is in line with the corresponding mark on the metal backing plate.

24 Align the timing mark bands on the belt with those on the sprockets. The crankshaft sprocket mark is in the form of a notch in its rear guide perimeter. The auxiliary shaft sprocket has no timing mark. Fit the timing belt over the crankshaft sprocket first, then the auxiliary shaft sprocket, followed by the camshaft sprocket. Be careful not to kink the belt.

25 Check that all the timing marks are still aligned, then temporarily tension the belt by turning the tensioner anti-clockwise and tightening the retaining nut. As a rough guide to the correct tension, it should just be possible to turn the belt through 90° using a finger and thumb placed approximately midway between the auxiliary shaft sprocket and the tensioner/idler wheel (photo). The tensioner position may also be adjusted using a bolt through the special hole next to the tensioner.

26 Remove the TDC locating rod from the cylinder block.

27 Refit the crankshaft pulley and retaining bolt, but do not fully tighten it at this stage.

28 Using a socket or spanner on the crankshaft pulley bolt, turn the crankshaft two complete turns in the normal direction of rotation. On regaining the TDC position with No 1 cylinder on compression, insert the TDC locating rod again.

29 Check that the timing marks are still aligned.

30 The belt deflection must now be checked. To do this, first make a mark on the engine in line with the outer surface of the timing belt, midway between the auxiliary shaft sprocket and idler wheel. A force of 30 N (7 lbf) applied to the timing belt must produce a deflection of 7.5 mm (0.3 in) with the engine cold. (Should the engine be hot, the deflection should be 5.5 mm/0.22 in.) Renault technicians use a special tool to do this, but an alternative arrangement can be made by using a spring balance and steel rule. Apply the force with the spring balance, and read off the deflection on the steel rule. Refer to Section 16 in Part B of this Chapter if necessary.

31 If the tension is incorrect, adjust the tensioner as necessary, then re-tighten the nut to the specified torque. This torque is critical, since if the nut were to come loose, considerable engine damage would result.

30.25 Checking the timing belt tension by twisting it through 90°

32 Remove the TDC locating rod, and refit the plug.

33 Prevent the crankshaft turning using the method given in paragraph 7, and tighten the crankshaft pulley bolt to the specified torque.

34 Refit the timing belt upper cover/mounting bracket to the cylinder head, and tighten the bolts.

35 Refit the timing belt lower cover.

36 Refit the engine mounting upper bracket. Check the mounting adjustment with reference to Section 43.

37 Refit the cover over the mounting, and tighten the screw. Remove the trolley jack.

38 Refit and adjust the alternator drivebelt with reference to Chapter 12, Section 6.

39 Refit the plastic cover to the inside of the right-hand wheel arch.

40 Refit the roadwheel, tighten the bolts and lower the car to the ground.

41 Refit the air cleaner assembly with reference to Chapter 4.

42 Refit the bonnet with reference to Chapter 11.

43 Reconnect both battery leads.

31 Timing belt (16-valve engine) - removal, inspection and refitting

⚠️ *Caution: If the timing belt breaks or slips in service, extensive engine damage may result. Renew the belt at the intervals specified in Chapter 1, or earlier if its condition is at all doubtful.*

Removal

1 Disconnect both battery leads.

2 For extra working room, remove the bonnet with reference to Chapter 11.

3 Remove the air cleaner assembly with reference to Chapter 4.

4 Apply the handbrake, then jack up the front of the car and support it on axle stands. Remove the right-hand roadwheel.

5 Remove the plastic cover from inside the wheel arch. Also unbolt the lower timing cover for access to the crankshaft pulley (photo).

6 Remove the right-hand headlight unit with reference to Chapter 12, Section 18.

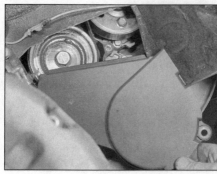

31.5 Removing the lower timing cover

7 Remove the alternator drivebelt with reference to Chapter 12, Section 6.

8 Slacken the crankshaft pulley bolt while holding the crankshaft stationary. To do this, have an assistant insert a screwdriver in the starter ring gear teeth, through the access hole in the top of the gearbox bellhousing. Take care not to damage the ignition timing sensor.

9 Support the weight of the engine on a trolley jack, with a piece of wood inserted between the jack head and sump.

10 Remove the screw and take off the cover from the right-hand engine mounting.

11 Unscrew the bolts and nuts, and remove the engine mounting upper bracket.

12 Unbolt and remove the timing belt plastic intermediate cover.

13 Unbolt the metal upper timing belt cover from the cylinder head.

14 Turn the engine in a clockwise direction, using a socket on the crankshaft pulley bolt, until the TDC marks on the camshaft sprockets are uppermost and in line with the corresponding marks or notches on the valve cover.

15 Lock the engine in the TDC position by inserting a rod through the TDC setting plug hole. See Section 28, paragraphs 13 and 14.

16 Fully unscrew the crankshaft pulley bolt and remove the crankshaft pulley from the nose of the crankshaft. If it is tight, use a puller.

17 Double-check that the camshaft sprocket timing marks are aligned with the corresponding marks on the valve cover.

18 Check if the belt is marked with arrows to indicate its running direction; mark it if necessary.

19 Loosen the nut, move the timing belt tensioner bracket out to release the tension from the belt, then retighten the nut.

20 Release the belt from the camshaft sprockets, idler wheel, auxiliary shaft sprocket and crankshaft sprocket, and remove it from the engine. Be careful not to kink or otherwise damage it, if it is to be re-used.

21 Clean the sprockets and tensioner wheels, and wipe them dry. Do not apply excessive amounts of solvent to the tensioner wheels, otherwise the bearing lubricant may be contaminated. Also clean the cylinder head and block behind the belt running area.

Inspection

22 Refer to paragraph 21 of the previous Section.

Refitting

23 Check that the crankshaft is still at the TDC position for No 1 cylinder, and locked in this position using the metal rod through the hole in the crankcase.

24 Check that the timing marks on the camshaft sprockets are in line with the corresponding marks or notches on the valve cover.

25 Align the timing mark bands on the belt with those on the sprockets. The crankshaft sprocket mark is in the form of a notch in its rear guide perimeter. The auxiliary shaft sprocket has no timing mark. Fit the timing belt over the crankshaft sprocket first, then the auxiliary shaft sprocket, followed by the camshaft sprockets.

26 Check that all the timing marks are still aligned, then temporarily tension the belt by pivoting the tensioner pulley on its bracket and tightening the retaining nut. As a rough guide to the correct tension, it should just be possible to turn the belt through 90°, using a finger and thumb placed approximately midway between the auxiliary shaft sprocket and the idler wheel. The tensioner bracket position may be adjusted using a bolt through the special hole next to the tensioner.

27 Remove the TDC locating rod from the cylinder block.

28 Refit the crankshaft pulley and retaining bolt, but do not fully tighten it at this stage.

29 Using a socket or spanner on the crankshaft pulley bolt, turn the crankshaft two complete turns in the normal direction of rotation. On regaining the TDC position with No 1 cylinder on compression, insert the TDC locating rod again.

30 Check that the timing marks are still aligned.

31 The belt deflection must now be checked. To do this, first make a mark on the valve cover in line with the outer surface of the timing belt, midway between the two camshaft sprockets. A force of approximately 100 N (22.5 lbf) applied to the timing belt must produce a deflection of 3.0 mm ± 0.5 mm (0.12 in ± 0.02 in) with the engine cold. Renault technicians use a special tool to do this, but an alternative arrangement can be made by using a spring balance and steel rule. Apply the force with the spring balance and read off the deflection on the steel rule. Refer to Section 16 in Part B of this Chapter if necessary.

32 If the tension is incorrect, adjust the tensioner as necessary, then re-tighten the nut to the specified torque. This torque is critical, since if the nut were to come loose, considerable engine damage would result.

33 Remove the TDC locating rod and refit the plug.

34 Prevent the crankshaft turning using the method given in paragraph 8, and tighten the crankshaft pulley bolt to the specified torque.

35 Refit the timing belt upper cover to the cylinder head, and tighten the bolts.

36 Refit the timing belt intermediate cover.

37 Refit the right-hand engine mounting upper bracket. Check the mounting adjustment with reference to Section 43.

38 Refit the cover over the mounting, and tighten the screw. Remove the trolley jack.

39 Refit and tension the alternator drivebelt with reference to Chapter 12, Section 6.

40 Refit the right-hand headlight unit with reference to Chapter 12, Section 18.

41 Refit the timing belt lower cover. Refit the plastic cover inside the wheel arch, then refit the roadwheel and lower the car to the ground.

42 Refit the air cleaner assembly with reference to Chapter 4.

43 Refit the bonnet with reference to Chapter 11.

44 Reconnect the battery leads.

32 Timing belt sprockets and tensioners - removal, inspection and refitting

Removal

1 Remove the timing belt as described in Section 30 or 31.

2 To remove a camshaft sprocket, hold the sprocket stationary using a metal bar with two bolts tightened onto it, inserted into the holes in the sprocket, then unscrew the bolt. Alternatively, an old timing belt may be wrapped around the sprocket and held firm with a pair of grips to hold the sprocket stationary, or a special gear-holding tool may be used. Do not allow the camshaft to turn while unscrewing the bolt, or the valves may contact the pistons.

3 Pull the sprocket from the end of the camshaft, if necessary using two levers or screwdrivers. Where applicable, check whether the Woodruff key is likely to drop out of its slot in the camshaft; if so, remove it and store it safely. (There are no keys on the 1764 cc/F7P engine, as the sprocket incorporates a tab to engage with the camshaft.)

4 The sprocket can be removed from the auxiliary shaft in the same manner (photo). Again check that the Woodruff key is firmly in the slot in the shaft.

5 A puller may be necessary to remove the crankshaft sprocket. It is a simple matter to make up a puller using two bolts, a metal bar and the existing crankshaft pulley bolt. By unscrewing the crankshaft pulley bolt, the sprocket is pulled from the end of the crankshaft. If necessary, remove the Woodruff key from the slot in the crankshaft (photo).

6 Unscrew the nut or bolt, and remove the timing belt upper idler wheel or tensioner as applicable. If necessary, the upper and lower rear timing covers may be unbolted at this stage (photo).

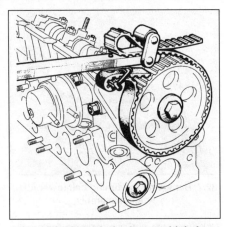

Fig. 2.30 Method of using an old timing belt to hold the camshaft sprocket stationary while the bolt is being loosened (Sec 32)

32.4 Removing the bolt from the auxiliary shaft

32.5 Removing the crankshaft sprocket

32.6 Removing the lower rear timing cover

2

32.7 Removing the timing belt lower idler wheel assembly

32.14 Using a special tool to hold the auxiliary shaft sprocket stationary while tightening the bolt

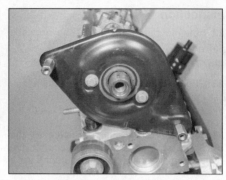

33.2 Metal backing plate located behind the camshaft sprocket

7 Unscrew the nut or bolts, and remove the timing belt lower idler wheel or tensioner as applicable (photo).

Inspection

8 Inspect the teeth of the sprockets for signs of nicks and damage. The teeth are not prone to wear, and should normally last the life of the engine.
9 Spin the tensioner and idler wheel by hand, and check it for any roughness or tightness. Do not attempt to clean them with solvent, as this may enter the bearings. If wear is evident, renew the tensioner and/or idler wheel.

Refitting

10 Refit the timing belt idler wheel, and tighten the nut or bolt to the specified torque.
11 Refit the tensioner, but do not tighten the nut at this stage.
12 Refit the timing belt upper and lower rear covers, and tighten the bolts.
13 Check that the Woodruff key is in the crankshaft slot, then slide on the sprocket. Use a piece of metal tube to tap it fully home.
14 Check that the Woodruff key is in the auxiliary shaft slot, then slide on the sprocket. Use a metal tube to tap it home, if necessary. Apply a little locking fluid to the bolt threads. Fit the bolt and washer, and tighten the bolt to the specified torque, holding the sprocket using one of the methods described in paragraph 2 (photo).
13 Fit and secure the camshaft sprocket in

the same way, being careful not to allow the camshaft to turn while tightening the bolt.
14 Refit the timing belt with reference to Section 30 or 31.

33 Camshaft oil seals - renewal

Sprocket end

1 Remove the camshaft sprocket with reference to Section 32.
2 On the 1721 cc (F2N) and 1794 cc (F3P) engines, unscrew the bolts securing the metal backing plate to the cylinder head, and remove the plate (photo).
3 Note the fitted depth of the oil seal, then prise it out using a small screwdriver. Be careful not to damage the seating or the camshaft seal rubbing surface.
4 Wipe clean the seating in the cylinder head.
5 Smear a little fresh oil on the outer surface of the new oil seal. Locate the seal squarely in the cylinder head, then drive it into position, using a metal tube of diameter slightly less than that of the bore in the cylinder head. Make sure that the oil seal is the correct way round, with the lips facing inwards.
6 When applicable, refit the cylinder head backing plate.
7 Refit the camshaft sprocket with reference to Section 32.

Distributor end

8 Remove the distributor cap and rotor arm (Chapter 5). Remove the rotor arm shield.
9 Prise out the old seal and fit the new one as described above for the sprocket end.
10 Refit the rotor arm shield, rotor arm and distributor cap.

34 Camshaft and tappets (8-valve engines) - removal, inspection and refitting

Removal

1 Remove the timing belt as described in Section 30.
2 On carburettor models, unbolt the fuel vapour separator from the front of the engine, leaving the hoses connected.
3 Disconnect the accelerator cable and move it aside (Chapter 4).
4 Unscrew the nuts from the valve cover, and withdraw the cover from the engine. On the 1721 cc (F2N) engine, lift the fuel pipe cluster slightly before removing the cover. Remove the gasket.
5 Disconnect the HT leads, and remove the distributor cap and rotor arm (Chapter 5). Remove the rotor arm shield.
6 On the 1721 cc (F2N) engine, remove the fuel pump (Chapter 4).
7 Remove the camshaft sprocket with reference to Section 32.
8 Unbolt and remove the backing plate from the cylinder head.
9 Using a dial gauge, measure the camshaft endfloat, and compare with the value given in the Specifications. This will give an indication of the amount of wear present on the thrust surfaces.
10 Make identifying marks on the camshaft bearing caps, so that they can be refitted in the same positions and the same way round.
11 Progressively slacken the bearing cap bolts until the valve spring pressure is released. Remove the bolts, and the bearing caps themselves (photos).
12 Note the position of the cam lobes. The lobes for No 1 cylinder (flywheel end) will be

34.11A Camshaft bearing cap and bolts

34.11B Removing the No 1 (flywheel end) camshaft bearing cap

pointing upwards. Lift out the camshaft together with the oil seals (photo).

13 Remove the tappets, each with its shim (photo). Place them in a compartment box, or on a sheet of card marked into eight sections, so that they may be refitted to their original locations. Write down the shim thicknesses - they will be needed later if any of the valve clearances are incorrect.

Inspection

14 Examine the camshaft bearing surfaces, cam lobes, and the fuel pump eccentric on the 1721 cc (F2N) engine, for wear ridges, pitting or scoring. Renew the camshaft if evident.

15 Renew the oil seals at the ends of the camshaft as a matter of course. Lubricate the lips of the new seals before fitting them, and store the camshaft so that its weight is not resting on the seals.

16 Examine the camshaft bearing surfaces in the cylinder head and bearing caps. Deep scoring or other damage means that the cylinder head must be renewed.

17 Inspect the tappet buckets and shims for scoring, pitting and wear ridges. Renew as necessary.

Refitting

18 Oil the tappets and fit them to the bores from which they were removed. Fit the correct shim, numbered side downwards, to each tappet.

19 Oil the camshaft bearings. Place the camshaft with its oil seals onto the cylinder head. The oil seals must be positioned so that they are flush with the cylinder head faces. The cam lobes must be positioned as noted before removal (paragraph 12).

 Caution: If the cam lobes are not positioned correctly, the valves may be forced into the pistons when the bearing caps are tightened.

20 Refit the camshaft bearing caps to their original locations, applying a little sealant to the end caps where they meet the cylinder head.

21 Apply sealant to the threads of the bearing cap bolts. Fit the bolts and tighten them progressively to the specified torque.

22 If a new camshaft has been fitted, measure the endfloat using a dial gauge, and check that it is within the specified limits.

23 Refit the cylinder head backing plate.

24 Refit the camshaft sprocket with reference to Section 32.

25 On the 1721 cc (F2N) engine, refit the fuel pump (Chapter 4).

26 Refit the timing belt with reference to Section 30.

27 Check and adjust the valve clearances as described in Section 29.

28 Refit the valve cover together with a new gasket, and tighten the nuts.

29 On carburettor models, refit the fuel

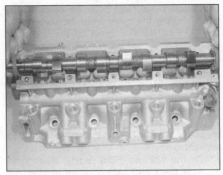
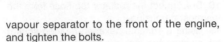

34.12 Lifting out the camshaft

vapour separator to the front of the engine, and tighten the bolts.

30 Reconnect the accelerator cable and adjust it if necessary (Chapter 4).

31 Refit the rotor arm shield, rotor arm and distributor cap with reference to Chapter 5.

32 Reconnect the HT leads.

35 Camshafts and tappets (16-valve engine) - removal, inspection and refitting

Removal

1 Remove the timing belt as described in Section 31.

2 Disconnect the accelerator cable and position it to one side, with reference to Chapter 4.

3 Remove the spark plugs and distributor with reference to Chapters 1 and 5.

4 Unbolt the spark plug tube housing from the top of the valve cover.

5 Disconnect the crankcase ventilation hoses, then unbolt the valve cover from the cylinder head and remove the double gasket (this may be in two sections or joined).

6 Remove both camshaft sprockets with reference to Section 32.

7 Using a dial gauge, measure the endfloat of the camshafts. No values were specified at the time of writing, but it is suggested that the value given for the 8-valve engines be used as a guide. This will give an indication of the amount of wear present on the thrust surfaces.

8 Progressively slacken the camshaft bearing housing bolts on both housings until the valve spring pressure is released. Remove the bolts and the housings.

9 Note the position of the cam lobes; the lobes for No 1 cylinder (flywheel end) will be pointing upwards. Remove the camshafts from the cylinder head, together with the oil seals. It is not necessary to identify them, as the exhaust camshaft has a slot in its end to drive the distributor. Recover the inlet camshaft plug from the cylinder head.

10 Remove the hydraulic tappets, noting which way round they are fitted, and place them in a compartment box, or on a sheet

34.13 Removing a tappet

of card marked into sixteen sections, so that they may be refitted to their original locations. Ideally, they should be placed in a compartment box filled with engine oil, to prevent the oil draining from them.

Inspection

11 Examine the camshaft bearing surfaces and cam lobes for wear ridges, pitting or scoring. Renew the camshafts if evident.

12 Renew the camshaft oil seals as a matter of course. Lubricate the lips of the new seals before fitting them, and store the camshaft so that its weight is not resting on the seals.

13 Examine the camshaft bearing surfaces in the cylinder head and bearing housings. Deep scoring or other damage means that the cylinder head and bearing housings must be renewed.

14 Examine the hydraulic tappets for scoring, pitting and wear ridges. The tappets should be renewed if they are obviously worn, or if they have been excessively noisy in operation.

15 If new hydraulic tappets are being fitted, or if the old ones have been allowed to drain, prime them with fresh engine oil before fitting them, as follows. Immerse each tappet in oil with the hole uppermost, and use flat-nosed pliers to move them up and down until all air is forced out.

Refitting

16 Fit the hydraulic tappets to the bores from which they were removed.

17 Oil the camshaft bearings. Place both camshafts, together with oil seals, onto the cylinder head. The oil seals must be positioned so that they are flush with the cylinder head faces. The cam lobes must be positioned as noted before removal (paragraph 9).

 Caution: If the cam lobes are not positioned correctly, the valves may be forced into the pistons when the bearing caps are tightened.

18 Apply a bead of sealing compound to the housing contact faces on the cylinder head. Also apply the compound to the inlet camshaft end plug, and locate it on the cylinder head.

2

19 Refit the camshaft bearing housings to their original locations, then insert the bolts and progressively tighten them to the specified torque. Note that the torque is different for 6 mm and 8 mm bolts.

20 If new camshafts have been fitted, measure the endfloat using a dial gauge.

21 Refit both camshaft sprockets with reference to Section 32.

22 Refit the valve cover together with new gaskets, and tighten the bolts to the specified torque.

23 Reconnect the crankcase ventilation hoses.

24 Refit the spark plug tube housing to the top of the valve cover, and tighten the bolts.

25 Refit the spark plugs and distributor with reference to Chapters 1 and 5.

26 Reconnect the accelerator cable with reference to Chapter 4.

27 Refit the timing belt with reference to Section 31.

36 Cylinder head (8-valve engines) - removal, inspection and refitting

Removal

1 Remove the battery with reference to Chapter 12.

2 Remove the bonnet with reference to Chapter 11.

3 Drain the cooling system with reference to Chapter 1. Also drain the cylinder block by unscrewing the drain plug located on the right-hand rear face of the engine. Refit the plug after draining.

4 Drain the engine oil with reference to Chapter 1. Refit the plug after draining, using a new washer.

5 Remove the air cleaner assembly or air inlet duct, with reference to Chapter 4.

6 Remove the timing belt with reference to Section 30 of this Chapter.

7 Remove the camshaft sprocket with reference to Section 32.

8 Unbolt the timing belt backplate from the right-hand end of the cylinder head.

9 Disconnect the accelerator cable and position it to one side. On carburettor models, also disconnect the choke cable. Refer to Chapter 4.

10 Disconnect the exhaust downpipe from the manifold (Chapter 4).

11 If wished, the inlet and exhaust manifolds can be removed at this stage. Otherwise, disconnect the various fuel, vacuum and coolant hoses and electrical feeds from the carburettor or throttle body, and from the inlet manifold (see Chapter 4). Be prepared for fuel spillage.

12 Remove the spark plugs and distributor cap with reference to Chapters 1 and 5.

13 On the 1721 cc (F2N) engine, unbolt the fuel vapour separator from the front of the engine, leaving the hoses connected.

14 Unscrew the nuts from the valve cover, and withdraw the cover from the engine. On the 1721 cc (F2N) engine, lift the fuel pipe cluster slightly before removing the cover. Remove the gasket.

15 Disconnect the lead from the temperature gauge sender on the cylinder head. On fuel injection models, also disconnect the injection system coolant temperature sensor.

16 Disconnect the multi-plug from the knock sensor on the front of the cylinder head, when applicable.

17 On the 1721 cc (F2N) engine, disconnect the fuel feed and return pipes at the fuel pump, and plug their ends.

18 Disconnect the radiator top hose from the thermostat housing.

19 Using a suitable hexagon-headed socket bit, slacken the cylinder head retaining bolts half a turn at a time in the reverse order to that shown in Fig. 2.32. When the tension has been relieved, remove all the bolts.

20 Lift the cylinder head upwards and off the cylinder block. If it is stuck, tap it upwards using a hammer and block of wood. Do not try to turn it (it is located by two dowels), nor attempt to prise it free using a screwdriver inserted between the block and head faces.

Inspection

21 The mating faces of the cylinder head and block must be perfectly clean before refitting the head. Use a scraper to remove all traces of gasket and carbon, and also clean the tops of the pistons. Take particular care with the aluminium cylinder head, as the soft metal is damaged easily. Also, make sure that debris is not allowed to enter the oil and water channels - this is particularly important for the oil circuit, as carbon could block the oil supply

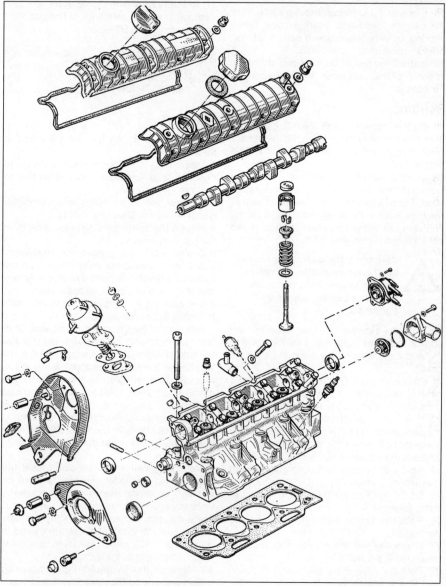

Fig. 2.31 Exploded view of the cylinder head on the 8-valve engine (Sec 36)

to the camshaft or crankshaft bearings. Using adhesive tape and paper, seal the water, oil and bolt holes in the cylinder block. Clean the piston crowns in the same way.

> **HAYNES HINT** *To prevent carbon entering the gap between the pistons and bores, smear a little grease in the gap. After cleaning the piston, rotate the crankshaft so that the piston moves down the bore, then wipe out the grease and carbon with a cloth rag.*

22 Check the block and head for nicks, deep scratches and other damage. If slight, they may be removed carefully with a file. More serious damage may be repaired by machining, but this is a specialist job.

23 If warpage of the cylinder head is suspected, use a straight-edge to check it for distortion. Refer to Part D of this Chapter if necessary.

24 Clean out the bolt holes in the block using a pipe cleaner, or a rag and screwdriver. Make sure that all oil is removed, otherwise there is a possibility of the block being cracked by hydraulic pressure when the bolts are tightened.

25 Examine the bolt threads and the threads in the cylinder block for damage. If necessary, use the correct-size tap to chase out the threads in the block, and use a die to clean the threads on the bolts.

Refitting

26 Ensure that the mating faces of the cylinder block and head are spotlessly clean, that the retaining bolt threads are also clean and dry, and that they screw easily in and out of their locations.

27 Check that No 1 piston is still at TDC, and that the camshaft is in the correct position (No 1 cylinder cam lobes pointing upwards). *If the camshaft position is wrong, there is a risk of valves being forced into pistons.*

28 Fit a new cylinder head gasket to the block, locating it over the dowels. Make sure it is the right way up.

29 Lower the cylinder head onto the block, engaging it over the dowels.

30 Lightly oil the cylinder head bolts, both on their threads and under their heads. Insert the bolts, and tighten them finger-tight.

31 Following the sequence in Fig. 2.32,

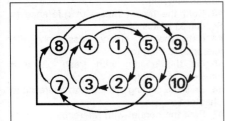

Fig. 2.32 Cylinder head bolt tightening sequence (Sec 36)

36.34 Angle-tightening the cylinder head bolts

tighten the bolts to the torque specified for Stage 1. In the same sequence, tighten to the torque specified for Stage 2.

32 Wait three minutes, then progressively slacken all the bolts in the reverse of the tightening sequence until completely loose. These are Stages 3 and 4.

33 Again following the tightening sequence, tighten the bolts to the torque specified for Stage 5.

34 Final tightening is carried out by turning each bolt through the angle specified for Stage 6. Measure the angle using a commercially-available gauge, or make up a cardboard template cut to the angle required (photo). There is no need for retightening after warm-up.

35 Reconnect the radiator top hose to the thermostat housing, and tighten the clip.

36 On the 1721 cc (F2N) engine, reconnect the fuel feed and return pipes at the fuel pump.

37 Reconnect the knock sensor multi-plug, when applicable.

38 Reconnect the temperature gauge sender and (when applicable) the fuel injection system coolant temperature sensor.

39 Refit the valve cover, using a new gasket, and tighten the nuts.

40 On the 1721 cc (F2N) engine, refit the fuel vapour separator, and tighten the bolts.

41 Refit the spark plugs and distributor cap with reference to Chapters 1 and 5.

42 If they were removed, refit the inlet and exhaust manifolds with reference to Chapter 4. Otherwise, reconnect the various services to the inlet manifold and to the carburettor or throttle body.

43 Reconnect the exhaust downpipe, using a new sealing ring.

44 Reconnect the choke cable (when applicable) and the accelerator cable, referring to Chapter 4.

45 Refit the timing belt backplate to the right-hand end of the cylinder head.

46 Refit the camshaft sprocket with reference to Section 32 of this Chapter.

47 Refit the timing belt with reference to Section 30 of this Chapter.

48 Check the valve clearances if necessary (Section 29).

49 Refit the air cleaner or air inlet duct.

50 Refill the engine with oil, with reference to Chapter 1.

51 Refill the cooling system with reference to Chapter 1.

52 Refit the bonnet with reference to Chapter 11.

53 Reconnect the battery with reference to Chapter 12.

37 Cylinder head (16-valve engine) - removal, inspection and refitting

Removal

1 Remove the battery with reference to Chapter 12.

2 Remove the bonnet with reference to Chapter 11.

3 Drain the cooling system, including the cylinder block, with reference to Chapter 1. The block drain plug is located on the right-hand rear face of the engine.

4 Drain the engine oil with reference to Chapter 1.

5 Remove the air cleaner assembly with reference to Chapter 4.

6 Remove the timing belt with reference to Section 31 of this Chapter.

7 Disconnect the exhaust downpipe from the manifold (Chapter 4).

8 Disconnect the crankcase ventilation hoses from the valve cover (photo).

9 Remove the distributor cap and spark plugs (Chapter 1).

10 Disconnect the accelerator cable and position it to one side, with reference to Chapter 4.

11 Remove the engine wiring harness as necessary (making notes to avoid confusion when refitting), and remove the fuel injection/ignition computer with reference to Chapter 4.

12 Unbolt the alternator mounting bracket from the inlet manifold.

13 Unbolt the starter motor mounting bracket from the exhaust manifold.

14 Unbolt the exhaust manifold heat shield.

15 Remove the radiator and cooling fan with reference to Chapter 3.

37.8 Crankcase ventilation hoses on the valve cover

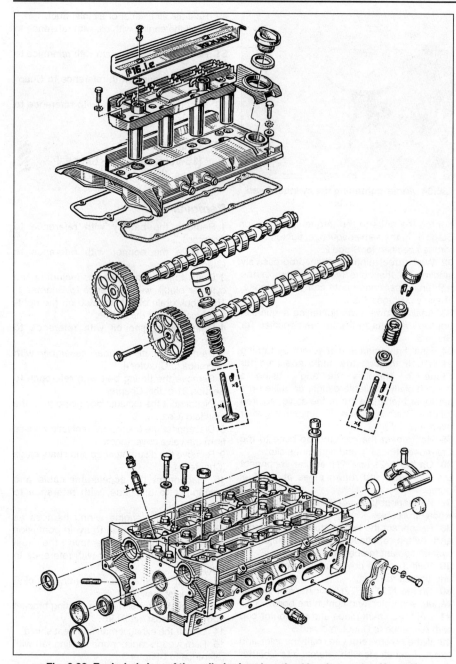

Fig. 2.33 Exploded view of the cylinder head on the 16-valve engine (Sec 37)

16 Unbolt the two bracing struts from the bottom of the inlet manifold.

17 Unbolt the spark plug tube housing from the top of the valve cover.

18 Unbolt the valve cover from the cylinder head, and remove the two gaskets.

19 Using a T55 Torx key, progressively unscrew the cylinder head bolts half a turn at a time, in the reverse order to that shown in Fig. 2.32. When the tension has been relieved, remove all the bolts.

20 Lift the cylinder head upwards and off the cylinder block. If it is stuck, tap it upwards using a hammer and block of wood. Do not try to turn it (it is located by two dowels), nor

attempt to prise it free using a screwdriver inserted between the block and head faces.

Inspection

21 Refer to paragraphs 21 to 25 of the previous Section.

Refitting

22 Ensure that the mating faces of the cylinder block and head are spotlessly clean, that the retaining bolt threads are also clean and dry, and that they screw easily in and out of their locations.

23 Check that No 1 piston is still at TDC. Temporarily refit the valve cover, and align the

timing marks on the camshaft sprockets with the marks on the cover.

 Caution: If the camshafts are wrongly positioned, the valves may be forced into the pistons.

24 Fit a new cylinder head gasket to the block, locating it over the dowels. Make sure it is the right way up.

25 Lower the cylinder head onto the block, engaging it over the dowels.

26 Lightly oil the cylinder head bolts, both on their threads and under their heads. Insert the bolts, and tighten them finger-tight.

27 Following the sequence in Fig. 2.32, tighten the bolts to the torque specified for Stage 1. Repeat the sequence by tightening to the torque specified for Stage 2.

28 Wait three minutes, then progressively slacken all the bolts in the reverse of the tightening sequence until completely loose. This is Stage 3 and Stage 4.

29 Again following the tightening sequence, tighten the bolts to the torque specified for Stage 5.

30 Final tightening is carried out by turning each bolt through the angle specified for Stage 6. Measure the angle using a commercially-available gauge, or make up a cardboard template cut to the angle required. No retightening after warm-up is required.

31 Refit the valve cover, together with new gaskets, and tighten the bolts to the specified torque.

32 Refit the spark plug tube housing to the valve cover, and tighten the bolts.

33 Secure the bracing struts to the inlet manifold.

34 Refit the radiator with reference to Chapter 3, but do not fill the cooling system at this stage.

35 Secure the starter motor mounting bracket to the exhaust manifold.

36 Refit the exhaust manifold heat shield.

37 Refit the alternator mounting bracket.

38 Refit the engine wiring harness and the fuel injection/ignition computer.

39 Reconnect the accelerator cable with reference to Chapter 4.

40 Refit the distributor cap and spark plugs with reference to Chapter 1.

41 Reconnect the crankcase ventilation hoses.

42 Reconnect the exhaust downpipe, using a new sealing ring.

43 Refit the timing belt with reference to Section 31 of this Chapter.

44 Refit the air cleaner assembly with reference to Chapter 4.

45 Fill the engine with fresh oil, with reference to Chapter 1.

46 Refill the cooling system with reference to Chapter 1.

47 Refit the bonnet with reference to Chapter 11.

48 Refit the battery with reference to Chapter 12.

39.12A Sump (viewed from the rear) showing drain plug

39.12B Sump front securing bolts (viewed from under the car)

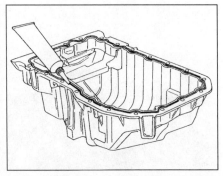

Fig. 2.34 Applying a bead of sealant to the sump mating faces (Sec 39)

38 Sump (8-valve engines) - removal and refitting

Note: *An engine lifting hoist is required during this procedure.*

Removal

1 Disconnect the battery negative and positive leads.
2 Drain the engine oil as described in Chapter 1. Refit and tighten the drain plug, using a new washer.
3 Jack up the front of the car and support it on axle stands. Remove the right-hand front roadwheel.
4 Remove the plastic cover from inside the wheel arch for access to the engine.
5 Remove the bonnet as described in Chapter 11.
6 Remove the air cleaner assembly with reference to Chapter 4.
7 Working beneath the car, loosen (but do not remove) the rear engine mounting link's rearmost bolt. Unscrew and remove the front bolt, and swivel the link downwards.
8 Unscrew the bolts securing the engine-to-gearbox tie-bar and cover assembly to the gearbox bellhousing and engine block, and lower it from the sump. A Torx key may be required to unscrew the bolts.
9 On models fitted with power steering, detach the hose from the right-hand side of the engine subframe.
10 Unscrew the bolts securing the sump to the cylinder block, but leave two bolts finger-tight at this stage.
11 Mark the position of the engine right-hand mounting stud in relation to the upper bracket, then unscrew the nut. Attach a hoist to the engine lifting eyes, and raise the engine as far as possible without straining the left-hand mounting and exhaust downpipe.
12 Remove the two remaining bolts, then break the joint by striking the sump with the palm of the hand. Lower the sump over the oil pump, and withdraw it.

Refitting

13 Clean all traces of sealing compound from the mating faces of the sump and crankcase.

14 Apply a bead of CAF 4/60 THIXO paste to the sump mating faces.
15 It is important that the sump is positioned correctly the first time, and not moved around after the paste has touched the crankcase. Long bolts or dowel rods may be fitted temporarily to help achieve this.
16 To prevent oil dripping from the oil pump and crankcase, wipe these areas clean before refitting the sump.
17 Lift the sump into position, insert the bolts and tighten them progressively to the specified torque.
18 Lower the engine onto the right-hand mounting stud, and refit the nut. Make sure that the nut is in the same position as when removed, then tighten it to the specified torque.
19 On models fitted with power steering, reconnect the hose to the right-hand side of the engine subframe.
20 Refit the engine-to-gearbox tie-bar and cover assembly, and tighten the bolts.
21 Refit the rear engine mounting link.
22 Refit the air cleaner with reference to Chapter 4.
23 Refit the bonnet with reference to Chapter 11.
24 Refit the plastic cover to the inside of the right-hand wheel arch.
25 Refit the roadwheel and lower the car to the ground.
26 Fill the engine with fresh oil, with reference to Chapter 1.
27 Reconnect the battery leads.
28 On models fitted with power steering, bleed the hydraulic system with reference to Chapter 10.

39 Sump (16-valve engine) - removal and refitting

Note: *An engine lifting hoist is required during this procedure.*

Removal

1 Disconnect the battery negative and positive leads.
2 Drain the engine oil as described in Chapter 1. Refit and tighten the drain plug, using a new washer.

3 Jack up the front of the car and support on axle stands. Remove the right-hand front roadwheel.
4 Remove the plastic cover from inside the wheel arch for access to the engine.
5 Remove the bonnet as described in Chapter 11.
6 Remove the air cleaner assembly with reference to Chapter 4.
7 Unbolt and remove the short tie-rod linking the engine subframe to the right-hand side inner wing panel.
8 Working beneath the car, loosen (but do not remove) the rear engine mounting link's rearmost bolt. Unscrew and remove the front bolt, and swivel the link downwards.
9 Remove the exhaust downpipe and heat shield with reference to Chapter 4.
10 Working on the right-hand side driveshaft, use a suitable punch to drive out the roll pins.
11 Loosen (but do not remove) the lower bolt securing the right-hand swivel hub to the bottom of the suspension strut. Unscrew and remove the upper bolt, then tilt the swivel hub and disconnect the driveshaft from the gearbox sun gear shaft. Support the driveshaft over to one side.
12 Unscrew the bolts securing the sump to the cylinder block, but leave two bolts finger-tight at this stage (photos).
13 Unscrew the nuts and remove the four studs attaching the sump to the gearbox. Use two nuts tightened against each other to remove the studs.
14 Mark the position of the engine right-hand mounting stud in relation to the upper bracket, then unscrew the nut. Attach a hoist to the engine lifting eyes, and raise the engine as far as possible without straining the left-hand mounting.
15 Remove the two remaining bolts, then break the joint by striking the sump with the palm of the hand. Lower the sump over the oil pump, and withdraw it.

Refitting

16 Clean all traces of sealing compound from the mating faces of the sump and crankcase.
17 Apply a bead of CAF 4/60 THIXO paste to the sump mating faces.
18 It is important that the sump is positioned

2

40.2 Removing the oil pump from the crankcase

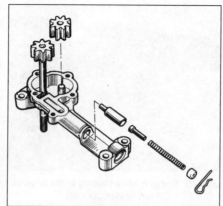

Fig. 2.35 Exploded view of the oil pump (Sec 40)

40.7A Checking the clearance between the oil pump gears and the body

correctly the first time, and not moved around after the paste has touched the crankcase. Long bolts or dowel rods may be fitted temporarily to help achieve this.

19 To prevent oil dripping from the oil pump and crankcase, wipe these areas clean before refitting the sump.

20 Lift the sump into position, then insert two diagonally-opposite bolts and finger-tighten them at this stage.

21 Insert and tighten the four studs attaching the sump to the gearbox. Use two nuts tightened against each other to tighten the studs.

22 Insert the remaining sump bolts, and tighten them progressively to the specified torque.

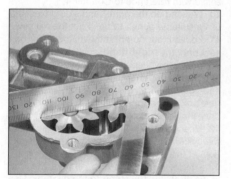

40.7B Checking the endfloat of the oil pump gears

23 Fit the four nuts to the studs, and tighten them to the specified torque.

24 Lower the engine onto the right-hand mounting stud, and refit the nut. Make sure that the nut is in the same position as when removed, then tighten it to the specified torque.

25 Reconnect the right-hand driveshaft to the gearbox sun gear shaft, and secure with new roll pins. Apply sealing compound to each end of the roll pins.

26 Refit the upper bolt securing the right-hand swivel hub to the bottom of the suspension strut, then tighten both bolts to the specified torque given in Chapter 10.

27 Refit the exhaust downpipe and heat shield with reference to Chapter 4.

28 Reconnect the rear engine mounting link, and tighten the bolts to the specified torque.

29 Refit the tie-rod to the subframe, and tighten the bolts.

30 Refit the air cleaner assembly with reference to Chapter 4.

31 Refit the bonnet with reference to Chapter 11.

32 Refit the plastic cover to the inside of the right-hand wheel arch.

33 Refit the roadwheel and lower the car to the ground.

34 Fill the engine with fresh oil, with reference to Chapter 1.

35 Reconnect the battery leads.

40 Oil pump - removal, inspection and refitting

Removal

1 Remove the sump as described in Section 38 or 39.

2 Unscrew the four retaining bolts at the ends of the pump body, and withdraw the pump from the crankcase and drivegear (photo).

Inspection

3 Unscrew the retaining bolts and lift off the pump cover.

4 Withdraw the idler gear and the drivegear/shaft. Mark the idler gear so that it can be refitted in its same position.

5 Extract the retaining clip, and remove the oil pressure relief valve spring retainer, spring, spring seat and plunger.

6 Clean the components, and carefully examine the gears, pump body and relief valve plunger for any signs of scoring or wear. Renew the pump complete if excessive wear is evident.

7 If the components appear serviceable, measure the clearance between the pump body and the gears using feeler blades. Also measure the gear endfloat, and check the flatness of the end cover (photos). If the clearances exceed the specified tolerances, the pump must be renewed.

8 If the pump is satisfactory, reassemble the

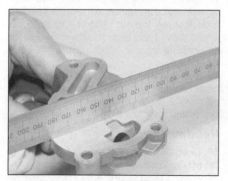

40.7C Checking the flatness of the oil pump cover

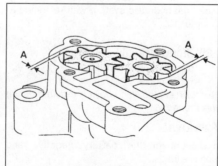

Fig. 2.36 Oil pump gear-to-body clearance measurement points (A) (Sec 40)

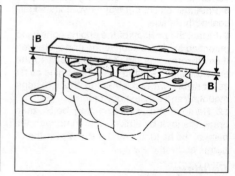

Fig. 2.37 Oil pump gear endfloat measurement points (B) (Sec 40)

40.8 Tightening the oil pump cover bolts

40.10 Tightening the oil pump mounting bolts

41.3 Removing the front plate and oil seal

components in the reverse order of removal. Fill the pump with oil, then refit the cover and tighten the bolts securely (photo). (A new pump should also be primed with oil.) Check that the locating dowel is in position where the driveshaft enters the oil pump body.

Refitting

9 Wipe clean the mating faces of the oil pump and crankcase.
10 Lift the oil pump into position with its shaft engaged with the drivegear. Insert and fully tighten the retaining bolts (photo).
11 Refit the sump as described in Section 38 or 39.

41 Crankshaft oil seals - renewal

Right-hand (pulley end) oil seal

1 Remove the timing belt and crankshaft sprocket with reference to Sections 30, 31 and 32 (as applicable).
2 Remove the sump as described in Section 38 or 39.
3 Unscrew the bolts securing the front plate to the cylinder block and withdraw the plate, noting that it is located by dowels in the two lower bolt hole locations (photo).
4 Clean the front plate, and scrape off all traces of sealant from the plate and cylinder block.
5 Prise out the old oil seal and clean the seating.
6 Fit a new seal so that it is flush with the outer face of the front plate, using a block of wood. Ensure that the open side of the seal is fitted towards the engine.
7 Lubricate the oil seal lips. Apply a bead of CAF 4/60 THIXO paste to the mating face of the front plate, making sure that the oilway cavity is not blocked.
8 Refit the front plate and insert the retaining bolts. The two bolts around the oil seal opening at the 2 o'clock and 8 o'clock positions should also have a small quantity of the sealant paste applied to their threads, as they protrude into the crankcase.

Progressively tighten the retaining bolts in a diagonal sequence.
9 Refit the sump as described in Section 38 or 39.
10 Refit the crankshaft sprocket and timing belt with reference to Sections 30, 31 and 32.

Left-hand (flywheel end) oil seal

11 Remove the flywheel as described in Section 42.
12 Prise out the old oil seal using a small screwdriver, taking care not to damage the surface of the crankshaft. Alternatively, the oil seal can be removed by drilling two small holes diagonally opposite each other and inserting self-tapping screws in them. A pair of grips can then be used to pull out the oil seal, by pulling on each side in turn.
13 Inspect the seal rubbing surface on the crankshaft. If it is grooved or rough in the area where the old seal was fitted, the new seal should be fitted slightly less deeply, so that it rubs on an unworn part of the surface.
14 Wipe clean the oil seal seating, then dip the new seal in fresh engine oil, and locate it over the crankshaft with its closed side facing outwards. Make sure that the oil seal lip is not damaged as it is located on the crankshaft.
15 Using a metal tube, drive the oil seal squarely into the bore until flush. A block of wood cut to pass over the end of the crankshaft may be used instead.
16 Refit the flywheel with reference to Section 42.

43.3A Remove the screw . . .

42 Flywheel - removal, inspection and refitting

Refer to Section 10 in Part A of this Chapter.

43 Engine/gearbox mountings - inspection and renewal

Note: *The engine mountings are positioned with a jig at the Renault factory, and it is important to mark the exact position of each mounting before removing it, to ensure correct alignment of the engine/gearbox.*

Inspection

1 Apply the handbrake, then jack up the front of the car and support it on axle stands.
2 Visually inspect the rubber pads on the engine/transmission mountings for signs of cracking and deterioration. Careful use of a lever will help to determine the condition of the rubber pads. If there is excessive movement in the mounting, or if the rubber has deteriorated, the mounting should be renewed.

Renewal

Right-hand mounting

3 Unscrew the screw and remove the plastic cover from the top of the mounting. On right-hand drive models with the 8-valve engine,

43.3B . . . and remove the cover (8-valve engine)

2

43.3C Remove the cover . . .

43.3D . . . for access to the upper mounting nuts (16-valve engine)

43.13 Gearbox mounting showing central nut and pad mounting nuts

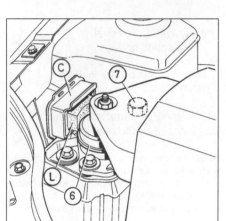

Fig. 2.38 Right-hand mounting on the F-type engine (Sec 43)

C Case
L Vertical movement limiter
6 Rubber mounting
7 Longitudinal movement limiter

note that the cover holds the accelerator cable in position (photos).

4 Mark the position of the mounting, using a scriber or dab of paint on the washer beneath the central nut, and around the lower mounting bracket.

5 Unscrew the nut and remove the washer.

6 Support the weight of the engine, using a hoist on the right-hand lifting eye, or a trolley jack and block of wood beneath the sump.

7 Unbolt the upper mounting bracket from the timing cover bracket on the end of the cylinder head.

8 Unbolt the lower mounting bracket and movement limiter from the body.

9 Fit the new components using a reversal of the removal procedure, but leave the mounting nuts and bolts loose. Finally, carry out the mounting adjustment procedure described later in this Section.

Gearbox mounting

10 Remove the battery as described in Chapter 12.

11 Take the weight of the transmission, using a trolley jack and block of wood, or by using a hoist on the left-hand engine lifting eye.

12 Mark the position of the mounting using a scriber or dab of paint.

13 Unscrew the central nut and the two nuts securing the mounting pad to the upper bracket (photo).

14 Unbolt the upper mounting bracket from the body, and remove the mounting pad (photos).

15 If necessary, unbolt the lower bracket from the gearbox.

16 Refitting is a reversal of the removal procedure, but leave the mounting nuts and bolts loose. Finally, carry out the mounting adjustment procedure described later in this Section.

Rear mounting/lower link

17 Working beneath the car, unscrew the bolts securing the rear mounting link to the

brackets on the subframe and cylinder block. Withdraw the link from under the car.

18 If necessary, unbolt and remove the brackets.

19 Refitting is a reversal of the removal procedure, but leave the mounting bolts loose. Finally, carry out the mounting adjustment procedure described later in this Section.

Adjustment

20 If just one mounting is being renewed, fit the new mounting in the same position as the old one, and check that the dimensions are correct. If more than one mounting has been renewed, take the weight of the engine/gearbox assembly on a hoist, loosen all of the mounting bolts and carry out the following procedure.

Gearbox mounting

21 Refer to Fig. 2.18 in Part B of this Chapter. Pre-tighten the turret nut (I) and lower mounting bolt (II) to 3 Nm (2 lbf ft). Tighten the two upper mounting bolts (III) to 20 Nm (15 lbf ft), then tighten the nut (I) and bolt (II) to the same torque.

22 With the gearbox mounting pad and lower bracket and the right-hand mounting lower bracket in place, check that the mounting bolts are loose.

23 Remove the right-hand front roadwheel, then remove the inner wheel arch plastic panel for access to the crankshaft pulley.

43.14A Removing the rear bolt from the gearbox upper mounting bracket

43.14B Gearbox upper mounting bracket front bolt (arrowed)

43.14C Gearbox upper mounting bracket lower bolt (arrowed)

43.23 Checking the clearance between the crankshaft pulley and strengthening rod when adjusting the engine mounting position

43.25 Checking the dimension between the outer edge of the movement limiter case and the centre of the stud

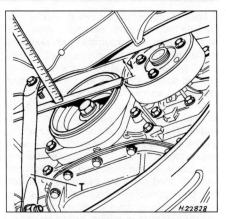

Fig. 2.39 Checking the clearance (Y) between the crankshaft pulley and the edge of the body on the F-type engine (Sec 43)

T Strengthening rod

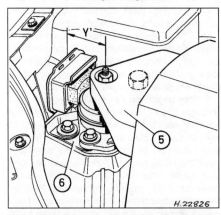

Fig. 2.40 Checking the clearance between the outer edge of the lower bracket and the centre of its stud. For Y' see text (Sec 43)

5 Upper bracket   6 Lower bracket

Refer to Fig. 2.39 and check that, on 8-valve engines, there is a minimum clearance of 36.5 mm (models without power steering) or 34 mm (models with air conditioning or power steering) between the crankshaft pulley and the edge of the body next to the strengthening rod (photo). On 16-valve engines, the minimum clearance is 29 mm, measured between the plastic cover over the crankshaft pulley and the edge of the body next to the strengthening rod. Note that this clearance does not take into consideration the thickness of the straight-edge on the pulley.

24 Maintaining this clearance, centre the left-hand mounting pad in the longitudinal direction in the middle of the upper bracket/battery tray gaps. Tighten the mounting nuts and bolts.

Right-hand mounting

25 Unbolt the upper bracket if it has already been fitted. Measure the dimension between the outer edge of the movement limiter case and the centre of the mounting stud (photo).

26 Refit the upper bracket, and tighten the bolts securing it to the specified torque.

27 With the lower mounting bracket and movement limiter bolts loose, lower the hoist supporting the engine/transmission, and allow it to rest on its mountings. Refer to Fig. 2.40 and measure again between the outer edge of the movement limiter case and the centre of the mounting stud. If necessary, adjust the mounting so that the dimension (Y') is the same as recorded previously (paragraph 25). Tighten the two lower bracket bolts to the specified torque.

28 With the remaining bolts loose, centre the limiter within the hole in the upper bracket so that the clearance is the same on both sides. Tighten the limiter mounting bolts to the specified torque.

29 Working under the car, refit the rear mounting link and tighten the bolts to the specified torque.

30 Recheck dimension Y' on the right-hand mounting, and adjust if necessary.

Part D: Engine removal and general overhaul procedures

44 General information

Included in this part of Chapter 2 are the general overhaul procedures for the cylinder heads, cylinder block/crankcase and internal engine components.

The information ranges from advice concerning preparation for an overhaul and the purchase of replacement parts, to detailed step-by-step procedures covering removal, inspection, renovation and refitting of internal engine parts.

The following Sections have been compiled based on the assumption that the engine has been removed from the car. For information concerning in-car engine repair, as well as the removal and refitting of the external components necessary for the overhaul, refer to

Part A (1108 cc/C1E engine), Part B (1171 cc/E5F and E7F engines, and 1390 cc/E6J and E7J engines), or Part C (1721 cc/F2N engine, 1764 cc/F7P engine, and 1794 cc/F3P engine) of this Chapter, and to Section 49 of this Part.

45 Engine overhaul - general information

It is not always easy to determine when, or if, an engine should be completely overhauled, as a number of factors must be considered.

High mileage is not necessarily an indication that an overhaul is needed, while low mileage does not preclude the need for an overhaul. Frequency of servicing is probably the most important consideration. An engine

which has had regular and frequent oil and filter changes, as well as other required maintenance, will most likely give many thousands of miles of reliable service. Conversely, a neglected engine may require an overhaul very early in its life.

Excessive oil consumption is an indication that piston rings, valve stem oil seals and/or valves and valve guides are in need of attention. Make sure that oil leaks are not responsible before deciding that the rings and/or guides are bad. Perform a cylinder compression check to determine the extent of the work required.

Check the oil pressure with a gauge fitted in place of the oil pressure sender, and compare it with the value given in the Specifications. If it is extremely low, the main and big-end bearings and/or the oil pump are probably worn out.

2

Loss of power, rough running, knocking or metallic engine noises, excessive valve gear noise and high fuel consumption may also point to the need for an overhaul, especially if they are all present at the same time. If a complete tune-up does not remedy the situation, major mechanical work is the only solution.

An engine overhaul involves restoring the internal parts to the specifications of a new engine. During an overhaul, the pistons and rings are renewed, and the cylinder bores are reconditioned. New main bearings, connecting rod bearings and camshaft bearings are generally fitted, and if necessary, the crankshaft may be reground to restore the journals. The valves are also serviced as well, since they are usually in less-than-perfect condition at this point. While the engine is being overhauled, other components, such as the distributor, starter and alternator, can be overhauled as well. The end result should be a like-new engine that will give many trouble-free miles. **Note:** *Critical cooling system components such as the hoses, drivebelts, thermostat and water pump MUST be renewed when an engine is overhauled. The radiator should be checked carefully, to ensure that it is not clogged or leaking. Also, it is a good idea to renew the oil pump whenever the engine is overhauled.*

Before beginning the engine overhaul, read through the entire procedure to familiarize yourself with the scope and requirements of the job. Overhauling an engine is not difficult if you follow all of the instructions carefully, have the necessary tools and equipment, and pay close attention to all specifications; however, it can be time-consuming. Plan on the vehicle being tied up for a minimum of two weeks, especially if parts must be taken to an engineering works for repair or reconditioning. Check on the availability of parts, and make sure that any necessary special tools and equipment are obtained in advance. Most work can be done with typical hand tools, although a number of precision measuring tools are required for inspecting parts to determine if they must be renewed. Often the engineering works will handle the inspection of parts, and offer advice concerning reconditioning and renewal. **Note:** *Always wait until the engine has been completely disassembled, and all components, especially the engine block, have been inspected before deciding what service and repair operations must be performed by an engineering works. Since the condition of the block will be the major factor to consider when determining whether to overhaul the original engine or buy a reconditioned unit, do not purchase parts or have overhaul work done on other components until the block has been thoroughly inspected.* As a general rule, time is the primary cost of an overhaul, so it does not pay to fit worn or substandard parts.

As a final note, to ensure maximum life and minimum trouble from a reconditioned engine, everything must be assembled with care, and in a spotlessly-clean environment.

46 Engine removal - methods and precautions

If you have decided that an engine must be removed for overhaul or major repair work, several preliminary steps should be taken.

Locating a suitable place to work is extremely important. Adequate work space, along with storage space for the vehicle, will be needed. If a garage is not available, at the very least a flat, level, clean work surface is required.

Cleaning the engine compartment and engine before beginning the removal procedure will help keep tools clean and organized.

An engine hoist or A-frame will also be necessary. Make sure the equipment is rated in excess of the combined weight of the engine and transmission. Safety is of primary importance, considering the potential hazards involved in lifting the engine out of the vehicle.

If the engine is being removed by a novice, an assistant should be available. Advice and aid from someone more experienced would also be helpful. There are many instances when one person cannot simultaneously perform all of the operations required when lifting the engine out of the vehicle.

Plan the operation ahead of time. Arrange for, or obtain, all of the tools and equipment you will need, prior to beginning the job. Some of the equipment necessary to perform engine removal and installation safely and with relative ease are (in addition to an engine hoist) a heavy-duty floor jack, complete sets of spanners and sockets as described in the front of this manual, wooden blocks, and plenty of rags and cleaning solvent for mopping up spilled oil, coolant and fuel. If the hoist must be hired, make sure that you arrange for it in advance, and perform all of the operations possible without it beforehand. This will save you money and time.

Plan for the vehicle to be out of use for quite a while. An engineering works will be required to perform some of the work which the do-it-yourselfer cannot accomplish without special equipment. These places often have a busy schedule, so it would be a good idea to consult them before removing the engine, in order to accurately estimate the amount of time required to rebuild or repair components that may need work.

Always be extremely careful when removing and refitting the engine. Serious injury can result from careless actions. Plan ahead, take your time, and you will find that a job of this nature, although major, can be accomplished successfully.

47 Engine (with manual gearbox) - removal and refitting

Note: *All engine types are removed upwards from the engine compartment together with the manual gearbox, and then separated on the bench. Downwards removal together with the engine subframe is possible, but this method is not described, as it requires the use of a special cradle and trolley.*

Removal

1 Remove the battery with reference to Chapter 12.

2 Remove the radiator with reference to Chapter 3. Also disconnect the top hose from the thermostat housing, and the bottom hose from the coolant pipe (photos).

3 Drain the manual gearbox oil with reference to Chapter 7, Section 7.

4 If necessary, drain the engine oil with reference to Chapter 1.

5 Remove the bonnet with reference to Chapter 11.

6 Remove the air cleaner assembly with reference to Chapter 4.

7 Apply the handbrake, then jack up the front of the car and support it on axle stands. Remove both front roadwheels.

8 Disconnect the clutch cable with reference to Chapter 6.

9 Working beneath the car, drive out the double roll pin securing the right-hand driveshaft to the differential sun gear shaft.

10 Unscrew the upper bolt securing the right-hand front suspension strut to the stub axle carrier. (Note that the nut is on the brake caliper side). Slacken the lower bolt, then pull out the driveshaft and disengage it from the sun gear. Tie the driveshaft to one side, taking

47.2A Disconnecting the top hose (arrowed) from the thermostat housing

47.2B Disconnecting the bottom hose from the coolant pipe

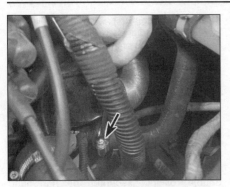

47.19A Disconnecting the heater hose from the coolant pipe . . .

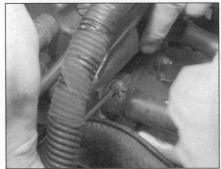

47.19B . . . and inlet manifold

47.21A Disconnecting the wiring from the oil pressure switch

care not to strain the flexible brake hydraulic hose.

11 Unscrew the two mounting bolts securing the left-hand brake caliper to the stub axle carrier. Tie the caliper to the coil spring, taking care not to strain the hose.

12 Unscrew the nut from the left-hand track rod end, and disconnect it from the stub axle carrier steering arm using a balljoint separator tool.

13 Unscrew the bolts securing the left-hand driveshaft inner rubber boot to the gearbox.

14 Unscrew and remove the pinch-bolt securing the front left lower balljoint to the bottom of the stub axle carrier.

15 Support the weight of the stub axle carrier and driveshaft on a trolley jack, then unscrew the two bolts and separate the stub axle carrier from the bottom of the suspension strut.

16 Withdraw the left-hand driveshaft and stub axle carrier from the gearbox. Make sure that the tripod components remain in position on the inner end of the driveshaft, otherwise they may fall into the gearbox. There may be some loss of oil from the gearbox, so position a small container on the floor to catch it.

17 Support the weight of the engine/gearbox assembly beneath the gearbox on a trolley jack and piece of wood.

18 Unbolt and remove the battery tray/gearbox mounting from the gearbox and body.

19 Disconnect the heater, fuel, charcoal canister and vacuum servo hoses (as applicable) (photos).

20 Disconnect the accelerator cable, and where applicable the choke cable, with reference to Chapter 4.

21 Either disconnect the wiring harness from the engine, or, if the engine is not being dismantled, disconnect the wiring from the connection box on the left-hand side of the engine compartment, and tie the loose end to the engine. Make a note of all connections to ensure correct refitting. The wiring includes connections to the following components according to model (photos).

(a) *Oil pressure switch.*
(b) *Temperature sensor(s).*
(c) *Oil level sender.*
(d) *Starter motor.*
(e) *Alternator.*
(f) *'King' HT lead from the distributor cap.*
(g) *Carburettor or fuel injection system.*
(i) *Reversing light switch.*
(j) *Ignition module.*
(k) *Knock sensor.*

22 Unbolt the earthing wire. Also unbolt the wiring harness from the engine-to-gearbox tie-bar (photo).

23 As a precaution against damage to the distributor cap, remove it with reference to Chapter 5.

24 Remove the ignition transistor assistance unit or ignition module (as applicable) with reference to Chapter 5.

25 Disconnect the speedometer cable with reference to Chapter 12.

26 Cut the plastic tie and pull back the rubber boot, then disconnect the gearchange rod from the lever on the gearbox by unscrewing the nut and removing the bolt. Recover the bush from inside the lever. Do not separate the gearchange rod at the clamp, otherwise it will be necessary to adjust the gear lever position on refitting. Tie the gearchange rod to one side.

27 On E- and F-type engines, remove the exhaust downpipe with reference to Chapter 4, disconnecting the oxygen sensor wiring when applicable. On C-type engines, it is only necessary to disconnect the exhaust downpipe from the exhaust manifold.

28 On models fitted with power-assisted steering, remove the pump with reference to Chapter 10. Leave the hose(s) attached, and position the pump to one side.

29 On models fitted with air conditioning, remove the compressor with reference to Chapter 3, but leave the hoses attached, and position the pump to one side. **Do not** open any refrigerant lines.

30 Disconnect the engine rear mounting/lower link, referring if necessary to the relevant Section in Part A, B or C.

31 Connect a hoist to the engine lifting eyes, and take the weight of the engine/gearbox.

32 Disconnect the engine right-hand mounting, referring if necessary to the relevant Section in Part A, B or C.

2

47.21B Disconnecting the wiring from the oil level sender

47.21C Disconnecting the 'king' HT lead from the distributor cap

47.22 Engine wiring harness support on the engine-to-gearbox tie-bar

47.34 Lifting the engine and gearbox assembly from the engine compartment

33 On E- and F-type engines, remove the remaining part of the gearbox mounting.
34 With the help of an assistant, slowly lift the engine/gearbox assembly from the engine compartment, taking care not to damage any components on the surrounding panels (photo). (On the 1764 cc/F7P engine, incline the engine/gearbox assembly forwards as far as possible before lifting it). When high enough, lift the engine/gearbox assembly over the front body panel, and lower to the ground.
35 To separate the gearbox from the engine, refer to the relevant paragraphs of Chapter 7, Section 7 or 8 .

Refitting

36 Refitting is a reversal of removal, noting the following additional points.
(a) Apply a little high melting-point grease to the splines of the gearbox input shaft.
(b) Tighten all nuts and bolts to the specified torque.
(c) Refer to the applicable Chapters and Sections as for removal.
(d) Check and if necessary adjust the engine mountings with reference to the relevant Section in Part A, B or C.
(e) Fill the engine and gearbox with oil, with reference to Chapter 1.
(f) Fill the cooling system with reference to Chapter 1.

48 Engine (with automatic transmission) - removal and refitting

Note: *The engine is removed upwards from the engine compartment together with the automatic transmission, and then separated on the bench. Downwards removal together with the engine subframe is possible, but this method is not described, as it requires the use of a special cradle and trolley. Before working on an engine fitted with an AD4 automatic transmission, read the note given in Chapter 7, Section 10. It is necessary to validate the full-throttle position using a special Renault test instrument after the transmission has been refitted.*

Removal

1 Remove the bonnet as described in Chapter 11.
2 Remove the battery as described in Chapter 12.
3 Remove the air cleaner housing assembly as described in Chapter 4.
4 Disconnect the oxygen sensor wiring at the connector (when applicable), and remove the exhaust downpipe with reference to Chapter 4. Recover the seal at the rear flange and the sealing ring at the exhaust manifold flange.
5 Apply the handbrake, then jack up the front of the car and support it on axle stands. Remove both front roadwheels.
6 Drain the cooling system as described in Chapter 1.
7 Drain the automatic transmission fluid as described in Chapter 1.
8 Loosen (but do not remove) the rear bolt from the engine rear mounting link. Unscrew and remove the front bolt, and swivel the link downwards.

AD4 transmission

9 Unscrew the bolts securing the driveshafts to the transmission flanges on both sides. An Allen key or 12-sided socket will be required for this (refer to Chapter 8).

10 Refer to Chapter 10, and disconnect the steering track rod ends from the steering arms on the swivel hubs on both sides.
11 Working on each side in turn, unscrew and remove the upper bolts securing the front suspension struts to the swivel hubs. Loosen (but do not remove) the lower bolts.
12 Pivot the swivel hubs outwards, and disconnect the driveshafts from the output flanges. Tie the driveshafts to one side.

MB1 transmission

13 Working beneath the car, drive out the double roll pin securing the right-hand driveshaft to the differential sun gear shaft.
14 Unscrew the upper bolt securing the right-hand front suspension strut to the swivel hub. (Note that the nut is on the brake caliper side). Slacken the lower bolt, then pull out the driveshaft and disengage it from the sun gear. Tie the driveshaft to one side, taking care not to strain the flexible brake hydraulic hose.
15 Unscrew the two mounting bolts securing the left-hand brake caliper to the swivel hub. Tie the caliper to the coil spring, taking care not to strain the hose.
16 Unscrew the nut from the left-hand track rod end, and disconnect it from the swivel hub steering arm using a balljoint separator tool.
17 Unscrew the bolts securing the left-hand driveshaft inner rubber boot to the transmission.
18 Unscrew and remove the pinch-bolt securing the front left lower balljoint to the bottom of the swivel hub.
19 Support the weight of the swivel hub and driveshaft on a trolley jack, then unscrew the two bolts and separate the swivel hub from the bottom of the suspension strut.
20 Withdraw the left-hand driveshaft and swivel hub from the transmission. Make sure that the tripod components remain in position on the inner end of the driveshaft, otherwise they may fall into the transmission. There may be some loss of oil from the transmission, so position a small container on the floor to catch it.

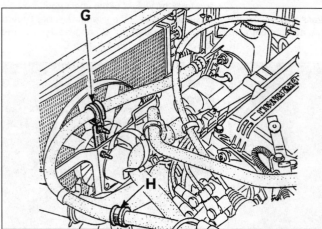

Fig. 2.41 The two brackets (G and H) retaining the power steering hose (Sec 48)

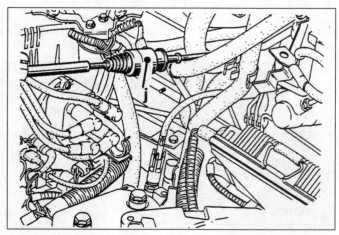

Fig. 2.42 Kickdown switch (J) on the accelerator cable (Sec 48)

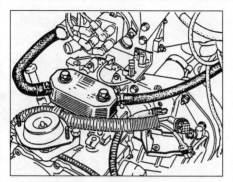

Fig. 2.43 Fluid cooler located on the top of the automatic transmission (Sec 48)

Models with power steering

21 Refer to Fig. 2.41, and unbolt the two brackets holding the power steering hose to the electric cooling fan bracket and side mounting bracket.

22 Remove the steering pump with reference to Chapter 10. Leave the hose(s) attached, and position the pump over to one side.

Models with air conditioning

23 Remove the compressor with reference to Chapter 3, leaving the hoses attached. **Do not** open any refrigerant lines. Position the compressor over to one side.

All models

24 Remove the expansion tank, the electric cooling fan assembly and the radiator, together with the folded deflector. Refer to Chapter 3.

25 Disconnect the fuel feed and return hoses with reference to Chapter 4.

26 Disconnect the brake servo vacuum hose from the inlet manifold.

27 Disconnect the hoses from the MAP sensor and the charcoal canister bleed valve (as applicable) with reference to Chapter 4, Sections 36 and 44 respectively.

28 Either disconnect the wiring harness from the engine, or, if the engine is not being dismantled, disconnect the wiring from the connection box on the left-hand side of the engine compartment, and tie the loose end to the engine. Make a note of all connections to ensure correct refitting. The wiring includes connections to the following components, according to model.

(a) Oil pressure switch.
(b) Temperature sensor(s).
(c) Oil level sender.
(d) Starter motor.
(e) Alternator.
(f) 'King' HT lead from the distributor cap.
(g) Carburettor or fuel injection system.
(h) Ignition module.
(i) The earth cable from the top of the crossmember and from the automatic transmission.
(j) On the MB1 automatic transmission, remove the vacuum capsule, solenoid valve wiring, speed sensor and multi-

function switch. Also where necessary remove the strap securing the electronic unit to the transmission.
(k) Knock sensor.

29 As a precaution against damage to the distributor cap, remove it with reference to Chapter 5.

30 Remove the ignition module with reference to Chapter 5.

31 On fuel injection models not equipped with air conditioning, remove the injection computer, and disconnect the multi-plugs from the fuel injection relays (Chapter 4).

32 Remove the mounting and detach the fuel injection harness from the scuttle.

33 Disconnect the speedometer cable by unscrewing it from the transmission.

34 Disconnect the accelerator cable with reference to Chapter 4 (including the wiring from the kickdown switch, when this is located on the cable). When applicable, also disconnect the choke cable.

35 Remove the evaporative emission control carbon canister, with reference to Chapter 4, Section 44.

36 Remove the automatic transmission computer, together with its mounting.

37 Disconnect the gear control cable.

38 Connect a hoist to the engine lifting eyes, and take the weight of the engine/transmission.

39 Unbolt and remove the battery mounting/left-hand transmission mounting from the transmission and body.

40 Disconnect the right-hand engine mounting, referring if necessary to the appropriate Section in Part B of this Chapter.

41 With the help of an assistant, slowly lift the engine/transmission assembly from the engine compartment, taking care not to damage any components on the surrounding panels. When high enough, lift the engine/transmission assembly over the front body panel, and lower onto a workbench.

42 To separate the transmission from the engine, first unbolt the engine-to-transmission tie-rod bracket.

43 Unscrew and remove the bolts securing the driveplate to the torque converter, at the same time holding the starter ring gear stationary with a wide-bladed screwdriver engaged with the ring gear teeth. Turn the ring gear as required to bring each of the bolts into view.

44 Attach a home-made retaining plate to the transmission bellhousing, in order to keep the torque converter in position while pulling the transmission from the engine. This is important because the torque converter could be seriously damaged if it is dropped. The retaining plate will also be necessary when refitting the transmission, in order to keep the torque converter engaged with the transmission pump.

45 Remove the starter motor with reference to Chapter 12.

46 Unbolt the fluid cooler from the top of the transmission, and remove the sealing rings.

47 Unscrew and remove the transmission-to-engine nuts and bolts, and withdraw the transmission from the engine.

Refitting

48 Refitting is a reversal of removal, noting the following additional points.

(a) Tighten all nuts and bolts to the specified torques (where applicable).
(b) Refer to Chapter 9 when refitting the front brake caliper.
(c) Make sure that the torque converter is fully engaged with the transmission pump before refitting the transmission to the engine.
(d) Make sure that the location dowels are correctly fitted to the engine cylinder block.
(e) Apply grease to the centre spigot of the torque converter.
(f) If necessary, adjust the selector cable with reference to Chapter 7, Section 11.
(g) When refitting the fluid cooler, check and if necessary renew the sealing O-rings.
(h) On the AD4 transmission, the full-throttle position must be validated - refer to the note in Chapter 7, Section 10.
(i) Fill the automatic transmission with fresh fluid, with reference to Chapter 1. On the AD4 transmission, also check that the final drive section of the transmission is filled with the correct quantity and grade of oil.
(j) Fill the cooling system with reference to Chapter 1.
(k) Adjust the accelerator and choke cables (as applicable) with reference to Chapter 4.

49 Engine overhaul - dismantling sequence

1 It is much easier to disassemble and work on the engine if it is mounted on a portable engine stand. These stands can often be hired from a tool hire shop. Before the engine is mounted on a stand, the flywheel/driveplate should be removed from the engine, so that the engine stand bolts can be tightened into the end of the cylinder block.

2 If a stand is not available, it is possible to disassemble the engine with it blocked up on a sturdy workbench or on the floor. Be extra-careful not to tip or drop the engine when working without a stand.

3 If you are going to obtain a reconditioned engine, all the external components must come off first, in order to be transferred to the replacement engine (just as they will if you are doing a complete engine overhaul yourself). Check with the engine supplier for details. Normally these components include:

(a) Alternator and brackets.
(b) Distributor, HT leads and spark plugs.
(c) Thermostat and cover.
(d) Carburettor or fuel injection equipment.
(e) Inlet and exhaust manifolds.

2

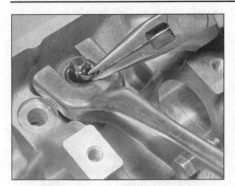

50.4A Removing the split collets

50.4B Removing a valve spring seat

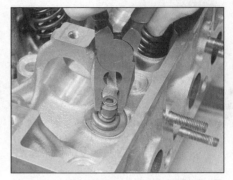

50.5 Removing the oil seal from the top of the valve guide

(f) Oil filter.
(g) Fuel pump (carburettor models).
(h) Engine mountings.
(i) Flywheel/driveplate.
  Note: *When removing the external components from the engine, pay close attention to details that may be helpful or important during refitting. Note the fitted position of gaskets, seals, spacers, pins, washers, bolts and other small items.*
4 If you are obtaining a 'short' motor (which, when available, consists of the engine cylinder block, crankshaft, pistons and connecting rods all assembled), then the cylinder head, sump, oil pump, and timing belt (where applicable) will have to be removed also.
5 If you are planning a complete overhaul, the engine can be disassembled and the internal components removed in the following order:
(a) Inlet and exhaust manifolds.
(b) Timing belt/chain and sprockets.
(c) Cylinder head.
(d) Flywheel/driveplate.
(e) Sump.
(f) Oil pump.
(g) Pistons.
(h) Crankshaft.
6 Before beginning the disassembly and overhaul procedures, make sure that you have all of the correct tools necessary. Refer to the introductory pages at the beginning of this manual for further information.

50 Cylinder head - dismantling

Note: *New and reconditioned cylinder heads are available from the manufacturers and from engine overhaul specialists. Due to the fact that some specialist tools are required for the dismantling and inspection procedures, and new components may not be readily available, it may be more practical and economical for the home mechanic to purchase a reconditioned head rather than dismantle, inspect and recondition the original head.*

1171 cc (E5F/E7F) and 1390 cc (E6J/E7J) engines

1 Remove the rocker shaft and retaining plate, if not already done.

1721 cc (F2N) and 1794 cc (F3P) engines

2 Withdraw the tappet buckets (if not already done), complete with shims, from their bores in the head. Place them on a sheet of cardboard numbered 1 to 8, with No 1 at the flywheel end. It is a good idea to write the shim thickness size on the card alongside each bucket, in case the shims are accidentally knocked off their buckets and mixed up. The size is etched on the shim bottom face.

1764 cc (F7P) engine

3 Remove the hydraulic tappets (if not already done), noting which way round they are fitted. Place them in a compartmented box, or on a sheet of card marked into sixteen sections, so that they may be refitted to their original locations.

> **HAYNES HiNT** *The hydraulic tappets should be placed in a compartmented box filled with engine oil, to prevent the oil draining from them.*

All engines

4 Using a valve spring compressor, compress each valve spring in turn until the split collets can be removed. Release the compressor and lift off the cap, spring(s) and spring seat (photos). If, when the valve spring compressor is screwed down, the valve spring cap refuses to free and expose the split collets, gently tap the top of the tool, directly over the cap, with a light hammer. This will free the cap.
5 Withdraw the oil seal from the top of the valve guide (photo), then remove the valve through the combustion chamber.
6 It is essential that the valves and associated components are kept in their correct sequence, unless they are so badly worn that they are to be renewed. If they are going to be kept and used again, place them in labelled polythene bags, or in a compartmented box (photos).

51 Cylinder head and valves - cleaning, inspection and renovation

1 Thorough cleaning of the cylinder head and valve components, followed by a detailed inspection, will enable you to decide how much valve service work must be carried out during the engine overhaul.

Cleaning

2 Scrape away all traces of old gasket material and sealing compound from the cylinder head.

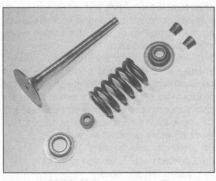

50.6A Valve components

50.6B Store the valve components in a polythene bag after removal

51.6 Checking the cylinder head surface for distortion

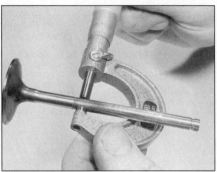

51.9 Measuring the valve stem diameter

51.11 Grinding-in the valves

3 Scrape away the carbon from the combustion chambers and ports, then wash the cylinder head thoroughly with paraffin or a suitable solvent.

4 Scrape off any heavy carbon deposits that may have formed on the valves, then use a power-operated wire brush to remove deposits from the valve heads and stems.

Inspection and renovation

Note: *Be sure to perform all the following inspection procedures before concluding that the services of an engine overhaul specialist are required. Make a list of all items that require attention.*

Cylinder head

5 Inspect the head very carefully for cracks, evidence of coolant leakage and other damage. If cracks are found, a new cylinder head should be obtained.

6 Use a straight-edge and feeler blade to check that the cylinder head surface is not distorted (photo). If it is, it may be possible to have it resurfaced ('skimmed') on the 1108 cc (C1E) engine. The manufacturers do not allow resurfacing on the other engine types.

7 Examine the valve seats in each of the combustion chambers. If they are severely pitted, cracked or burned, then they will need to be renewed or recut by an engine overhaul specialist. If they are only slightly pitted, this can be removed by grinding the valve heads and seats together with coarse, then fine, grinding paste as described below.

8 If the valve guides are worn, indicated by a side-to-side motion of the valve in the guide, new guides must be fitted. A dial gauge may be used to determine the amount of side play of the valve. Recheck the fit using a new valve if in doubt, to decide whether it is the valve or the guide which is worn. If new guides are to be fitted, the valves must be renewed in any case. Valve guides may be renewed using a press and a suitable mandrel, making sure that they are at the correct height. The work is best carried out by an engine overhaul specialist, since if it is not done skilfully, there is a risk of damaging the cylinder head.

Valves

> ⚠️ **Warning: The exhaust valves on the 16-valve 1764 cc (F7P) engine are filled with sodium, and must be disposed of carefully. Ideally, they should be cut in half and then immersed in water (approximately 10 litres for four exhaust valves). It is important to take adequate safety precautions during this operation by wearing goggles and suitable clothing, since the sodium reacts violently when it contacts the water. When the reaction has subsided, the water will contain sodium hydroxide (caustic soda), which must in turn be disposed of safely.**

9 Examine the heads of each valve for pitting, burning, cracks and general wear, and check the valve stem for scoring and wear ridges. Rotate the valve, and check for any obvious indication that it is bent. Look for pits and excessive wear on the end of each valve stem. If the valve appears satisfactory at this stage, measure the valve stem diameter at several points using a micrometer (photo). Any significant difference in the readings obtained indicates wear of the valve stem. Should any of these conditions be apparent, the valve(s) must be renewed. If the valves are in satisfactory condition, or if new valves are being fitted, they should be ground (lapped) into their respective seats to ensure a smooth gas-tight seal.

10 Valve grinding is carried out as follows. Place the cylinder head upside-down on a bench, with a block of wood at each end to give clearance for the valve stems.

11 Smear a trace of coarse carborundum paste on the seat face, and press a suction grinding tool onto the valve head. With a semi-rotary action, grind the valve head to its seat, lifting the valve occasionally to redistribute the grinding paste (photo). When a dull-matt even surface is produced on both the valve seat and the valve, wipe off the paste and repeat the process with fine carborundum paste. A light spring placed under the valve head will greatly ease this operation. When a smooth unbroken ring of light grey matt finish is produced on both the valve and seat, the grinding operation is complete. Be sure to remove all traces of grinding paste, using paraffin or a suitable solvent, before reassembly of the cylinder head.

Valve components

12 Examine the valve springs for signs of damage and discoloration, and also measure their free length using vernier calipers (photo) or by comparing the existing spring with a new component.

13 Stand each spring on a flat surface, and check it for squareness. If any of the springs are damaged, distorted or have lost their tension, obtain a complete new set of springs. It is normal to renew the springs as a matter of course during a major overhaul.

14 On the 1721 cc (F2N) and 1794 cc (F3P) engines, inspect the tappet buckets and their

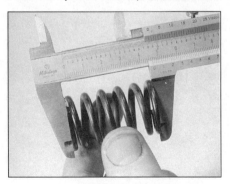

51.12 Checking the valve spring free length

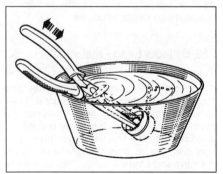

Fig. 2.44 Priming the hydraulic tappets with oil (Sec 51)

2

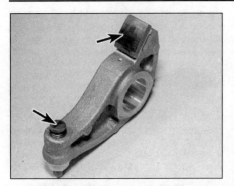

51.16A Check the rocker arm contact surfaces (arrowed) for wear

51.16B Checking the internal diameter of the rocker arms

51.16C Clean the oil spill holes using a length of wire

shims for scoring, pitting (especially on the shims), and wear ridges. Renew any components as necessary. Note that some scuffing is to be expected, and is acceptable provided that the tappets are not scored.

15 On the 1764 cc (F7P) engine, examine the hydraulic tappets for scoring, pitting and wear ridges. Renew as necessary. If new hydraulic tappets are being fitted, or if the old ones have been allowed to drain, prime them with fresh engine oil before fitting them, as follows; immerse each tappet in oil with the hole uppermost, and use flat-nosed pliers to move them up and down until all air is forced out.

Rocker arm components - except 1721 cc (F2N), 1764 cc (F7P) and 1794 cc (F3P) engines

16 Check the rocker arm contact surfaces for pits, wear, score marks or any indication that the surface-hardening has worn through. Dismantle the rocker shaft, and check the rocker arm and rocker shaft pivot and contact areas in the same way. Measure the internal diameter of each rocker, and check their fit on the shaft. Clean out the oil spill holes in each rocker using a length of wire (photos). Renew the rocker arm or the rocker shaft itself if any are suspect.

17 On the 1108 cc (C1E) engine, inspect the pushrod ends for scuffing and excessive wear. Roll each pushrod on a flat surface, such as a piece of plate glass, and check for straightness.

Valve stem oil seals

18 The valve stem oil seals should be renewed as a matter of course.

52 Cylinder head - reassembly

1 Lubricate the stems of the valves, and insert them into their original locations. If new valves are being fitted, insert them into the locations to which they have been ground.

2 Working on the first valve, dip the new oil seal in engine oil, then carefully locate it over the valve and onto the guide. Take care not to damage the seal as it is passed over the valve stem. Use a suitable socket or metal tube to press the seal firmly onto the guide (photo).

3 Locate the spring seat on the guide, followed by the spring(s) and cap. Where applicable, the spring should be fitted with its closest coils towards the head.

4 Compress the valve spring and locate the split collets in the recess in the valve stem. Release the compressor, then repeat the procedure on the remaining valves.

Use a little grease to hold the collets in place.

5 With all the valves installed, place the cylinder head flat on the bench and, using a hammer and interposed block of wood, tap the end of each valve stem to settle the components.

6 On the 1764 cc (F7P) engine, fit the hydraulic tappets to the bores from which they were removed.

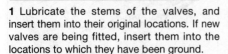

52.2 Pressing a valve seal onto its guide

7 On the 1721 cc (F2N) and 1794 cc (F3P) engines, lubricate the tappet buckets and insert them into their respective locations as noted during removal. Make sure that each bucket has its correct tappet shim in place on its upper face, and that the shim is installed with its etched size number facing downwards.

53 Auxiliary shaft (F-type engine) - removal, inspection and refitting

Removal

1 Remove the timing belt with reference to Section 30 or 31 in Part C of this Chapter.

2 Remove the auxiliary shaft sprocket with reference to Section 32 in Part C.

3 Unbolt the timing belt rear lower cover from the cylinder block.

4 Unscrew the four bolts and withdraw the auxiliary shaft housing (photo), then remove the gasket (if fitted).

5 From the top, unscrew the two bolts and withdraw the oil pump drivegear cover plate and O-ring. Screw a suitable bolt into the oil pump drivegear, or use a tapered wooden shaft, and withdraw the drivegear from its location (photos).

6 Unscrew the two bolts and washers, and lift out the auxiliary shaft thrustplate and the auxiliary shaft (photos).

53.4 Removing the auxiliary shaft housing

53.5A Removing the oil pump drivegear cover plate . . .

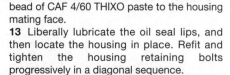

53.5B . . . and drivegear

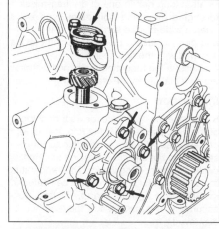

Fig. 2.45 Auxiliary shaft housing retaining bolt locations, oil pump driveshaft and cover plate (Sec 53)

Inspection

7 Examine the auxiliary shaft and oil pump driveshaft for pitting, scoring or wear ridges on the bearing journals, and for chipping or wear of the gear teeth. Renew as necessary. Check the auxiliary shaft bearings in the cylinder block for wear and, if worn, have these renewed by your Renault dealer or suitably-equipped engineering works. Wipe them clean if they are still serviceable.

8 Temporarily fit the thrustplate to its position on the auxiliary shaft, and use a feeler gauge to check that the endfloat is as given in the Specifications. If it is greater than the upper tolerance, a new thrustplate should be obtained, but first check the thrust surfaces on the shaft to ascertain if wear has occurred here.

Refitting

9 Clean off all traces of the old gasket or sealant from the auxiliary shaft housing, and prise out the oil seal with a screwdriver. Install the new oil seal using a block of wood, and tap it in until it is flush with the outer face of the housing. The open side of the seal must be towards the engine (photos).

10 Liberally lubricate the auxiliary shaft, and slide it into its bearings.

11 Place the thrustplate in position with its curved edge away from the crankshaft, and refit the two retaining bolts, tightening them securely.

12 Place a new housing gasket in position over the dowels of the cylinder block (photo).

If a gasket was not used previously, apply a bead of CAF 4/60 THIXO paste to the housing mating face.

13 Liberally lubricate the oil seal lips, and then locate the housing in place. Refit and tighten the housing retaining bolts progressively in a diagonal sequence.

14 Lubricate the oil pump drivegear, and lower the gear into its location.

15 Refit the rear lower timing belt cover to the cylinder block, and tighten the bolts.

16 Position a new O-ring seal on the drivegear cover plate, fit the plate and secure with the two retaining bolts.

17 Refit the auxiliary shaft sprocket with reference to Section 32.

18 Refit the timing belt with reference to Section 30 or 31.

53.6A Removing the auxiliary shaft thrustplate . . .

54 Driveplate (E-type engine with automatic transmission) - removal, inspection and refitting

Removal

1 Remove the automatic transmission as described in Section 48.

2 Mark the driveplate in relation to the crankshaft.

3 The driveplate must now be held stationary while the bolts are loosened. To do this,

53.6B . . . and the auxiliary shaft itself

53.9A Prising the auxiliary shaft oil seal from the housing

53.9B Driving in the new oil seal

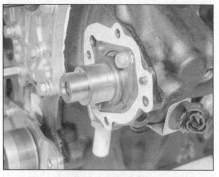

53.12 Auxiliary shaft housing gasket positioned on the dowels

2

locate a long bolt in one of the transmission-to-engine mounting bolt holes, and either insert a wide-bladed screwdriver in the starter ring gear, or use a piece of bent metal bar engaged with the ring gear.

4 Unscrew the mounting bolts, and withdraw the driveplate from the crankshaft.

Inspection

5 Examine the driveplate for wear or chipping of the ring gear teeth. If the ring gear is worn or damaged, it may be possible to renew it separately, but this job is best left to a Renault dealer or engineering works. The temperature to which the new ring gear must be heated for installation is critical and, if not done accurately, the hardness of the teeth will be destroyed.

6 Check the driveplate carefully for signs of distortion, and for hairline cracks around the bolt holes, or radiating outwards from the centre. If damage of this sort is found, the driveplate must be renewed.

Refitting

7 Clean the driveplate and crankshaft faces, then locate the unit on the crankshaft, making sure that any previously-made marks are aligned.

8 Apply a few drops of locking fluid to the mounting bolt threads, fit the bolts and tighten them in a diagonal sequence to the specified torque.

9 Refit the automatic transmission as described in Section 48.

55 Left-hand (driveplate end) crankshaft oil seal (E-type engine with automatic transmission) - renewal

1 Remove the driveplate as described in Section 54.

2 The procedure is now as described in Part B of this Chapter, Section 23.

3 Refit the driveplate with reference to Section 54.

56 Camshaft and followers (1108 cc/C1E engine) - removal, inspection and refitting

Removal

1 Remove the cylinder head, timing chain and camshaft sprocket with reference to Part A.

2 Remove the distributor with reference to Chapter 5.

3 Using a suitable bolt screwed into the distributor drivegear, or a tapered wooden rod, extract the drivegear from the distributor aperture.

4 Withdraw the cam followers from the top of the cylinder block, keeping them in strict order of removal. Identify them for location.

5 Unscrew and remove the two bolts securing the camshaft retaining plate to the cylinder block, and carefully withdraw the camshaft from its location.

Inspection

6 Examine the camshaft bearing surfaces, cam lobes and skew gear for wear ridges, pitting, scoring or chipping of the gear teeth. Renew the camshaft if any of these conditions are apparent.

7 If the camshaft is serviceable, temporarily refit the sprocket, and secure with the retaining bolt. Using a feeler blade, measure the clearance between the camshaft retaining plate and the outer face of the bearing journal. If the clearance (endfloat) exceeds the specified dimension, renew the retaining plate. To do this, remove the sprocket, and draw off the plate and retaining collar using a suitable puller. Fit the new plate and a new collar, using a hammer and tube to drive the collar into position.

8 Examine the condition of the camshaft bearings. If renewal is necessary, have this work carried out by a Renault dealer or engineering works.

9 Inspect the cam followers for wear ridges and pitting of their camshaft lobe contact faces, and for scoring on the sides of the follower body. Light scuff marks are normal, but there should be no signs of scoring or ridges. If the followers show signs of wear, renew them. Note that they must all be renewed if a new camshaft is being fitted.

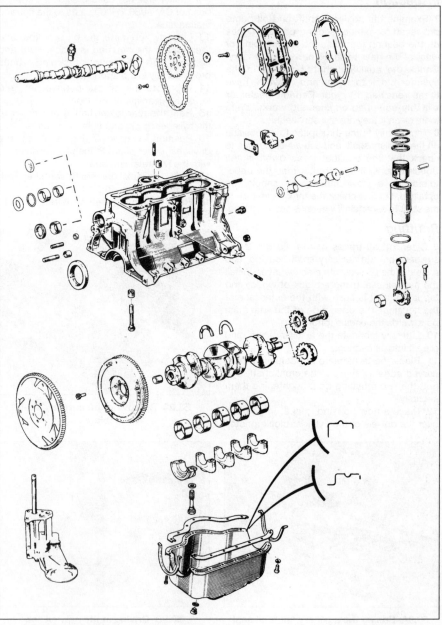

Fig. 2.46 Cylinder block components on the 1108 cc (C1E) engine (Secs 56 to 59)

Refitting

10 Lubricate the camshaft bearings, and carefully insert the camshaft from the timing gear end of the engine.

11 Refit the retaining plate, then insert and tighten the bolts securely. Check that the camshaft rotates smoothly.

12 Lubricate the cam followers, and insert them into their original bores in the cylinder block.

13 Refit the camshaft sprocket, timing chain and cylinder head, with reference to Part A.

14 Turn the crankshaft until No 1 piston (flywheel end) is at the top of its compression stroke.

15 Without moving the crankshaft, position the distributor drivegear so that its slots are at the 2 o'clock and 8 o'clock positions, with the larger offset side facing away from the engine (photo).

16 Lower the drivegear into mesh with the camshaft and oil pump driveshaft. As the gear meshes with the camshaft, it will rotate anti-clockwise. It should end up with its slot at right-angles to the crankshaft centre-line and with the larger offset towards the flywheel. It will probably be a tooth out on the first attempt; it may take two or three attempts to get it just right (photo).

17 Refit the distributor with reference to Chapter 5.

57 Piston/connecting rod assemblies - removal

1 With the cylinder head, sump and oil pump removed, proceed as follows.

2 Rotate the crankshaft so that No 1 big-end cap (nearest the flywheel position) is at the lowest point of its travel. If the big-end cap and rod are not already numbered, mark them with a centre-punch (photo). Mark both cap and rod to identify the cylinder they operate in.

3 Before removing the big-end caps, use a feeler blade to check the amount of side play between the caps and the crankshaft webs (photo).

4 Unscrew and remove the big-end bearing cap nuts (C- and E-type engines) or bolts (F-type engine). Withdraw the cap, complete with shell bearing, from the connecting rod. Strike the cap with a wooden or copper mallet if it is stuck.

5 If only the bearing shells are being attended to, push the connecting rod up and off the crankpin, and remove the upper bearing shell. Keep the bearing shells and cap together in their correct sequence if they are to be refitted.

C- and E-type engines

6 Remove the liner clamps and withdraw each liner, together with piston and connecting rod, from the top of the cylinder block. Mark the liners using masking tape, so that they may be refitted in their original locations.

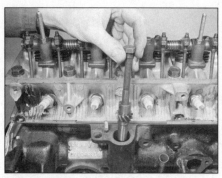

56.15 Fit the distributor drivegear . . .

7 Withdraw the piston from the bottom of the liner. Keep each piston with its respective liner if they are to be re-used.

F-type engines

8 Push the connecting rod up, and remove the piston and rod from the bore. Note that if there is a pronounced wear ridge at the top of the bore, there is a risk of damaging the piston as the rings foul the ridge. However, it is reasonable to assume that a rebore and new pistons will be required in any case if the ridge is so big.

58 Crankshaft - removal

1 With the timing belt/chain, rear timing cover and front plate (E- and F-type engines), flywheel/driveplate and pistons removed, proceed as follows.

2 Before the crankshaft is removed, check the endfloat using a dial gauge in contact with the end of the crankshaft (photo). Push the crankshaft fully one way, and then zero the gauge. Push the crankshaft fully the other way, and check the endfloat. The result can be compared with the specified amount, and will give an indication as to whether new thrustwashers are required.

3 If a dial gauge is not available, feeler blades can be used. First push the crankshaft fully towards the flywheel end of the engine, then slip the feeler blade between the web of No 2 crankpin and the thrustwasher of the centre main bearing (C- and E-type engines) (photo),

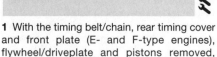

56.16 . . . so that the slot is at right-angles to the crankshaft centre-line, with its larger offset side facing the flywheel when fitted

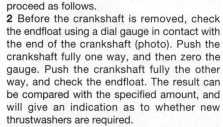

57.2 Big-end caps marked with a centre-punch

57.3 Checking the big-end cap side play

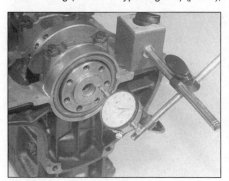

58.2 Checking the crankshaft endfloat with a dial gauge

58.3 Checking the crankshaft endfloat with a feeler blade - C- and E-type engines

2

58.4 Identification numbers (arrowed) on the main bearings

58.5 Removing No 1 main bearing cap

58.7A Removing the thrustwashers . . .

or between the web of No 1 crankpin and the thrust washer of No 2 main bearing (F-type engine).

4 Identification numbers should already be cast onto the base of each main bearing cap. If not, number the cap and crankcase using a centre-punch, as was done for the connecting rods and caps (photo).

5 Unscrew and remove the main bearing cap retaining bolts, and withdraw the caps, complete with bearing shells (photo). Tap the caps with a wooden or copper mallet if they are stuck.

6 Carefully lift the crankshaft from the crankcase.

7 Remove the thrustwashers at each side of the centre main bearing (C- and E-type

engines) or No 2 main bearing (F-type engine), then remove the bearing shell upper halves from the crankcase (photos). Place each shell with its respective bearing cap.

8 Remove the oil seal from the rear of the crankshaft.

59 Cylinder block/crankcase and bores - cleaning and inspection

Cleaning

1 For complete cleaning, the core plugs should be removed. Drill a small hole in them, then insert a self-tapping screw and pull out the plugs using a pair of grips or a slide-hammer. Also remove all external

components and senders, and where applicable unbolt the drivebelt tensioner bracket from the cylinder block. On F-type engines, unbolt the coolant pipe, oil cooler and alternator drivebelt tensioner bracket. Remove the oil sprayers (when fitted) from the bottom of each bore by unscrewing the retaining bolts (photos).

2 Scrape all traces of gasket from the cylinder block, taking care not to damage the head and sump mating faces.

3 Remove all oil gallery plugs. The plugs are usually very tight - they may have to be drilled out and the holes re-tapped. Use new plugs when the engine is reassembled.

4 If the block is extremely dirty, it should be steam-cleaned.

58.7B . . . and main bearing shell upper halves

59.1A Removing the coolant pipe from the cylinder block (E-type engine)

59.1B Oil pressure switch location on the cylinder block (F-type engine)

59.1C Oil sprayer location on the bottom of the bore

59.1D Removing the oil level sender from the cylinder block

59.1E Coolant pipe and mounting bolt - F-type engine

59.21A Checking the liner protrusion with a dial gauge . . .

59.21B . . . and with feeler blades

60.2 Using an old feeler blade to remove the piston rings

5 After the block has been steam-cleaned, clean all oil holes and oil galleries one more time. Flush all internal passages with warm water until the water runs clear, dry the block thoroughly and wipe all machined surfaces with a light rust-preventative oil. If you have access to compressed air, use it to speed up the drying process and to blow out all the oil holes and galleries.

 Warning: Wear eye protection when using compressed air!

6 If the block is not very dirty, you can do an adequate cleaning job with hot soapy water and a stiff brush. Take plenty of time, and do a thorough job. Regardless of the cleaning method used, be sure to clean all oil holes and galleries very thoroughly, dry the block completely and coat all machined surfaces with light oil.

7 The threaded holes in the block must be clean to ensure accurate torque wrench readings during reassembly. Run the proper-size tap into each of the holes to remove rust, corrosion, thread sealant or sludge, and to restore damaged threads. If possible, use compressed air to clear the holes of debris produced by this operation. Now is a good time to clean the threads on the head bolts and the main bearing cap bolts as well.

8 Refit the main bearing caps, and tighten the bolts finger-tight.

9 After coating the mating surfaces of the new core plugs with suitable sealant, refit them in the cylinder block. Make sure that they are driven in straight and seated properly, or leakage could result. Special tools are available for this purpose, but a large socket, with an outside diameter that will just slip into the core plug, will work just as well.

10 Apply suitable sealant to the new oil gallery plugs, and insert them into the holes in the block. Tighten them securely.

11 If the engine is not going to be reassembled right away, cover it with a large plastic bag to keep it clean and prevent it rusting.

Inspection

12 Visually check the block for cracks, rust and corrosion. Look for stripped threads in the threaded holes. If there has been any

history of internal water leakage, it may be worthwhile having an engine overhaul specialist check the block with special equipment. If defects are found, have the block repaired, if possible, or renewed.

13 Check the cylinder bores/liners for scuffing and scoring. Normally, bore wear will be evident in the form of a wear ridge at the top of the bore. This ridge marks the limit of piston travel.

14 Measure the diameter of each cylinder at the top (just under the ridge area), centre and bottom of the cylinder bore, parallel to the crankshaft axis.

15 Next measure each cylinder's diameter at the same three locations across the crankshaft axis. If the difference between any of the measurements is greater than 0.20 mm, indicating that the cylinder is excessively out-of-round or tapered, then remedial action must be considered.

16 Repeat this procedure for the remaining cylinders.

17 If the cylinder walls are badly scuffed or scored, or if they are excessively out-of-round or tapered, have the cylinder block rebored (F-type engine) or obtain new cylinder liners (C- and E-type engines). New pistons (oversize in the case of a rebore) will also be required.

18 If the cylinders are in reasonably good condition, then it may only be necessary to renew the piston rings.

19 If this is the case, the bores should be honed in order to allow the new rings to bed in correctly and provide the best possible seal. The conventional type of hone has spring-loaded stones, and is used with a power drill. You will also need some paraffin or honing oil and rags. The hone should be moved up and down the cylinder to produce a crosshatch pattern, and plenty of honing oil should be used. Ideally, the crosshatch lines should intersect at approximately a 60° angle. Do not take off more material than is necessary to produce the required finish. If new pistons are being fitted, the piston manufacturers may specify a finish with a different angle, so their instructions should be followed. Do not withdraw the hone from the cylinder while it is still being turned, but stop it first. After honing a cylinder, wipe out all

traces of the honing oil. If equipment of this type is not available, or if you are not sure whether you are competent to undertake the task yourself, an engine overhaul specialist will carry out the work at a moderate cost.

20 Before refitting the cylinder liners, their protrusions must be checked as follows, and new base seals/O-rings fitted. On the 1108 cc (C1E) engine, the seals are a flat ring-type gasket of special material. The seals are available in three thicknesses, colour-coded blue, red and green. On the 1171 cc and 1390 cc (E-type) engines, the seals are in the form of a rubber O-ring.

21 Place the liner with a blue base seal (C-type engine) or without a base O-ring (E-type engine) in the cylinder block, and press down to make sure that it is seated correctly. Using a dial gauge or straight-edge and feeler blade, check that the protrusion of the liner above the upper surface of the cylinder block is within the specified limits (photos). Check all of the liners in the same manner, and record the protrusions. If the protrusions are incorrect on the C-type engine, try again using a red or green seal instead of the blue one. Note that there is also a limit specified for the difference of protrusion between two adjacent liners. If new liners are being fitted, it is permitted to interchange them to bring this difference within limits. The protrusions may be stepped upwards or downwards from the flywheel end of the engine.

22 Refit all external components and senders.

60 Piston/connecting rod assemblies - inspection and reassembly

Inspection

1 Before the inspection process can begin, the piston/connecting rod assemblies must be cleaned, and the original piston rings removed from the pistons.

2 Carefully expand the old rings over the top of the pistons. The use of two or three old feeler blades will be helpful in preventing the rings dropping into empty grooves (photo).

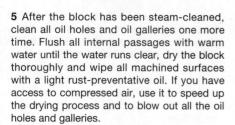

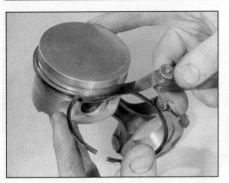

60.10 Measuring the piston ring-to-groove clearance

60.11 Measuring the pistons for ovality

60.13 Measuring the piston ring end gap

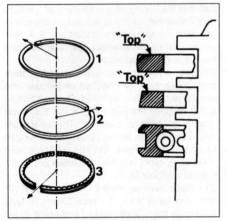

Fig. 2.47 Piston ring identification on the E-type engine (Sec 60)

3 Scrape away all traces of carbon from the top of the piston. A hand-held wire brush or a piece of fine emery cloth can be used once the majority of the deposits have been scraped away.

4 Remove the carbon from the ring grooves in the piston by cleaning them using an old ring. Break the ring in half to do this. Be very careful to remove only the carbon deposits; do not remove any metal, nor nick or scratch the sides of the ring grooves. Protect your fingers - piston rings are sharp.

5 Once the deposits have been removed, clean the piston/connecting rod assembly with paraffin or a suitable solvent, and dry thoroughly. Make sure the oil return holes in the back sides of the ring grooves are clear.

6 If the pistons and cylinder bores are not damaged or worn excessively, and if the cylinder block does not need to be rebored, the original pistons can be re-used. Normal piston wear appears as even vertical wear on the piston thrust surfaces, and slight looseness of the top ring in its groove. New piston rings, however, should always be used when the engine is reassembled.

7 Carefully inspect each piston for cracks around the skirt, at the gudgeon pin bosses, and at the piston ring lands (between the piston ring grooves).

8 Look for scoring and scuffing on the sides of the skirt, holes in the piston crown, and burned areas at the edge of the crown. If the skirt is scored or scuffed, the engine may have been suffering from overheating and/or abnormal combustion, which caused excessively-high operating temperatures. The cooling and lubricating systems should be checked thoroughly. A hole in the piston crown is an indication that abnormal combustion (pre-ignition) was occurring. Burned areas at the edge of the piston crown are usually evidence of knocking (detonation). If any of the above problems exist, the causes must be corrected, or the damage will occur again. The causes may include inlet air leaks, incorrect fuel/air mixture or (when applicable) incorrect ignition timing.

9 Corrosion of the piston, in the form of small pits, indicates that coolant is leaking into the combustion chamber and/or the crankcase. Again, the cause must be corrected, or the problem may persist in the rebuilt engine.

10 If new rings are being fitted to old pistons, measure the piston ring-to-groove clearance by placing a new piston ring in each ring groove and measuring the clearance with a feeler blade (photo). Check the clearance at three or four places around each groove. No values are specified, but if the measured clearance is excessive - say greater than 0.10 mm - new pistons will be required. If the new ring is excessively tight, the most likely cause is dirt remaining in the groove.

11 Check the piston-to-bore/liner clearance by measuring the cylinder bore/liner (see Section 59) and the piston diameter. Measure the piston across the skirt, at a 90° angle to the gudgeon pin, approximately half way down the skirt (photo). Subtract the piston diameter from the bore/liner diameter to obtain the clearance. On F-type engines, if this is greater than the figures given in the Specifications, the block will have to be rebored and new pistons and rings fitted. On C- and E-type engines, new pistons and liners are supplied in matched pairs.

12 Check the fit of the gudgeon pin by twisting the piston and connecting rod in opposite directions. Any noticeable play indicates excessive wear, which must be corrected. If the pistons or connecting rods

are to be renewed, the work should be carried out by a Renault garage or engine overhaul specialist. (In the case of the 1764 cc/F7P engine, the gudgeon pins are secured by circlips, so the pistons and connecting rods can be separated without difficulty. Note the position of the piston relative to the rod before dismantling, and use new circlips on reassembly.)

13 Before refitting the rings to the pistons, check their end gaps by inserting each of them in their cylinder bores (photo). Use the piston to make sure that they are square. No values are specified, but typical gaps would be of the order of 0.50 mm for compression rings, perhaps somewhat greater for the oil control rings. Renault rings are supplied pre-gapped; no attempt should be made to adjust the gaps by filing.

Reassembly

14 Install the new rings by fitting them over the top of the piston, starting with the oil control scraper ring. Use feeler blades or strips of tin in the same way as when removing the old rings. Note that the second compression ring is tapered. Both compression rings must be fitted with the word 'TOP' uppermost. Be careful when handling the compression rings; they will break if they are handled roughly or expanded too far.

15 With all the rings in position, space the ring gaps at 120° to each other.

16 Note that if new piston and liner assemblies have been obtained, each piston is matched to its respective liner, and they must not be interchanged.

61 Crankshaft - inspection

1 Clean the crankshaft and dry it with compressed air if available.

 Warning: Wear eye protection when using compressed air! Be sure to clean the oil holes with a pipe cleaner or similar probe.

61.3 Using a penny to check the crankshaft journals for scoring

61.5 Using a micrometer to check the crankshaft journals

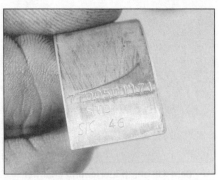

62.1 Typical marking on the back metal of a bearing shell

2 Check the main and big-end bearing journals for uneven wear, scoring, pitting and cracking.

3 Rub a penny across each journal several times (photo). If a journal picks up copper from the penny, it is too rough.

4 Remove all burrs from the crankshaft oil holes with a stone, file or scraper.

5 Using a micrometer, measure the diameter of the main bearing and connecting rod journals, and compare the results with the Specifications at the beginning of this Chapter (photo). By measuring the diameter at a number of points around each journal's circumference, you will be able to determine whether or not the journal is out-of-round. Take the measurement at each end of the journal, near the webs, to determine if the journal is tapered. If any of the measurements vary by more than 0.025 mm, the crankshaft will have to be reground, and undersize bearings fitted.

6 Check the oil seal journals as applicable at each end of the crankshaft for wear and damage. If the seal has worn an excessive groove in the journal, consult an engine overhaul specialist, who will be able to advise if a repair is possible, or whether a new crankshaft is necessary.

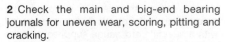

62 Main and big-end bearings - inspection

1 Even though the main and big-end bearings should be renewed during the engine overhaul, the old bearings should be retained for close examination, as they may reveal valuable information about the condition of the engine. The size of the bearing shells is stamped on the back metal (photo), and this information should be given to the supplier of the new shells.

2 Bearing failure occurs because of lack of lubrication, the presence of dirt or other foreign particles, overloading the engine, and corrosion. Regardless of the cause of bearing failure, it must be corrected before the engine is reassembled, to prevent it from happening again.

3 When examining the bearings, remove them from the engine block, the main bearing caps, the connecting rods and the rod caps, and lay them out on a clean surface in the same general position as their location in the engine. This will enable you to match any bearing problems with the corresponding crankshaft journal.

4 Dirt and other foreign particles get into the engine in a variety of ways. Dirt may be left in the engine during assembly, or it may pass through filters or the crankcase ventilation system. It may get into the oil, and from there into the bearings. Metal chips from machining operations and normal engine wear are often present. Abrasives are sometimes left in engine components after reconditioning, especially when parts are not thoroughly cleaned using the proper cleaning methods. Whatever the source, these foreign objects often end up embedded in the soft bearing material, and are easily recognized. Large particles will not embed in the bearing, and will score or gouge the bearing and journal. The best prevention for this cause of bearing failure is to clean all parts thoroughly, and keep everything spotlessly-clean during engine assembly. Frequent and regular engine oil and filter changes are also recommended.

5 Lack of lubrication (or lubrication breakdown) has a number of interrelated causes. Excessive heat (which thins the oil), overloading (which squeezes the oil from the bearing face) and oil leakage (from excessive

bearing clearances, worn oil pump or high engine speeds) all contribute to lubrication breakdown. Blocked oil passages, which usually are the result of misaligned oil holes in a bearing shell, will also oil-starve a bearing and destroy it. When lack of lubrication is the cause of bearing failure, the bearing material is wiped or extruded from the steel backing of the bearing. Temperatures may increase to the point where the steel backing turns blue from overheating.

6 Driving habits can have a definite effect on bearing life. Full-throttle, low-speed operation (labouring the engine) puts very high loads on bearings, which tends to squeeze out the oil film. These loads cause the bearings to flex, which produces fine cracks in the bearing face (fatigue failure). Eventually, the bearing material will loosen in pieces and tear away from the steel backing. Short-trip driving leads to corrosion of bearings, because insufficient engine heat is produced to drive off the condensed water and corrosive gases. These products collect in the engine oil, forming acid and sludge. As the oil is carried to the engine bearings, the acid attacks and corrodes the bearing material.

7 Incorrect bearing installation during engine assembly will lead to bearing failure as well. Tight-fitting bearings leave insufficient bearing oil clearance, and will result in oil starvation. Dirt or foreign particles trapped behind a bearing shell result in high spots on the bearing which lead to failure.

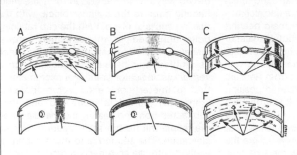

Fig. 2.48 Typical bearing failures (Sec 62)

A Scratched by dirt; dirt embedded into bearing material
B Lack of oil; overlay wiped out
C Improper seating; bright (polished) sections
D Tapered journal; overlay gone from entire surface
E Radius ride
F Fatigue failure; craters or pockets

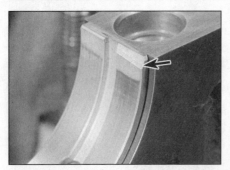

64.4 Oil pump lubricating chamfer (arrowed) on No 5 upper main bearing shell on the E-type engine

64.7 Thread of Plastigage (arrowed) placed on a crankshaft main journal

64.11 Measuring the Plastigage width with the special gauge

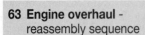

63 Engine overhaul - reassembly sequence

1 Before reassembly begins, ensure that all new parts have been obtained, and that all necessary tools are available. Read through the entire procedure to familiarise yourself with the work involved, and to ensure that all items necessary for reassembly of the engine are at hand. In addition to all normal tools and materials, a thread-locking compound will be needed. A tube of RTV sealing compound will also be required for the joint faces that are fitted without gaskets; it is recommended that CAF 4/60 THIXO paste (obtainable from Renault dealers) is used, as it is specially-formulated for this purpose.

2 In order to save time and avoid problems, engine reassembly can be carried out in the following order.

(a) *Crankshaft.*
(b) *Pistons/connecting rod assemblies.*
(c) *Oil pump.*
(d) *Sump.*
(e) *Flywheel/driveplate.*
(f) *Cylinder head.*
(g) *Timing chain/belt and sprockets.*
(h) *Engine external components.*

64 Crankshaft - main bearing running clearance check and refitting

1 Before fitting the crankshaft and main bearings on the F-type engine, a decision has to be made whether the No 1 main bearing cap is to be sealed using butyl seals or silicone sealant. If butyl seals are to be used, it is necessary to determine the correct thickness of the seals to obtain from Renault. To do this, place the bearing cap in position without any seals, and secure it with the two retaining bolts. Locate a twist drill, dowel rod or any other suitable implement which will just fit in the side seal groove. Now measure the implement - this dimension is the side seal groove size. If the dimension is less than or equal to 5 mm, a 5.10 mm thick side seal is needed. If the dimension is more than 5 mm, a

5.4 mm thick side seal is required. Having determined the side seal size and obtained the necessary seals, proceed as follows for the other types of engine as well. Note that silicone sealant is used instead of side seals when the engine is originally assembled at the factory.

Main bearing running clearance check

2 Clean the backs of the bearing shells and the bearing recesses in both the cylinder block and main bearing caps.

3 Press the bearing shells without oil holes into the caps, ensuring that the tag on the shell engages in the notch in the cap.

4 Press the bearing shells with the oil holes/grooves into the recesses in the cylinder block. Note the following points.

(a) *On the C-type engine, the upper shells for Nos 1 and 3 main bearings are identical, as are the shells for Nos 2, 4 and 5 main bearings.*
(b) *On the E-type engine, No 5 upper main bearing shell is different from the rest, in that it has an oil pump pinion lubricating chamfer which must face the timing belt end of the crankshaft (photo).*
(c) *On all engines, if the original main bearing shells are being re-used, these must be refitted to their original locations in the block and caps.*

5 Before the crankshaft can be permanently installed, the main bearing running clearance should be checked; this can be done in either of two ways. One method is to fit the main bearing caps to the cylinder block, with the bearing shells in place. With the cap retaining bolts tightened to the specified torque, measure the internal diameter of each assembled pair of bearing shells using a vernier dial indicator or internal micrometer. If the diameter of each corresponding crankshaft journal is measured and then subtracted from the bearing internal diameter, the result will be the main bearing running clearance. The second (and more accurate) method is to use an American product known as 'Plastigage'. This consists of a fine thread of perfectly-round plastic which is compressed between the bearing cap and the

journal. When the cap is removed, the deformation of the plastic thread is measured with a special card gauge supplied with the kit. The running clearance is determined from this gauge. Plastigage is sometimes difficult to obtain in the UK, but enquiries at one of the larger specialist chains of quality motor factors should produce the name of a stockist in your area. The procedure for using Plastigage is as follows.

6 With the upper main bearing shells in place, carefully lay the crankshaft in position. Do not use any lubricant; the crankshaft journals and bearing shells must be perfectly clean and dry.

7 Cut several pieces of the appropriate-size Plastigage (they should be slightly shorter than the width of the main bearings), and place one piece on each crankshaft journal axis (photo).

8 With the bearing shells in position in the caps, fit the caps to their numbered or previously-noted locations. Take care not to disturb the Plastigage.

9 Starting with the centre main bearing and working outward, tighten the main bearing cap bolts progressively to their specified torque setting. Don't rotate the crankshaft at any time during this operation.

10 Remove the bolts and carefully lift off the main bearing caps, keeping them in order. Don't disturb the Plastigage or rotate the crankshaft. If any of the bearing caps are difficult to remove, tap them from side-to-side with a soft-faced mallet.

11 Compare the width of the crushed Plastigage on each journal to the scale printed on the Plastigage envelope to obtain the main bearing running clearance (photo).

12 If the clearance is not as specified, the bearing shells may be the wrong size (or excessively-worn if the original shells are being re-used). Before deciding that different size shells are needed, make sure that no dirt or oil was trapped between the bearing shells and the caps or block when the clearance was measured. If the Plastigage was wider at one end than at the other, the journal may be tapered.

13 Carefully scrape away all traces of the Plastigage material from the crankshaft and

64.17 Rear oil seal located on the crankshaft

64.18A Applying sealant to No 1 main bearing cap on the E-type engine

64.18B Injecting sealant into the side grooves of No 1 main bearing cap

bearing shells, using a fingernail or something similar which is unlikely to score the shells.

Final refitting

14 Carefully lift the crankshaft out of the cylinder block once more.

15 Using a little grease, stick the thrustwashers to each side of the centre main bearing (C- and E-type engines) or No 2 main bearing (F-type engine). Ensure that the oilway grooves on each thrustwasher face outwards from the bearing.

16 Lubricate the lips of the new crankshaft rear oil seal, and carefully slip it over the crankshaft rear journal. Do this carefully, as the seal lips are very delicate. Ensure that the open side of the seal faces the engine.

17 Liberally lubricate each bearing shell in the cylinder block, and lower the crankshaft into position. Check that the rear oil seal is positioned correctly (photo).

18 Lubricate the bearing shells, then fit the bearing caps in their numbered or previously-noted locations. Note the following points (photos).

(a) On the C-type engine, apply Loctite Frenetanch to the bearing faces of No 1 main bearing cap.

(b) On the E-type engine, apply a thin coating of CAF 4/60 THIXO sealant to the outer

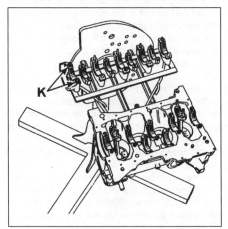

Fig. 2.49 Apply a thin coating of CAF 4/60 THIXO sealant to the outer corners of No 1 main bearing cap (K) on the E-type engine (Sec 64)

corners of No 1 main bearing cap, as shown in Fig. 2.49.

(c) When fitting the butyl seals to No 1 main bearing cap on the F-type engine, fit the seals with their grooves facing outwards. Position the seals so that approximately 0.2 mm (0.008 in) of seal protrudes at the bottom-facing side (the side towards the crankcase). Lubricate the seals with a little oil, and apply a little CAF 4/60 THIXO sealant to the bottom corners of the cap prior to fitting it. When the cap is being fitted, use the bolts as a guide by just starting them in their threads, then pressing the cap firmly into position.

(d) Where silicone sealant is to be used on the No 1 main bearing on the F-type engine, do not press the cap right down onto the crankshaft, but leave it raised so that the first few threads of the main bearing bolts can just be entered. Now inject the sealant into each of the cap side grooves until it enters the space below the cap and completely fills the grooves.

19 Fit the main bearing cap bolts, and tighten them progressively to the specified torque. Where butyl seals have been fitted on the F-type engine, trim the protruding ends flush with the surface of the crankcase. Where silicone sealant has been used, wipe away any excess sealant from the main bearing cap, and smooth it flush with the surface of the crankcase.

20 Check that the crankshaft is free to turn. Some stiffness is normal if new components

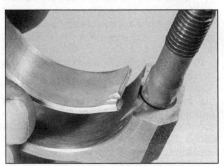

65.2 Engaging the shell tab with the cut-out in the connecting rod

have been fitted, but there must be no jamming or tight spots.

21 Check the crankshaft endfloat with reference to Section 58.

65 Piston/connecting rod assemblies - big-end bearing running clearance check and refitting

1 Clean the backs of the big-end bearing shells and the recesses in the connecting rods and big-end caps. If new shells are being fitted, ensure that all traces of the protective grease are cleaned off using paraffin. Wipe the shells and connecting rods dry with a lint-free cloth.

2 Press the big-end bearing shells into the connecting rods and caps in their correct positions. Make sure that the location tabs are engaged with the cut-outs in the connecting rods (photo).

C-type (1108 cc) and E-type (1171 cc and 1390 cc) engines
Big-end bearing running clearance check

3 Place the four liners face down in a row on the bench, in their correct order. Turn them as necessary so that the flats on the edges of liners 1 and 2 are towards each other, and the flats on liners 3 and 4 are towards each other also. Fit the previously-selected base seals (C-type engine) or O-rings (E-type engine) to the base of each liner (photo).

65.3 Fitting a liner base O-ring (E-type engine)

2

65.6 The arrow on the piston crown must be facing the flywheel end of the engine

65.11 Liner clamp to hold the liners in place

65.14 Checking the gap between the liners

4 Lubricate the pistons and piston rings, then place each piston and connecting rod assembly with its respective liner.

5 Starting with assembly No 1, make sure that the piston ring gaps are still spaced at 120° to each other. Clamp the piston rings using a piston ring compressor.

6 Insert the piston and connecting rod assembly into the bottom of the liner, ensuring that the arrow on the piston crown will be facing the flywheel/driveplate end of the engine (photo). Using a block of wood or a hammer handle against the end of the connecting rod, tap the piston into the liner until the top of the piston is approximately 25 mm away from the top of the liner.

7 Repeat the procedure for the remaining three piston-and-liner assemblies.

8 Turn the crankshaft so that No 1 crankpin is at the bottom of its travel.

9 With the liner seal/O-ring in position, place No 1 liner, piston and connecting rod assembly into its location in the cylinder block. Ensure that the arrow on the piston crown faces the flywheel/driveplate end of the engine, and that the flat on the liner is positioned as described previously.

10 To measure the big-end bearing running clearance, refer to the information contained in Section 64; the same general procedures apply. If the Plastigage method is being used, ensure that the crankpin journal and the big-end bearing shells are clean and dry, then pull the connecting rod down and engage it with the crankpin. Place the Plastigage strip on the crankpin, check that the marks made on the cap and rod during removal are next to each other, then refit the cap and retaining nuts. Tighten the nuts to the specified torque. Do not rotate the crankshaft during this operation. Remove the cap and check the running clearance by measuring the Plastigage as previously described.

11 With the liner/piston assembly installed, retain the liner using a bolt and washer screwed into the cylinder head bolt holes, or using liner clamps (photo).

12 Repeat the above procedures for the remaining piston-and-liner assemblies.

Final refitting

13 Having checked the running clearance of all the crankpin journals and taken any

corrective action necessary, clean off all traces of Plastigage from the bearing shells and crankpin.

14 Liberally lubricate the crankpin journals and big-end bearing shells. Refit the bearing caps once more, ensuring correct positioning as previously described. Tighten the bearing cap nuts to the specified torque, and turn the crankshaft each time to make sure that it is free before moving on to the next assembly. On completion, check that all the liners are positioned relative to each other, so that a 0.1 mm feeler blade can pass freely through the gaps between the liners (photo). If this is not the case, it may be necessary to interchange one or more of the complete piston/liner assemblies to achieve this clearance.

F-type (1721 cc, 1764 cc and 1794 cc) engines

Big-end bearing running clearance check

15 Lubricate No 1 piston and piston rings, and check that the ring gaps are spaced at 120° intervals to each other.

16 Fit a ring compressor to No 1 piston, then insert the piston and connecting rod into No 1 cylinder. With No 1 crankpin at its lowest point, drive the piston carefully into the cylinder with the wooden handle of a hammer, at the same time guiding the connecting rod onto the crankpin. Make sure that the arrow on the piston crown faces the flywheel end of the engine.

17 To measure the big-end bearing running clearance, refer to the information contained in Section 64; the same general procedures apply. If the Plastigage method is being used, ensure that the crankpin journal and the big-end bearing shells are clean and dry, then engage the connecting rod with the crankpin. Place the Plastigage strip on the crankpin, fit the bearing cap in its previously-noted position, then tighten the bolts to the specified torque. Do not rotate the crankshaft during this operation. Remove the cap and check the running clearance by measuring the Plastigage as previously described.

18 Repeat the above procedures on the remaining piston/connecting rod assemblies.

Final refitting

19 Having checked the running clearance of all the crankpin journals and taken any corrective action necessary, clean off all traces of Plastigage from the bearing shells and crankpin.

20 Liberally lubricate the crankpin journals and big-end bearing shells. Refit the bearing caps once more, ensuring correct positioning as previously described. Tighten the bearing cap bolts to the specified torque, and turn the crankshaft each time to make sure that it is free before moving on to the next assembly.

66 Engine -
initial start-up after overhaul

1 With the engine refitted in the vehicle, double-check the engine oil and coolant levels.

2 With the spark plugs removed and the ignition system disabled by connecting the coil HT lead to ground with a jumper lead, crank the engine on the starter motor until the oil pressure light goes out.

3 Refit the spark plugs and connect all the HT leads.

4 Start the engine, noting that this may take a little longer than usual, due to the fuel pump and carburettor (where applicable) being empty.

5 While the engine is idling, check for fuel, water and oil leaks. Do not be alarmed if there are some odd smells and smoke from parts getting hot and burning off oil deposits.

6 Keep the engine idling until hot water is felt circulating through the top hose, then switch it off.

7 After a few minutes, recheck the oil and water levels, and top-up as necessary.

8 There is no requirement to retighten the cylinder head bolts on E- and F-type engines. On the 1108 cc (C1E) engine, the bolts must be retightened as described in Section 6, Part A of this Chapter.

9 If new pistons, rings or crankshaft bearings have been fitted, the engine must be run-in for the first 500 miles (800 km). Do not operate the engine at full-throttle, nor allow it to labour in any gear during this period. It is recommended that the oil and filter be changed at the end of this period.

Chapter 3
Cooling, heating and air conditioning systems

Contents

Degrees of difficulty

Easy, suitable for novice with little experience	**Fairly easy,** suitable for beginner with some experience	**Fairly difficult,** suitable for competent DIY mechanic	**Difficult,** suitable for experienced DIY mechanic	**Very difficult,** suitable for expert DIY or professional

Specifications

General

Cooling system type	Pressurised, with belt-driven pump, front-mounted radiator and electric cooling fan. Auxiliary electric pump on 16-valve models
Coolant type	See Chapter 1 (*'Lubricants, fluids and capacities'*)
Coolant capacity	See Chapter 1 (*'Lubricants, fluids and capacities'*)
System pressure:	
Brown cap	1.2 bars
Blue cap	1.6 bars

Thermostat

Type	Wax	
Opening temperatures:	**Starts to open**	**Fully open**
1108 cc (C1E) engine	83°C	95°C
1171 cc (E5F/E7F) and 1390 cc (E6J/E7J) engines	86°C	98°C
1721 cc (F2N), 1794 cc (F3P) and 1764 cc (F7P) engines	89°C	101°C
Travel (closed to fully open):		
1721 cc (F2N), 1794 cc (F3P) and 1764 cc (F7P) engines	8.0 mm	
All other engines	7.5 mm	

Temperature gauge sender

Resistance values (typical):	
20°C	3550 ± 500 ohms
80°C	335 ± 35 ohms
90°C	240 ± 30 ohms

3

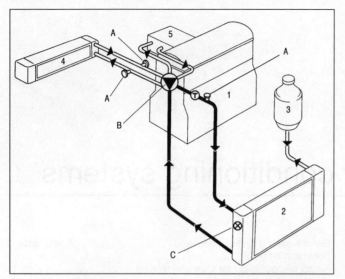

Fig. 3.1 Cooling system schematic - 1108 cc (C1E) engine (Sec 1)

1 Cylinder block
2 Radiator
3 Expansion tank
4 Heater radiator
5 Inlet manifold
6 Restrictors
7 Auxiliary water pump

8 Engine oil cooler
14 Restrictor
15 Restrictor
A Bleed screw
B Water pump
C Fan thermostatic switch
T Thermostat

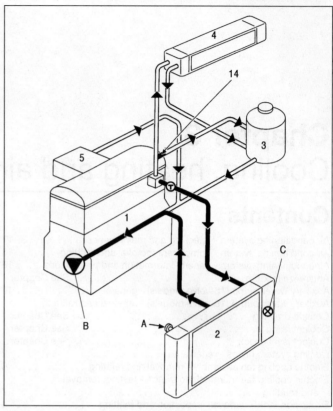

Fig. 3.2 Cooling system schematic - 1171 and 1390 cc (E-type) engines (Sec 1)

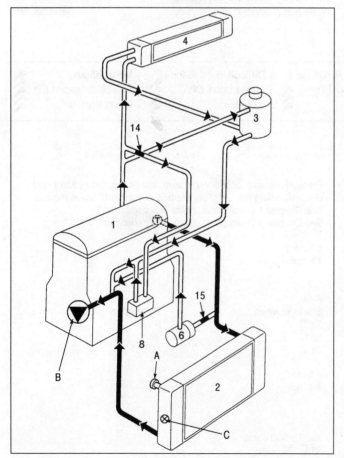

Fig. 3.3 Cooling system schematic - 1764 cc (F7P) engine (Sec 1)

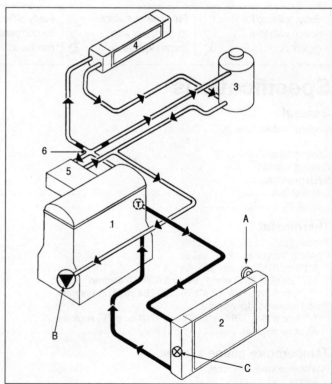

Fig. 3.4 Cooling system schematic - 1721 and 1794 cc (F2N and F3P) engines (Sec 1)

1 General information

The cooling system is of the pressurised type. The main components are a belt-driven pump, an aluminium crossflow radiator, an expansion tank, an electric cooling fan, a thermostat, and the associated hoses.

The system functions as follows. When the engine is cold, coolant is pumped around the cylinder block and head passages. After cooling the cylinder bores, combustion surfaces and valve seats, the coolant passes through the heater and inlet manifold, and is returned to the water pump.

When the coolant reaches a predetermined temperature, the thermostat opens, and the hot coolant passes through the top hose to the radiator. As the coolant circulates through the radiator, it is cooled by the inrush of air when the car is in motion. The airflow is supplemented by the action of the electric cooling fan when necessary. Upon reaching the bottom of the radiator, the coolant returns to the pump via the radiator bottom hose, and the cycle is repeated.

As the coolant warms up, it expands; the increased volume is accommodated in an expansion tank. On the 1108 cc (C1E) engine, the expansion tank is 'cold', simply receiving a small volume of coolant as the temperature increases, and returning it to the system as it cools down. On all other models, the tank is 'hot', coolant circulating through the tank all the time that the engine is running.

The electric cooling fan, mounted behind the radiator, is controlled by a thermostatic switch located in the side of the radiator. At a predetermined coolant temperature, the switch contacts close, actuating the fan via a relay.

The cooling system on 16-valve models has two principal additional components: an engine oil cooler (mounted under the oil filter) and an auxiliary water pump (described later in this Chapter).

For details of the air conditioning system (when fitted) and the precautions associated with it, refer to Section 16.

2 Cooling system hoses - renewal

1 The number, routing and pattern of hoses will vary according to model, but the same basic procedure applies. Before commencing work, make sure that the new hoses are to hand, along with new hose clips if needed. It is good practice to renew the hose clips at the same time as the hoses.

2 Drain the cooling system, saving the coolant if it is fit for re-use (Chapter 1, Section 33). Squirt a little penetrating oil onto the hose clips if they are rusty.

3 Release the hose clips from the hose concerned. Three clip types are used: worm-

2.3 Releasing a spring hose clip using self-locking pliers

drive, spring and 'sardine-can'. The worm-drive clip is released by turning its screw anti-clockwise. The spring clip is released by squeezing its tags together with pliers, at the same time working the clip away from the hose stub (photo). The 'sardine-can' clip is not re-usable, and is best cut off with snips or side cutters.

4 Unclip any wires, cables or other hoses which may be attached to the hose being removed. Make notes for reference when reassembling if necessary.

5 Release the hose from its stubs with a twisting motion. Be careful not to damage the stubs on delicate components such as the radiator. If the hose is stuck fast, the best course is often to cut it off using a sharp knife, but again be careful not to damage the stubs.

6 Before fitting the new hose, smear the stubs with washing-up liquid or a suitable rubber lubricant to aid fitting. **Do not** use oil or grease, which may attack the rubber.

7 Fit the hose clips over the ends of the hose, then fit the hose over its stubs. Work the hose into position. When satisfied, locate and tighten the hose clips.

8 Refill the cooling system (Chapter 1, Section 33). Run the engine, and check that there are no leaks.

9 Recheck the tightness of the hose clips on any new hoses after a few hundred miles.

3.4 Disconnecting the radiator top hose

3 Radiator - removal, inspection, cleaning and refitting

Note: *If the radiator is to be removed for a period of more than 48 hours, precautions must be taken against internal corrosion. Either rinse the radiator with clean water and dry it thoroughly by blowing air through it, or fill it with coolant and plug the hose stubs. If the reason for removing the radiator is concern over coolant loss, note that minor leaks may be repaired by using a radiator sealant with the radiator in situ.*

Removal

1 Disconnect the battery earth lead.

2 Remove the cooling fan (Section 6).

3 Drain the cooling system by disconnecting the radiator bottom hose. Save the coolant in a clean container if it is fit for re-use.

4 Release the hose clips and disconnect the remaining hoses from the radiator (photo).

5 Remove the two radiator mounting bolts from the top crossmember. Note the position of each mounting (they are not the same) and remove them (photos).

6 If wished, the bonnet and the top crossmember may be removed to improve access. This is essential on 16-valve models, where the inlet manifold prevents the radiator

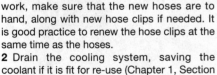

3.5A Removing a radiator mounting bolt

3.5B Right-hand mounting has a locating tag (arrowed) . . .

3

3.5C ... left-hand mounting does not

3.6 Top crossmember must be removed on 16-valve models

3.8A Carefully lift out the radiator ...

being moved back far enough to clear the crossmember (photo).

7 On models with air conditioning, separate the condenser from the radiator by removing the two screws. **Do not** disconnect the refrigerant pipes.

8 Carefully lift the radiator off its bottom mountings and remove it. Recover the mountings; renew them if they are in poor condition (photos).

Inspection and cleaning

9 Radiator repair is best left to a specialist, but minor leaks may be sealed without removing the radiator, using a sealant. The area of a leak can usually be identified by a rusty or crystalline deposit.

10 Clear the radiator matrix of flies and leaves with a soft brush, or by hosing.

3.8B ... and recover the bottom mountings

Refitting

11 Refit by reversing the removal operations. Refill the cooling system on completion (Chapter 1, Section 33).

4 Expansion tank - removal, inspection and refitting

1 With the engine cold, drain some coolant from the system until the expansion tank is empty.

2 Release the strap which secures the tank. Disconnect the hose (or hoses) and remove the tank.

3 Clean the tank and inspect it for cracks and other damage. Renew it if necessary. Also inspect the cap; if there is evidence that

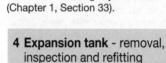

5.1 Cylinder head outlet elbow (arrowed) contains the thermostat on E-type engines

coolant has been vented through the cap, renew it.

4 Refit by reversing the removal operations. Refill and bleed the cooling system as described in Chapter 1, Section 33.

5 Thermostat - removal, testing and refitting

1 On the 1171 cc (E5F/E7F) and 1390 cc (E6J/E7J) engines, the thermostat is located in the cylinder head outlet elbow on the left-hand side of the engine (photo).

2 On the 1108 cc (C1E) engine, the thermostat is located in the end of the radiator top hose at the water pump, and is retained by a hose clip.

3 On the 1721 cc (F2N), 1794 cc (F3P) and 1764 cc (F7P) engines, the thermostat is located in a housing bolted to the left-hand side of the cylinder head beneath the distributor cap. Access can be improved by removing the distributor cap and the air cleaner or air inlet trunking.

Removal

4 Partially drain the cooling system, so that the coolant level is below the thermostat location.

5 Loosen the clip and disconnect the hose.

6 On the 1108 cc (C1E) engine, withdraw the thermostat from inside the hose. On other engines, unbolt the cover and remove the thermostat, then remove the sealing ring (photos).

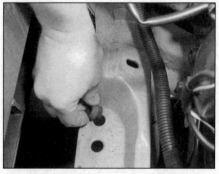

5.6A Removing the thermostat from the top hose on the 1108 cc (C1E) engine

5.6B Removing the thermostat from a 1390 cc (E6J) engine

5.6C Removing the thermostat from a 1721 cc (F2N) engine

6.3A Remove the retaining screws . . .

6.3B . . . and withdraw the fan assembly. Motor wiring is on the underside

7.1A Free the multi-plug from its bracket . . .

Testing

7 To test whether the unit is serviceable, suspend it on a string in a saucepan of cold water, together with a thermometer. Heat the water, and note the temperature at which the thermostat begins to open. Continue heating the water until the thermostat is fully open, and then remove it from the water.

8 The temperature at which the thermostat should start to open is stamped on the unit. If the thermostat does not start to open at the specified temperature, does not fully open in boiling water, or does not fully close when removed from the water, then it must be discarded and a new one fitted.

Refitting

9 Refitting is a reversal of removal, but where applicable renew the sealing ring. On the 1108 cc engine, make sure that the thermostat bleed hole is in the slot on the end of the water pump outlet. Refill the cooling system as described in Chapter 1.

6 Electric cooling fan assembly - removal and refitting

1 Disconnect the battery earth lead.
2 Release the multi-plugs from each side of the fan frame, and disconnect them.
3 Remove the four retaining screws and withdraw the fan assembly. Note which way up it is fitted (photos).
4 Refit by reversing the removal operations.

7 Electric cooling fan thermostatic switch - testing, removal and refitting

Testing

1 Free the switch multi-plug and disconnect it (photos).
2 With the ignition on, bridge the multi-plug terminals on the wiring harness with a paper clip or a short length of wire. Be careful not to let the bridging link touch earth (bare metal). The fan should run.
3 If there has been a problem of overheating due to the fan not operating, but it runs with the bridge in place, this suggests that the thermostatic switch is defective.
4 If the fan does not run with the bridge in place, there is a fault in the fan itself or in its supply circuit (including the fuse and, if applicable, the relay).
5 The switch can be tested further after removal by immersing it in a heated water bath (refer to Section 5, testing the thermostat), and using a continuity tester to check the temperature at which the switch contacts open and close.

Removal

6 Drain the cooling system (Chapter 1, Section 33).
7 Free the switch multi-plug and disconnect it.

8 Unscrew the switch from the radiator and remove it.

Refitting

9 Apply a little sealant to the threads of the switch, and screw it into position.
10 Reconnect and secure the multi-plug.
11 Refill the cooling system (Chapter 1, Section 33).

8 Temperature gauge and warning light sender unit - testing, removal and refitting

1 The location of the temperature gauge/warning light sender unit varies according to model. On the 1108 cc (C1E) engine, it is on top of the water pump. On the 1764 cc (F7P) engine, it is screwed into the thermostat housing (photo). On the other engines, it is on the front of the cylinder head at the left-hand end. In all cases, the procedures are the same.

Testing

2 Disconnect the multi-plug from the sender unit (photo). Using an ohmmeter, measure the resistance of the sender, and compare it with the values given in the Specifications. If the value obtained is greatly different from that specified, the sender is probably defective.
3 For accurate testing across the temperature range, the sender unit will have to be removed.

7.1B . . . and disconnect it. Fan switch itself (arrowed) screws into radiator

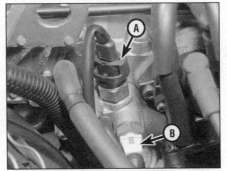

8.1 Temperature gauge sender (A) - 1764 cc (F7P) engine. Fuel injection system temperature sensor (B) is also visible

8.2 Disconnecting the multi-plug from the temperature gauge sender unit - 1794 cc (F3P) engine

3

9.7 Work through the holes in the pulley for access to one of the water pump bolts - 1108 cc (C1E) engine

9.8 Removing the water pump - 1108 cc (C1E) engine

9.11A With the timing belt removed, undo the bolts . . .

Removal

4 Drain the cooling system (Chapter 1, Section 33). Alternatively, remove the expansion tank cap to depressurise the system, and have the new sender unit or a suitable bung to hand.
5 Disconnect the multi-plug and unscrew the sender unit.

Refitting

6 Apply a little sealant to the sensor threads, and screw it into position. Reconnect the multi-plug.
7 Top-up or refill the cooling system as necessary.

9 Water pump -
removal and refitting

1 If the water pump is leaking, or is noisy in operation, it must be renewed.
2 Disconnect the battery earth lead. Drain the cooling system (Chapter 1, Section 33).

1108 cc (C1E) engine

3 Remove the alternator drivebelt (Chapter 12, Section 6).

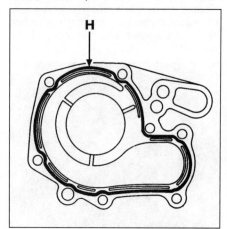

Fig. 3.5 Apply a bead of sealant (H) as shown when refitting the water pump - 1171 and 1390 cc (E-type engines (Sec 9)

4 Unscrew the bolt which secures the alternator adjusting arm to the water pump body. Remove the bolt, and swing the arm clear.
5 Slacken the hose clips and disconnect the hoses from the pump.
6 Disconnect the lead from the temperature gauge sender on top of the pump body.
7 Unscrew and remove the bolts securing the water pump to the cylinder head. Access to the bolt behind the pulley can be gained by inserting a socket and extension bar through the holes in the pulley (photo).
8 With all the bolts removed, withdraw the pump from the cylinder head (photo). If it is stuck, strike it sharply with a plastic or hide mallet. Recover the gasket.
9 Refit by reversing the removal operations, using a new gasket.

1171 cc (E5F/E7F) and 1390 cc (E6J/E7J) engines

10 Remove the timing belt (Chapter 2).
11 Unbolt the water pump from the front of the cylinder block (photos).
12 Clean the mating faces of the water pump and cylinder block, then apply a bead of sealant, 0.6 to 1.0 mm wide, around the inner perimeter of the water pump sealing face on the water pump (Fig. 3.5).
13 Locate the water pump on the cylinder block, then insert the bolts and tighten them evenly.
14 Refit the timing belt (Chapter 2).

9.16 Three bolts (arrowed) which secure the water pump pulley - F-type engine

9.11B . . . and withdraw the water pump (E-type engine)

1721 cc (F2N), 1794 cc (F3P) and 1764 cc (F7P) engines

15 Remove the alternator drivebelt (Chapter 12, Section 6) and the alternator itself (Chapter 12, Section 7). Access will be improved by removing the right-hand headlight.
16 Unscrew the three bolts and remove the water pump pulley (photo).
17 Unscrew the bolts which secure the water pump to the cylinder block, and remove the pump (photo). If it is stuck, strike it sharply with a plastic or hide mallet. Recover the gasket.
18 Refit by reversing the removal operations, using a new gasket.

All engines

19 Refill and bleed the cooling system (Chapter 1, Section 33).

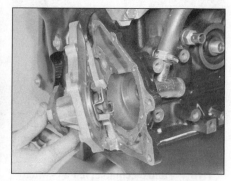

9.17 Removing the water pump - F-type engine

11.3 Auxiliary water pump multi-plug (arrowed)

10 Auxiliary water pump (16-valve models) - general

The auxiliary water pump fitted to 16-valve models is electrically operated. It forms part of the anti-percolation system, which is designed to reduce under-bonnet temperatures when the car is stopped after a run. This system is described in Chapter 4, Section 39.

11 Auxiliary water pump (16-valve models) - removal and refitting

1 Disconnect the battery earth lead. Drain the cooling system as described in Chapter 1, Section 33.
2 The pump is located on the left-hand side of the engine bay, towards the front. Access can be improved by removing the air cleaner.
3 Disconnect the pump multi-plug (photo). Disconnect the coolant hoses, release the pump from its mounting and remove it.
4 Refit by reversing the removal operations. Refill and bleed the cooling system as described in Chapter 1, Section 33.

12 Heating system - general information and checks

General information

The heater and fresh air ventilation unit works on the principle of mixing hot and cold air in the proportions selected by means of the central (temperature) control knob. Coolant flows through the heater radiator all the time that the engine is running, regardless of the temperature selected.

Air distribution is selected by the left-hand control knob. Additional control is possible by opening, closing or redirecting individual vents in the facia panel.

A three-speed blower is controlled by the right-hand knob.

For details of the air conditioning system fitted to some models, refer to Section 16.

Checks

Periodically check that all the controls operate as intended. Problems related to the temperature and air distribution controls may be due to cables being broken or disconnected (Section 15).

If the blower does not operate at all, check the fuse and the blower multi-plug before condemning the motor. If one or two speeds do not work, the fault is almost certainly in the dropper resistor (Section 13).

Check the condition and security of the coolant hoses which feed the heater radiator. The radiator-to-hose joints are at the bulkhead under the bonnet. If water leaks inside the car seem to be coming from the heater, establish whether the leak is of coolant (indicating a leaking heater radiator) or of rainwater (indicating a defective scuttle seal). Cooling system antifreeze has a distinctive sweet smell.

13 Heater components - removal and refitting

Blower assembly

1 Disconnect the battery earth lead.
2 Remove the windscreen wiper arms (Chapter 12) and the windscreen cowl panel (Chapter 11). Remove the vehicle jack.
3 On right-hand drive models only, remove the windscreen wiper motor and linkage (Chapter 12).
4 On fuel injection models, remove the computer from the right-hand scuttle (Chapter 4, Section 36).
5 Remove the heater blower cover, which is secured by two screws (photo).
6 Disconnect the multi-plug from the resistor on the side of the blower (photo).
7 Remove the two bolts which secure the

13.5 Removing a heater blower cover screw

13.6 Disconnecting the multi-plug from the heater blower resistor

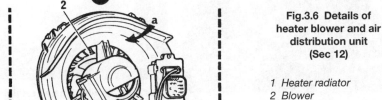

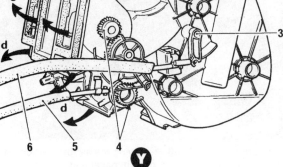

Fig.3.6 Details of heater blower and air distribution unit (Sec 12)

1 Heater radiator
2 Blower
3 Temperature control flap
4 Air distribution flaps
5 Air distribution control cable
6 Temperature control cable
a Air inlet
b Outlet to demisting ducts
c Outlet to facia vents
d Outlet to lower vents
X Scuttle (under-bonnet)
Y Passenger compartment

3

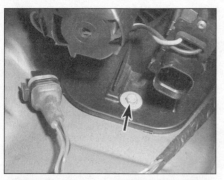

13.7A One of the heater blower securing bolts (arrowed)

13.7B Removing the heater blower

13.12 Disconnecting the blower motor multi-plug

blower. Manipulate the blower and remove it through the right-hand scuttle (photos).

8 If the motor is to be renewed, release the clips and separate the half-housings. Use new clips (supplied with a new motor) on reassembly.

9 Check the condition of the seal at the base of the blower. Renew it if its condition is in doubt. **Note:** *If the seal is defective, rainwater entering the scuttle will leak into the passenger compartment.*

10 Refit by reversing the removal operations.

Blower motor dropper resistor

11 This resistor is located in the right-hand side of the blower casing. It is switched into the circuit at low and intermediate speeds. If it

fails, one or more speeds will not be operative.

12 Gain access to the blower assembly as described earlier, and disconnect the multi-plug from the resistor. Also disconnect the multi-plug from the blower motor (photo).

13 Remove the two screws, release the clips and withdraw the resistor (photos).

14 Refit by reversing the removal operations.

Air distribution unit

15 Remove the complete facia assembly (Chapter 11).

16 Remove the blower assembly as described earlier.

17 Remove the single retaining screw now accessible from the scuttle (photo).

18 Clamp the coolant hoses where they pass

through the bulkhead. Alternatively, drain the cooling system. Disconnect the hoses from the heater radiator stubs (photo).

19 Remove the two screws and free the duct panel from the air distribution unit, bending the facia central support bracket rearwards if necessary (photo).

20 Remove the air distribution unit, complete with control panel and cables, from inside the car (photo). Be prepared for coolant spillage from the heater radiator.

21 Refit by reversing the removal operations. Refill and bleed the cooling system (Chapter 1, Section 33).

Heater radiator

22 Remove the air distribution unit as previously described.

13.13A Remove the screws . . .

13.13B . . . release the clips (arrowed) and withdraw the resistor

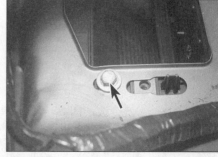

13.17 Air distribution unit retaining screw (arrowed) is only accessible after removing the blower

13.18 Disconnecting the heater hoses at the bulkhead. Hoses are different sizes, so they cannot get mixed up

13.19 Removing the duct panel from the air distribution unit

13.20 Removing the air distribution unit, control panel and cables

13.23 Removing the foam seal and the closing plate from the heater radiator stubs

13.24A Remove the heater radiator securing screws, if fitted . . .

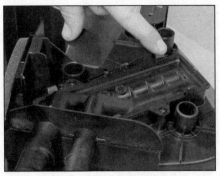

13.24B . . . release the clips . . .

23 Remove the foam seal and the closing plate from the heater radiator stubs (photo).
24 Remove the two screws (if fitted), release the clips and withdraw the heater radiator (photos). Be careful not to damage the fins.
25 Refit by reversing the removal operations. If the clips were damaged during removal, secure the radiator using two screws in the holes provided.

14 Heater ducts and vents - removal and refitting

1 The facia side vents are each secured by three Torx screws - one at the side, and two at the bottom. Remove the screws and withdraw the vent (photos).

2 The centre vent cannot be removed independently of the facia.
3 The removal of the central duct panel is included in the operation to remove the air distribution unit (Section 13).

15 Heater controls - removal and refitting

Control panel and bulb

1 Disconnect the battery earth lead. Remove the radio (Chapter 12).
2 Remove the radio housing, which is secured by four screws (Chapter 11, Section 26).
3 Remove the two screws which secure the

heater control panel. Withdraw the control panel (photos).
4 If the reason for removing the panel is to renew the bulb, this can be done without further dismantling (photos). To remove the panel completely, disconnect the multi-plugs and the cables.
5 Refit by reversing the removal operations. Check that the controls operate over their full range before securing the panel.

Control cables

6 Remove the control panel as just described.
7 Remove the two screws which secure the duct panel to the air distribution unit. Withdraw the duct panel for access to the cables.

13.24C . . . and remove the heater radiator

14.1A Remove the screws (two at the bottom and one at the side) . . .

14.1B . . . to free the facia side vent

15.3A Removing a heater control panel screw

15.3B Withdrawing the heater control panel

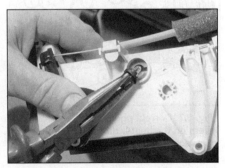

15.4A To renew the heater control panel bulb, free the bulbholder by twisting it a quarter of a turn with pliers . . .

3

15.4B . . . then separate the bulb and holder

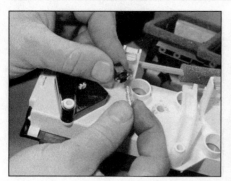

15.8A Remove the heater control cable securing clip . . .

15.8B . . . and disconnect the cable from the lever

8 Remove the cable securing clips. Disconnect each cable from its lever by turning the cable through 90° (photos). Also disconnect the return spring from the temperature control lever.

9 Remove the cables. Note that they are of different lengths; the longer cable controls the temperature flap.

10 Before refitting, position the flap levers on the distribution unit in the 'cold' and 'ventilation' positions (Fig. 3.7). Turn the control panel knobs fully anti-clockwise to the corresponding positions.

11 Connect the cables and the return spring. Fit the cable securing clips, and check the operation of the controls.

12 Refit the duct panel and the control panel.

16 Air conditioning system - general information and precautions

General information

1 An air conditioning system is available on some models. It enables the temperature of incoming air to be lowered; it also dehumidifies the air, which makes for rapid demisting and increased comfort.

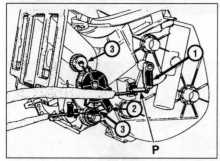

Fig. 3.7 Correct positions of control flap levers on the heater distribution unit when reconnecting the cables (Sec 15)

1 Temperature control flap
2 Air distribution flap
3 Gear alignment marks
P Spring

2 The cooling side of the system works in the same way as a domestic refrigerator. Refrigerant gas is drawn into a belt-driven compressor, and passes into a condenser in front of the radiator, where it loses heat and becomes liquid. The liquid passes through an expansion valve to an evaporator, where it changes from liquid under high pressure to gas under low pressure. This change is accompanied by a drop in temperature, which cools the evaporator. The refrigerant returns to the compressor and the cycle begins again.

3 Air blown through the evaporator passes to the air distribution unit, where it is mixed with hot air blown through the heater radiator, to achieve the desired temperature in the passenger compartment.

4 The heating side of the system works in the same way as on models without air conditioning.

Precautions

5 The refrigerant (Freon R12) is potentially dangerous, and should only be handled by qualified persons. If it is splashed onto the skin, it can cause frostbite. It is not itself

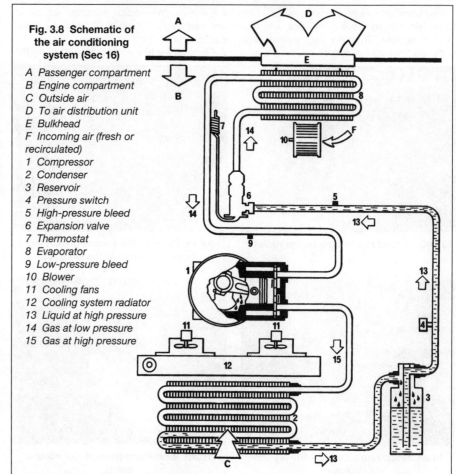

Fig. 3.8 Schematic of the air conditioning system (Sec 16)

A Passenger compartment
B Engine compartment
C Outside air
D To air distribution unit
E Bulkhead
F Incoming air (fresh or recirculated)
1 Compressor
2 Condenser
3 Reservoir
4 Pressure switch
5 High-pressure bleed
6 Expansion valve
7 Thermostat
8 Evaporator
9 Low-pressure bleed
10 Blower
11 Cooling fans
12 Cooling system radiator
13 Liquid at high pressure
14 Gas at low pressure
15 Gas at high pressure

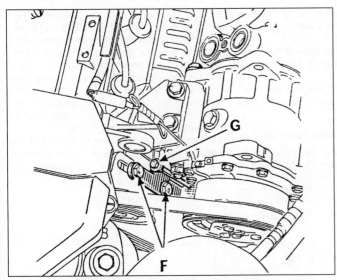

Fig. 3.9 Air conditioning compressor drivebelt adjustment points -
1390 cc (E6J/E7J) engine shown (Sec 18)

F Adjusting strap bolts   G Tensioning screw

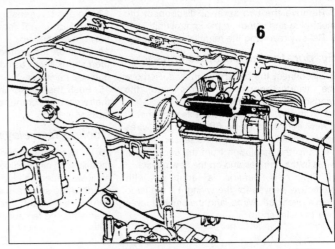

Fig. 3.10 Air conditioning system electronic control unit (6)
(Sec 18)

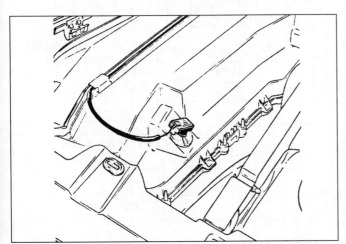

Fig. 3.11 Air conditioning evaporator temperature sensor
(Sec 18)

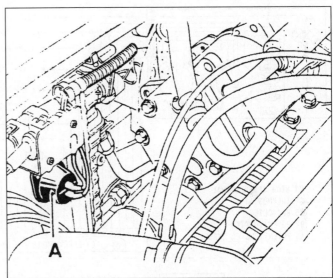

Fig. 3.12 Cooling fan resistor (A) - models with air conditioning
(Sec 18)

poisonous, but in the presence of a naked flame (including a cigarette) it forms a poisonous gas.

6 Uncontrolled discharging of the refrigerant is dangerous, and potentially damaging to the environment. It follows that any work on the air conditioning system which involves opening the refrigerant circuit must only be carried out by a Renault dealer or an air conditioning specialist.

7 Do not operate the air conditioning system if it is known to be short of refrigerant; the compressor may be damaged.

17 Air conditioning system - checking and maintenance

1 Routine maintenance is limited to checking the tension and condition of the compressor drivebelt, and checking the refrigerant sight

glass for bubbles. Refer to Chapter 1, Section 6.

2 Periodic recharging of the system will be required, since there is inevitably a slow loss of refrigerant. It is suggested that the system be inspected by a specialist every 2 years, or at once if a loss of performance is noticed.

18 Air conditioning system - component removal and refitting

 Warning: Do not attempt to open the refrigerant circuit. Refer to the precautions at the end of Section 16.

1 The only operations described here are those which can be carried out without discharging the refrigerant. All other operations must be referred to a specialist.

2 If necessary, the compressor can be unbolted and moved aside, without disconnecting its flexible hoses, after removing the drivebelt.

Compressor drivebelt

3 Disconnect the battery earth lead. For improved access, remove the bonnet.

4 Slacken the two bolts on the compressor adjusting strap. Back off the tensioning screw to slacken the drivebelt, and slip the drivebelt off the pulleys.

5 Refit by reversing the removal operations. Tension the drivebelt in the same way as described for the alternator drivebelt (Chapter 1).

Electronic control unit

6 Disconnect the battery earth lead. Remove the windscreen wiper arms and the windscreen cowl panel.

3

7 Remove the two screws, disconnect the multi-plug and withdraw the control unit.
8 Refit by reversing the removal operations.

Evaporator temperature sensor

9 Disconnect the battery earth lead. Remove the windscreen wiper arms and the windscreen cowl panel.
10 Withdraw the sensor from its grommet. If a new sensor is to be fitted, cut the wires approximately 15 cm from the sensor.
11 Strip the insulation from the ends of the wires in the harness and on the new sensor.
12 The new sensor is supplied with protective sleeves for the wiring. Fit the large sleeve over both wires, and the small sleeves over the wire ends.

13 Solder the sensor wires to the harness wires. It does not matter which way round they are connected.
14 Position the protective sleeves correctly, then shrink them using hot air from a hot air gun or a hairdryer.
15 Push the sensor into its grommet. Refit the remaining components and reconnect the battery.

Cooling fan relays and resistor

16 A two-speed cooling fan is fitted to models with air conditioning. The fan operates at low speed all the time that the air conditioning system is in use. If pressure rises in the refrigerant circuit or if the engine overheats, the fan operates at high speed.

17 The two fan speeds are obtained using two relays and a resistor of 0.28 ohms. For low-speed operation, the resistor is switched in series with the fan motor. For high-speed operation, the resistor is bypassed.
18 The fan relays are located in the relay box on the left-hand side of the engine bay. To remove them, open the box, disconnect the multi-plugs and withdraw the relays. Refit by reversing the removal operations.
19 The resistor is located to the left of the radiator. Its resistance can be checked without removing it, after disconnecting the multi-plug.

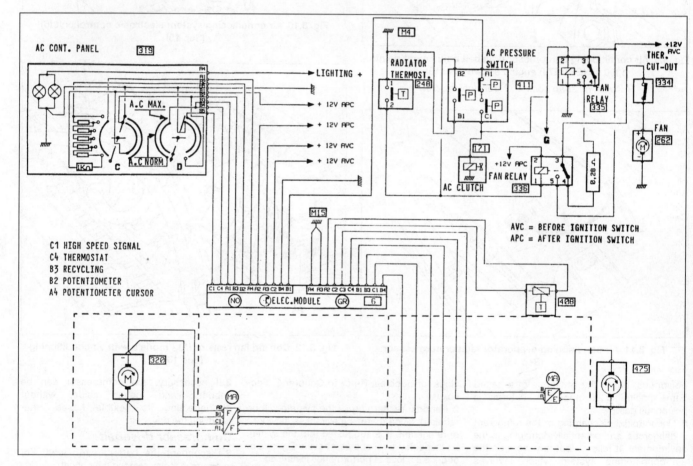

Fig. 3.13 Air conditioning system wiring diagram (Sec 18)

+APC	Positive feed (ignition-controlled)	
+AVC	Positive feed (full-time)	
E	Recycling motor connector	
F	Blower connector	
G	Signal to idle speed compensator	
M4	Earth point for relay 336	
M15	Earth point for blower	
6	Electronic control unit	

171	Compressor
248	Thermostatic switch in radiator
262	Engine cooling fan
319	Control panel
320	Blower
334	Thermal cut-out
335	Fan relay
336	Fan relay (high speed)

408	Evaporator temperature sensor
411	Pressure switch
475	Recycling motor

Connector colours

GR	Grey
MA	Brown
NO	Black

Chapter 4
Fuel, exhaust and emission control systems

Contents

Degrees of difficulty

Easy, suitable for novice with little experience	**Fairly easy,** suitable for beginner with some experience	**Fairly difficult,** suitable for competent DIY mechanic	**Difficult,** suitable for experienced DIY mechanic	**Very difficult,** suitable for expert DIY or professional

Specifications

Part A: Carburettor models

Carburettor – general

Make and type:
1108 cc (C1E) engine	Solex 32 BIS, single-barrel
1171 cc (E5F) engine	Pierburg 32 1B1, single-barrel
1390 cc (E6J) engine	Weber 32 TLDR, twin-barrel
1721 cc (F2N) engine	Solex 32-34 Z13, twin-barrel
Idle speed	See Chapter 1 Specifications
CO content	See Chapter 1 Specifications

Recommended fuel

All carburettor models	Leaded or unleaded petrol, 95 RON minimum

Fuel pump

Delivery pressure (static, with engine idling – see text):
Minimum	0.170 bars
Maximum	0.325 bars

Solex 32 BIS carburettor data

Identification number	963
Barrel diameter	23 mm
Main jet	115
Idle jet	42
Air correction jet	150
Pneumatic enrichment device	40
Accelerator pump injector	40
Needle valve	1.3 mm
Needle valve washer thickness	1.0 mm
Float level	Not adjustable

Pierburg 32 1B1 carburettor data

Identification number	7.17625.21
Barrel diameter	23 mm
Main jet	102.5
Air correction jet	100
Idle fuel/air jet	45/112.5
Auxiliary idle fuel/air jet	40/150
Pneumatic enrichment device	62.5
Accelerator pump injectors	30/40
Accelerator pump delivery	1.3 ± 0.15 ml per 10 strokes
Auxiliary jet	102.5
Needle valve	1.5 mm
Float level (not adjustable)	28.5 ± 1.0 mm
Cold enrichment device coolant switch:	
Closes below	45°C
Opens above	55°C

Weber 32 TLDR carburettor data

	Primary	Secondary
Barrel diameter	23 mm	24 mm
Main jet	122	122
Air correction jet	175	200
Idle jet	52	40
Emulsifier	F3	F24
Enrichment device	50	75/110
Accelerator pump injector	50	
Needle valve	1.5 mm	
Float level	31 mm	

Solex 32-34 Z13 carburettor data

	Primary	Secondary
Barrel diameter	24 mm	26 mm
Main jet	112.5	135
Air correction jet	155	175
Idle jet	42	50
Econostat	—	110
Enrichment device	40	—
Accelerator pump injectors	40	35
Needle valve	1.6 mm	
Float level	33.5 ± 0.5 mm	

Inlet manifold heater (1721 cc/F2N engine)

Thermal switch operating temperatures:
Switches on below	56°C
Switches off above	63°C

Part B: Fuel injection models

General

System type:
1171 cc (E7F), 1390 cc (E7J) and 1794 cc (F3P) engines	Single-point (throttle body injection)
1764 cc (F7P) engines	Multi-point
Idle speed (not adjustable)	See Chapter 1 Specifications
CO content	See Chapter 1 Specifications

Recommended fuel

Engine types E7F-706 and F3P-714 (vehicle codes X57R and X57U)	Unleaded petrol only, 91 RON minimum
Engine type F7P-720 (vehicle code C575)	Leaded or unleaded petrol, 95 RON minimum
All other models	Unleaded petrol only, 95 RON minimum

Fuel pump

Type	Electric, immersed in fuel tank
Pressure (unregulated):	
Single-point system	3 bars approx
Multi-point system	5 bars approx
Delivery (with 12-volt supply) – minimum quantity:	
Single-point system	0.83 litres per minute
Multi-point system	1.08 litres per 30 seconds

Regulated fuel pressure

Single-point system	1.06 ± 0.05 bars
Multi-point system:	
Manifold vacuum zero	3.00 ± 0.15 bars
Manifold vacuum 500 mbars	2.50 ± 0.15 bars

Fuel injectors

Operating voltage	12
Resistance:	
Single-point injector	1.2 ohms approx
Multi-point injectors	2.5 ± 0.5 ohms
Cold start injector (multi-point system only)	10 ± 0.5 ohms

Inlet air temperature sensor

Resistance at temperature of:	Single-point system	Multi-point system
0°C	5890 ± 600 ohms	9720 ± 2250 ohms
20°C	2500 ± 100 ohms	3550 ± 500 ohms
40°C	1170 ± 100 ohms	1470 ± 180 ohms

Coolant temperature sensor

Resistance at temperature of:	Single-point system	Multi-point system
20°C	3500 ± 500 ohms	3500 ± 500 ohms
40°C	1460 ± 140 ohms	Not stated
80°C	333 ± 33 ohms	335 ± 35 ohms
90°C	243 ± 30 ohms	240 ± 30 ohms

Torque wrench settings

	Nm	lbf ft
Fuel pump ring nut:		
Single-point system	30	22
Multi-point system	70	52

4

Part A: Carburettor models

1 General information and precautions

The fuel system consists of a fuel tank mounted under the rear of the car, a mechanical fuel pump, and a single- or twin-barrel downdraught carburettor. The mechanical fuel pump is operated by an eccentric on the camshaft, and is mounted on the forward side of the cylinder block on the 1108 cc (C1E) engine, or on the rear side of the cylinder head on the 1171 cc (E5F), 1390 cc (E6J) and 1721 cc (F2N) engines. All models have a fuel filter and a vapour separator (defuming device) fitted between the fuel pump and the carburettor. Excess fuel is returned from the vapour separator to the fuel tank.

The air cleaner contains a disposable paper filter element, and incorporates a flap valve air temperature control system. This system allows cold air from the outside of the car and warm air from the exhaust manifold to enter the air cleaner in the correct proportions according to ambient air temperatures. The flap is controlled automatically by a temperature-sensitive wax capsule.

Carburettors may be of Solex, Pierburg or Weber manufacture, according to model. The various carburettors are described in Section 12. Mixture enrichment for cold starting is by a manually-operated choke control.

Most models are equipped with an anti-percolation system, the function of which is to reduce carburettor temperature quickly after the car is stopped. For full details, see Section 15.

The exhaust system is in three sections, connected by clamped or spring-loaded flange joints. The system is suspended throughout its entire length by rubber mountings.

⚠️ **Warning: Many of the procedures in this Chapter require the removal of fuel lines and connections, which may result in some fuel spillage. Before carrying out any** operation on the fuel system, refer to the precautions given in 'Safety first!' at the beginning of this manual, and follow them implicitly. Petrol is a highly-dangerous and volatile liquid, and the precautions necessary when handling it cannot be overstressed.

2 Air cleaner housing assembly - removal and refitting

Removal

1108 cc (C1E) engine

1 Remove the air cleaner filter element as described in Chapter 1.
2 Disconnect the air inlet hose and hot air hose from the air cleaner inlet.
3 Unscrew and remove the mounting bolts, noting the arrangement of the rubber spacers, washers and sleeves.
4 Withdraw the air cleaner body from the engine.

1171 cc (E5F) and 1390 cc (E6J) engines

5 Remove the air cleaner element as described in Chapter 1.
6 Disconnect the hot and cold air inlet hoses from the air cleaner.
7 Lift the air cleaner body off the carburettor.

1721 cc (F2N) engine

8 Disconnect the coil HT lead from the distributor cap.
9 Release the clip, and disconnect the hot air pick-up hose from the elbow on the cylinder head. Also disconnect the outlet hose from the air cleaner housing.
10 Remove the two bolts which secure the air cleaner to the battery tray.
11 Lift out the air cleaner assembly, freeing it from the cold air pick-up hose at the base.

Refitting

12 Refit by reversing the removal operations.

3 Accelerator cable - removal, refitting and adjustment

Removal

Refer to Section 24 in Part B of this Chapter, disregarding references to the idle speed control motor. It will be necessary to remove the air cleaner or the carburettor air inlet duct for access to the throttle quadrant.

Refitting and adjustment

Refitting is a reversal of removal, but if necessary adjust the cable as follows. With the pedal fully released, check that there is a small amount of slack in the cable with the throttle sector on its stop. Have an assistant fully depress the accelerator pedal, then check that the throttle quadrant is in its fully-open position. If adjustment is required, remove the spring clip from the adjustment ferrule, reposition the ferrule as necessary, then insert the clip in the next free groove on the ferrule.

4 Accelerator pedal - removal and refitting

Refer to Section 25 in Part B of this Chapter.

5 Choke cable - removal, refitting and adjustment

1 Disconnect the battery negative lead.
2 Remove the air cleaner or the carburettor air inlet duct as necessary for access to the carburettor.
3 Using a screwdriver or pair of pliers, disconnect the coiled end of the choke cable from the lever on the carburettor (photo).
4 Prise out the clip securing the choke outer cable to the bracket on the side of the carburettor (photo).
5 Working inside the car, remove the single screw which secures the choke control assembly to the facia. Withdraw the control assembly.
6 Disconnect the warning light lead, and the inner and outer cable. Remove the control assembly.
7 Prise out the bulkhead grommet, and withdraw the choke cable into the engine compartment.
8 Refitting is a reversal of removal. Adjust the cable so that with the control knob raised, the choke valve plate on the carburettor is closed. Move the control knob to the position shown (Fig. 4.1), and check that the choke valve plate is fully open. New control assemblies are supplied with a temporary limit stop in the correct position; after adjustment, the control knob must be pushed downwards to break the stop.

5.3 Disconnecting the choke cable inner from the carburettor

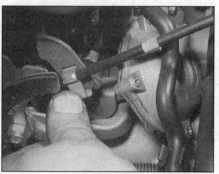

5.4 Releasing the choke cable outer securing clip

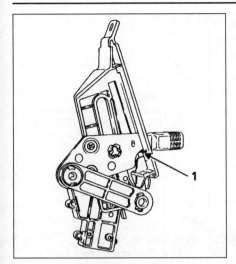

Fig. 4.1 With the choke control knob in this position (1), choke valve plate must be fully open (Sec 5)

6 Unleaded petrol - general information

All carburettor-engined models covered by this manual can run on premium unleaded fuel (octane rating 95), super unleaded (octane rating 98) or super (4-star) leaded fuel (octane rating 97).

7 Fuel pump - testing, removal and refitting

Note: *Refer to the warning note in Section 1 before proceeding.*

Testing

1 To test the fuel pump on the engine, temporarily disconnect the outlet pipe which leads to the carburettor, and hold a wad of rag over the pump outlet while an assistant spins the engine on the starter. *Keep the hands away from the electric cooling fan.* Regular spurts of fuel should be ejected as the engine turns.
2 The pump can also be tested by removing it. With the pump outlet pipe disconnected but the inlet pipe still connected, hold the wad of rag at the outlet. Operate the pump lever or plunger by hand; if the pump is in a satisfactory condition, a strong jet of fuel should be ejected. On the 1108 cc (C1E), 1171 cc (E5F) and 1390 cc (E6J) engines, the pump lever should be moved up and down, whereas on the 1721 cc (F2N) engine, the plunger should be pushed in and out.
3 If a suitable pressure gauge is available, a more accurate test may be carried out. Before connecting the gauge to the fuel system, run the engine at idle speed for several minutes, in order to completely fill the carburettor float chamber. With the engine switched off, disconnect the fuel supply pipe at the carburettor end, and then connect the

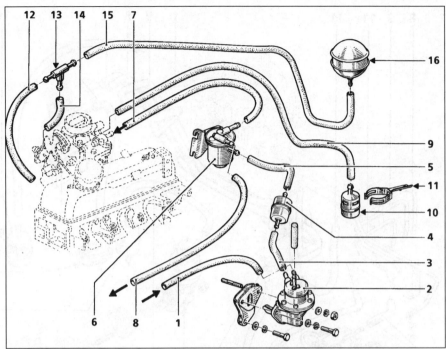

Fig. 4.2 Fuel pump and associated components - 1108 cc (C1E) engine (Sec 7)

1 *Fuel inlet hose*	8 *Fuel return hose*	13 *T-piece*
2 *Fuel pump*	9 *Vent hose*	14 *Vacuum hose (to choke*
3 *Fuel outlet hose*	10 *Vapour reservoir*	*vacuum capsule)*
4 *Fuel filter*	11 *Clip*	15 *Vacuum hose (to*
5 *Connecting hose*	12 *Vacuum hose*	*reservoir)*
6 *Vapour separator*	*(from manifold)*	16 *Vacuum reservoir*
7 *Carburettor inlet hose*		

pressure gauge to it. The gauge connection pipe should be transparent and short. Using a hose clamp, pinch the return pipe leading to the fuel tank. Hold the gauge as high as possible with the pipe vertical, then start the engine and allow it to idle. Lower the gauge until the level of fuel in the transparent pipe is level with the fuel pump diaphragm, then check that the pump static pressure is as given in the Specifications. (Note that the engine is idling, using the fuel already in the float chamber, but there is no fuel movement as the gauge is connected to the pump outlet). Check the return pipe for obstruction by removing the clamp from the return hose and checking that the pressure then drops by 0.01 to 0.02 bar - if not, blow through the return pipe to clear the obstruction.

Removal

4 Disconnect the battery negative lead.
5 Identify the fuel pump inlet and outlet hoses for position, then disconnect and plug them.
6 Unscrew the nuts or bolts securing the pump to the cylinder block (1108 cc/C1E) or cylinder head (all other models), and remove the washers.
7 Withdraw the fuel pump from the engine, and remove the gasket block. On the 1721 cc (F2N) engine, note the number and location of the gaskets.

Refitting

8 Refitting is a reversal of removal, but clean the mating surfaces, and fit a new gasket block. Tighten the nuts/bolts securely on completion.

8 Fuel flow meter - general

1 A fuel flow meter is fitted to some 1171 cc (E5F) and 1390 cc (E6J) models. No information on this item was available at the time of writing.

9 Fuel gauge sender unit - removal and refitting

Refer to Section 31 in Part B of this Chapter. The hose and wiring connections are different (see Fig. 4.5), but the procedure is the same.

10 Fuel tank - removal and refitting

Refer to Section 32 in Part B of this Chapter, ignoring references to the fuel pump and filter.

4

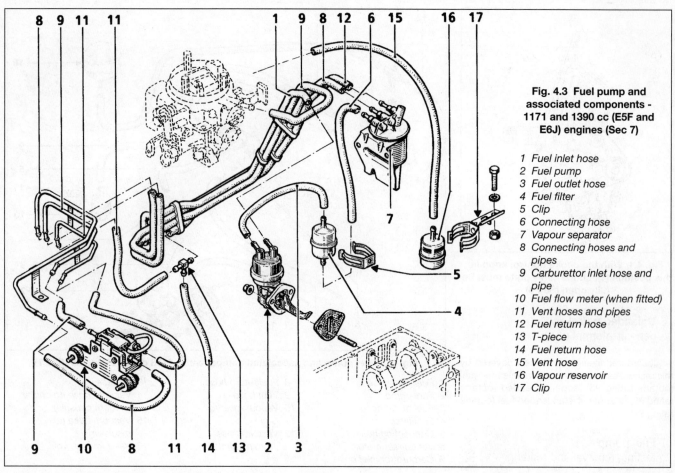

Fig. 4.3 Fuel pump and associated components - 1171 and 1390 cc (E5F and E6J) engines (Sec 7)

1 Fuel inlet hose
2 Fuel pump
3 Fuel outlet hose
4 Fuel filter
5 Clip
6 Connecting hose
7 Vapour separator
8 Connecting hoses and pipes
9 Carburettor inlet hose and pipe
10 Fuel flow meter (when fitted)
11 Vent hoses and pipes
12 Fuel return hose
13 T-piece
14 Fuel return hose
15 Vent hose
16 Vapour reservoir
17 Clip

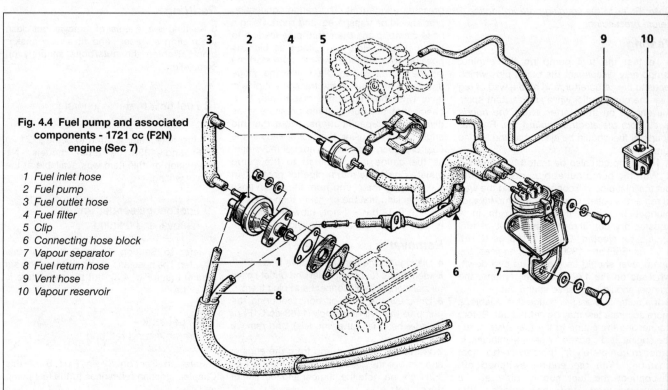

Fig. 4.4 Fuel pump and associated components - 1721 cc (F2N) engine (Sec 7)

1 Fuel inlet hose
2 Fuel pump
3 Fuel outlet hose
4 Fuel filter
5 Clip
6 Connecting hose block
7 Vapour separator
8 Fuel return hose
9 Vent hose
10 Vapour reservoir

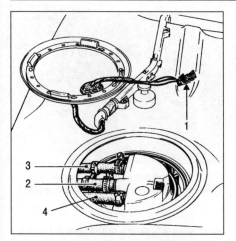

Fig. 4.5 Fuel gauge sender unit connections - carburettor models (Sec 9)

1 *Electrical connector* 3 *Fuel return hose*
2 *Fuel feed hose* 4 *Vent hose*

11 Fuel tank filler pipe - removal and refitting

Refer to Section 34 in Part B of this Chapter.

12 Carburettors - general information

The various carburettors fitted to the models in the Clio range are listed in the Specifications. They are all downdraught carburettors with manual choke. The two smaller engines have single-barrel carburettors (Solex 32 BIS and Pierburg 32 1B1). The larger engines have twin-barrel carburettors (Weber 32 TLDR and Solex 32-34 Z13), the opening of the secondary throttle butterfly being controlled by the vacuum developed in the primary barrel.

To improve fuel vaporisation, the Solex 32 BIS carburettor has a coolant-heated base. The other carburettors have an electric heater in the idle mixture circuit; the 1721 cc (F2N) engine also has a heater in the inlet manifold.

On models with air conditioning and/or power steering, an idle speed compensator is fitted, so that the engine does not stall under increased load from one of these systems.

The Pierburg 32 1B1 carburettor is fitted with an external cold enrichment device, consisting of a solenoid valve which controls a vacuum feed to a mixture corrector diaphragm. When the engine is cold and/or the choke is in operation, the solenoid is energised and the mixture corrector cannot operate. When the engine is warm and the choke control is pushed home, the solenoid circuit is interrupted. The vacuum feed to the mixture corrector is opened, weakening the fuel/air mixture at idle and low-load conditions. This improves fuel economy and reduces exhaust emissions.

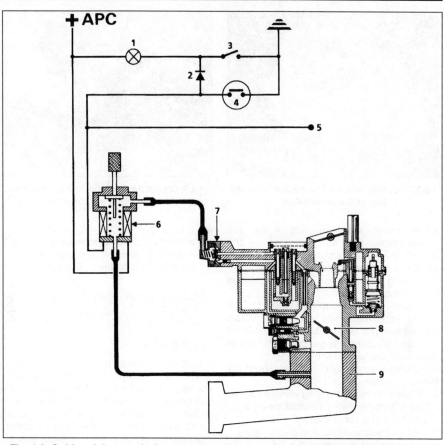

Fig. 4.6 Cold enrichment device operating diagram - Pierburg 1B1 carburettor (Sec 12)

1 *Choke warning light*
2 *Diode*
3 *Choke warning light switch*
4 *Coolant temperature switch*
5 *Test lead (to be earthed when adjusting idle mixture)*
6 *Solenoid valve*
7 *Mixture corrector*
8 *Throttle butterfly*
9 *Inlet manifold*

13 Carburettor - removal and refitting

1 Disconnect the battery earth lead. On models with an anti-percolation system, remove the carburettor cooling duct.
2 On the 1108 cc (C1E) engine, drain the cooling system as described in Chapter 1, then disconnect the coolant hoses from the carburettor.

3 Remove the air cleaner or inlet duct from the top of the carburettor, and place to one side. Remove the gasket (photos).
4 Disconnect the accelerator and choke cables from the carburettor, as described in the relevant Sections of this Chapter.
5 Disconnect the fuel inlet hose, and plug its end (photo).
6 On the 1390 cc (E6J) and 1721 cc (F2N) engines, disconnect the crankcase ventilation hose (photo).

13.3A Remove the bolts . . .

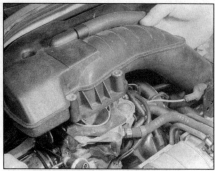

13.3B . . . lift off the air inlet duct . . .

4

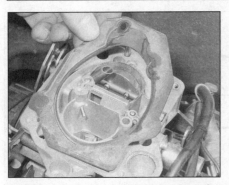

13.3C . . . and recover the gasket from the top of the carburettor

7 Disconnect any remaining wires and vacuum hoses from items such as the anti run-on solenoid, carburettor temperature sender and carburettor heater (photos). Depending on model and equipment, not all these items will be found.

8 Unscrew the mounting nuts, bolts or Torx screws, remove the washers, and withdraw the carburettor from the inlet manifold (photos). Recover the gaskets and, where applicable, the heat shield.

9 Refitting is a reversal of removal, noting the following points:
(a) Make sure that the mating surfaces of the carburettor and inlet manifold are clean, and fit a new set of gaskets.
(b) Adjust the accelerator and choke cables as described in Sections 3 and 5.

13.7A Disconnecting the wiring from the anti run-on solenoid . . .

13.8A Four Torx screws (arrowed) securing the carburettor - 1390 cc (E6J) engine shown

13.5 Disconnecting the fuel inlet hose - 1721 cc (F2N) engine shown

(c) On the 1108 cc (C1E) engine, refill the cooling system as described in Chapter 1.
(d) Adjust the idle speed and mixture as described in Chapter 1, Section 3.

14 Carburettor - fault finding, overhaul and adjustments

Fault finding

1 Problems such as high fuel consumption, uneven idling and poor accelerator response are not necessarily due to faults in the carburettor. Haphazard tinkering with carburettor adjustments is unlikely to do any good. Check first that the ignition system components are in good condition, that the

13.7B . . . and from the carburettor temperature sender. Details will vary with model

13.8B Removing the carburettor

13.6 Disconnecting the crankcase ventilation hose - 1721 cc (F2N) engine shown

valve clearances are correctly adjusted and that the general mechanical condition of the engine is good. Also check that the fuel and air filters are clean.

2 Faults with the carburettor are usually associated with dirt entering the float chamber and blocking the jets, causing a weak mixture or power failure within a certain engine speed range. If this is the case, then a thorough clean will normally cure the problem. If the carburettor is generally worn, uneven running may be caused by air entering through the throttle valve spindle bearings. All the carburettors fitted to the Renault Clio are fitted with manually-operated chokes, which do not normally cause any problems.

3 On carburettors fitted with an anti run-on solenoid, failure of the solenoid or its power supply will prevent the engine from idling smoothly, or at all. Check for a 12-volt supply at the solenoid terminal when the ignition is switched on. There should be an audible click from the solenoid when the ignition is switched on and off.

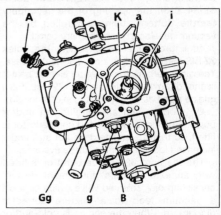

Fig. 4.7 Solex 32 BIS carburettor with cover removed (Sec 14)

a Air correction jet
g Idle jet
i Accelerator pump injector
A Idle speed adjustment screw
B Idle mixture (CO) adjustment screw
Gg Main jet
K Barrel

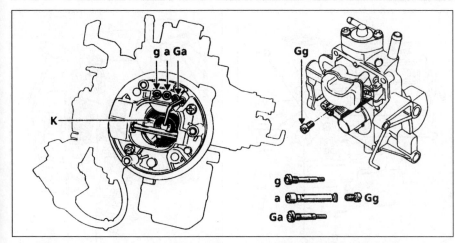

Fig. 4.8 Jet identification and location - Pierburg 32 1B1 carburettor (Sec 14)

a Air correction jet	*Ga Auxiliary jet*
g Idle jet	*Gg Main jet*

K Barrel

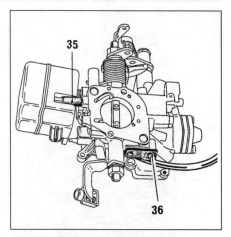

Fig. 4.9 Rubber hose (35) joining vacuum reservoir to carburettor base - Pierburg 1B1 carburettor (Sec 14)

36 Heating element securing screw

Overhaul and adjustments

4 The following paragraphs describe cleaning and adjustment procedures which can be carried out by the home mechanic, after the carburettor has been removed from the inlet manifold. On some carburettors, partial cleaning can be carried out after removal of the top cover, without removing the carburettor itself. This is useful if it is simply a question of clearing a blocked jet, but be careful not to drop small parts into the carburettor barrel.

5 If the carburettor is worn or damaged, it should either be renewed, or overhauled by a specialist who will be able to restore the carburettor to its original calibration.

6 Before commencing work, obtain a repair or overhaul kit for the carburettor in question. This will contain the various gaskets, seals, etc, which must be renewed once they have been disturbed.

7 If it is necessary to remove adjuster screws or nuts, or to disturb adjustable items such as an accelerator pump cam, make alignment marks or take notes so that the original setting can be regained.

Solex 32 BIS

8 Remove the securing screws and lift off the carburettor cover, at the same time

disengaging the choke valve operating linkage.

9 The various jets are shown in the accompanying illustration (Fig. 4.7). Each jet should be removed and identified for position, then the float chamber can be cleaned of any sediment. Clean the main body and the cover thoroughly with fuel, and blow through the carburettor internal channels and jets using air from an air line or foot pump.

10 No adjustment of float level is possible on this carburettor, except by changing the thickness of the washer under the needle valve. The thickness of washer required is given in the Specifications.

11 Reassembly is a reversal of dismantling.

Pierburg 32 1B1

12 Remove the securing screws and lift off the carburettor cover, at the same time disengaging the choke valve operating linkage.

13 The various jets are shown in the accompanying illustration (Fig. 4.8). Each jet

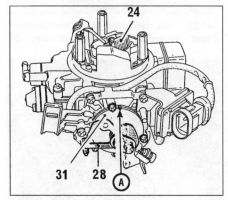

Fig. 4.11 Choke cover and spring casing alignment marks (A) - Pierburg 1B1 carburettor (Sec 14)

24 Choke valve plate   31 Choke cover
28 Choke lever

should be removed and identified for position, then the float chamber can be cleaned of any sediment. Clean the main body and the cover thoroughly with fuel, and blow through the carburettor internal channels and jets using air from an air line or foot pump.

14 Check the condition of the rubber hose which joins the vacuum reservoir to the carburettor base (Fig. 4.9). Renew the hose if necessary.

15 Check the float level by inverting the carburettor cover and measuring the height of the float above the gasket face (Fig. 4.10). No adjustment is possible; if the height is not within the specified limits, inspect the float, needle valve and cover, and renew as necessary. The needle valve seat is pressed into the carburettor cover; if the seat is defective, the cover must be renewed.

16 If the choke spring cover is disturbed for any reason, note the alignment marks between the cover and the spring casing, and ensure that they are correctly positioned on reassembly (Fig. 4.11).

17 Reassembly is a reversal of dismantling.

Weber 32 TLDR

18 Disconnect the vacuum hose from the choke vacuum capsule (photo).

14.18 Choke vacuum capsule hose (arrowed) - Weber 32 TLDR

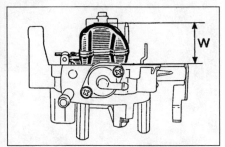

Fig. 4.10 Float level checking dimension - Pierburg 1B1 carburettor (Sec 14)

W = 28.5 ± 1 mm

4

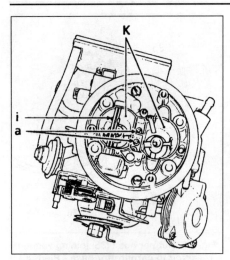

Fig. 4.12 Identification of jets in the carburettor body - Weber 32 TLDR carburettor (Sec 14)

a Air correction jets
i Accelerator pump injector
K Barrels

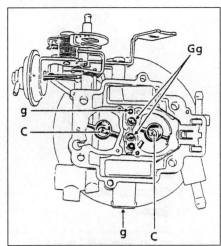

Fig. 4.13 Identification of jets in the carburettor cover - Weber 32 TLDR carburettor (Sec 14)

C Mixture centralising tubes
g Idle jets   Gg Main jets

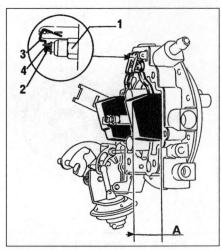

Fig. 4.14 Float level checking dimension - Weber 32 TLDR carburettor (Sec 14)

1 Needle valve
2 Valve ball
3 Bend tag here to adjust
4 End of tag to be square to valve
A = 31 mm

19 Unscrew and remove the two slotted screws from the top of the carburettor cover, and lift the cover from the main body (photo). Remove the gasket from the cover.
20 The various jets are shown in the accompanying illustrations (Fig. 4.12 and 4.13) (photo). Each jet should be removed and

identified for position, then the float chamber can be cleaned of any sediment. Clean the main body and the cover thoroughly with fuel, and blow through the carburettor internal channels and jets using air from an air line or foot pump.
21 With the jets refitted and a new gasket

located on the cover, check the float level setting as follows. Hold the cover vertical so that the floats hang down, and close the needle valve without causing the valve ball to be depressed. Measure the distance between the gasket and the tip of the float, and compare with the dimension given in the Specifications (Fig. 4.14). If adjustment is necessary, bend the tag on the float arm and make the check again.
22 Reassembly is a reversal of dismantling.

Solex 32-34 Z13

23 Remove the screws securing the carburettor cover to the main body (photo).
24 Lift the cover, and at the same time disengage the degassing valve plunger from the operating lever (photo).
25 Refer to the accompanying illustration for the location of the various jets (Fig. 4.15) (photo). Remove each jet and identify its location. Using fuel, thoroughly clean the float chamber, main body and cover. Blow through

14.19 Carburettor cover retaining screws (arrowed) - Weber 32 TLDR

14.20 Underside view of the carburettor cover - Weber 32 TLDR. Main jets arrowed

14.23 Removing the cover retaining screws - Solex 32-34 Z13 carburettor

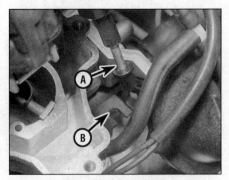

14.24 Disengage the degassing valve plunger (A) from the operating lever (B) - Solex 32-34 Z13 carburettor

14.25 Carburettor body viewed with the cover removed - Solex 32-34 Z13. Main jets arrowed

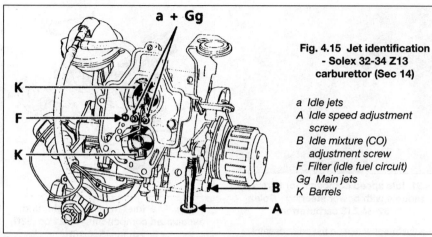

Fig. 4.15 Jet identification - Solex 32-34 Z13 carburettor (Sec 14)

a Idle jets
A Idle speed adjustment screw
B Idle mixture (CO) adjustment screw
F Filter (idle fuel circuit)
Gg Main jets
K Barrels

14.26 Underside view of the carburettor cover - Solex 32-34 Z13. Float fulcrum pin arrowed

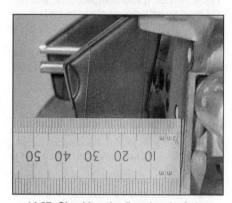

14.27 Checking the float level - Solex 32-34 Z13 carburettor

the carburettor internal channels and jets using air from an air line or foot pump.

26 Note how the needle valve is attached to the float arm, then push out the fulcrum pin, remove the float assembly and valve, and remove the gasket. Renew the gasket and refit the float assembly (photo).

27 To check the float level, hold the cover vertically so that the floats hang down. The needle valve should be closed, but the spring-tensioned ball in the end of the valve should not be depressed. Measure the distance between the gasket and the floats, and compare with the dimension given in the Specifications (photo). If adjustment is necessary, bend the tag on the float arm and make the check again.

28 Reassembly is a reversal of dismantling.

Idle speed compensator adjustments (models with power steering and/or air conditioning)

29 On models equipped with power-assisted steering, air conditioning, or both, an idle speed compensator is fitted, to prevent stalling under the extra load of the steering pump and/or air conditioning compressor. Each compensator receives a vacuum feed controlled by a solenoid valve (Fig. 4.16). If adjustment is necessary, this is done after refitting the carburettor. Proceed as follows.

30 Carry out the normal idle speed adjustment as described in Chapter 1.

31 On models with power-assisted steering, make sure that the front wheels are pointing straight-ahead. Disconnect the vacuum hose from the compensator, then apply a vacuum

4

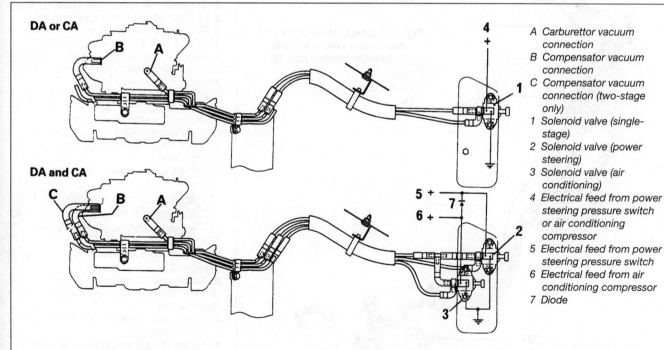

A Carburettor vacuum connection
B Compensator vacuum connection
C Compensator vacuum connection (two-stage only)
1 Solenoid valve (single-stage)
2 Solenoid valve (power steering)
3 Solenoid valve (air conditioning)
4 Electrical feed from power steering pressure switch or air conditioning compressor
5 Electrical feed from power steering pressure switch
6 Electrical feed from air conditioning compressor
7 Diode

Fig. 4.16 Idle speed compensator vacuum hose connections for models with power steering (DA) and/or air conditioning (CA) (Sec 14)

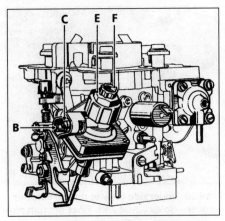

Fig. 4.17 Idle speed compensator details - models with power steering and air conditioning (Sec 14)

B Vacuum hose connection - power steering
C Vacuum hose connection - air conditioning
E Adjustment screw - power steering
F Adjustment screw - air conditioning

of 600 mbars (or manifold vacuum) to the connector. The engine speed must rise to 975 ± 50 rpm. If not, turn the adjustment screw on the compensator (photo). Remake the original vacuum hose connections when adjustment is correct.

32 On models with air conditioning, switch on the air conditioner to its maximum position; the engine speed must rise to 950 rpm. If not, turn

14.31 Idle speed compensator screw for models with power steering - Solex 32-34 Z13 carburettor

the adjustment screw on the compensator.
33 On models with both power steering and air conditioning, there are two adjustment screws on the compensator and two vacuum hose connections (Fig. 4.17). Make the power steering adjustment first, followed by the adjustment for the air conditioning.

15 Anti-percolation system (carburettor models) - general

1 The anti-percolation system is fitted to most carburettor models. Its purpose is to reduce carburettor temperature when the car is stopped after a run. This prevents fuel percolation occurring in the carburettor, so

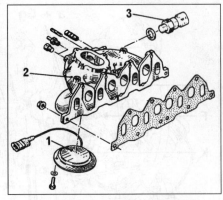

Fig. 4.19 Inlet manifold heater and associated components - 1721 cc (F2N) engine (Sec 16)

1 Heater   2 Manifold   3 Thermal switch

avoiding hot-start problems and fuel vapour emissions.
2 The main components of the system are an electric fan, air ducts leading to the carburettor, and the control circuitry. A combined temperature switch/timer unit actuates the fan for a set period after switch-off, if the temperature exceeds a certain value.
3 No specific testing or repair procedures for the anti-percolation system were available at the time of writing.

16 Inlet manifold heater (1721 cc/F2N engine) - removal and refitting

1 Disconnect the battery earth lead.
2 Remove the carburettor cooling duct.
3 Remove the air inlet duct from the top of the carburettor, and move it to one side.
4 Separate the connector which joins the main wiring harness to the manifold heater wiring.
5 Unbolt the heater from the bottom of the inlet manifold, and lower it through the aperture in the exhaust manifold.
6 The thermal switch which controls the manifold heater is located in the side of the inlet manifold.
7 Refitting is a reversal of removal.

17 Inlet and exhaust manifold assembly (1108 cc/C1E engine) - removal and refitting

1 Remove the carburettor as described in Section 13.
2 Disconnect the brake servo vacuum hose and the crankcase ventilation hose from the inlet manifold.
3 Unbolt and remove the starter motor protective shield.
4 Remove the hot air ducting, then unbolt the metal tube and air cleaner support bracket.
5 Unscrew the nut, and withdraw the heat shield from the exhaust manifold.

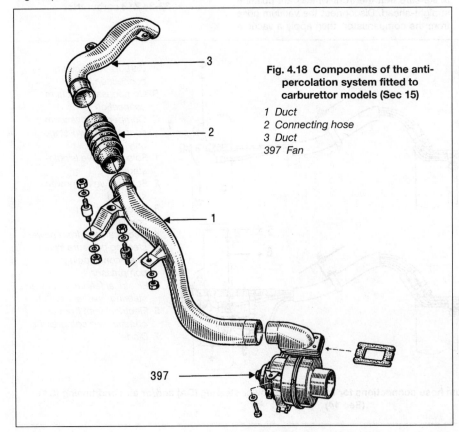

Fig. 4.18 Components of the anti-percolation system fitted to carburettor models (Sec 15)

1 Duct
2 Connecting hose
3 Duct
397 Fan

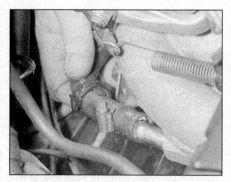

18.2 Disconnecting a coolant hose from the inlet manifold - 1390 cc (E6J) engine shown

18.3 Disconnecting the brake servo vacuum hose

18.4A Removing the inlet manifold complete with carburettor - 1390 cc (E6J) engine shown

6 Unscrew and remove the two nuts, then remove the washers, tension springs and sleeves securing the exhaust front pipe to the manifold. Slide the flange plate off the manifold studs, and separate the joint.

7 Progressively unscrew the nuts securing the manifold assembly to the cylinder head, and remove the washers.

8 Withdraw the manifold assembly from the studs on the cylinder head, then remove the gasket.

9 Refitting is a reversal of removal. Ensure that the cylinder head and manifold mating faces are clean. Use new gaskets on the manifold and carburettor.

18 Inlet manifold (1171 cc/E5F and 1390 cc/E6J engines) - removal and refitting

1 Remove the carburettor as described in Section 13. Alternatively, the inlet manifold may be removed together with the carburettor, but if this method is used, it will still be necessary to disconnect the accelerator and choke cables and the various hoses.

2 Drain the cooling system as described in Chapter 1, then disconnect the coolant hoses from the inlet manifold (photo).

3 Disconnect the brake servo vacuum hose (photo).

4 Progressively unscrew the nuts, then withdraw the inlet manifold from the studs on the cylinder head. Remove the gasket (photos).

5 Refitting is a reversal of removal. Ensure that the cylinder head and manifold mating surfaces are clean, and use a new gasket. Where applicable, refit the carburettor with reference to Section 13. Refill the cooling system with reference to Chapter 1.

19 Exhaust manifold (1171 cc/E5F and 1390 cc/E6J engines) - removal and refitting

1 Unscrew and remove the two nuts, then remove the washers, tension springs and sleeves securing the exhaust front pipe to the manifold (photo). Slide the flange plate off the manifold studs, and separate the joint.

2 Loosen the clip and disconnect the hot air hose for the air cleaner from the exhaust manifold hot air shroud (photo).

3 Unscrew the nuts and remove the hot air shroud from the manifold (photo).

4 Progressively unscrew the nuts and bolts securing the exhaust manifold, then withdraw it from the cylinder head. Recover the manifold gasket (photos).

5 Refitting is a reversal of removal. Ensure that the cylinder head and manifold mating surfaces are clean, and use a new gasket.

18.4B Recover the manifold gasket. Use a new gasket on reassembly

19.1 Unscrewing the exhaust front pipe-to-manifold nuts - note the arrangement of the tension springs and sleeves

4

19.2 Disconnecting the hot air hose from the exhaust manifold shroud - 1390 cc (E6J) engine shown

19.3 Removing the hot air shroud from the manifold

19.4A Unscrew the nuts and bolts (arrowed) . . .

19.4B . . . remove the exhaust manifold . . .

19.4C . . . and recover the gasket. Use a new gasket on reassembly

20 Inlet and exhaust manifolds (1721 cc/F2N engine) - removal and refitting

1 Remove the carburettor as described in Section 13.
2 Drain the cooling system as described in Chapter 1, then disconnect the coolant hoses from the inlet manifold.

3 Disconnect the brake servo vacuum hose from the inlet manifold.
4 Unscrew and remove the two nuts, then remove the washers, tension springs and sleeves securing the exhaust front pipe to the manifold. Slide the flange plate off the manifold studs, and separate the joint.
5 Unscrew the nuts securing the hot air shroud to the manifold, and remove the shroud.

6 Unbolt and remove the manifold support brackets.
7 Progressively unscrew the nuts and bolts securing the inlet and exhaust manifolds, and withdraw them from the cylinder head. Although the manifolds are separate, they are retained by the same bolts, since the bolt holes are split between the manifold flanges. Recover the manifold gasket.
8 Refitting is a reversal of removal. Ensure that the cylinder head and manifold mating surfaces are clean, and use a new gasket. Refit the carburettor with reference to Section 13. Refill the cooling system with reference to Chapter 1.

21 Exhaust system - general information and component renewal

Refer to Section 42 in Part B of this Chapter, but disregard references to the catalytic converter.

Part B: Fuel injection models

22 General information and precautions

This part of Chapter 4 deals with the fuel injection systems and associated components fitted to the Renault Clio. There are two fuel injection systems: a single-point system fitted to the 1171, 1390 and 1794 cc (E7F, E7J and F3P) engines, and a multi-point system fitted to the 1764 cc (F7P) 16-valve engine. The systems are described in detail in Section 27.

The fuel pump on all injection models is electric; it is located inside the fuel tank. As with the carburettor models, fuel circulates continuously when the pump is operating, excess fuel being returned to the tank. Sixteen-valve models have an auxiliary fuel tank and pump - see Section 33.

The exhaust system is conventional. On nearly all models, it incorporates a catalytic converter, to reduce emissions of toxic gases. The catalytic converter and other emission control components are described in the final part of this Chapter.

⚠️ **Warning: Many of the procedures in this Chapter require the removal of fuel lines and connections, which may result in some fuel spillage (on fuel injection models, refer to Section 28 and depressurise the fuel system before disconnecting any fuel lines or hoses). Before carrying out any operation on the fuel system, refer to the precautions given in 'Safety first!' at the beginning of this manual, and follow them implicitly. Petrol is a highly-dangerous and volatile liquid, and the precautions necessary when handling it cannot be overstressed.**

23 Air cleaner housing assembly - removal and refitting

1171 cc (E7F) and 1390 cc (E7J) engines

1 Disconnect the hot and cold air intake ducts from the air cleaner housing.
2 Remove the three screws which secure the air cleaner to the throttle housing. Lift off the air cleaner, and recover the sealing ring.
3 Refit by reversing the removal operations, using a new sealing ring if necessary.

1794 cc (F3P) engine

4 Disconnect the coil HT lead from the distributor cap.
5 Release the clip, and disconnect the hot air pick-up hose from the elbow on the cylinder head. Also disconnect the outlet hose from the air cleaner housing (photos).
6 Disconnect the temperature control vacuum hose from the valve on the air intake housing.
7 Remove the two bolts which secure the air cleaner to the battery tray (photo).
8 Lift out the air cleaner assembly, freeing it from the cold air pick-up hose at the base (photo).
9 If it is necessary to remove the cold air pick-up hose, access is gained by removing the front part of the wheel arch splash shield. The hose can then be unclipped (photo).
10 Refit by reversing the removal operations.

23.5A Disconnect the hot air pick-up hose from the elbow . . .

23.5B . . . release the clip (arrowed) and disconnect the outlet hose

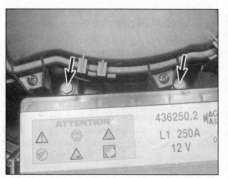

23.7 Remove the two bolts (arrowed) which secure the air cleaner to the battery tray

23.8 Removing the air cleaner assembly

23.9 The end of the cold air pick-up hose is clipped in position inside the front wheel arch

1764 cc (F7P) engine

11 Remove the air cleaner element (Chapter 1, Section 21).

12 Remove the nut which secures the air cleaner housing. Free the housing from the air intake duct, and withdraw it.

13 Refit by reversing the removal operations.

24 Accelerator cable - removal, refitting and adjustment

Note: *Correct adjustment of the accelerator cable is essential to the operation of the fuel injection system. In particular, a stable idle speed will not be obtained if there is insufficient slack in the cable.*

Removal and refitting

1 Working under the bonnet, turn the throttle quadrant by hand to release the cable tension. Disconnect the inner cable from the quadrant (photo).

2 Pull the outer cable and grommet from the locating bracket (photo).

3 On automatic transmission models, disconnect the kickdown switch.

4 Working inside the car, disconnect the cable from the accelerator pedal by squeezing the lugs of the cable end fitting (photo).

5 Withdraw the cable into the engine bay, and remove it. On some right-hand-drive models, the cable is routed through the engine right-hand mounting cover; in this case, the cover

24.1 Disconnecting the inner cable from the throttle quadrant

must be unbolted in order to free the cable (photo).

6 Refit by reversing the removal operations. Adjust the cable if necessary as follows.

Adjustment

7 Cable adjustment is carried out by altering the position of the hairpin clip on the outer cable. Adjustment is correct when the throttle is fully open with the pedal depressed, with some slack in the cable when the pedal is released.

8 On single-point injection models, the idle speed control motor opens the throttle slightly when the ignition is switched off. There must be enough slack in the cable to allow the throttle to close past this position, otherwise a stable idle speed will not be obtained.

24.2 Withdraw the outer cable and grommet from the locating bracket

8 On automatic transmission models, the kickdown switch should only operate when the pedal is depressed beyond the wide-open throttle position. Correct adjustment of this switch requires the use of special test equipment. If its function is in doubt, consult a Renault dealer.

25 Accelerator pedal - removal and refitting

1 Remove the nut which secures the accelerator pedal pivot bush (photo).

2 Disconnect the accelerator cable from the pedal as described in the previous Section.

3 Remove the pedal and bush.

4 Refit by reversing the removal operations.

24.4 Accelerator cable connection to pedal

24.5 On this model, the engine right-hand mounting cover must be removed to free the accelerator cable

25.1 Remove the nut (arrowed) which secures the accelerator pedal pivot bush

4

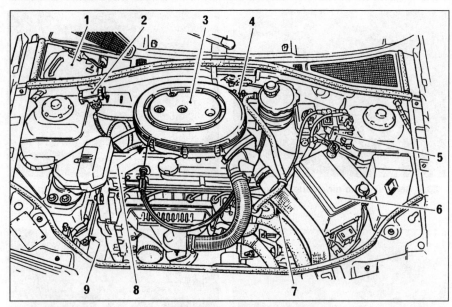

Fig. 4.20 Location of major components of the single-point injection system. 1390 cc (E7J) engine shown; other versions similar (Sec 27)

1 Computer and relays
2 MAP sensor
3 Air cleaner
4 Evaporative emission control solenoid valve
5 Ignition module
6 Battery
7 Engine speed and position (flywheel) sensor
8 Throttle housing
9 Evaporative emission control canister

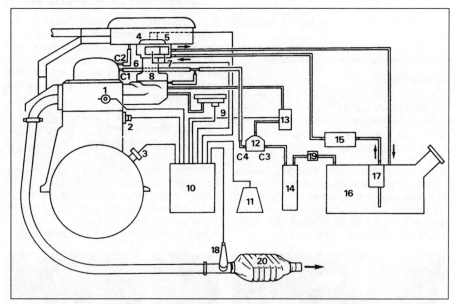

Fig. 4.21 Schematic showing relationship of the components of the single-point injection system (Sec 27)

1 Coolant temperature sensor
2 Knock sensor (when fitted)
3 Engine speed/position (flywheel) sensor
4 Injector
5 Air temperature sensor
6 Throttle potentiometer
7 Idle speed control motor
8 Throttle housing
9 MAP sensor
10 Computer
11 Ignition module
12 Vacuum bleed valve
13 Evaporative emission control solenoid valve
14 Evaporative emission control canister
15 Fuel filter
16 Fuel tank
17 Fuel pump
18 Oxygen sensor
19 Non-return valve
20 Catalytic converter
C1 Restrictor
C2 Restrictor
C3 Restrictor
C4 Restrictor

26 Unleaded petrol - general information

All fuel injection models sold in the UK must only use unleaded petrol. Leaded fuel must **not** be used, as it will damage the catalytic converter. The fuel filler neck is designed to accept only unleaded pump nozzles.

All models can use the 95 octane unleaded fuel available throughout the UK. Super unleaded (98 octane) can also be used, though there is no advantage in doing so. Some models can use 91 octane unleaded fuel, where available. The minimum octane rating is given in the Specifications.

A 16-valve model without a catalytic converter (vehicle code C575, engine code F7P-720) is sold in some markets. This model can be run on 97 octane leaded fuel if wished.

27 Fuel injection systems - general information

Single-point system

This system is fitted to models with the 1171, 1390 and 1794 cc (E7F, E7J and F3P) engines. It is essentially the same on all three engine sizes; detail differences are given in the Specifications.

The system can be considered as a halfway stage between a conventional fixed-jet carburettor and an electronically-controlled multi-point injector system. As with a carburettor, fuel metering and vaporisation take place in one unit (the throttle housing), and the fuel/air mixture is then distributed to the cylinders via the inlet manifold. As with any modern petrol injection system, fuel metering is controlled electronically by a computer, which receives information from

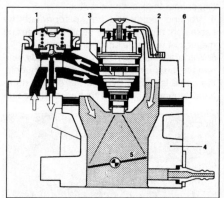

Fig. 4.22 Throttle housing components - single-point injection system (Sec 27)

1 Fuel pressure regulator
2 Air temperature sensor
3 Injector
4 Lower half-casing
5 Throttle butterfly
6 Upper half-casing

various sensors. The computer also controls the idle speed, and manages the emission control and ignition systems.

The main advantage of single-point injection over a carburettor is that the injection system allows much more precise control over the air-fuel ratio (mixture) under all operating conditions. This control is necessary to keep exhaust emissions to a minimum, and for the correct operation and long life of the catalytic converter. Additional benefits from the driver's point of view are improved fuel economy under all conditions, and better driveability when cold.

The system is best understood if considered in two parts: the fuel supply and the control system.

Fuel supply

The fuel supply sub-system consists of an electric fuel pump immersed in the tank, a fuel filter, a fuel pressure regulator and a fuel injector. These last two items are located in the throttle body.

Fuel passes from the pump through the filter to the throttle body, where it enters the fuel pressure regulator. The pressure regulator maintains a constant pressure in the injector, relative to the pressure in the throttle body upstream of the butterfly. Excess fuel is returned to the tank. Fuel is therefore circulating all the time that the pump is running, which ensures a constant supply of cool fuel, and avoids problems with vapour-locks.

The fuel injector consists of a needle valve, which is opened by an electromagnet and closed by a spring. The electromagnet is controlled by the injection computer. When the valve is open, fuel is sprayed from the injector onto the throttle plate. Because the relative fuel pressure is constant, the quantity of fuel injected varies directly according to the opening time.

The electrical feed to the fuel pump is via a relay controlled by the injection computer. The pump is only energised when the engine is running or when the starter motor is operating. Some procedures in this Chapter require the pump to be running when the engine is stopped; this is achieved by bridging two terminals in the relay socket (see Section 29).

Control

The injection control sub-system consists of the injection computer and its associated sensors and actuators. As mentioned earlier, the computer also controls the emission control and ignition systems; there is some overlap between these functions.

The computer is located in the engine bay, in the right-hand scuttle. (On models with air conditioning, the computer is located inside the car, below the glovebox.) It receives signals from sensors which monitor the following functions:

Coolant temperature.
Inlet air temperature.
Engine speed (flywheel sensor).
Road speed (speedometer cable sensor) - not on all models.
Inlet manifold pressure (MAP sensor).
No-load condition (idle switch).
Throttle position and full-load condition (throttle potentiometer).
Exhaust gas oxygen (oxygen sensor).

On the two larger engines, the computer also receives signals from a knock sensor which detects pre-ignition (pinking). This belongs to the ignition side of the system, and is covered in Chapter 5. The oxygen sensor is properly part of the emission control system, and is covered in detail in Part C of this Chapter.

Outputs from the computer control the operation of the following components:

Fuel pump (via the fuel pump relay).
Fuel injector.
Idle speed control motor.
Evaporative emission control solenoid valve.
Ignition power module.

The computer processes the information from the various sensors, to determine the quantity of fuel which must be injected to suit operating conditions at any given moment. For this purpose, the volume of inlet air is calculated from the engine speed and throttle position signals, with a correction being made for inlet air temperature. Extra enrichment is provided at full-load and during warm-up. Compensation is also made for the effect of changes in battery voltage.

In the no-load condition (throttle lever pressing on the idle switch), the computer cuts off injection at engine speeds above idle. At idle, the engine speed is controlled by the computer sending signals to the idle speed control motor incorporated in the throttle housing. The motor opens or closes the throttle butterfly, to maintain a constant idle speed regardless of engine temperature or additional loads (alternator, steering pump, etc). There is no need for periodic adjustment of idle speed or mixture, even to compensate for long-term factors such as engine wear. The system is also self-correcting for changes in altitude.

When stopping the engine, the idle speed control motor first closes the throttle completely, then opens it again in readiness for the next start. This means that with the engine stopped there will be more slack in the accelerator cable than with the engine idling.

Multi-point system

This system is fitted to models with the 1764 cc (F7P) engine. Two versions of the system exist: one with an oxygen sensor and catalytic converter, the other without. On the version without a catalytic converter, adjustment of the exhaust gas CO content (mixture) is possible. Refer to Chapter 1, Section 9, for details.

In many respects, the multi-point system is similar to the single-point system just described. The major differences are as follows.

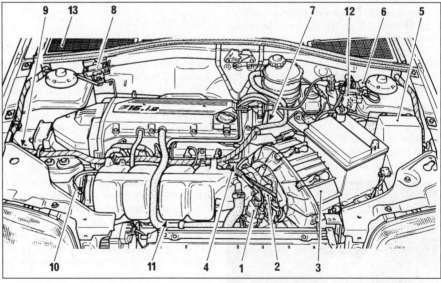

Fig. 4.23 Location of major components of the multi-point injection system (Sec 27)

1 Idle speed control valve
2 Throttle potentiometer
3 Air cleaner
4 Injector rail and pressure regulator
5 Relay box
6 CO (mixture) adjustment potentiometer (when fitted)
7 Coolant temperature sensor
8 MAP sensor
9 Evaporative emission control solenoid valve (when fitted)
10 Air temperature sensor
11 Cold start injector (when fitted)
12 Ignition module
13 Computer (under cowl panel)

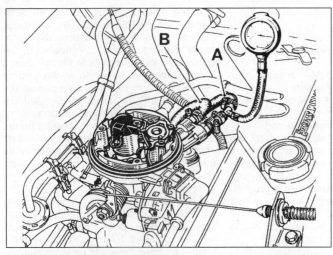

Fig. 4.24 Fuel system pressure check - single-point injection
(Sec 29)

A Fuel feed pipe   B Fuel return pipe

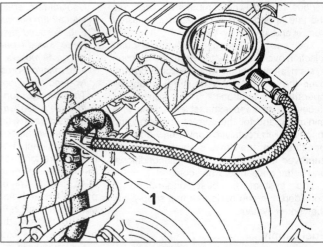

Fig. 4.25 Fuel system pressure check - multi-point injection
(Sec 29)

1 T-piece

Fuel supply

Four fuel injectors, one per cylinder, are used. The injectors are located in the inlet manifold, downstream of the throttle butterfly. On versions with a catalytic converter, a single cold start injector provides additional fuel when the starter motor is operating and coolant temperature is below 20°C. All the injectors are fed from a common rail, which also carries the fuel pressure regulator. As with the single-point system, excess fuel is returned to the tank.

Control

The computer receives information from various sensors, as detailed for the single-point system.

The idle speed regulation system is slightly different. An idle speed control valve opens or closes an auxiliary air circuit, bypassing the throttle butterfly. There is no idle switch.

Fault finding - all systems

In the event of a fault developing, the amount of diagnostic work possible for the home mechanic is limited. Renault technicians use dedicated test equipment which can find

29.4 Fuel pump relay (arrowed) is in the relay box on this model. On other models, it may be in the computer case

a fault quickly and precisely. Initial fault-finding checks should concentrate on external factors such as a blocked fuel filter, loose or corroded electrical connections and loose or damaged vacuum hoses. Remember also that incorrect adjustment of the accelerator cable can affect operation of the fuel injection system, especially at idle.

If it appears that the fault is in one of the sensors or in the computer, it will probably be more satisfactory to consult a Renault dealer or other specialist than to embark on a hit-and-miss programme of component renewal.

28 Fuel system - depressurisation

Note: Refer to the warning at the end of Section 22 before proceeding.

1 Fuel lines may contain fuel under pressure, even though the engine is not running. Depressurisation of the fuel system is therefore necessary before undertaking any work which will involve opening a fuel line or union, otherwise fuel under pressure could spray out onto hot engine components.
2 Remove the fuel pump fuse, which is No 14 in the under-bonnet (auxiliary) fusebox.
3 Try to start the engine. It may run briefly and die, or it may not start at all. Operate the starter motor a couple more times, to ensure that all fuel pressure has been relieved.
4 Switch off the ignition and refit the fuel pump fuse.

29 Fuel system - pressure check

Note: Refer to the warning at the end of Section 22 before proceeding.

1 Depressurise the fuel system (Section 28), remembering to refit the fuel pump fuse.

Single-point system

2 Remove the air cleaner housing or the air inlet housing, as applicable.
3 Connect a pressure gauge (range 0 to 3 bars) into the fuel feed pipe where it enters the throttle housing, using a T-piece.
4 Make the fuel pump run by bridging terminals 3 and 5 of the fuel pump relay (located in the computer case or in the relay box in front of the left-hand suspension turret) (photo). The fuel pump relay can be distinguished by the thick leads connected to terminals 3 (red or red/white lead) and 5 (brown or white lead).
5 Note the pressure shown on the gauge. It should correspond to the regulated fuel pressure given in the Specifications.
6 Momentarily pinch the fuel return pipe. Note the pressure registered on the gauge: this must correspond to the unregulated pressure given in the Specifications.

Multi-point system

7 Connect a pressure gauge (range 0 to 6 bars) into the fuel feed pipe where it joins the injector rail, using a T-piece.
8 Make the fuel pump run by bridging terminals 3 and 5 of the fuel pump relay (located in the relay box in front of the left-hand suspension turret).
9 Note the pressure shown on the gauge. It should correspond to the regulated fuel pressure given in the Specifications.
10 Momentarily pinch the fuel return pipe. Note the pressure registered on the gauge: this must correspond to the unregulated pressure given in the Specifications.
11 If a vacuum pump and gauge are available, disconnect the vacuum hose from the pressure regulator. Apply vacuum of approximately 500 mbars to the pressure regulator, and check that the fuel pressure drops by the same amount.

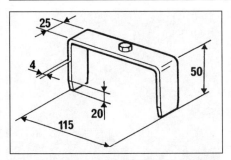

Fig. 4.26 Ring nut spanner tool dimensions (in mm) (Secs 30 and 31)

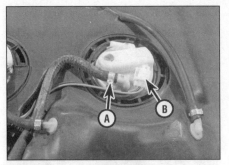

30.10 Disconnect the fuel hose (A) and the multi-plug (B) from the top of the fuel pump

30.11A Unscrewing the fuel pump ring nut using an adjustable peg spanner

All systems

12 If the pressure readings are low, check the pump delivery (Section 30) to see whether the fault is in the pump or the pressure regulator. If the readings are high, the fault can only be in the pressure regulator.

13 Remove the bridge from the fuel pump relay. Depressurise the fuel system and disconnect the pressure gauge, then remake the original connections.

30 Fuel pump -
testing, removal and refitting

Note: *Refer to the warning at the end of Section 22 before proceeding.*

Testing

1 Carry out a pressure check as described in the previous Section.

2 Disconnect the fuel return pipe from the throttle housing or injector rail. Connect a length of flexible hose to the return union, and place the other end of the hose in a measuring jar (capacity 2 litres approx).

3 Make the fuel pump run by bridging terminals 3 and 5 of the fuel pump relay as described in the previous Section.

4 Note the quantity of fuel delivered in exactly one minute (single-point system) or 30 seconds (multi-point system). Correct values are given in the Specifications.

5 Remove the bridge from the fuel pump relay. Return the pumped fuel to the tank, or put it in a suitable sealed container.

6 If delivery is low, check that full battery voltage is available at the pump relay contacts, and at the pump itself. (The pump wiring connector is accessible through the fuel gauge sender hatch in the floor.) A 1 volt drop in supply voltage will result in a drop of approximately 10% in delivery.

7 If the supply voltage is correct but delivery is still low, there are two possible reasons. Either there is a blockage somewhere in the supply line (including the main fuel filter and the pump pick-up filter), or the pump is defective. If the pump is defective, it must be renewed.

8 On completion, remake the original fuel pipe connections.

Removal

9 Remove the fuel tank (Section 32).

10 Disconnect the fuel hose and the multi-plug from the top of the pump (photo).

11 Unscrew the ring nut, using a tool made to the dimensions shown in Fig. 4.26 or an adjustable peg spanner such as that shown. The tool must fit the slots snugly to avoid damage. Remove the ring nut and withdraw the pump. Recover the seal (photos).

12 Release the retaining clips. Note which way round the wires are connected, then disconnect them and remove the fuel pump.

Refitting

13 Refit by reversing the removal operations, making sure that the wires are connected the right way round. Use a new seal and (if

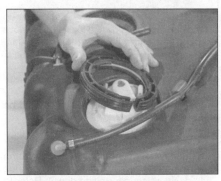

30.11B Remove the ring nut . . .

possible) tighten the ring nut to the specified torque.

31 Fuel gauge sender unit -
removal, testing and refitting

Note: *Refer to the warning at the end of Section 22 before proceeding.*

Removal

1 Disconnect the battery earth lead. Tilt the rear seat, and remove the access cover from the fuel gauge sender hatch.

2 Disconnect the hose(s) and the multi-plugs from the top of the sender unit, making marks for reference when refitting if there is any possibility of confusion (photo).

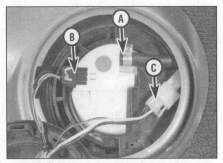

31.2 Fuel gauge sender unit connections on a fuel injection model - breather hose (A), sender multi-plug (B) and fuel pump connector (C)

30.11C . . . withdraw the pump . . .

30.11D . . . and recover the seal. Note the position of the cut-out (arrowed)

4

31.3 Removing the fuel gauge sender unit (fuel tank removed)

32.5 Fuel filler pipe, breather and vent hoses

32.7 Removing the fuel tank

3 Unscrew the ring nut, using a tool made up to the dimensions shown in Fig. 4.26. Remove the ring nut and withdraw the sender unit, allowing any fuel contained in it to drain back into the tank (photo).

Testing

4 Invert the sender unit: the float should be heard to move.
5 Connect an ohmmeter across the terminals, two at a time, and observe the changes in resistance between the 'empty' position (sender upright and the right way up) and the 'full' position (sender upside-down). Refer to Fig. 4.27 for terminal identification. The following results were obtained in the workshop:

Terminals	'Empty'	'Full'
A - B	10 ohms	Infinity (open-circuit)
A - C	290 ohms	0.7 ohms
B - C	300 ohms	Infinity (open-circuit)

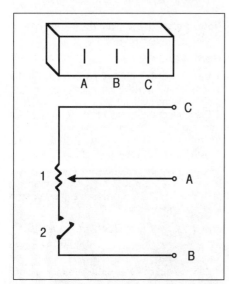

Fig. 4.27 Identification of fuel gauge sender unit terminals (viewed with sender in upright position) and schematic of internal connections (Sec 31)

A Warning light switch   1 Potentiometer
B Common   2 Switch
C Gauge

6 If the readings obtained do not change as described, and if erratic operation of the fuel gauge or warning light has been experienced, the sender unit should be renewed.

Refitting

7 Refit by reversing the removal operations. Use a new seal and (if possible) tighten the ring nut to the specified torque.

32 Fuel tank - removal and refitting

Note: *Refer to the warning at the end of Section 22 before proceeding.*
1 Disconnect the battery earth lead. Disconnect the hose(s) and the multi-plug(s) from the fuel gauge sender unit (Section 31).
2 Siphon the fuel from the tank into a suitable container. (On 16-valve models, the fuel must be pumped out as described in the next Section.)
3 Raise and support the car. Remove the exhaust system (Section 42) and the centre section heat shield. On models with ABS and/or an open-bar rear axle, the rear section of the exhaust cannot be removed without cutting it; in this case, remove the downpipe and the centre section.
4 Disconnect the handbrake cables from the equaliser. Unclip the cables and move them out of the way.
5 Disconnect the fuel filler pipe from the tank. Also disconnect the breather and vent hoses

33.1 Auxiliary fuel tank and pump fitted to 16-valve models

(photo). If crimped hose clips have been used, cut them off and obtain worm-drive clips for reassembly.
6 Disconnect the outlet hose from the fuel filter. Be prepared for fuel spillage.
7 Have an assistant support the fuel tank. Remove the five mounting bolts and lower the tank, freeing it from the brake pipe clips. Disconnect any remaining wires or hoses and remove the tank (photo).
8 The tank is made of plastic. If it is damaged, it should be renewed. Proprietary repair kits are available, but check that they are suitable for use on plastic tanks. A fuel leak is not just expensive, it is dangerous.
9 Refit by reversing the removal operations. Adjust the handbrake as described in Chapter 1.

33 Auxiliary fuel tank and pump (16-valve models) - description, removal and refitting

Description

1 There are two fuel tanks on 16-valve models: a main tank and an auxiliary tank (photo). The auxiliary tank is at a lower level than the main tank. When filling with fuel, the fuel flows down the filler pipe into the auxiliary tank, then through a connecting pipe into the main tank. A flap at the end of the connecting pipe prevents fuel running back out of the main tank.
2 In normal operation, fuel is drawn from the main tank by the main fuel pump. When the level in the main tank falls to a certain point, the auxiliary fuel pump runs for 40 seconds to transfer fuel from the auxiliary to the main tank. At the same time, the 'low level' warning light illuminates for 5 seconds. The auxiliary pump will only operate when the engine is running, or when the starter motor is cranking.
3 If both tanks are allowed to run dry, it will be necessary to pour at least 10 litres (over 2 gallons) of fuel in through the filler before any will reach the main tank. If this quantity of fuel is not available, it will be necessary to crank the engine on the starter motor for about 60 seconds until the auxiliary pump has

33.8 Auxiliary fuel pump electrical connector (arrowed)

transferred the fuel. (A similar problem can be avoided after intentional draining of the main tank by adding a few litres of fuel via the fuel gauge sender hole.)

Removal

4 Disconnect the fuel feed pipe from the injector rail; be prepared for fuel spillage. Connect a length of flexible fuel hose to the feed pipe, and place the other end of the hose in a suitable container.

5 Activate the fuel pumps by bridging terminals 3 and 5 on the fuel pump relay: fuel will be pumped into the container. Remove the bridging wire when fuel stops flowing.

 Caution: Do not allow the main fuel pump to run dry for more than a few seconds.

6 Disconnect the battery earth lead. Reconnect the fuel feed pipe to the injector rail.

7 Raise and support the vehicle. Remove the spare wheel and both rear wheels.

8 Remove the two bolts which secure the auxiliary fuel pump. Disconnect the electrical feed and the hoses from the auxiliary pump (photo). Be prepared for fuel spillage.

 Warning: If the auxiliary pump has not been working, the auxiliary tank may still be full of fuel.

9 Disconnect the pipe which joins the auxiliary and main tanks. Also disconnect the filler and breather pipes.

10 Remove the auxiliary tank mounting bolts (Fig. 4.28) and lower the tank from under the vehicle.

Refitting

11 Refit by reversing the removal operations, using new hose clips when necessary.

12 Prime the main fuel tank if necessary by pouring a few litres of fuel in through the fuel gauge sender unit hole.

34 Fuel tank filler pipe - removal and refitting

Note: *Refer to the warning note in Section 22 before proceeding.*

1 A drain plug is not provided on the fuel tank, and it is therefore preferable to carry out the removal operation when the tank is nearly

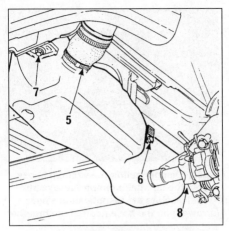

Fig. 4.28 Auxiliary fuel tank fastenings (Sec 33)

5 *Fuel filler pipe*   7 *Mounting bolt*
6 *Mounting bolt*   8 *Mounting bolt*

empty. Before proceeding, disconnect the battery earth lead, then syphon or pump the remaining fuel from the tank.

2 Chock the front wheels, then jack up the rear of the car and support it on axle stands. Remove the spare wheel and the right-hand rear wheel. Also remove the mudflap.

3 Loosen the clip and disconnect the filler hose from the filler pipe.

4 Remove the two screws which secure the breather pipe assembly.

5 Mark the positions of the breather pipes, then disconnect them from the filler pipe.

6 Open the fuel filler flap and remove the cap.

7 Unscrew the cross-head screws located inside the filler flap recess, and remove the filler pipe assembly.

8 Refit by reversing the removal operations, using new hose clips when necessary.

35 Single-point fuel injection system - component testing

1 Renault dealers use dedicated test equipment (test box XR25) which plugs into the diagnostic socket provided on the main fuse/relay carrier in the passenger compartment. This equipment provides a rapid and comprehensive diagnosis of the complete system. It is the only means of checking the operation of the computer.

2 Limited testing of individual components is still possible. The information in the following paragraphs is given as a guide to the owner who wishes to carry out preliminary checks if a fault develops. However, it may still be necessary to enlist the help of a Renault dealer or other specialist for positive diagnosis of a fault.

3 Where the use of an electrical multi-meter is called for, this should be a modern digital type, with an internal resistance of at least 10 Megohms. **Do not** use analogue (moving pointer) meters, which may have an unacceptably-low internal resistance.

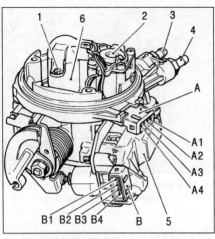

Fig. 4.29 Terminal and component identification - single-point injection system (Sec 35)

1 *Air temperature sensor*
2 *Fuel pressure regulator*
3 *Fuel return connection*
4 *Fuel feed connection*
5 *Idle speed control motor*
6 *Injector*
A *Air temperature sensor/injector connector*
A1 *Air temperature sensor*
A2 *Injector +*
A3 *Injector -*
A4 *Air temperature sensor*
B *Idle speed motor/idle switch connector*
B1 *Motor feed*
B2 *Motor feed*
B3 *Idle switch*
B4 *Idle switch*

4 No meaningful testing is possible on the idle speed control motor, the throttle potentiometer, the manifold pressure (MAP) sensor, the road speed sensor (when fitted), or the computer itself.

5 Before testing individual components, make sure that all system electrical connections are clean and secure, and that all fuel, air and vacuum hoses are in good condition. Also remember that problems with idle speed regulation may simply be due to incorrect accelerator cable adjustment.

6 Where a component is found to be defective, it must be renewed as described in the following Section.

Inlet air temperature sensor

7 Remove the air cleaner or the air intake housing, as applicable.

8 Disconnect the inlet air temperature sensor multi-plug.

9 Use an ohmmeter on terminals A1 and A4 (Fig. 4.29) to measure the resistance of the temperature sensor. Desired values are given in the Specifications. Use a hairdryer or a fan heater to raise the sensor temperature, and check that the resistance of the sensor falls as it warms up.

4

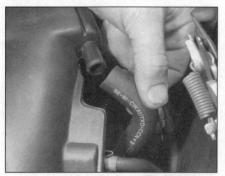

36.1 Disconnecting the breather hose from the air inlet housing

36.2 Disconnecting the vacuum hoses from the air intake temperature control valve. Note the identification sleeve (arrowed) on the hose nearest the breather

36.3 Remove the three Torx screws which secure the air inlet housing

Coolant temperature sensor

10 Refer to Chapter 3, Section 8 for testing, and to the Specifications at the start of this Chapter for resistance values.

Fuel injector

11 The resistance of the injector winding can be checked at the injector connector (terminals A2 and A3 in Fig. 4.29).

12 If the resistance is not as specified, remove the inlet air temperature sensor (Section 36) and repeat the measurement at the terminals of the injector itself.

13 If the resistance is now correct, there was a fault in the wiring between the multi-plug

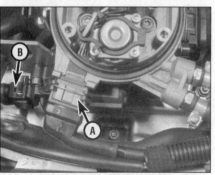

36.4 Air inlet trunking is secured to the housing by a hose clip (arrowed)

and the injector. If the resistance is still incorrect, the injector is defective.

14 If it is suspected that dirt or gum in the injector is impairing performance, the use of a proprietary cleaning additive in the fuel tank may be effective.

Fuel pressure regulator

15 Testing of the fuel pressure regulator is included in the fuel pressure check (Section 29).

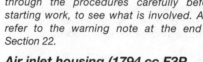

36 Single-point fuel injection system - component removal and refitting

Note: *Special test equipment may be required for initial setting-up after renewal of some components described in this Section. Read through the procedures carefully before starting work, to see what is involved. Also refer to the warning note at the end of Section 22.*

Air inlet housing (1794 cc F3P engine only)

1 Disconnect the breather hose from the air inlet housing (photo).

2 Disconnect the vacuum hoses from the air intake temperature control valve. Mark the hoses if there is any possibility of mixing them up (photo).

3 Remove the three Torx screws which secure the air inlet housing to the throttle housing (photo).

4 Release the hose clip and disconnect the air inlet trunking from the housing (photo).

5 Lift off the air inlet housing and recover the gasket.

6 Refit by reversing the removal operations, making sure that the vacuum hoses are connected correctly. Use a new gasket if necessary on the top of the throttle housing.

Throttle housing

7 Disconnect the battery earth lead. Depressurise the fuel lines (Section 28).

8 Remove the air cleaner housing or the air inlet housing, as applicable, from the top of the throttle housing.

9 Disconnect the three multi-plugs from the throttle housing (photos).

10 Disconnect the fuel supply and return lines, making identification marks for reference when refitting. Be prepared for fuel spillage.

11 Disconnect the accelerator cable.

12 Remove the four Torx screws and lift off the throttle housing. Recover the gasket.

13 The two halves of the housing can be separated if necessary, after releasing the press-stud fastenings which hold the halves together (photos). Renew the seals between the housing halves when reassembling.

36.9A Disconnect the injector/air temperature sensor multi-plug (A), the idle speed motor/idle switch multi-plug (B) . . .

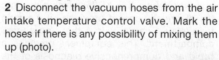

36.9B . . . and the throttle potentiometer multi-plug (C)

36.13A Release the studs by squeezing them with pliers . . .

36.13B . . . to separate the two halves of the throttle housing

36.14 Fitting a new gasket to the top of the throttle housing

36.18 Removing the Torx screw (size TX25) which secures the air temperature sensor cover

14 Refit by reversing the removal operations, using new gaskets (photo).

15 If the lower half of the throttle housing has been renewed, the initial setting of the throttle potentiometer must be verified using Renault test box XR25.

Inlet air temperature sensor

16 Remove the air cleaner housing or the air inlet housing, as applicable.

17 Disconnect the sensor multi-plug.

18 Remove the single securing screw, and free the sensor cover (photo).

19 Free the connector lugs, and withdraw the sensor together with the cover, wiring and connector (photo).

20 Refit by reversing the removal operations.

Coolant temperature sensor

21 The location of the coolant temperature sensor varies according to model. On the 1171 cc and 1390 cc (E7F and E7J) engines, it is located on the left-hand end of the cylinder head. On the 1794 cc (F3P) engine, it is located on the rear of the cylinder block (photo).

22 The procedure is as described in Chapter 3, Section 8.

Fuel injector

23 Remove the air cleaner housing or the air inlet housing, as applicable.

24 Remove the inlet air temperature sensor as described previously.

25 Note the fitted position of the injector, then withdraw it (photo).

26 Refit by reversing the removal operations. Use new O-ring seals on the injector, and lubricate them with silicone grease.

Idle speed control motor

27 Remove the air cleaner housing or the air inlet housing, as applicable.

28 Remove the throttle housing securing screws. Carefully lift the throttle housing to improve access to the motor.

29 Disconnect the motor multi-plug.

30 Remove the three mounting screws and withdraw the motor (photo).

31 Refit by reversing the removal operations, then perform an initial setting operation as follows.

32 Place a shim or feeler blades of thickness approximately 5 mm between the throttle lever and the motor plunger. Switch the ignition on for a few seconds, then switch it off again. Remove the shim or blades, then switch on and off again.

33 Run the engine and check that the idle speed and quality are satisfactory.

Fuel pressure regulator

34 The fuel pressure regulator is an integral part of the upper half of the throttle housing. If it is defective, the upper half of the housing must be renewed.

Throttle potentiometer

35 The throttle potentiometer is an integral part of the lower part of the throttle housing. If it is defective, the lower half of the housing must be renewed. The idle speed control motor can be transferred from the old half-housing to the new one.

36 After renewal, the initial setting of the potentiometer must be verified using Renault test box XR25.

Computer

Note: *The following paragraphs describe the procedure for models without air conditioning. On models with air conditioning, the computer is located inside the car, below the glovebox.*

36.19 Removing the air temperature sensor assembly. The cover also serves as a connector for the fuel injector

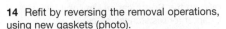

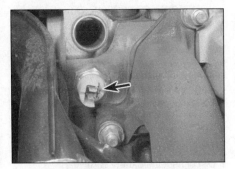

36.21 Coolant temperature sensor (arrowed) on the rear of the cylinder block - 1794 cc (F7P) engine

36.25 Removing the fuel injector

36.30 Three screws (arrowed) securing the idle speed control motor

4

36.38A To remove the computer mounting bracket, undo the single nut here . . .

36.38B . . . and the nut and the bolt here (arrowed)

36.39 Unhook the retaining strap to release the computer case

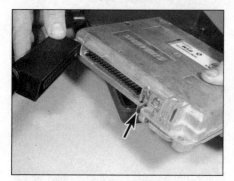

36.40 Depress the tag (arrowed) to release the computer multi-plug

36.42 Removing the MAP sensor - disconnect the multi-plug (A) and the vacuum hose (B)

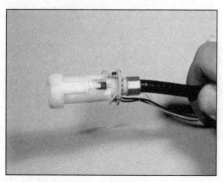

36.44 Road speed sensor is at the speedometer end of the cable

37 Disconnect the battery earth lead. Open the hatch in the right-hand side of the windscreen cowl panel and remove the jack. For improved access, remove the cowl panel completely (Chapter 11).
38 Remove the mounting bracket, which is secured by two nuts and a single bolt (photos).
39 Unhook the retaining strap. Lift out the mounting bracket and the computer with its protective case (photo).
40 Open the protective case by prising with a screwdriver in the slots in the side of the case. Disconnect the multi-plug and withdraw the computer (photo).
41 Refit by reversing the removal operations.

Manifold pressure (MAP) sensor

42 Unclip the MAP sensor from its bracket on the bulkhead. Disconnect the multi-plug and the vacuum hose from it (photo).
43 Refit by reversing the removal operations.

Road speed sensor

44 The road speed sensor is an integral part of the speedometer cable (photo). For cable renewal procedures, see Chapter 12.

37 Multi-point fuel injection system - component testing

1 Refer to Section 35, paragraphs 1 to 6.
2 Do not attempt to adjust the stop screw on the throttle housing in an attempt to correct idle speed problems, or for any other reason. Note that idle speed problems on this system may be caused by erroneous or missing signals from the road speed sensor.

Inlet air temperature sensor

3 Refer to Section 35. The air temperature sensor on multi-point injection models is located in the right-hand end of the inlet manifold.

Coolant temperature sensor

4 The coolant temperature sensor is located on the thermostat housing, next to the temperature gauge sensor. Refer to Chapter 3, Section 8 - the procedure is the same;

refer to the Specifications at the start of this Chapter for resistance values.

Fuel injectors

5 Withdraw the injector rail complete with injectors, but without disconnecting the fuel hoses (see Section 38).
6 Position each injector in a clean glass jar. Run the fuel pump (Section 29) and check that no fuel drips from any injector.
7 With the fuel pump still running, apply 12 volts to the terminals of each injector in turn. With voltage applied, the injector must spray fuel into the glass jar.

Cold start injector

8 The cold start injector is tested in the same way as just described for the main injectors.

Idle speed control valve

9 A quick check of the idle speed control valve can be made by pinching or plugging the air hose on the inlet side of the valve while the engine is idling. The idle speed should drop, perhaps to the point where the engine stalls. When the airflow is restored, the idle speed should rise and then stabilise.

Fuel pressure regulator

10 Testing of the fuel pressure regulator is included in the fuel pressure check (Section 29).

38 Multi-point fuel injection system - component removal and refitting

Note: *Special test equipment may be required for initial setting-up after renewal of some components described in this Section. Read through the procedures carefully before starting work, to see what is involved. Also refer to the warning note at the end of Section 22.*

Throttle housing

1 Disconnect the battery earth lead.
2 Disconnect the accelerator cable, the electrical multi-plugs and any breather or vacuum hoses from the throttle housing.

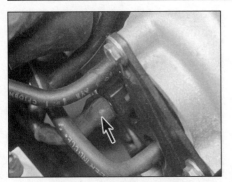

38.6 Inlet air temperature sensor multi-plug (arrowed) - multi-point injection system

3 Release the air cleaner housing from the throttle housing.
4 Unbolt the throttle housing from the inlet manifold, and remove it.
5 Refit by reversing the removal operations.

Inlet air temperature sensor

6 Disconnect the sensor multi-plug (photo).
7 Unscrew the sensor from the end of the inlet manifold, and remove it.
8 Refit by reversing the removal operations.

Coolant temperature sensor

9 The coolant temperature sensor is located on the thermostat housing, next to the temperature gauge sensor. Refer to Chapter 3, Section 8; the procedure is the same.

Fuel injectors

10 Depressurise the fuel system (Section 28). Disconnect the battery earth lead.
11 Remove the cover from the top of the inlet manifold.
12 Disconnect the fuel feed and return hoses from the injector rail. Be prepared for fuel spillage. Plug or clamp the hoses.
13 Remove the breather hoses and pipes which run from the engine valve cover across the injector rail. Also disconnect the vacuum hose from the fuel pressure regulator.
14 Unplug the electrical connectors from the injectors (photo). (If preferred, this can be done as the injector rail is withdrawn).
15 Remove the single nut which secures the

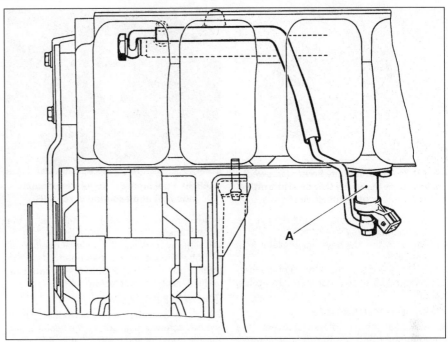

Fig. 4.30 Location of the cold start injector (A) on the underside of the inlet manifold (Sec 38)

injector rail (photo). Withdraw the rail and the injectors.
16 The injectors can be separated from the rail after removing the retaining clips.
17 Refit by reversing the removal operations. Use new injector sealing rings if necessary. Apply a little silicone grease or a similar lubricant to the sealing rings.
18 Check that there are no fuel leaks on completion.

Cold start injector

19 Disconnect the battery earth lead.
20 Remove the bonnet and the bonnet hinge crossmember.
21 Tilt the radiator forwards. Disconnect the crankcase ventilation hose from the inlet manifold, and unbolt the bracing strut.
22 Disconnect the fuel union from the cold start injector. Be prepared for fuel spillage.
23 Remove the two screws which secure the cold start injector to the manifold.

24 Refit by reversing the removal operations. Before refitting the bonnet hinge crossmember, reconnect the battery and run the fuel pump by bridging the relay connectors (Section 29). With the pump running, check that there are no fuel leaks from the fuel union on the cold start injector.

Idle speed control valve

25 Disconnect the battery earth lead.
26 Disconnect the multi-plug from the top of the valve (photo).
27 Remove the clamp which secures the valve. Disconnect the air hoses and withdraw the valve.
28 Refit by reversing the removal operations.

Fuel pressure regulator

29 Depressurise the fuel system (Section 28). Disconnect the battery earth lead.
30 Remove the cover from the top of the inlet manifold. Disconnect the vacuum hose from

4

38.14 Unplugging an injector connector

38.15 The injector rail is secured by a single nut (arrowed)

38.26 Idle speed control valve multi-plug (arrowed)

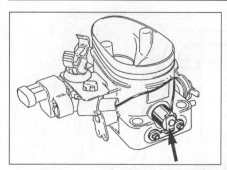

Fig. 4.31 Throttle body heater (arrowed) located on the underside of the throttle housing (Sec 38)

the fuel pressure regulator, and remove the two securing screws (photo).

31 Withdraw the fuel pressure regulator from the fuel rail.

32 Refit by reversing the removal operations. On completion, check that there are no fuel leaks.

Throttle potentiometer

33 Make alignment marks between the potentiometer and the throttle housing, for reference when refitting.

34 Disconnect the multi-plug, and remove the two screws which secure the throttle

39.2 Coolant temperature sensor (arrowed) for the anti-percolation system

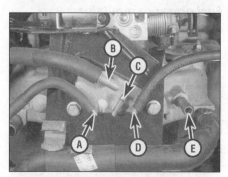

40.4 Vacuum hose connections on the rear of the inlet manifold, viewed with the engine removed - 1794 cc (F3P) engine

A Air cleaner temperature control
B Crankcase ventilation system
C MAP sensor
D Evaporative emission control canister
E Brake servo

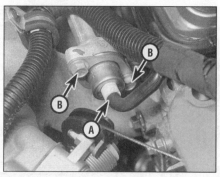

38.30 Fuel pressure regulator vacuum hose (A) and securing screws (B)

potentiometer (photo). Remove the potentiometer.

35 Refit by reversing the removal operations. If a new potentiometer has been fitted, the initial setting of the potentiometer must be verified using Renault test box XR25.

Throttle body heater

36 Disconnect the battery earth lead. Undo the two screws which secure the heater to the throttle housing, and remove it (Fig. 4.31).

37 Refit by reversing the removal operations.

Computer

38 Refer to Section 36. The procedure is the same.

Manifold pressure (MAP) sensor

39 Refer to Section 36. The procedure is the same.

Road speed sensor

40 Refer to Section 36. The procedure is the same.

39 Anti-percolation system (1764 cc/ F7P engine only) - general

1 The anti-percolation system comes into operation only when the engine is stopped and the coolant temperature is 105°C or greater. Under these conditions, the auxiliary water pump and the cooling fan are energised for approximately 8 minutes. Operation continues over this period even though the coolant temperature falls below 105°C.

2 The coolant temperature sensor for the anti-percolation system is the lowest of the three on the thermostat housing (photo).

3 For details of the auxiliary water pump, refer to Chapter 3.

40 Inlet manifold - removal and refitting

1171 cc and 1390 cc (E7F and E7J) engines

1 Remove the throttle housing (Section 36).

2 The procedure is now as described in Part A of this Chapter, Section 18, disregarding references to the carburettor.

38.34 Two screws (arrowed) secure the throttle potentiometer

1794 cc (F3P) engine

3 Remove the throttle housing (Section 36) and the heat shield which is below it.

4 The inlet and exhaust manifolds are now removed together as described in Part A of this Chapter, Section 20, disregarding references to the carburettor. Note the various vacuum hose connections on the rear of the manifold (photo).

1764 cc (F7P) engine

5 Remove the radiator (Chapter 3).

6 Remove the fuel injectors, including the cold start injector (Section 38).

7 Remove the air cleaner (Section 23).

8 Remove the alternator (Chapter 12) and the bolts which secure the alternator bracket to the manifold.

9 Disconnect the remaining hoses and electrical connectors from the manifold. Unbolt the manifold bracing struts, undo the nuts and withdraw the manifold.

10 Refit by reversing the removal operations, using a new manifold gasket.

41 Exhaust manifold - removal and refitting

1171 cc and 1390 cc (E7F and E7J) engines

1 Proceed as described in Part A of this Chapter, Section 19.

1794 cc (F3P) engine

2 The inlet and exhaust manifolds are removed together as described in the preceding Section and in Part A of this Chapter, Section 20.

1764 cc (F7P) engine

3 Remove the heat shield. Disconnect the exhaust front pipe from the manifold.

4 Remove the securing nuts and bolts, and withdraw the manifold from the cylinder head. Recover the gasket.

5 Refit by reversing the removal operations, using a new gasket.

All models

6 The fitting of a new exhaust sealing ring is recommended.

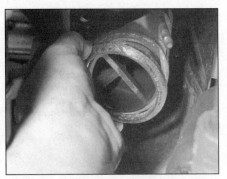

42.7 Renew the exhaust downpipe-to-manifold sealing ring

42.8 One of the exhaust rear section rubber mountings (arrowed)

42.10 Withdraw the heat shield when separating the catalytic converter from the exhaust intermediate section

42 Exhaust system - general information and component renewal

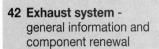

1 Typical exhaust system components are shown in Fig. 4.32. The system consists of three sections, plus (except on C575 16-valve models - see Section 26) a catalytic converter.

2 On models with ABS and/or a open-bar rear axle, the rear section of the exhaust system cannot be removed intact, but must be cut. (Alternatively, the rear axle and shock absorbers can be unbolted and lowered.) Repair sections have an additional joint in the pipe which passes over the axle. The downpipe and the intermediate section can be removed separately, leaving the rear section in place if wished.

3 On all other models, both the downpipe and the rear section can be removed leaving the remaining exhaust system in position. To remove the intermediate section, it is recommended that the rear section is removed first, then the intermediate section can be disconnected from the catalytic converter or downpipe. In practice, it is often easier to remove the whole system and then separate the sections on the bench.

4 Exhaust system clamps and connections will often be found to be badly rusted. Apply penetrating oil or releasing fluid before starting work, or cut them off if they are to be renewed. The use of a high-temperature anti-seize compound on reassembly will avoid future problems. Exhaust jointing compound should be used on the joints between sections.

5 To remove the system or part of the system, first jack up the front or rear of the car and support it on axle stands. Alternatively, position the car over an inspection pit or on car ramps. Disconnect the oxygen sensor multi-plug, when applicable - refer to Part C of this Chapter.

6 To remove the downpipe, unscrew and remove the two nuts, then remove the washers, tension springs and sleeves securing the downpipe to the manifold. Slide the flange plate off the manifold studs, then

separate the joint, recovering the sealing ring. Remove the nuts and bolts which secure the downpipe to the catalytic converter. (On C575 16-valve models, release the clamp which secures the intermediate section to the downpipe.) Also unbolt the heat guard if necessary.

7 Refitting of the downpipe is a reversal of removal. Use a new sealing ring at the exhaust manifold (photo). When applicable, also use a new gasket at the catalytic

converter. Transfer the oxygen sensor to the new pipe if necessary (refer to Part C of this Chapter). Tighten the manifold nuts to compress the springs until the nuts butt against the spacers on the studs.

8 To remove the rear section, release the clamp which secures it to the intermediate section. Slide the rubber mountings off the bars, and if necessary, remove them from the underbody by turning them through 90° (photo). Separate the rear and intermediate

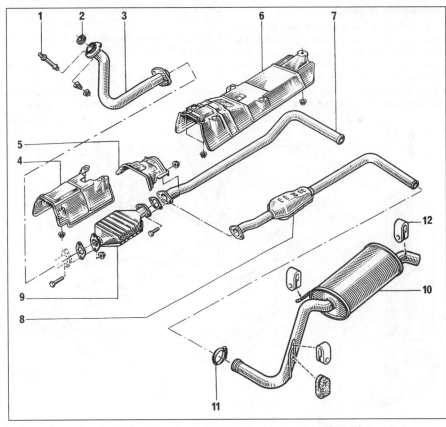

Fig. 4.32 Typical exhaust system components (Sec 42)

1 Stud	6 Heat shield	9 Catalytic converter
2 Sealing ring	7 Intermediate section	10 Rear section (silencer)
3 Downpipe (front section)	(straight-through)	11 Clamp
4 Heat shield	8 Intermediate section (with	12 Rubber mounting
5 Heat shield	expansion chamber)	

4

sections, cutting the intermediate section if necessary to clear the rear axle (see paragraph 2).

9 Refitting of the rear section is a reversal of removal, using a repair section with an additional joint if necessary. Tighten the clamp(s) securely, and renew the rubber mountings if necessary.

10 To remove the intermediate section, first remove or disconnect the rear section as just described. Remove the nuts and bolts which secure the catalytic converter to the

intermediate section. (On C575 16-valve models, release the clamp which secures the intermediate section to the downpipe.) Slide the rubber mountings off the bars, and if necessary remove them from the underbody by turning them through 90°. Separate the intermediate section from the catalytic converter, at the same time removing the heat shield (photo).

11 Refitting is a reversal of removal; tighten the nuts and bolts securely, and renew the rubber mountings if necessary.

12 Refer to Part C of this Chapter for details of precautions relating to the catalytic converter. It is removed in the same way as the intermediate section.

13 When the whole system has been removed, refit it with the various clamps and connections loose. Begin tightening at the exhaust manifold, and work towards the tail pipe, ensuring that the system hangs easily and without strain.

Part C: Emission control systems

43 General information and precautions

Evaporative emission control

1 The purpose of the evaporative emission control system is to minimise emissions of fuel vapour. The fuel tank filler cap and hoses form part of this system. The rest of the system varies according to model, as follows.

Carburettor engines

2 All carburettor models are equipped with a vapour separator (or defuming device) and a fuel vapour reservoir. Typical pipe runs and connections are shown earlier in this Chapter, in connection with the fuel pump (Section 7).

3 Fuel vapour which forms in the carburettor float chamber when the engine is not running

is stored in the vapour reservoir. The vapour trap intercepts vapour formed in the fuel supply line, and returns it to the fuel tank.

Fuel injection engines - 1171 and 1390 cc (E7F and E7J)

4 Fuel vapour which forms when the engine is not running is stored in a canister located under the right-hand headlight. With the engine running, vapour is drawn into the inlet manifold, and fresh air is drawn into the canister. A solenoid valve and a calibrated bleeding valve, both located on the bulkhead, control the rate at which this happens.

Fuel injection engines - 1794 and 1764 cc (F3P and F7P)

5 The system is as just described, but a simple restrictor takes the place of the calibrated bleeding valve. The solenoid valve is located behind the right-hand headlight.

Crankcase emission control

6 This system ensures that there is always a partial vacuum in the crankcase, and so prevents pressure which could cause oil contamination, fume emission and oil leakage past seals.

7 The layout of the crankcase ventilation system, according to engine type, is shown in the accompanying illustrations.

8 When the engine is idling, or under partial load conditions, the high depression in the inlet manifold draws the crankcase fumes (diluted by air from the air cleaner side of the throttle valve) through the calibrated restrictor and into the combustion chambers.

9 The crankcase ventilation hoses and restrictors should be periodically cleaned to ensure correct operation of the system.

Exhaust emission control

Carburettor models - 1171 cc (E5F) engine

10 This is the only carburettor engine with an exhaust emission control system. It consists of a cold enrichment device, which reduces exhaust emissions at idle and light load once the engine has warmed up. Refer to Section 12 in Part A of this Chapter for details. Testing of this system is included in the normal idle adjustments (Chapter 1, Section 9).

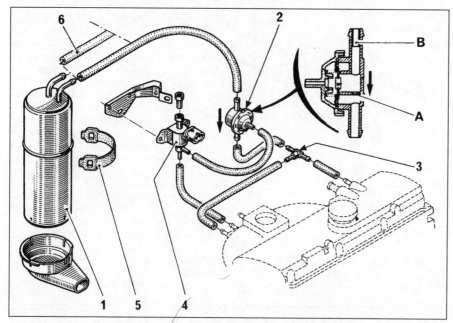

Fig. 4.33 Evaporative emission control components - E7F/E7J engines (Sec 43)

1 Canister
2 Bleeding valve
3 T-piece
4 Solenoid valve
5 Strap
6 Hose
A Restrictor (0.8 mm diameter)
B Restrictor (2 mm diameter)

Arrows show direction of flow with engine running

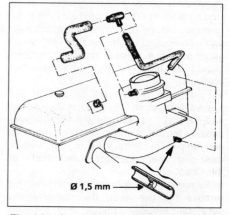

Fig. 4.34 Crankcase ventilation hoses and restrictor - 1108 cc (C1E) engine (Sec 43)

Ø 1,5 mm

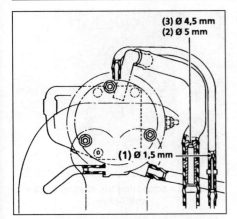

Fig. 4.35 Crankcase ventilation hoses and restrictors - 1171 cc (E5F) and 1390 cc (E6J) engines (Sec 43)

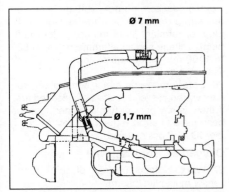

Fig. 4.36 Crankcase ventilation hoses and restrictors - 1721 cc (F2N) engine (Sec 43)

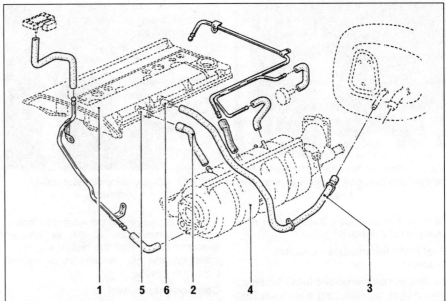

Fig. 4.37 Crankcase ventilation hoses and restrictors - 1764 cc (F7P) engine (Sec 43)

1 Cylinder head cover	3 Hose (upstream of throttle)	5 Restrictor (1.7 mm diameter)
2 Hose (downstream of throttle)	4 Inlet manifold	6 Restrictor (5.5 mm diameter)

Fuel injection models

11 All fuel injection models (except model C575, 1764 cc engine code F7P-720) are fitted with an oxygen sensor and a catalytic converter. These components ensure a very low level of toxic emissions in the exhaust gas.

12 The oxygen sensor (also known as a 'lambda' sensor, from the symbol used to indicate air/fuel ratio) is located in the exhaust downpipe. It provides a signal to the injection computer which varies according to the oxygen content of the exhaust gas. The computer uses the signal to calculate the amount of fuel which must be injected to ensure efficient combustion and low emission levels.

13 The oxygen sensor contains an electric heater, which is energised when the ignition is switched on. This enables it to reach its operating temperature (in excess of 250°C) more quickly, so reducing exhaust emissions during the warm-up phase.

14 The catalytic converter is located in the exhaust system at the end of the downpipe furthest from the manifold. It contains a ceramic honeycomb structure coated with a precious metal compound. This compound acts as a catalyst in the conversion of carbon

monoxide (CO) to carbon dioxide (CO_2), and unburned hydrocarbons (HC_x) to water and carbon dioxide (H_2O and CO_2).

Catalytic converter - precautions

15 For long life and satisfactory operation of the catalytic converter, certain precautions must be observed. These are as follows.

16 Only use unleaded fuel. Leaded fuel will poison the catalyst and the oxygen sensor.

17 Do not run the engine for long periods if it is misfiring. Unburnt fuel entering the catalytic converter can cause it to overheat, resulting in permanent damage. For the same reason, do not try to start the engine by pushing or towing the car, nor crank it on the starter motor for long periods.

18 Do not strike or drop the catalytic converter. The ceramic honeycomb which forms part of its internal structure may be damaged.

19 Always renew seals and gaskets upstream of the catalytic converter whenever they are disturbed.

20 Remember that the catalytic converter gets hotter than the rest of the exhaust system. Avoid parking in long grass, undergrowth or similar material which could catch fire or be damaged.

44 Emission control components - testing and renewal

Evaporative emission control
Carburettor models

1 There are no specific testing procedures for the carburettor evaporative emission control components. Check periodically that the

various hoses are in good condition and securely connected.

2 Should component renewal be necessary, it is simply a case of disconnecting the hoses, and unbolting or unclipping the vapour trap or reservoir. Dispose of the old component safely, bearing in mind that it may contain liquid fuel and/or fuel vapour.

Fuel injection models - testing

3 The operating principle of the system is that the solenoid valve be open only when the engine is warm and at part-throttle conditions. The following procedure is based on information relating to the 1764 cc (F7P) engine, but it is broadly applicable to the others.

4 Bring the engine to normal operating temperature, then switch it off. Connect a vacuum gauge (range 0 to 1000 mbars) into the hose between the canister and the restrictor. Connect a voltmeter to the solenoid valve terminals.

5 Start the engine and allow it to idle. There should be no vacuum shown on the gauge, and no voltage present at the solenoid.

6 If manifold vacuum is indicated although no voltage is present, check that the restrictor is of the correct size. If it is, the solenoid valve is stuck open.

7 If voltage is present at idle, there is a fault in the wiring or the computer.

8 Depress the accelerator slightly. Voltage should appear momentarily at the solenoid terminals, and manifold vacuum be indicated on the gauge.

9 If vacuum is not indicated even though voltage is present, either there is a leak in the hoses, or the valve is not opening.

4

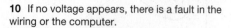

44.11 Slacken the screws (arrowed) to release the evaporative canister mounting strap

44.12 Evaporative canister hoses (arrowed)

44.27 Disconnecting the oxygen sensor multi-plug

10 If no voltage appears, there is a fault in the wiring or the computer.

Fuel injection models - canister renewal

11 Remove the two screws which secure the front part of the right-hand wheel arch liner, and pull the liner out of the way. Release the mounting strap and free the canister (photo).

12 Disconnect the hoses from the canister, noting how they are connected (photo). (If preferred, access to the hoses can be gained by removing the right-hand headlight.)

13 Dispose of the old canister safely, bearing in mind that it may contain liquid fuel and/or fuel vapour.

14 Refit by reversing the removal operations. Make sure that the hoses are connected correctly.

Solenoid valve renewal

15 Disconnect the hoses from the solenoid valve, marking them if necessary for reference when reassembling.

16 Disconnect the multi-plug from the valve. Release the valve from its mounting bracket, and remove it.

17 Refit by reversing the removal operations. Make sure that the hoses are connected correctly.

Crankcase emission control

Testing

18 There is no specific test procedure for the crankcase emission control system. If

problems are suspected (sometimes indicated by oil contamination of the air cleaner element), check that the hoses are clean internally, and that the restrictors are not blocked or missing.

Component renewal

19 This is self-evident. Mark the various hoses before disconnecting them, if there is any possibility of confusion on reassembly.

Exhaust emission control (fuel injection models only)

20 The remainder of this Section only applies to fuel injection models equipped with oxygen sensor and catalyst.

Testing

21 An exhaust gas analyser (CO meter) will be needed. The ignition system must be in good condition, the air cleaner element must be clean, and the engine must be in good mechanical condition.

22 Bring the engine to normal operating temperature, then connect the exhaust gas analyser in accordance with the equipment maker's instructions.

23 Allow the engine to idle, and check the CO level (Chapter 1 Specifications). Repeat the check at 2500 to 3000 rpm. If the CO level is within the specified limits, the system is operating correctly.

24 If the CO level is higher than specified, try the effect of disconnecting the oxygen sensor wiring. If the CO level rises when the sensor is disconnected, this suggests that the oxygen

sensor is OK and that the catalytic converter is faulty. If disconnecting the sensor has no effect, this suggests a fault in the sensor.

25 If a digital voltmeter is available, the oxygen sensor output voltage can be measured (terminal C of the sensor multi-plug, or terminal 35 of the computer multi-plug). Voltage should alternate between 625 to 1100 mV (rich mixture) and 0 to 100 mV (lean mixture).

26 Before renewing either component, run the car using several tankfuls of unleaded fuel, if there is any possibility that leaded fuel has been used in error. Repeat the test to see if the CO level has reduced.

Oxygen sensor - renewal

27 Working under the car, disconnect the sensor multi-plug (photo).

28 Unclip the heat shield. Unscrew the sensor from the exhaust downpipe, and remove it (photos).

29 Clean the threads in the exhaust pipe, and the threads of the sensor (if it is to be refitted).

30 Note that if the sensor wires are broken, the sensor must be renewed. No attempt should be made to repair them.

31 Apply high-temperature anti-seize compound to the sensor threads. Screw the sensor in by hand, then tighten it fully.

32 Reconnect the sensor multi-plug.

Catalytic converter - renewal

33 The catalytic converter is renewed in the same way as any other part of the exhaust system. Refer to Part B of this Chapter.

44.28A Unclip the heat shield . . .

44.28B . . . unscrew the oxygen sensor . . .

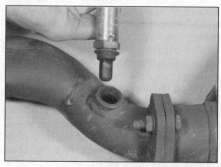

44.28C . . . and remove it from the exhaust pipe

Chapter 5 Ignition system

Contents

Degrees of difficulty

Easy, suitable for novice with little experience	**Fairly easy,** suitable for beginner with some experience	**Fairly difficult,** suitable for competent DIY mechanic	**Difficult,** suitable for experienced DIY mechanic	**Very difficult,** suitable for expert DIY or professional

Specifications

Part A: Transistor-assisted contact breaker ignition system

General

System type .... Contact breaker and coil, with electronic assistance in low-tension circuit
Application .... 1108 cc (C1E) engine
Firing order .... 1—3—4—2
Location of No 1 cylinder .... Flywheel end

Distributor

Type .... Conventional, with contact breaker points and condenser
Direction of rotation .... Clockwise

Ignition coil

Type .... Conventional
Primary resistance (typical) .... 1.5 ohms
Secondary resistance (typical) .... 5000 ohms
For ignition timing and spark plug specifications, see Chapter 1.

Part B: Electronic ignition system

General

System type .... Fully-electronic, computer-controlled
Application .... All engines except 1108 cc (C1E)
Firing order .... 1—3—4—2
Location of No 1 cylinder .... Flywheel end

Distributor

Type .... Rotor arm and cap only
Direction of rotation .... Anti-clockwise

Ignition coil

Type .... Renix, mounted on ignition module
Primary r esistance (typical) .... 0.4 to 0.8 ohms
Secondary resistance (typical) .... 6500 ohms
Ignition timing is not adjustable. For spark plug specifications, see Chapter 1

5

Part A: Transistor-assisted contact breaker ignition system

1 General information

This type of ignition system is fitted to the 1108 cc (C1E) engine. It consists of a conventional contact breaker ignition system, but with the addition of a transistor assistance unit.

The ignition system functions as follows. Low-tension voltage from the battery is fed to the ignition coil, where it is converted into high-tension voltage. The high-tension voltage is powerful enough to jump the spark plug gap in the cylinder many times a second under high compression pressure, provided that the ignition system is in good working order and that all adjustments are correct.

The ignition system consists of two individual circuits, known as the low-tension (LT) circuit and high-tension (HT) circuit.

The low-tension circuit (sometimes known as the 'primary' circuit) consists of the battery, the lead to ignition switch, the lead to the low-tension or primary coil windings, and the lead from the low-tension coil windings to the contact breaker points and condenser in the distributor.

The high-tension circuit (sometimes known as the 'secondary' circuit) consists of the high-tension or secondary coil winding, the heavily-insulated lead from the centre of the coil to the centre of the distributor cap, the rotor arm, the spark plug leads and the spark plugs.

The complete ignition system operation is as follows. Low-tension voltage from the battery is changed within the ignition coil to high-tension voltage by the opening and closing of the contact breaker points in the low-tension circuit. High-tension voltage is then fed, via a contact in the centre of the distributor cap, to the rotor arm of the distributor. The rotor arm revolves inside the distributor cap, and each time it passes one of the four metal segments in the cap, the opening and closing of the contact breaker points causes the high-tension voltage to build up, jump the gap from the rotor arm to the appropriate metal segment and so, via the spark plug lead, to the spark plug, where it finally jumps the gap between the two spark plug electrodes, one being earthed.

In order that the engine may run correctly, it is necessary for the spark to ignite the fuel/air mixture in the combustion chamber at exactly the right moment in relation to engine speed and load. The ignition timing is advanced and retarded automatically to ensure this, as follows.

The automatic advance and retard is controlled both mechanically and by a vacuum-operated system. The mechanical governor mechanism consists of two weights, which move out under centrifugal force from the central distributor shaft as the engine speed rises. As they move outwards, they rotate the cam relative to the distributor shaft, and so advance the spark. The weights are held in position by two light springs, and it is the tension of these springs which is largely responsible for correct spark advancement.

The vacuum control consists of a diaphragm, one side of which is connected, via a small-bore tube, to the carburettor and the other side to the contact breaker plate. Depression in the induction manifold and carburettor, which varies with engine speed and throttle opening, causes the diaphragm to move, so rotating the contact breaker plate and advancing or retarding the spark.

The function of the transistor assistance unit is to relieve the contact points of carrying the full primary current, thus extending their service life and providing a more reliable system. The contact points are used to switch a transistor on and off, and it is the transistor which carries the full primary current.

On some models, the unit has two sockets on it; the top one is for normal use, and includes transistor assistance, and the bottom one is for emergency use, bypassing the transistor unit and converting the system to a conventional contact breaker ignition system. The wiring plug is simply moved from one socket to the other as required. Some models have a sealed unit without the two socket positions.

The transistor operates as follows, noting that the components mentioned relate to those shown in area F of Fig. 5.1. With the points open, the base (B) and emitter (E) are of the same potential, and therefore no current flows. When the points close, the base (B) becomes negative due to the voltage drop at point (A), and current then flows through the transistor collector (C). The resistances R1 and R2 are fitted in the circuit to provide a low control voltage, and this also greatly increases the life of the contact points, as they only carry a small current. When the points open again, the voltage increases at point (A) and the base (B), and the transistor is switched off.

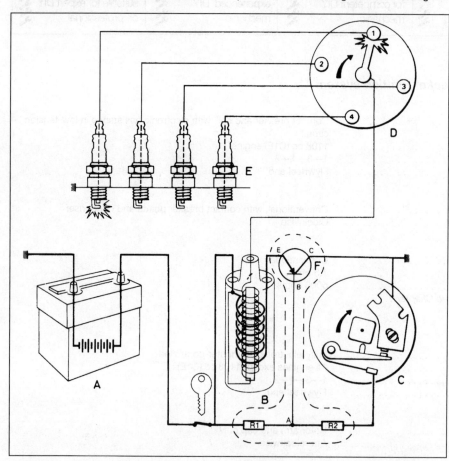

Fig. 5.1 Schematic of transistor-assisted contact breaker ignition system (Sec 1)

A Battery
B Ignition coil
C Distributor contact breaker

D Distributor cap and rotor
    arm

E Spark plugs
F Transistor assistance unit

2 Ignition system - testing

1 Should a fault occur in the ignition system, first bypass the transistor assistance unit by pulling the connector from the top of the unit and plugging it into the lower socket. (On models without a lower socket, the same effect can be achieved by joining pins 1 and 3 in the connector - see paragraph 9). If the fault disappears, the fault is proved to be in the transistor assistance unit. If the fault remains, check the basic conventional system with reference to the following paragraphs.

2 There are two main symptoms indicating faults in the ignition system. Either the engine will not start or fire, or the engine is difficult to start and misfires. If it is a regular misfire (ie the engine is running on only two or three cylinders), the fault is almost sure to be in the secondary (high-tension) circuit. If the misfiring is intermittent, the fault could be in either the high- or low-tension circuits. If the car stops suddenly, or will not start at all, it is likely that the fault is in the low-tension circuit. Loss of power and overheating, apart from faulty carburation settings, are normally due to faults in the distributor, or to incorrect ignition timing.

Engine fails to start

3 If the engine fails to start and the engine was running normally when it was last used, first check that there is fuel in the fuel tank. If the engine turns normally on the starter motor and the battery is fully charged, then the fault may be in either the high- or low-tension circuits. First check the high-tension circuit. (If the starter motor turns slowly or not at all, this is not an ignition system fault. Refer to Chapter 12.)

4 One of the most common reasons for bad starting is wet or damp spark plug leads and distributor. Remove the distributor cap. If condensation is visible internally, dry the cap with a rag, and also wipe over the leads. Refit the cap. Alternatively, using a moisture-dispersant such as Holts Wet Start can be very effective in starting the engine. To prevent the problem recurring, Holts Damp Start can be used to provide a sealing coat, so excluding any further moisture from the ignition system. In extreme difficulty, Holts Cold Start will help to start a car when only a very poor spark occurs.

5 If the engine still fails to start, check that the current is reaching the plugs, by disconnecting each plug lead in turn at the spark plug end, and holding the end of the cable about 5 mm away from the cylinder block. Hold the lead with rubber-insulated pliers to avoid electric shocks. Spin the engine on the starter motor.

6 Sparking between the end of the cable and the block should be fairly strong with a good, regular blue spark. If current is reaching the plugs, then remove them and check their gaps. When the plugs are refitted, the engine should start - if not, try new plugs.

7 If there is no spark at the plug leads, take off the HT lead from the centre of the distributor cap, and hold it to the block as before. Spin the engine on the starter once more. A rapid succession of blue sparks between the end of the lead and the block indicates that the coil is in order, and that either the distributor cap is cracked, the rotor arm faulty, or the carbon brush in the distributor cap is not making good contact with the rotor arm.

8 If there are no sparks from the end of the lead from the coil, check the connections at the coil end of the lead. If it is in order, start checking the low-tension circuit.

9 Either transfer the connector to the lower socket on the transistor-assistance unit, or remove the connector and interconnect pins 1 and 3 in the connector, as shown in Fig. 5.2. The 5 amp fuse in the figure protects the Renault diagnostic equipment, and does not need to be fitted if diagnostic equipment is not being used. As a precaution against a faulty condenser at the distributor, a known good condenser should be connected to earth as shown.

10 Use a 12-volt voltmeter, or a 12-volt bulb and two lengths of wire. With the ignition switched on and the points open, test between the low-tension wire to the coil positive (+) terminal and earth. No reading indicates a break in the supply from the ignition switch. Check the connections at the switch to see if any are loose; refit them, and the engine should run.

11 If a reading is indicated but the engine does not run, this shows a faulty coil or condenser, or a broken lead between the coil and the distributor. Take the condenser wire off the points assembly, and with the points open, test between the moving point and earth. If there is now a reading, then the fault is in the condenser. Fit a new one, as described in Chapter 1, and the fault should clear.

12 With no reading from the moving point to earth, take a reading between earth and the coil negative (-) terminal. A reading here shows a broken wire between the coil and distributor. No reading confirms that the coil has failed and must be renewed, after which the engine should run. Remember to refit the condenser wire to the points assembly. For these tests, it is sufficient to separate the points with a piece of paper while testing with the points open.

Engine misfires

13 If the engine misfires regularly, run it at a fast idle speed. Pull off each of the plug caps in turn, and listen to the note of the engine. Hold the plug cap in a dry cloth or with a rubber glove, as additional protection against a shock from the HT supply.

14 No difference in engine running will be noticed when the lead from the defective cylinder is removed. Removing the lead from one of the good cylinders will accentuate the misfire.

15 Remove the plug lead from the end of the defective plug, and hold it about 5 mm away from the block. Restart the engine. If the sparking is fairly strong and regular, the fault must lie in the spark plug.

16 The plug may be loose, the insulation may be cracked, or the electrodes may have burnt away, giving too wide a gap for the spark to jump. Worse still, one of the electrodes may have broken off. Renew the spark plugs in this case.

17 If there is no spark at the end of the plug lead, or if it is weak and intermittent, check the ignition lead from the distributor to the plug. If the insulation is cracked or perished, renew the lead. Check the connections at the distributor cap.

18 If there is still no spark, examine the distributor cap carefully for tracking. This can be recognised by a very thin black line running between two or more electrodes, or between an electrode and some other part of the distributor. These lines are paths which now conduct electricity across the cap, thus letting it run to earth. The only answer is a new distributor cap.

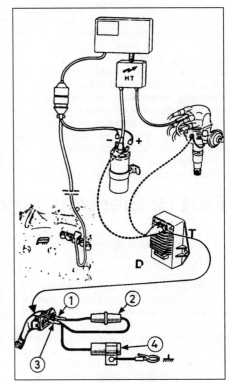

Fig. 5.2 Testing the transistor-assisted ignition system (Sec 2)

1 Terminal 1   4 Condenser
2 Fuse (see text) D Connector
3 Terminal 2   T Transistor assistance unit

Note that Renault diagnostic equipment is shown in this figure

5

19 Apart from the ignition timing being incorrect, other causes of misfiring have already been dealt with under the section dealing with the failure of the engine to start. To recap, these are that:

(a) *The coil may be faulty, giving an intermittent misfire.*
(b) *There may be a damaged wire or loose connection in the low-tension circuit.*
(c) *The condenser may be short-circuiting.*
(d) *There may be a mechanical fault in the distributor (broken driving spindle or contact breaker spring).*

20 If the ignition is too far retarded, it should be noted that the engine will tend to overheat, and there will be quite a noticeable drop in power. If the engine is overheating and the power is down, and the ignition timing is correct, then the carburettor should be checked, as it is likely that this is where the fault lies.

3 Distributor -
removal, overhaul and refitting

Removal

1 Mark the spark plug HT leads to aid refitting, and pull them off the ends of the plugs. Release the distributor cap retaining clips, and place the cap and leads to one side.
2 Remove No 1 spark plug (nearest the flywheel end of the engine).
3 Place a finger over the plug hole, and turn the engine in the normal direction of rotation (clockwise, viewed from the crankshaft pulley end) until pressure is felt in No 1 cylinder. This indicates that the piston is commencing its compression stroke. The engine can be turned with a socket or spanner on the crankshaft pulley bolt.
4 Continue turning the engine until the mark on the flywheel is aligned with the TDC notch on the clutch bellhousing.
5 Using a dab of paint or a small file, make a reference mark between the distributor base and the cylinder block.
6 Detach the vacuum advance pipe, and disconnect the LT lead at the wiring connector. Release the wiring loom from the support clip on the distributor body.
7 Unscrew the distributor clamp retaining nut, and lift off the clamp. Withdraw the distributor from the engine, and recover the seal.

Overhaul

8 Renewal of the contact breaker assembly, condenser, rotor and distributor cap should be regarded as the limit of overhaul on these units, as few other spares are available separately. Refer to Chapter 1 for these procedures. It is possible to renew the vacuum unit, but this must then be set up to suit the advance curve of the engine by adjustment of the serrated cam on the baseplate; this is best left to a Renault dealer or automotive electrician.
9 When the distributor has seen extended service, and the shaft, bushes and centrifugal mechanism become worn, it is advisable to purchase a new distributor.

Refitting

10 To refit the distributor, first check that the engine is still at the TDC position with No 1 cylinder on compression. If the engine has been turned while the distributor was removed, return it to the correct position as previously described. Also make sure that the seal is in position on the base of the distributor.
11 With the rotor arm pointing directly away from the engine and the vacuum unit at approximately the 5 o'clock position, slide the distributor into the cylinder block, and turn the rotor arm slightly until the offset peg on the distributor drive dog positively engages with the drivegear.
12 Align the previously-made reference marks on the distributor base and cylinder block. If a new distributor is being fitted, position the distributor body so that the rotor arm points toward the No 1 spark plug HT lead segment in the cap. With the distributor in this position, refit the clamp and secure with the retaining nut.
13 Reconnect the LT lead at the connector, refit the vacuum advance pipe and secure the wiring loom in the support clip.
14 Refit the No 1 spark plug, the distributor cap and the spark plug HT leads.
15 Check and if necessary adjust the ignition timing, as described in Chapter 1.

4 Ignition coil -
removal and refitting
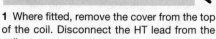

1 Where fitted, remove the cover from the top of the coil. Disconnect the HT lead from the coil.
2 Identify the LT leads for position, then disconnect them from the terminals on the coil.
3 Loosen the mounting nut, and slide the coil and bracket from the mounting stud.
4 Refitting is a reversal of removal, but if necessary wipe clean the top of the coil to prevent any tracking of the HT current.

5 Transistorised assistance unit
- removal and refitting
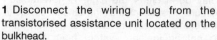

1 Disconnect the wiring plug from the transistorised assistance unit located on the bulkhead.
2 Unscrew the mounting nuts and remove the unit.
3 Refitting is a reversal of removal.

Part B: Electronic ignition system

6 General information and precautions

This part of Chapter 5 deals with the fully-electronic ignition system fitted to most Clio models. The system fitted to carburettor models is known as AEI (Allumage Electronique Intégrale). A modified version of this system is fitted to models with fuel injection, one computer having control over both the injection and the ignition systems.

Operation of both systems is similar. An engine speed sensor, bolted to the flywheel housing, provides signals indicating engine speed and crankshaft angular position. A vacuum sensor provides a signal varying with inlet manifold vacuum, which is an indication of load. The signals are processed by a computer, and used to determine the moment when ignition must occur. The computer controls an amplifier which drives the ignition coil.

Fuel injection models (except the 1171 cc engine) also have a knock sensor, which provides a signal if pre-ignition (pinking) occurs. This enables the computer to advance ignition timing to the most efficient point, without risking engine damage through pre-ignition. It also means that the system is self-correcting for small variations in fuel octane rating.

High tension passes from the coil to the distributor rotor arm, and thence to the appropriate spark plug. The distributor has no other function than the distribution of HT, speed- and load-related advance being handled by the computer. There is no means of adjusting ignition timing, short of changing the computer or the sensors.

On carburettor models, the computer, amplifier, ignition coil and vacuum sensor are all mounted together on a single module. The only item which can be renewed separately is the ignition coil.

On fuel injection models, the ignition module consists of the amplifier and the ignition coil, mounted together. The computer is shared with the fuel injection system, and is described in detail in Chapter 4; the same applies to the manifold vacuum (MAP) sensor.

The following precautions must be observed, to prevent damage to the ignition system components and to reduce risk of personal injury.

(a) *Ensure that the ignition is switched off before disconnecting any of the ignition wiring.*
(b) *Ensure that the ignition is switched off before connecting or disconnecting any*

ignition test equipment, such as a timing light.

(c) *Do not connect a suppression condenser or test light to the ignition coil negative terminal.*

(d) *Do not connect any test appliance or stroboscopic timing light requiring a 12-volt supply to the ignition coil positive terminal.*

(e) *Do not allow an HT lead to short out or spark against the computer control unit body.*

(f) *Do not earth the coil primary or secondary circuits.*

 Warning: The voltages produced by the electronic ignition system are considerably higher than those produced by conventional systems. Extreme care must be taken when working on the system with the ignition switched on. Persons with surgically-implanted cardiac pacemaker devices should keep well clear of the ignition circuits, components and test equipment.

7 Ignition system - testing

Carburettor models

1 There are two main symptoms indicating faults in the ignition system. Either the engine will not start or fire, or the engine is difficult to start and misfires. Each of these possibilities is covered separately in the following paragraphs.

Engine turns normally but fails to start

2 First check the HT leads, distributor cap and spark plugs.

3 Remove the two connectors from the bottom of the ignition module, and check the terminals for corrosion. If necessary, remove and refit the connectors several times in order to clean the terminals. If they are very dirty, scrape the terminals with a suitable instrument.

4 Using a voltmeter, and with the ignition switched on, check that the voltage between the coil positive (+) terminal at the interference suppression condenser output and earth, is at least 9.5 volts.

5 Disconnect the supply connector (A in Fig. 5.3) and connect a voltmeter between terminal 1 and earth. Switch on the ignition and attempt to start the engine. The reading on the voltmeter should be at least 9.5 volts. If not, check the battery voltage and recharge if necessary. Also check the ignition module supply wiring.

6 With connector (A) still disconnected and the ignition switched off, connect an ohmmeter between terminal 2 and the vehicle earth. If the reading is not zero, check the earth wiring from the connector for a possible loose connection.

7 With connector (A) still disconnected and

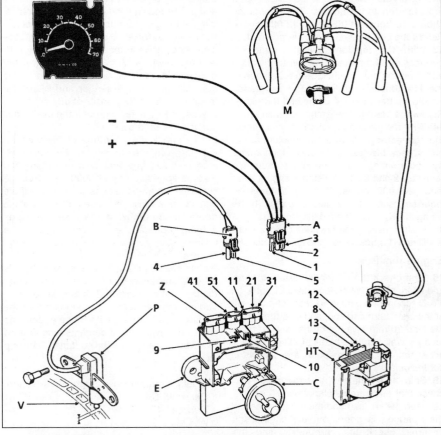

Fig. 5.3 Electronic ignition system components - carburettor models (Sec 6)

1 *Positive supply (+)*
2 *Earth (-)*
3 *Tachometer connector*
4 *Engine position/speed sensor terminal*
5 *Engine position/speed sensor terminal*
7 *Coil positive (+) and interference suppression terminals*
8 *Coil negative (-) terminal*
9 *Coil positive (+) terminal on the control unit*
10 *Coil negative (-) terminal on the control unit*

11 *Control unit positive (+) supply*
12 *Coil HT terminal*
13 *Coil positive (+) terminal for interference suppression capacitor*
21 *Control unit negative (-) terminal*
31 *Tachometer output*
41 *Engine position/speed signal*
51 *Engine position/speed signal*

A *Supply/tachometer connector*
B *Position/speed signal connector*
C *Vacuum capsule*
E *Control unit*
HT *Ignition coil*
M *Distributor cap*
P *Engine position/speed sensor*
V *Flywheel*
Z *Socket for temperature-related timing advance (1721 cc/F2N engine)*

the ignition switched off, connect an ohmmeter between terminals 9 and 11. If the reading is not zero, renew the ignition module.

8 Reconnect connector (A) and switch on the ignition. Connect a voltmeter between terminal 13 and the vehicle earth. If the reading is not at least 9.5 volts, shake the connector and observe if the reading increases. If it is still incorrect, check the coil terminal contacts for corrosion. If still incorrect, renew the connector (A).

9 With the ignition switched off, disconnect the position/speed sensor signal connector (B in Fig. 5.3). Check the resistance of the sensor by connecting an ohmmeter across terminals 4 and 5. The reading should be 200 ± 50

ohms. If the reading is incorrect, renew the sensor.

10 Using a feeler blade, check that the clearance between the end of the sensor (P) and the flywheel (V) is 1.0 ± 0.5 mm. If the clearance is incorrect, renew the sensor.

11 Reconnect both connectors (A) and (B), then remove the ignition coil (Section 9). Connect a 12-volt test light (2 watts maximum) between the coil feed terminals (9 and 10) on the ignition module. Spin the engine on the starter motor, and check that the test light flashes. If the light does not flash, renew the ignition module.

12 With the ignition coil still removed, connect an ohmmeter between terminals 7

5

and 12 on the coil to check the resistance of the high-tension windings. The resistance should be between 2000 and 12 000 ohms. Renew the coil if the resistance is incorrect.

13 With the ignition coil still removed, connect an ohmmeter between terminals 6 and 7 on the coil, to check the resistance of the low-tension windings. The resistance should be between 0.4 ohms and 0.8 ohms. Renew the coil if the resistance is incorrect.

14 With the ignition switched off, disconnect the connector (A) from the computer control unit. Check the resistance of the tachometer by connecting an ohmmeter between terminals 2 and 3. The resistance should be at least 20 000 ohms. If the resistance is incorrect, check the wiring for a fault. If the wiring is OK, renew the tachometer.

15 If after making the previous checks there is still no HT spark, renew the ignition module.

Engine misfires

16 First check the HT leads, distributor cap and spark plugs, with reference to Chapter 1.

17 Disconnect the main HT lead from the coil at the distributor cap end, and hold the end of the lead 20 mm away from the cylinder head using a well-insulated pair of pliers. *Do not allow the HT lead to touch the computer control unit.*

18 Spin the engine on the starter motor, and check that there are regular strong HT sparks between the HT lead and the cylinder head. If the sparking is good, but the engine still misfires, check the carburation and the mechanical condition of the engine (valve clearances, valve timing, compression) for possible faults. If there are no sparks, continue with the following checks.

19 Disconnect the supply connector (A in Fig. 5.3) and connect a voltmeter between terminal 1 and earth. Switch on the ignition and attempt to start the engine. The reading on the voltmeter should be at least 9.5 volts. If not, check the battery voltage and recharge if necessary. Also check the computer control unit supply wiring.

20 With the ignition switched off, disconnect the connector (B). Check the resistance of the engine speed and position sensor by connecting an ohmmeter across terminals 4 and 5. The reading should be 200 ± 50 ohms. If the reading is incorrect, renew the sensor.

21 Using a feeler blade, check that the clearance between the end of the sensor (P) and the flywheel (V) is 1.0 ± 0.5 mm. If the clearance is incorrect, renew the sensor. If it is correct, remove the sensor and clean the magnetic end of the sensor of any oil, dirt or grease. Refit the sensor, but if the misfire still persists, renew the sensor.

22 To check the condition of the vacuum capsule on the computer control unit, reconnect all plugs, then start the engine and hold its speed steady at 3000 rpm. Pull the vacuum pipe from the capsule, and note if the engine speed falls. If it does, the capsule is operating correctly; if the speed remains constant, check the vacuum pipes. If the misfire still persists, renew the ignition module.

23 On 1721 cc (F2N) engine models, there is an additional connector fitted to the computer control unit at position Z; this is linked to a coolant temperature switch, located on the cylinder head at the heater hose outlet. The switch has two sets of contacts, and is used to advance the ignition timing during warm-up as follows:

Coolant temperature	Ignition advance
Below 45°C	*15°*
Between 45° and 77°C	*6°*
Above 77°C	*0*

Fuel injection models

24 As mentioned in Chapter 4, there is little that can be done by way of testing the control circuitry on fuel injection models without special test equipment. The following paragraphs provide suggestions for preliminary checks which can be made without such equipment.

Engine turns normally but fails to start

25 Check the HT leads, distributor cap and spark plugs as described in Chapter 1.

26 Remove the connectors from the ignition amplifier, and check the terminals for corrosion. If necessary, remove and refit the connectors several times in order to clean the terminals. If they are very dirty, scrape the terminals with a suitable instrument.

27 Check the engine speed and position sensor as described in paragraphs 9 and 10, using an ohmmeter at the sensor connector on top of the gearbox.

28 Disconnect the coil terminals and check the resistance of the windings using an ohmmeter. Typical values are given in the Specifications.

Engine misfires

29 Carry out the HT checks as described for carburettor models (paragraphs 16 to 18).

30 Check the engine speed and position sensor as described earlier.

31 If no improvement is produced, consult a Renault dealer or other specialist.

8 Distributor - removal, overhaul and refitting

Removal

1 On models with the 1764 cc (F7P) engine, remove the engine top cover for access to the spark plug leads.

2 On all models, check that the spark plug leads and distributor cap are marked to aid refitting. Pull the leads off the plugs or distributor cap, as wished. Unscrew the retaining screws, and place the distributor cap and leads to one side (photos).

1171 cc and 1390 cc (E-type) engines

3 Pull the rotor arm off the distributor shaft, and remove the plastic shield. Use a Torx key to unscrew the two mounting bolts, and withdraw the distributor from the cylinder head. Note that although the upper mounting bolt hole in the distributor body may be slotted, the lower bolt hole is not slotted, so it is not possible to adjust the position of the distributor. Remove the seal from the base of the distributor (photos).

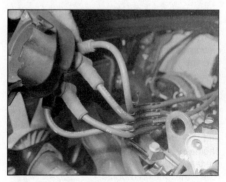

8.2A Removing the distributor cap. Note that spark plug leads are numbered . . .

8.2B . . . and so is the distributor cap cover (No 1 cylinder is at flywheel end)

8.3A Removing the rotor arm (E-type engine)

8.3B Distributor mounting bolts (arrowed) - E-type engine

8.3C Removing the distributor (E-type engine)

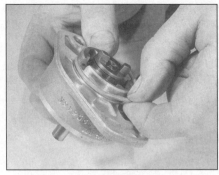

8.3D Removing the seal from the base of the distributor

1721 cc (F2N) and 1794 cc (F3P) engines

4 The rotor arm is connected directly to the camshaft. It may be attached by a circlip, or alternatively, it may be bonded. To remove the circlip type, carefully prise it off. To remove the bonded type, twist it using a pair of grips - this should break the bond, and enable the arm to be removed. (It is easy to damage the rotor arm when doing this, so a spare should be available.) Remove the plastic shield (photos).

1764 cc (F7P) engine

5 The procedure is similar to that just described. The rotor arm is fitted to the exhaust camshaft.

Overhaul
All models

6 Check the rotor arm and distributor cap as described in Chapter 1, and renew them if necessary. Note that on some models, a cover is fitted to the distributor cap; this should be removed for inspection and cleaning (photo).

1171 cc and 1390 cc (E-type) engines

7 Check the amount of play between the distributor shaft and body. This should not be excessive unless the engine has covered a high mileage. If it is excessive, the complete distributor should be renewed.
8 Check the condition of the distributor base seal, and renew it if necessary.

Refitting

9 Refitting is a reversal of removal. Do not attempt to bond a rotor arm to a camshaft designed to accept a rotor arm with a circlip.

9 Ignition coil (carburettor models) - removal and refitting

1 Disconnect the HT lead from the coil.
2 Remove the four screws and withdraw the coil from the ignition module, at the same time disconnecting the LT terminals.
3 Refitting is a reversal of removal, making sure that the wires are fitted securely.

10 Ignition module (carburettor models) - removal and refitting

1 Disconnect the battery earth lead. Disconnect the HT lead from the coil.
2 Disconnect the two or three multi-plugs (according to model) from the bottom of the unit.
3 Disconnect the hose from the vacuum advance capsule (photo).
4 Unbolt and remove the ignition module from the bulkhead.
5 The coil is the only part of the module which can be renewed separately. If a fault develops in the computer, the amplifier or the vacuum unit, the complete module must be renewed.
6 Refitting is a reversal of removal.

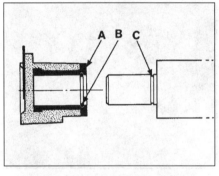

Fig. 5.4 Sectional view of rotor arm connection to the camshaft - 1721 cc (F2N) and 1794 cc (F3P) engines (Sec 8)

A Insert   B Circlip   C Groove in camshaft

8.4A Removing the rotor arm . . .

5

8.4B . . . and the plastic shield - 1794 cc (F3P) engine shown

8.6 Separating the distributor cap and cover

10.3 Vacuum hose (arrowed) connected to the vacuum advance capsule (carburettor models)

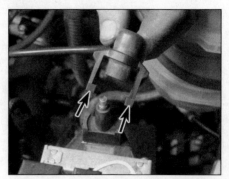

11.1 Disconnecting the HT lead from the coil. Note securing lugs (arrowed)

11.2 Disconnecting the multi-plugs from the ignition module. Third socket is not used on this version

11.4 Removing the ignition module

12.3 Removing an engine speed and position sensor bolt. Only the special shouldered bolts may be used

11 Ignition module (fuel injection models) - removal and refitting

Note: *The ignition coil is not renewable separately on fuel injection models.*

1 Disconnect the battery earth lead. Release the securing lugs and disconnect the HT lead from the coil (photo).
2 Disconnect the two multi-plugs from the module (photo).
3 Remove the two nuts which secure the module. Note that one of them also secures the radio interference suppression condenser.
4 Remove the ignition module (photo).
5 Refit by reversing the removal operations.

12 Engine speed and position sensor - removal and refitting

1 Remove the air cleaner or the air inlet trunking if necessary for access.
2 Disconnect the sensor wire from the bottom of the ignition module (carburettor models) or at the multi-plug on top of the gearbox (fuel injection models).
3 Unbolt and remove the sensor from the aperture at the top of the gearbox bellhousing (photo).

4 Refit by reversing the removal operations. It is important to use only the special shouldered bolts to attach the sensor to the gearbox bellhousing, as these determine its correct distance from the flywheel.

13 Knock sensor - removal and refitting

1 The location of the knock sensor varies according to engine type. On the 1390 cc (E7J) engine, it is on the rear of the engine, below the inlet manifold; access is from below. On the 1794 cc (F3P) engine, it is on the front of the cylinder head. On the 1764 cc (F7P) engine, the sensor is again on the front of the cylinder head, but it is obscured by the inlet manifold. A deep 24 mm socket will be needed to reach it, and it may also be necessary to remove the radiator for access.
2 Disconnect the sensor multi-plug. Unscrew the sensor and remove it (photo).
3 Refit by reversing the removal operations. Make sure that the multi-plug is securely connected.

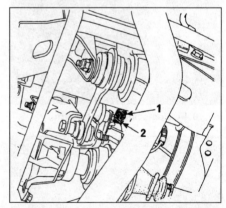

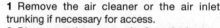
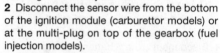

Fig. 5.5 Knock sensor (2) on the 1390 cc (E7J) engine viewed from below (Sec 13)

1 Multi-plug

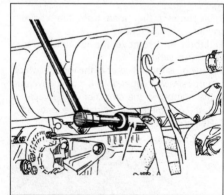

Fig. 5.6 Knock sensor removal on the 1764 cc (F7P) engine requires a deep socket (arrowed) (Sec 13)

13.2 Disconnecting the knock sensor multi-plug - 1794 cc (F3P) engine shown

Chapter 6 Clutch

Contents

Degrees of difficulty

Easy, suitable for novice with little experience	**Fairly easy,** suitable for beginner with some experience	**Fairly difficult,** suitable for competent DIY mechanic	**Difficult,** suitable for experienced DIY mechanic	**Very difficult,** suitable for expert DIY or professional

Specifications

General

Clutch type . . . . Single dry plate, diaphragm spring, cable-operated
Adjustment . . . . Automatic

Clutch disc

Diameter:
  1108 cc (C1E), 1171 cc (E5F and E7F), 1390 cc (E6J and E7J) . . . . . 181.5 mm
  1721 cc (F2N), 1764 cc (F7P) and 1794 cc (F3P) . . . . 200.0 mm
Lining thickness (new, and in compressed position) . . . . 7.7 mm
Number of springs:
  8-valve engine models 6
  16-valve engine models 4

Torque wrench settings

	Nm	lbf ft
Clutch cover bolts:		
7 mm diameter	25	18
8 mm diameter	30	22

1 General information

All manual gearbox models are equipped with a cable-operated clutch. The unit consists of a steel cover which is dowelled and bolted to the rear face of the flywheel, and contains the pressure plate and diaphragm spring.

The clutch disc is free to slide along the gearbox splined input shaft. The disc is held in position between the flywheel and the pressure plate by the pressure of the diaphragm spring. Friction lining material is riveted to the clutch disc, which has a spring-cushioned hub to absorb transmission shocks and help ensure a smooth take-up of the drive.

The clutch is actuated by a cable, controlled by the clutch pedal. The clutch release mechanism consists of a release arm and bearing which are in permanent contact with the fingers of the diaphragm spring.

Depressing the clutch pedal actuates the release arm by means of the cable. The arm pushes the release bearing against the diaphragm fingers, so moving the centre of the diaphragm spring inwards. As the centre of the spring is pushed in, the outside of the spring pivots out, so moving the pressure plate backwards and disengaging its grip on the clutch disc.

When the pedal is released, the diaphragm spring forces the pressure plate into contact with the friction linings on the clutch disc. The disc is now firmly sandwiched between the pressure plate and the flywheel, thus transmitting engine power to the gearbox.

Wear of the friction material on the clutch disc is automatically compensated for by a self-adjusting mechanism attached to the clutch pedal. The mechanism consists of a serrated quadrant, a notched cam and a tension spring. One end of the clutch cable is attached to the quadrant, which is free to pivot on the pedal, but is kept in tension by a spring. When the pedal is depressed, the notched cam contacts the quadrant, thus locking it and allowing the pedal to pull the cable and operate the clutch. When the pedal is released, the tension spring causes the notched cam to move free of the quadrant; at the same time tension is maintained on the cable, keeping the release bearing in contact with the diaphragm spring. As the friction material on the disc wears, the self-adjusting quadrant will rotate when the pedal is released, and the pedal free play will be maintained between the notched cam and the quadrant.

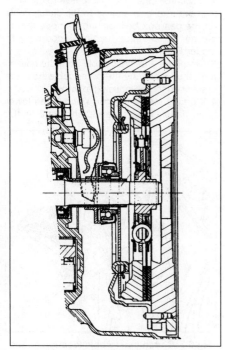

Fig. 6.1 Cross-section of the clutch components (Sec 1)

6

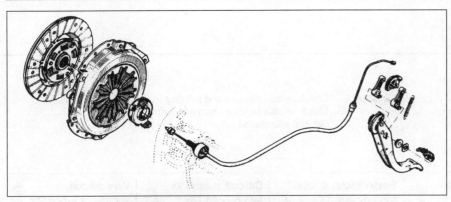

Fig. 6.2 Clutch cable and pedal components (Sec 2)

2.3B Withdrawing the clutch outer cable from the bracket on the bellhousing

2.4 Disconnecting the inner cable end from the quadrant

2.3A Disconnecting the clutch inner cable from the release fork

2 Clutch cable - removal and refitting

1 Disconnect the battery negative terminal.
2 In order to gain access to the clutch cable on the gearbox bellhousing, remove the air cleaner or air inlet ducting, according to model. Refer to Chapter 4.
3 Disengage the inner cable from the release fork, then withdraw the outer cable from the bracket on the bellhousing (photos).
4 Working inside the car, remove the lower facia panel. Press the clutch pedal to the floor. Release the pedal, and free the inner cable end from the serrated quadrant on the self-adjusting mechanism (photo).
5 Using a screwdriver, tap out the inner cable guide from the top of the pedal.
6 With the inner cable released, push the outer cable out of its location in the bulkhead.
7 Pull the cable through into the engine compartment, detach it from the support clips and remove it from the car.
8 To refit the cable, thread it through from the engine compartment, place it over the self-adjusting cam, and connect the inner cable end to the quadrant. Make sure that the self-adjusting cam support arms return to their rest position freely under the tension of the return spring.

9 Refit the inner cable guide to the top of the pedal.
10 Working in the engine compartment, slip the other end of the cable through the bellhousing bracket, and connect the inner cable to the release fork. Refit the cable to the support clips.
11 Depress the clutch pedal to draw the outer cable into its locating hole in the bulkhead, ensuring that it locates properly. At the same time, the inner cable guide will be automatically pulled onto the top of the pedal to hold the inner cable in position.
12 Depress the clutch pedal several times in order to allow the self-adjusting mechanism to set the correct free play.
13 When the self-adjusting mechanism on the clutch pedal is functioning correctly, there should be a minimum of 20 mm slack in the cable. To check this dimension, pull out the inner cable near the release fork on the gearbox as shown in Fig. 6.4. If there is less than the minimum slack in the cable, the self-adjusting quadrant should be checked for seizure or possible restricted movement.
14 Depress the clutch pedal fully, and check that the total movement at the top of the release fork is between 17 and 18 mm, as shown in Fig. 6.5. This movement ensures

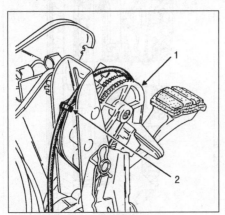

Fig.6.3 Clutch inner cable location on the self-adjusting quadrant (Sec 2)

1 Quadrant 2 Guide

Fig.6.4 Checking the clutch inner cable slack at the release fork end (Sec 2)

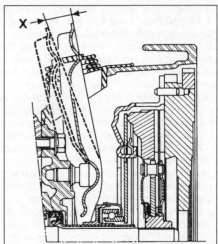

Fig.6.5 Checking the clutch release fork movement (X) (Sec 2)

For dimension X, see text

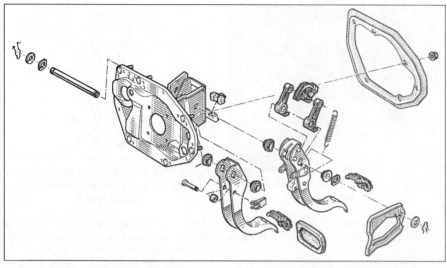

Fig. 6.6 Pedal components and bracket (right-hand drive) (Sec 3)

that the clutch pedal stroke is correct. If not, make sure that the quadrant and support arms are free to turn on their respective pivots, and that the spring has not lost its tension. If necessary, check the free length of the spring against a new one. Also check that the inner cable is not seizing in the outer cable.

15 Finish by refitting the lower facia panel.

3 Clutch pedal -
removal and refitting

Removal

1 Disconnect the battery negative terminal.
2 Proceed as described for cable removal in the previous Section, but without actually removing the cable from the bulkhead.
3 Extract the retaining clips from both ends of the clutch/brake pedal cross-shaft, and recover the washers. Note that the clip ends engage with the pedal bracket (photos).
4 Unscrew the mounting nuts, and remove the pedal shaft left-hand support bracket from the left-hand side of the clutch pedal. Remove the spring washer (photo).
5 Partially withdraw the clutch/brake pedal

cross-shaft to the right-hand side, and remove the pedal together with the self-adjusting mechanism and bushes (photo). Note that the thicker bush is on the right-hand side.
6 With the pedal removed, unhook and remove the self-adjusting quadrant return spring. Note the fitted position of the self-adjusting components for correct refitting.
7 Remove the bushes and withdraw the self-adjusting support arms and quadrant (photo). Inspect these components and renew them if worn.

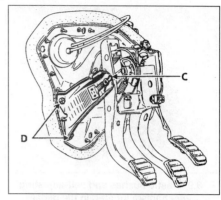

Fig. 6.7 Support bracket removal (Sec 3)
Left-hand drive shown
1 Mounting nuts   2 Retaining clips

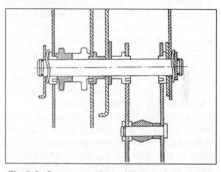

Fig.6.8 Cross-section of the pedal bushes (Sec 3)

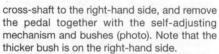

3.3A Remove the pedal cross-shaft retaining clip (arrowed) . . .

3.3B . . . and washer

3.4 Removing the pedal shaft left-hand support bracket (right-hand drive shown)

3.5 Removing the clutch pedal

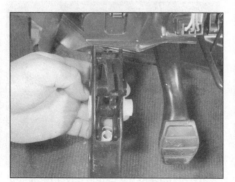

3.7 Removing the self-adjusting mechanism from the pedal

6

3.9 Plastic bushes and self-adjusting mechanism located in the pedal

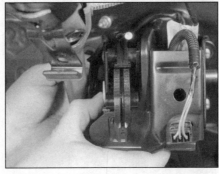

3.11 Locating the pedal assembly in the bracket

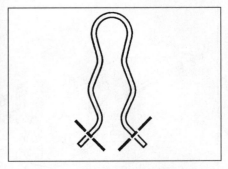

Fig.6.9 The ends of the retaining clip may be shortened as shown to make fitting easier (Sec 3)

Refitting

8 Apply some multi-purpose grease to the bushes, and to the bearing surfaces of the support arms, quadrant and pedal shaft.

9 Locate the plastic bushes in the clutch pedal, making sure that the largest bush is at the brake pedal side, and ensuring that the lugs on the bushes locate in the cut-outs in the pedal (photo). Reconnect the self-adjusting quadrant return spring.

10 To facilitate the refitting procedure, and to hold the bushes and support arms together, it is helpful to assemble the pedal on a dummy shaft. If a suitable shaft is not available, the pedal can still be refitted, but it will be necessary to hold the bushes together until the retaining clips are in place.

11 Locate the pedal assembly in the bracket, and push the pedal shaft through from the right-hand side (photo). Refit the right-hand retaining spring clip and washer, making sure that the clip engages correctly in the bracket aperture.

12 Fit the spring washer and support bracket on the left-hand end of the shaft. Screw on the bracket mounting nuts hand-tight only at this stage.

13 Using a G-clamp or similar tool, clamp the sides of the bracket together and fit the retaining clip in its groove, also making sure that the ends of the clip engage the aperture in the support bracket. The ends of the clip may be shortened by 2.0 mm if necessary in

order to make it easier to fit, but make sure that this has not already been done. Refer to Fig. 6.9.

14 Remove the G-clamp and fully tighten the support bracket mounting nuts. Note that the support bracket has a locking notch to prevent the shaft from rotating.

15 Reconnect the cable, and check the operation of the self-adjusting mechanism as described in the previous Section.

4 Clutch assembly - removal, inspection and refitting

> **Warning: Dust created by clutch wear and deposited on the clutch components may contain asbestos which is a health hazard. DO NOT blow it out with compressed air or inhale any of it. DO NOT use petrol or petroleum-based solvents to clean off the dust. Brake system cleaner or methylated spirit should be used to flush the dust into a suitable receptacle. After the clutch components are wiped clean with rags, dispose of the contaminated rags and cleaner in a sealed, marked container.**

Removal

1 Access to the clutch may be gained in one of two ways. Either the gearbox may be removed independently, as described in

Chapter 7, or the engine/gearbox unit may be removed as described in Chapter 2, and the gearbox separated from the engine on the bench. If the gearbox is being removed independently on the 1721 cc (F2N), 1764 cc (F7P) and 1794 cc (F3P) engines, it need only be moved to one side for access to the clutch.

2 Having separated the gearbox from the engine, unscrew and remove the clutch cover retaining bolts. Work in a diagonal sequence and slacken the bolts only a few turns at a time. Hold the flywheel stationary by positioning a screwdriver over the dowel on the cylinder block and engaging it with the starter ring gear (photos).

3 Ease the clutch cover off its locating dowels. Be prepared to catch the clutch disc, which will drop out as the cover is removed. Note which way round the disc is fitted (photo).

Inspection

4 With the clutch assembly removed, clean off all traces of asbestos dust using a dry cloth. This is best done outside or in a well-ventilated area; refer to the warning at the beginning of this Section.

5 Examine the linings of the clutch disc for wear or loose rivets, and the disc rim for distortion, cracks, broken torsion springs and worn splines (photo). The surface of the friction linings may be highly glazed, but as long as the friction material pattern can be

4.2A Unscrewing the clutch cover retaining bolts, showing a screwdriver engaged with the starter ring gear

4.2B Removing the clutch cover retaining bolts. Note that the gearbox has not been removed, just moved to one side

4.3 Removing the clutch cover and disc from the flywheel

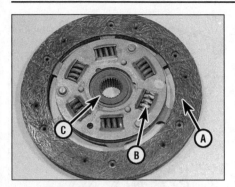

4.5 Inspect the clutch disc linings (A), springs (B) and splines (C)

4.6A Check the machined face of the pressure plate (arrowed) . . .

4.6B . . . and the diaphragm spring, paying particular attention to the tips (arrowed)

clearly seen, this is satisfactory. If there is any sign of oil contamination, indicated by shiny black discoloration, the disc must be renewed and the source of the contamination traced and rectified. This will be a leaking crankshaft oil seal, gearbox input shaft oil seal, or both. The renewal procedure for the former is given in Chapter 2. Renewal of the gearbox input shaft oil seal should be entrusted to a Renault garage, as it involves dismantling the gearbox and the renewal of the clutch release bearing guide tube using a press. The disc must also be renewed if the linings have worn down to, or just above, the level of the rivet heads.

6 Check the machined faces of the flywheel and pressure plate. If either is grooved, or heavily scored, renewal is necessary. The pressure plate must also be renewed if any cracks are apparent, or if the diaphragm spring is damaged or its pressure suspect (photos).

7 Take the opportunity to check the condition of the release bearing, as described in Section 5.

8 It is good practice to renew the clutch disc, pressure plate and release bearing as an assembly. Renewing the disc alone is not always satisfactory. A clutch repair kit can be obtained containing the new components.

9 Renault clutch kits for the F-type engine contain a special dummy bush which may be fitted in the crankshaft bore to enable the use of the clutch centring tool also supplied in the kit. To fit this bush, first clean the bore in the end of the crankshaft and apply locking fluid to the outer surface of the bush. Press the bush fully into the crankshaft using a length of tubing of 38 mm outside diameter, making sure that the open end of the bush faces outwards as shown in Fig. 6.10.

Refitting

10 Before commencing the refitting procedure, apply a little high-melting-point grease to the splines of the gearbox input shaft. (A sachet of suitable grease may be supplied with the clutch kit.) Distribute the grease by sliding the clutch disc on and off the splines a few times. Remove the disc and wipe away any excess grease.

11 It is important that no oil or grease is allowed to come into contact with the friction material of the clutch disc or the pressure plate and flywheel faces. It is advisable to refit the clutch assembly with clean hands, and to wipe the pressure plate and flywheel faces with a clean dry rag before assembly begins.

12 Begin reassembly by placing the clutch disc against the flywheel, with the side having the larger offset facing away from the flywheel.

13 Place the clutch cover over the dowels.

Refit the retaining bolts and tighten them finger-tight so that the clutch disc is gripped, but can still be moved.

14 The clutch disc must now be centralised so that, when the engine and gearbox are mated, the splines of the gearbox input shaft will pass through the splines in the centre of the clutch disc hub. If this is not done accurately, it will be impossible to refit the gearbox.

15 Centralisation can be carried out quite easily by inserting a round bar through the hole in the centre of the clutch disc, so that the end of the bar rests in the hole in the end of the crankshaft. Note that a plastic centralising tube is supplied with Renault clutch kits, making the use of a bar unnecessary (photo).

16 If a bar is being used, move it sideways or up and down until the clutch disc is centralised. Centralisation can be judged by removing the bar and viewing the clutch disc hub in relation to the bore in the end of the crankshaft. When the bore appears exactly in the centre of the clutch disc hub, all is correct.

17 If a non-Renault clutch is being fitted, an alternative and more accurate method of centralisation is to use a commercially-available clutch aligning tool obtainable from most accessory shops (photo).

18 Once the clutch is centralised, progressively tighten the cover bolts in a diagonal sequence to the torque setting given

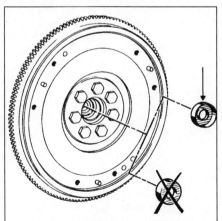

Fig.6.10 Fit the dummy bush in the crankshaft bore with the open end facing outwards (Sec 4)

4.15 Centralising the clutch disc using the special tube supplied with Renault clutch kits

4.17 Using a clutch alignment tool to centralise the clutch disc

6

4.18 Tightening the clutch cover bolts

in the Specifications (photo). Remove the centralising device.

19 The gearbox can now be refitted to the engine, referring to the appropriate Chapter of this manual. Check the functioning of the clutch pedal as described in Section 2.

5 Clutch release bearing - removal, inspection and refitting

Removal

1 To gain access to the release bearing, it is necessary to separate the engine and gearbox as described at the beginning of the previous Section.

2 With the gearbox removed from the engine, tilt the release fork and slide the bearing assembly off the gearbox input shaft guide tube.

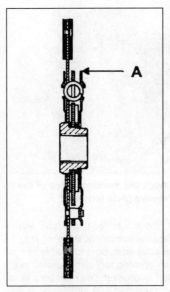

Fig.6.11 Clutch disc offset (A) faces away from flywheel (Sec 4)

3 To remove the release fork, disengage the rubber cover and then pull the fork off its pivot ball stud.

Inspection

4 Check the bearing for smoothness of operation. Renew it if there is any roughness or harshness as the bearing is spun. It is good practice to renew the bearing as a matter of course during clutch overhaul, regardless of its apparent condition.

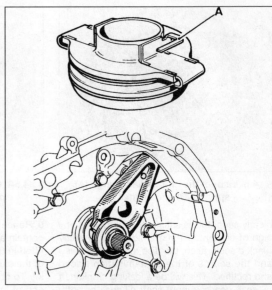

Fig. 6.12 Clutch release components. Clip (A) on bearing carrier must engage with release fork (Sec 5)

Refitting

5 Refitting the release fork and release bearing is the reverse sequence to removal, but note the following points.

(a) Lubricate the release fork pivot ball stud and the release bearing-to-diaphragm spring contact areas sparingly with molybdenum disulphide grease.

(b) Ensure that the clip on the bearing carrier engages with the release fork (Fig. 6.12).

Chapter 7 Transmission

Contents

Degrees of difficulty

Easy, suitable for novice with little experience	Fairly easy, suitable for beginner with some experience 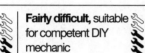	Fairly difficult, suitable for competent DIY mechanic	Difficult, suitable for experienced DIY mechanic	Very difficult, suitable for expert DIY or professional

Specifications

Part A: Manual gearbox

Type .... Four or five forward speeds (all synchromesh) and reverse. Final drive differential integral with main gearbox

Designation

Four-speed units .... JB0 or JB4
Five-speed units .... JB1, JB3 or JB5

Gear ratios (typical)

	4-speed	5-speed
1st	3.7 : 1	3.7 : 1 (JB3 048 - 3.1 : 1)
2nd	2.1 : 1	2.1 : 1 (JB3 048 - 1.8 : 1)
3rd	1.3 : 1	1.3 : 1
4th	0.9 : 1	1.0 : 1
5th	-	0.8 : 1
Reverse	3.6 : 1	3.6 : 1

Final drive ratios

JB0 (031) and JB4 (004)	3.4 : 1
JB0 (032) and JB4 (008)	3.6 : 1
JB1 (038 and 043)	3.5 : 1
JB1 (046)	4.3 : 1
JB3 (041)	3.3 : 1
JB3 (045)	3.4 : 1
JB3 (046)	3.9 : 1
JB3 (048)	4.2 : 1
JB5 (005 and 010)	3.6 : 1
JB5 (015)	4.2 : 1

Torque wrench settings

	Nm	lbf ft
Gearbox mounting nuts/bolts	40 to 50	30 to 37
Clutch bellhousing to engine	50	37
Gearchange casing nuts	15	11
Gearchange link rod clamp nuts and bolts	30	22

7

Part B: Automatic transmission

Type

MB1	Three forward speeds and reverse. Final drive differential integral with transmission
AD4	Four forward speeds and reverse. Final drive differential integral with transmission

Application

Carburettor models (E6J engine)	MB1
Fuel injection models (E7J engine)	AD4

Ratios (typical)

	MB1	**AD4**
1st	2.50 : 1	2.71 : 1
2nd	1.50 : 1	1.55 : 1
3rd	1.00 : 1	1.00 : 1
4th	-	0.68 : 1
Reverse	2.00 : 1	2.11 : 1
Final drive	3.87 : 1	3.76 : 1 (E7J 719 - 4.12 : 1)

Torque wrench settings

	Nm	**lbf ft**
Driveplate to crankshaft	50 to 55	37 to 41
Driveplate to torque converter:		
MB1	25	18
AD4	15	11
Gauze filter:		
MB1	9	7
AD4	5	4
Sump pan:		
MB1	6	4
AD4	10	7
Fluid cooler:		
MB1	40	30
AD4	25	18
Transmission mounting bolts	40	30

Part A: Manual gearbox

1 General information

The manual gearbox is either of four-speed (JB0 or JB4) or five-speed (JB1, JB3 or JB5) type, with one reverse gear. Baulk ring synchromesh gear engagement is used on all the forward gears. The final drive (differential) unit is integral with the main gearbox, and is located between the mechanism casing and the clutch and differential housing. The gearbox and differential both share the same lubricating oil.

Gearshift is by means of a floor-mounted lever, connected by a remote control housing and gearchange rod to the gearbox fork contact shaft.

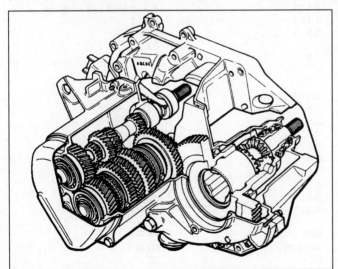

Fig. 7.1 Cutaway view of the four-speed gearbox (Sec 1)

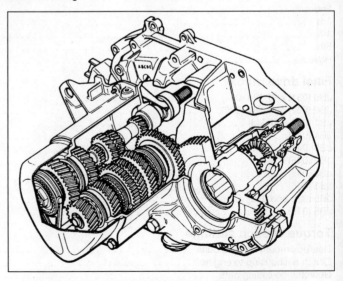

Fig. 7.2 Cutaway view of the five-speed gearbox (Sec 1)

2 Gearchange linkage/ mechanism - adjustment

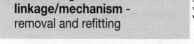

1 Apply the handbrake, then jack up the front of the car and support it on axle stands.

2 Unbolt and remove the splash guard from under the gearbox.

3 Select 1st gear on the gearbox by moving the lever to the correct position (Fig. 7.6). Renault technicians use a special tool to hold the lever in position and take up any free play (Fig. 7.7); a suitable alternative tool can be made from flat metal bar or wood.

4 Using a feeler blade, check that the clearance between the reverse stop-ring on the gear lever and the inclined plane on the right-hand side of the gear lever housing is between 2 and 5 mm (Fig. 7.8).

5 If adjustment is necessary, unhook the return spring from the gear lever end of the link rod, then loosen the clamp bolt at the gearbox end of the link rod so that the rod can be moved on the clevis.

6 Move the gear lever so that the reverse stop-ring is against the inclined plane on the housing, then insert a 2 mm feeler blade between the ring and plane. Hold the lever in this position, then tighten the clamp bolt.

7 Remove the holding tool and refit the return spring.

8 Recheck the clearance shown in Fig. 7.8.

9 Check that all gears can be selected, then refit the splash guard and tighten the bolts. Lower the car to the ground.

3 Gearchange linkage/mechanism - removal and refitting

1 Working inside the car, prise the gear lever gaiter from the centre console.

2 Apply the handbrake, then jack up the front of the car and support it on axle stands.

3 Working beneath the car, disconnect the exhaust pipe flexible mountings. Unbolt and remove the splash guard from under the gearbox.

4 Unhook the return spring from the link rod.

5 Pull back the rubber boot from the front end of the link rod. Remove the bolt and disconnect the rod from the gearbox lever. Recover the bush and sleeve. Note that the clevis at the front end of the rod is offset, and must be refitted correctly.

6 Remove the nuts securing the casing assembly to the underbody. Lower the assembly, at the same time pulling the exhaust system to one side.

7 Mark the link rod and gear lever clevis in relation to each other. Unscrew the pinch-bolt and remove the rod from the clevis.

8 Grip the gear lever in a vice, then remove the knob and gear lever gaiter. The knob is bonded to the lever, and may be hard to remove.

9 Extract the circlip from the bottom of the

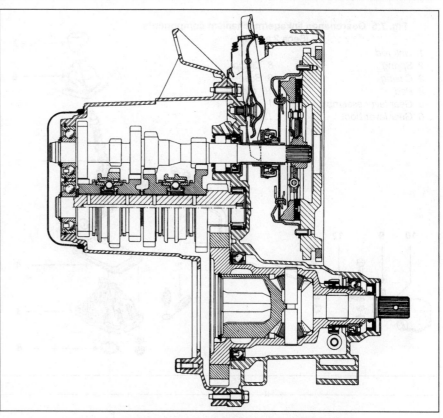

Fig. 7.3 Sectional view of the four-speed gearbox (Sec 1)

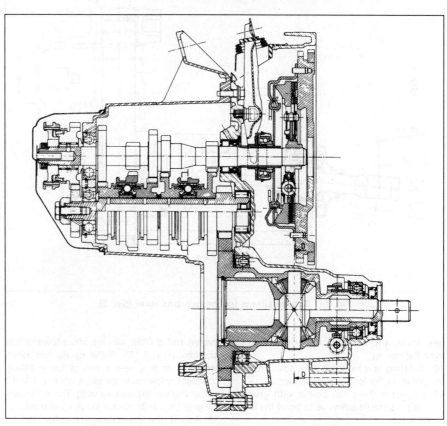

Fig. 7.4 Sectional view of the five-speed gearbox (Sec 1)

7

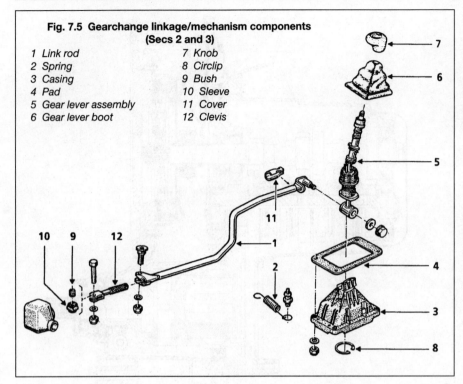

**Fig. 7.5 Gearchange linkage/mechanism components
(Secs 2 and 3)**

1 Link rod	7 Knob
2 Spring	8 Circlip
3 Casing	9 Bush
4 Pad	10 Sleeve
5 Gear lever assembly	11 Cover
6 Gear lever boot	12 Clevis

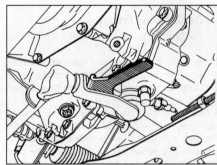

Fig. 7.7 Using the special Renault tool to hold the gearbox lever in 1st gear position (Sec 2)

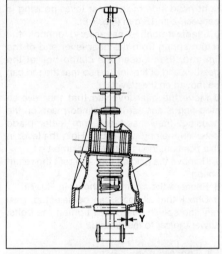

Fig. 7.8 The adjustment gap (Y) should be between 2 and 5 mm (Sec 2)

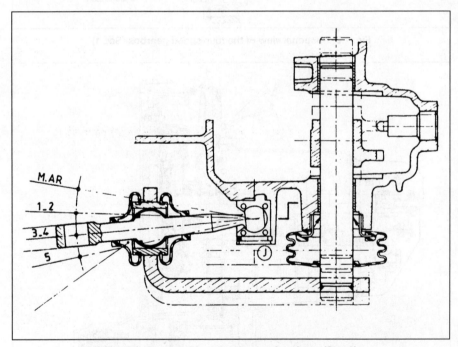

Fig. 7.6 Gear positions for the gearbox lever (Sec 2)

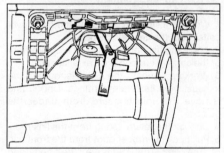

Fig. 7.9 Adjusting the gear lever mechanism with a 2 mm feeler blade (Sec 2)

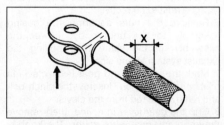

Fig. 7.10 Area of the clevis to be showing when reconnecting the link rod (Sec 3)

X = 10 to 12 mm

gear lever, and withdraw the lever and latch from the casing.

10 Refitting is a reversal of removal, noting the points in the following paragraphs.

11 Lubricate the pivot points with grease, and use a suitable adhesive to bond the knob to the lever.

12 Make sure that the clevis on the front end of the link rod is fitted with the offset towards the gearbox (Fig. 7.10). If the clevis has been removed or if a new clevis is being fitted, connect the link rod to the clevis leaving 10 to 12 mm of the knurled area showing. This will locate the gear lever in its correct longitudinal position.

13 Adjust the gearchange mechanism as described in Section 2.

4.0 Speedometer drive plastic connection (arrowed) on later models

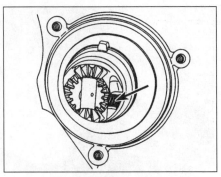

Fig. 7.11 View of the speedometer drivegear (arrowed) with the differential sun wheel removed (Sec 4)

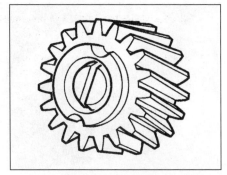

Fig. 7.12 Notches in the speedometer drivegear engage with the shaft (Sec 4)

4 Speedometer drive - removal and refitting

Note: *On some later models, the speedometer drive is taken from the right-hand side of the gearbox, just above the driveshaft. It is not possible to remove the drive on this type. The later type can be identified by the plastic cable connection to the gearbox instead of the clip type connection on earlier models (photo).*

Removal

1 Disconnect the left-hand driveshaft at the gearbox end - refer to Chapter 8. There is no need to disturb the hub end of the driveshaft; the driveshaft/swivel hub assembly can be removed together, as described for engine removal (Chapter 2, Part D).
2 Extract the circlip and thrustwasher, then withdraw the left-hand sun wheel from the differential. The sun wheel also acts as the driveshaft spider housing.
3 Turn the differential until the planet wheels are in a vertical plane so that the speedometer drivegear is visible.
4 Pull out the clip and disconnect the speedometer cable from the outside of the gearbox.
5 Using long-nosed pliers, extract the speedometer drivegear shaft vertically from the outside of the gearbox.
6 Using the same pliers, extract the speedometer drivegear from inside the differential housing, being very careful not to drop it.
7 Examine the drivegear teeth for wear and damage. Renew it if necessary. Note that if the drivegear teeth on the differential are worn or damaged, it will be necessary to dismantle the gearbox - this work should be entrusted to a Renault dealer.

Refitting

8 Using long-nosed pliers, insert the speedometer drivegear into its location.
9 From outside the gearbox, refit the drivegear shaft. Make sure that it engages with the gear location notches correctly, as shown in Fig. 7.12.

10 Refit the speedometer cable, and secure with the clip.
11 Insert the differential sun wheel, then refit the thrustwasher and circlip.
12 Reconnect the left-hand driveshaft with reference to Chapter 8.

5 Differential output oil seal (right-hand side) - renewal

1 Apply the handbrake, then jack up the front of the car and support it on axle stands. Remove the right-hand wheel.
2 Position a suitable container beneath the gearbox, then unscrew the drain plug and allow the oil to drain. (On some models, it may be necessary to remove a splash guard from the bottom of the gearbox first.) When most of the oil has drained, clean and refit the drain plug, tightening it securely.
3 Using a pin punch (5 mm diameter), drive out the double roll pins securing the inner end of the right-hand driveshaft to the differential side gear. New pins will be required when reassembling.
4 Unscrew the nut securing the steering track rod end to the steering arm. Use a balljoint removal tool to separate the balljoint taper.
5 Refer to Chapter 9 and unbolt the brake caliper from the swivel hub. Do not disconnect the hydraulic hose from the caliper. Tie the caliper to the suspension coil spring without straining the hydraulic hose.

6 Loosen (but do not remove) the lower bolt securing the swivel hub to the bottom of the suspension strut. Unscrew and remove the upper bolt, then tilt the swivel hub and disconnect the driveshaft. Take care not to damage the driveshaft rubber bellows.
7 Recover the O-ring from the side gear shaft.
8 Wipe clean the old oil seal, and measure its fitted depth below the casing edge. This is necessary to determine the correct fitted position of the new oil seal if the special Renault fitting tool is not being used.
9 Free the old oil seal, using a small drift to tap the outer edge of the seal inwards so that the opposite edge of the seal tilts out of the casing (photo). A pair of pliers or grips can then be used to pull out the oil seal. Take care not to damage the splines of the differential side gear.
10 Wipe clean the oil seal seating in the casing.
11 Before fitting the new oil seal, it is necessary to cover the splines on the side gear to prevent damage to the oil seal lips. Ideally, a close-fitting plastic cap should be located on the splines. If this is not available, wrap some adhesive tape over the splines.
12 Smear a little grease on the lips of the new oil seal and on the protective cap or tape.
13 Carefully locate the new oil seal over the side gear, and enter it squarely into the casing. Using a piece of metal tube or a socket, tap the oil seal into position to its correct depth, as noted previously (photos).

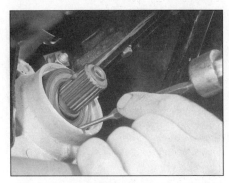

5.9 Tap the old differential output oil seal with a small drift to remove it

5.13A Position the new differential output oil seal on the gearbox . . .

7

5.13B . . . and drive it in with a socket or metal tube

Renault use a special tool to ensure that the oil seal is fitted to the correct depth; it may be possible to hire this tool from a Renault garage or tool hire shop.

13 Remove the plastic cap or adhesive tape, and apply a little grease to the splines of the side gear. Fit a new O-ring to the side gear shaft.

14 Engage the driveshaft with the splines on the side gear so that the roll pin holes are correctly aligned. Tilt the swivel hub and slide the driveshaft onto the side gear, making sure that it enters the oil seal centrally.

15 With the holes aligned, tap the new roll pins into position. Seal the ends of the roll pins with a suitable sealant.

16 Refit the upper bolt securing the swivel hub to the bottom of the suspension strut, then tighten both upper and lower bolts to the specified torque (see Chapter 10).

17 Refit the brake caliper to the swivel hub, and tighten the bolts to the specified torque with reference to Chapter 9.

18 Clean the track rod end balljoint taper and the steering arm, then refit the balljoint to the arm and tighten the nut to the specified torque (see Chapter 10).

19 Refill the gearbox with the correct quantity and grade of oil, with reference to Chapter 1. Refit the splash guard where necessary.

20 Refit the roadwheel and lower the car to the ground.

6 Reversing light switch - removal and refitting

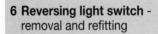

Removal

1 Apply the handbrake, then jack up the front of the car and support it on axle stands.

2 Where applicable, unbolt and remove the splash guard from the bottom of the gearbox.

3 Position a suitable container beneath the gearbox, then unscrew the drain plug and allow the oil to drain. When all of the oil has drained, clean and refit the drain plug, tightening it securely.

4 The switch is located on the left-hand side of the gearbox, next to the driveshaft (see photo 8.21). Disconnect the wiring from the switch.

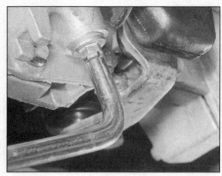

7.5 Unscrewing the gearbox oil drain plug

5 Unscrew the switch from the gearbox, and remove the washer.

Refitting

6 Clean the location in the gearbox and the threads of the switch.

7 Insert the switch together with a new washer, and tighten it securely.

8 Reconnect the wiring.

9 Refill the gearbox with the correct quantity and grade of oil, with reference to Chapter 1.

10 Refit the gearbox splash guard, where applicable.

11 Lower the car to the ground.

7 Manual gearbox (C- and E-type engines) - removal and refitting

Note: *This Section describes the removal of the gearbox, leaving the engine in position in the car. If adequate lifting gear is available, it may be easier to remove the engine and the gearbox together, as described in Chapter 2, and to separate them on the bench.*

Removal

1 The gearbox is removed upwards from the engine compartment, after disconnecting it from the engine. Due to the weight of the unit, it will be necessary to have suitable lifting equipment (such as an engine crane or hoist) available.

2 Apply the handbrake, then jack up the front

of the car and support it on axle stands. Remove both the front roadwheels.

3 Remove the battery with reference to Chapter 12.

4 Remove the bonnet with reference to Chapter 11.

5 Remove the plastic cover from the bottom of the gearbox. Position a suitable container beneath the gearbox, then unscrew the drain plug and allow the oil to drain (photo). When all of the oil has drained, clean and refit the drain plug, tightening it securely.

6 Unscrew the nut securing the left-hand steering track-rod end to the steering arm, then use a balljoint removal tool to separate the balljoint taper.

7 Working in the engine compartment, unscrew the three bolts securing the left-hand driveshaft inner rubber boot to the gearbox.

8 Refer to Chapter 9 and unbolt the left-hand brake caliper from the swivel hub. Do not disconnect the hydraulic hose from the caliper. Tie the caliper to the suspension coil spring without straining the hydraulic hose.

9 Unscrew and remove the pinch-bolt securing the front left lower balljoint to the bottom of the swivel hub.

10 Support the weight of the swivel hub and driveshaft on a trolley jack, then unscrew the two bolts and separate the swivel hub from the bottom of the suspension strut.

11 Withdraw the left-hand driveshaft and swivel hub from the gearbox. Make sure that the tripod components remain in position on the inner end of the driveshaft, otherwise they may fall into the gearbox.

12 Working on the right-hand side driveshaft, use a 5 mm diameter pin punch to drive out the roll pins.

13 Loosen (but do not remove) the lower bolt securing the right-hand swivel hub to the bottom of the suspension strut. Unscrew and remove the upper bolt, then tilt the swivel hub and disconnect the driveshaft from the gearbox sun gear shaft.

14 Cut the plastic tie and pull back the rubber boot, then disconnect the gearchange rod from the lever on the gearbox by unscrewing the nut and removing the bolt (photos). Recover the bush from inside the lever. Do not separate the gearchange rod at

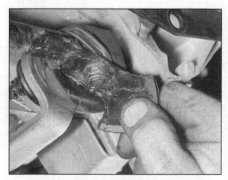

7.14A Pull back the rubber boot . . .

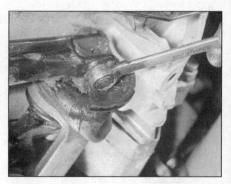

7.14B . . . then unbolt the gearchange rod from the lever on the gearbox

the clamp, otherwise it will be necessary to adjust the gear lever position on refitting.

15 Unbolt the front left-hand inner wing protective cover (where necessary, drill out the retaining rivets).

16 On C-type engines, unbolt and remove the clutch cover from the bottom of the gearbox bellhousing.

17 On E-type engines, unscrew the engine-to-gearbox tie-rod bracket mounting bolts and remove the bracket (photos).

18 On all engines, unscrew the engine-to-gearbox mounting nut located near the right-hand output shaft.

19 Pull out the clip and disconnect the speedometer cable from the rear of the gearbox, near the rear mounting.

20 Disconnect the clutch cable from the gearbox (see Chapter 6, Section 2).

21 Unbolt the earth cable from the gearbox casing.

22 Pull the wiring connector from the reversing light switch on the gearbox casing.

C-type engines

23 Unbolt and remove the engine-to-gearbox tie-rod.

E-type engines

24 Refer to Fig. 7.13. Loosen (but do not remove) bolt (A), then unscrew and remove bolt (B) from the rear mounting link.

25 Drain the cooling system (Chapter 1).

26 Loosen the clips and remove the radiator top hose. Also remove the thermostat.

27 Remove the distributor cap with reference to Chapter 5.

28 Remove the engine speed (flywheel) sensor and the ignition module with reference to Chapter 5.

29 Loosen (but do not remove) the bolt securing the rear mounting bracket to the gearbox, and swivel the mounting down.

All engines

30 Refer to Fig. 7.14 and remove stud (A). Use a nut and locknut to remove the stud.

31 Connect a suitable hoist to the engine, and lift it slightly. Alternatively, the engine may be supported on a trolley jack so that the hoist can be used to remove the gearbox.

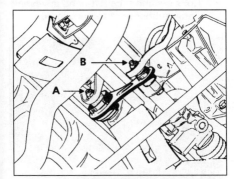

Fig. 7.13 Engine/gearbox rear mounting link - E-type engine (Sec 7)

A Rear mounting bolt
B Front mounting bolt

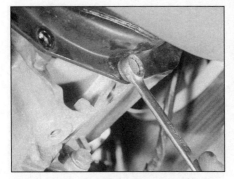

7.17A Unscrew the engine-to-gearbox tie-rod bracket bolts at the gearbox . . .

32 Remove the battery mounting bracket, complete with gearbox mounting pad (when applicable).

C-type engines

33 Remove the air cleaner housing assembly with reference to Chapter 4.

34 Unscrew the nut securing the gearbox front mounting to the subframe. Unbolt the front mounting from the gearbox.

35 Unbolt and remove the rear mounting from the rear of the gearbox.

36 Note the routing of the engine wiring harness, then disconnect it and remove it from the gearbox.

All engines

37 Unscrew and remove the gearbox-to-engine nuts and bolts from around the gearbox and from the starter motor. There is no need to remove the starter motor.

38 Support the weight of the gearbox. Disconnect the gearbox from the engine, sliding the gearbox end housing between the engine subframe and the front wing side panel. Careful use of a wide-bladed screwdriver may be necessary to free the bellhousing from the location dowels. Do not allow the weight of the gearbox to hang on the input shaft.

39 Pivot the gearbox forwards to release the final drive section. Lift the gearbox upwards from the engine compartment, bellhousing

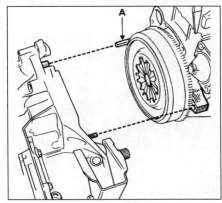

Fig. 7.14 Stud (A) locating gearbox on engine (Sec 7)

7.17B . . . and at the engine

end first. If the hoist is being used, attach it to the bellhousing mounting bolt holes.

Refitting

40 Refitting is a reversal of removal, noting the following additional points.

(a) Before assembling the gearbox to the engine, position a length of wood or a similar distance piece between the end of the clutch release fork and the outer cable bracket on the gearbox casing, in order to hold the fork in its released position. This will prevent the release bearing from becoming detached from the end of the release fork during the refitting procedure.

(b) Make sure that the location dowels are correctly positioned in the gearbox.

(c) Apply a little high-melting-point grease to the splines of the gearbox input shaft. Do not apply too much, otherwise there is the possibility of the grease contaminating the clutch friction disc.

(d) Make sure that the centring bush for the starter motor is correctly fitted. Refer to Chapter 12 if necessary.

(e) Use new roll pins when reconnecting the right-hand driveshaft, and seal the ends using a suitable sealant.

(f) Refit and tighten the brake caliper mounting bolts, with reference to Chapter 9.

(g) When applicable, check the engine mounting adjustment dimensions, as given in Chapter 2.

(h) Refill the gearbox with oil, and check the level with reference to Chapter 1.

(i) Tighten all nuts and bolts to the specified torque.

(j) Refill the cooling system with reference to Chapter 1.

7

8 Manual gearbox (F-type engine) - removal and refitting

Note: *This Section describes the removal of the gearbox, leaving the engine in the car. If adequate lifting gear is available, it may be easier to remove the engine and gearbox together, as described in Chapter 2, and to separate them on the bench.*

1 The gearbox is removed upwards from the

8.5A Removing the plastic cover from the bottom of the gearbox

8.5B Unscrew the drain plug . . .

8.5C . . . and drain the oil

engine compartment, after disconnecting it from the engine. Due to the weight of the unit, it will be necessary to have suitable lifting equipment available, such as an engine crane or hoist, to enable the unit to be removed in this way.

2 Apply the handbrake, then jack up the front of the car and support it on axle stands. Remove both the front roadwheels.

3 Remove the battery with reference to Chapter 12, and the air cleaner assembly with reference to Chapter 4.

4 Remove the bonnet with reference to Chapter 11.

5 Remove the plastic cover from the bottom of the gearbox. Position a suitable container beneath the gearbox, then unscrew the drain and filler plugs and allow the oil to drain (photos). When all of the oil has drained, clean and refit the drain plug, tightening it securely.

6 Unscrew the nut securing the left-hand steering track-rod end to the steering arm, then use a balljoint removal tool to separate the balljoint taper. On models fitted with ABS, unbolt the sensor from the swivel hub (photo).

7 Working in the engine compartment, unscrew the three bolts securing the left-hand driveshaft inner rubber boot to the gearbox (photo).

8 Refer to Chapter 9 and unbolt the left-hand brake caliper from the swivel hub (photo). Do not disconnect the hydraulic hose from the caliper. Tie the caliper to the suspension coil spring without straining the hydraulic hose.

9 Unscrew and remove the pinch-bolt securing the front left lower balljoint to the bottom of the swivel hub (photo).

10 Support the weight of the swivel hub and driveshaft on a trolley jack. Unscrew the two bolts, and separate the swivel hub from the bottom of the suspension strut and from the lower balljoint (photos).

11 Withdraw the left-hand driveshaft and swivel hub from the gearbox. Make sure that the tripod components remain in position on the inner end of the driveshaft, otherwise they may fall into the gearbox.

12 Working on the right-hand side driveshaft, use a 5 mm diameter pin punch to drive out the roll pins.

13 Loosen (but do not remove) the lower bolt

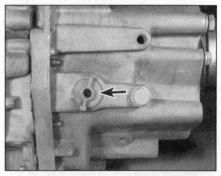

8.5D Oil filler plug (arrowed) on the front of the gearbox

8.6 Unbolt the ABS sensor from the swivel hub

8.7 Bolts securing the left-hand driveshaft inner rubber boot to the gearbox

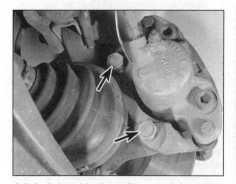

8.8 Left-hand brake caliper retaining bolts (arrowed)

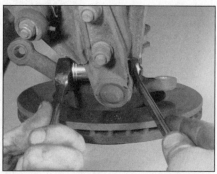

8.9 Removing the pinch-bolt securing the lower balljoint to the swivel hub

8.10A Unscrewing the bolts securing the swivel hub to the bottom of the suspension strut

8.10B Disconnecting the swivel hub from the lower balljoint

8.14A Pull back the rubber boot on the gearchange rod . . .

8.14B . . . unscrew and remove the bolt . . .

securing the right-hand swivel hub to the bottom of the suspension strut. Unscrew and remove the upper bolt, then tilt the swivel hub and disconnect the driveshaft from the gearbox.
14 Cut the plastic tie and pull back the rubber boot, then disconnect the gearchange rod from the lever on the gearbox by unscrewing the nut and removing the bolt. Recover the bush from inside the lever (photos). Do not separate the gearchange rod at the clamp, otherwise it will be necessary to adjust the gear lever position on refitting.
15 Unbolt the front left-hand inner wing protective cover (where necessary, drill out the retaining rivets) (photos).
16 On 1721 cc (F2N) and 1794 cc (F3P) engines, unscrew the engine-to-gearbox tie-

rod bracket mounting bolts, and remove the bracket (photo). No tie-rod bracket is fitted to the 1764 cc (F7P) engine.
17 Unscrew the engine-to-gearbox mounting nut located near the right-hand gearbox output shaft (photo).
18 Squeeze together the plastic clip, and disconnect the speedometer cable from the rear of the gearbox, near the rear mounting (photo).
19 Refer to Chapter 6, Section 2 and disconnect the clutch cable from the gearbox.
20 Unbolt the earth cable from the gearbox casing and from the cylinder block. Also unbolt the wiring harness supports and the earth strap from the gearbox and gearbox mountings, as applicable (photos).

8.14C . . . and recover the bush from inside the lever

8.15A Rivets (arrowed) holding the inner wing protective cover

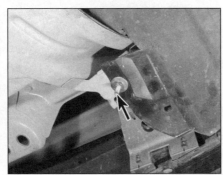

8.15B Inner wing protective cover bottom bolt (arrowed)

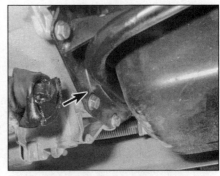

8.16 Engine-to-gearbox tie-rod bracket (arrowed) on the F-type engine

7

8.17 Engine-to-gearbox mounting nut (arrowed) located on the rear of the engine

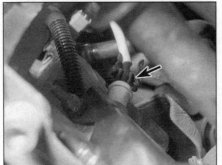

8.18 Speedometer cable connection (arrowed) at the rear of the gearbox

8.20A Wiring harness support on the gearbox (16-valve model)

8.20B Earth strap attachment to the gearbox casing

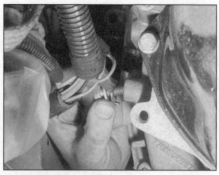

8.20C Wiring harness earth wire attachment to the gearbox casing

8.21 Disconnecting the wiring from the reversing light switch

21 Pull the wiring connector from the reversing light switch on the gearbox casing (photo).

22 Refer to Fig. 7.13. Loosen (but do not remove) bolt (A), then unscrew and remove bolt (B) from the rear mounting link. Swivel the link down (photo).

23 Drain the cooling system and remove the radiator, referring to Chapters 1 and 3. (On the 1764 cc/F7P engine, it will be necessary to unbolt the crossmember from the front of the engine compartment).

24 Disconnect the radiator top hose from the engine.

25 Remove the distributor cap with reference to Chapter 5.

26 Remove the ignition module and the engine speed (flywheel) sensor with reference to Chapter 5.

27 Where applicable, unclip the wiring harness from the ABS unit and from the gearbox mounting (photo).

28 Disconnect the wiring for the engine from inside the plastic box on the left-hand side of the engine compartment.

29 Disconnect the wiring from the starter motor and (when applicable) from the oxygen sensor. Detach the wiring bracket from the rear of the cylinder block, and feed it through the starter cover. Remove the starter motor (see Chapter 12), and pull the engine wiring loom through the hole in the bellhousing (photo).

30 Loosen (but do not remove) the bolt securing the rear mounting bracket to the gearbox, and swivel the mounting down.

31 Connect a suitable hoist to the engine,

and lift it slightly. Alternatively, the engine may be supported on a trolley jack so that the hoist can be used to remove the gearbox.

32 Disconnect and remove the bottom hose from the coolant pipe on the cylinder block.

33 On models with ABS, unbolt the ABS unit bracket from the right-hand inner wing panel. Unscrew the uppermost nut securing the gearbox mounting to the left-hand suspension turret. Tie the ABS unit to one side, being careful not to strain the hydraulic pipes (photos).

34 Support the weight of the gearbox.

35 Remove the remaining nuts and bolts which secure the gearbox mounting/battery tray to the bodywork. Remove the bolts which secure the mounting bracket to the gearbox, and lift out the mounting assembly (photo).

8.22 Rear mounting link swivelled downwards

8.27 ABS wiring harness attachment to the gearbox mounting

8.29 Removing a starter motor mounting bolt

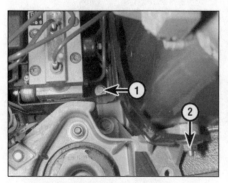

8.33A ABS unit bracket bolt (1) and gearbox mounting nut (2)

8.33B Using a cable tie (arrowed) to hold the ABS unit away from the gearbox

8.35 Removing the gearbox mounting assembly

8.36 Short brace on the left-hand side of the subframe

36 Unbolt the short brace from the left-hand side of the engine subframe (photo).

37 Remove the air inlet duct from the left-hand front corner of the engine compartment.

38 Loosen (but do not remove) the front bumper lower mounting bolt on the left-hand side.

39 Loosen (but do not remove) the two left-hand subframe nuts, and lower the subframe to the extent permitted by the length of the bolts (photo). This will give the extra room necessary to manoeuvre the gearbox.

40 Unscrew and remove all of the gearbox-to-engine bolts and nuts, noting the location of the washers (photos).

41 Disconnect the gearbox from the engine, sliding the gearbox end housing between the engine subframe and the front wing side panel. Careful use of a wide-bladed screwdriver may be necessary to free the

8.40B . . . and nuts

8.43 Lifting the gearbox out of the engine compartment

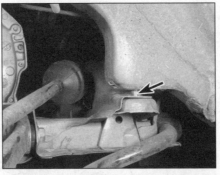

8.39 Subframe lowered to the extent of the mounting bolts (exposed thread arrowed)

bellhousing from the location dowels. Do not allow the weight of the gearbox to hang on the input shaft.

42 With the gearbox moved as far to the left-hand side of the engine compartment as possible, refer to Chapter 6 and remove the clutch. This is necessary in order to give additional room to manoeuvre the gearbox from the engine compartment.

43 Pivot the gearbox forwards to release the final drive section. Lift the gearbox upwards

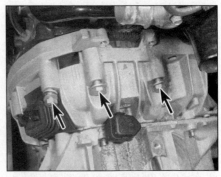

8.40A Removing the gearbox-to-engine mounting bolts (arrowed) . . .

from the engine compartment, bellhousing end first. If the hoist is being used, attach it to the bellhousing mounting bolt holes (photo).

44 Refitting is a reversal of removal, noting the following additional points.

(a) Before assembling the gearbox to the engine, position a length of wood or similar distance piece between the end of the clutch release fork and the outer cable bracket on the gearbox casing, in order to hold the fork in its released position. This

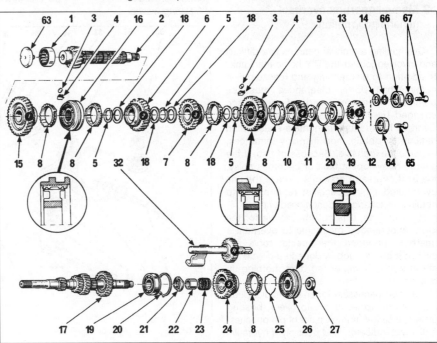

Fig. 7.15 Gearbox internal components (Sec 9)

1 Roller race	13 Washer	24 5th speed gear (primary)
2 Output shaft	14 5th speed circlip	25 5th speed spring
3 Roller	15 1st speed gear	26 5th speed gear hub
4 Spring	16 1st/2nd gear hub	27 5th speed nut
5 Circlip	17 Input shaft	32 Reverse shaft and gear
6 2nd speed gear	18 Splined ring	63 Oil baffle
7 3rd speed gear	19 Ball race	64 Thrustwasher
8 Synchro-ring	20 Circlip	65 5th speed end bolt on
9 3rd/4th gear hub	21 Washer	output shaft
10 4th speed gear	22 5th speed ring	66 Shouldered washer
11 Washer	23 Needle race	67 Retaining bolt and washer
12 5th speed gear		

7

will prevent the release bearing from becoming detached from the end of the release fork during the refitting procedure.

(b) Make sure that the location dowels are correctly positioned in the gearbox.

(c) Apply a little high-melting-point grease to the splines of the gearbox input shaft. Do not apply too much, otherwise there is the possibility of the grease contaminating the clutch friction disc.

(d) Make sure that the centring bush for the starter motor is correctly fitted. Refer to Chapter 12 if necessary.

(e) Use new roll pins when reconnecting the right-hand driveshaft, and seal the ends using a suitable sealant.

(f) Refit and tighten the brake caliper mounting bolts with reference to Chapter 9.

(g) Check the engine mounting adjustment dimensions, as given in Chapter 2.

(h) Refill the gearbox with oil, and check the level with reference to Chapter 1.

(i) Tighten all nuts and bolts to the specified torque.

(j) Refill the cooling system with reference to Chapter 1.

9 Manual gearbox overhaul - general information

Overhauling a manual gearbox is a difficult and involved job for the DIY home mechanic. In addition to dismantling and reassembling many small parts, clearances must be precisely measured and, if necessary, changed by selecting shims and spacers. Gearbox internal components are also often difficult to obtain, and in many instances, extremely expensive. Because of this, if the gearbox develops a fault or becomes noisy, the best course of action is to have the unit overhauled by a specialist repairer, or to obtain an exchange reconditioned unit.

Nevertheless, it is not impossible for the more experienced mechanic to overhaul a gearbox, provided the special tools are available and the job is done in a deliberate step-by-step manner so that nothing is overlooked.

The tools necessary for an overhaul include internal and external circlip pliers, bearing pullers, a slide-hammer, a set of pin punches, a dial test indicator, and possibly a hydraulic press. In addition, a large, sturdy workbench and a vice will be required.

During dismantling of the gearbox, make careful notes of how each component is fitted, to make reassembly easier and more accurate.

Before dismantling the gearbox, it will help if you have some idea what area is malfunctioning. Certain problems can be closely related to specific areas in the gearbox, which can make component examination and replacement easier. Refer to the *Fault diagnosis* Section at the beginning of this manual for more information.

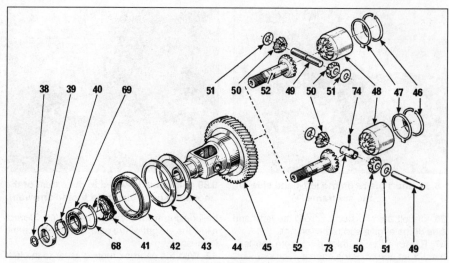

Fig. 7.16 Differential components (Sec 9)

38 O-ring	45 Differential housing	51 Planet wheel washer
39 Oil seal	46 Circlip	52 Sun wheel with tail shaft
40 Circlip	47 Shim	68 Circlip
41 Speedometer drivegear	48 Spider-type sun wheel	69 Ball race
42 Ball race	49 Planet wheel shaft	73 Sleeve
43 Spacer washer	50 Planet wheels	74 Pin
44 Spring washer		

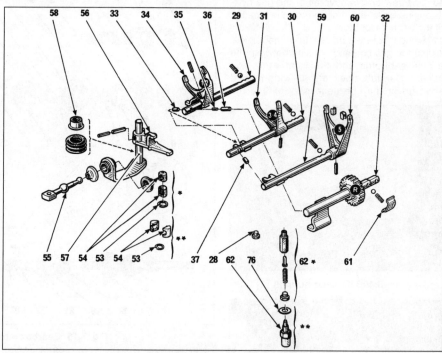

Fig. 7.17 Gear selector components (Sec 9)

Note: * = 1st type,
     ** = 2nd type

28 Threaded stop (four-speed)	34 1st/2nd shift fork	57 Input shaft
	35 1st/2nd plunger	58 Bush
29 1st/2nd shift rod	36 Plunger between 1st/2nd and reverse	59 5th speed rod (five-speed)
30 3rd/4th shift rod	37 5th speed plunger (five-speed)	60 5th speed shift fork (five-speed)
31 3rd/4th gear fork	53 Circlip	61 Reverse stirrup
32 Reverse shaft	54 Link support	62 5th speed detent assembly (five-speed)
33 Plunger between 1st/2nd and 3rd/4th	55 Link	
	56 Selector finger	76 5th speed detent shim washer

Part B: Automatic transmission

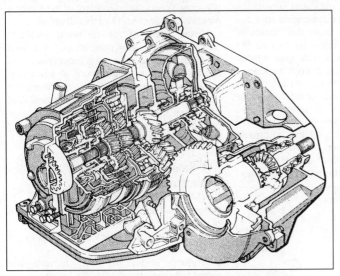

7.18 Cutaway view of the MB1 automatic transmission (Sec 10)

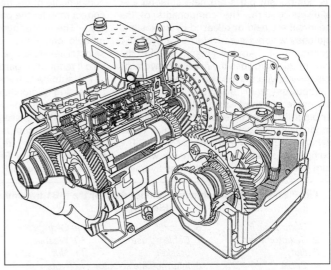

Fig. 7.19 Cutaway view of the AD4 automatic transmission (Sec 10)

10 General information

Automatic transmission is available as an option on 1390 cc (E6J and E7J) engines. Carburettor models (E6J engine) are fitted with the MB1 three-speed automatic transmission, and fuel injection models (E7J engine) are fitted with the AD4 four-speed automatic transmission.

The transmission consists of a torque converter, an epicyclic geartrain, hydraulically-operated clutches and brakes, and a computer control unit.

The torque converter provides a fluid coupling between engine and transmission, which acts as an automatic clutch, and also provides a degree of torque multiplication when accelerating.

Fig. 7.20 MB1 automatic transmission electronic control layout (Sec 10)

1 Fuse - reversing light (5 amp)
2 Fuse (1.5 amp)
3 Ignition switch
4 Starter relay
5 Reversing lights
6 Starter
7 Automatic transmission warning lamp
8 Automatic transmission earth
22 Vacuum capsule
BE Computer module
CM Multi-function switch
CV Speed sensor
EL Solenoid valves
RC Kickdown switch
P Load potentiometer

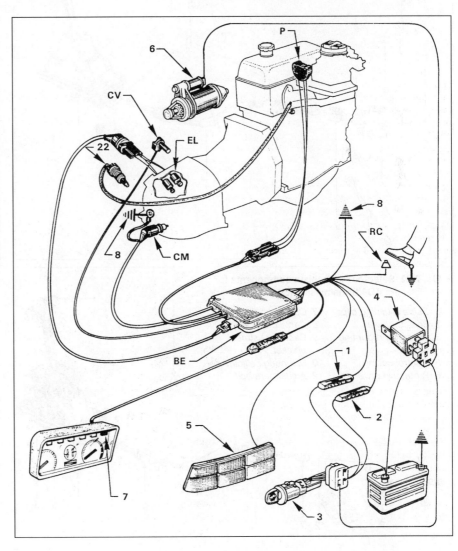

7

The epicyclic geartrain provides the forward gears or reverse gear, depending on which of its component parts are held stationary or allowed to turn. The components of the geartrain are held or released by brakes and clutches which are activated by a hydraulic control unit. An oil pump within the transmission provides the necessary hydraulic pressure to operate the brakes and clutches.

Impulses from switches and sensors connected to the transmission throttle and selector linkages are directed to a computer module, which determines the ratio to be selected from the information received. The computer activates solenoid valves, which in turn open or close ducts within the hydraulic control unit. This causes the clutches and brakes to hold or release the various components of the geartrain, and provide the correct ratio for the particular engine speed or load. The information from the computer module can be overridden by use of the selector lever, and a particular gear can be held if required, regardless of engine speed.

The automatic transmission fluid is cooled by passing it through a cooler located on the top of the transmission. Coolant from the cooling system passes through the cooler.

Due to the complexity of the automatic transmission, any repair or overhaul work must be left to a Renault dealer with the necessary special equipment for fault diagnosis and repair. In the event of a fault developing, begin by checking the fluid level (Chapter 1) and the adjustment of the selector mechanism (Section 11 of this Chapter).

Note: *Some of the following Sections dealing with the AD4 transmission require a full-throttle validation setting procedure using special Renault test equipment. Where this is the case, the work should not be attempted unless the necessary equipment is available, otherwise the transmission will not change gear at the correct speeds. Consult a Renault dealer for more information.*

11 Selector mechanism - adjustment

1 Apply the handbrake, then jack up the front of the car and support it on axle stands.
2 Move the selector lever inside the car to the 'D' position.
3 Working under the car, remove the protective cover from under the selector lever. Disconnect the cable end fitting from the lever.
4 Check that the lever on the transmission is in the 'D' position, and if necessary move the lever accordingly.
5 Refer to Fig. 7.23 or 7.24, and check that the dimension between the cable end fitting

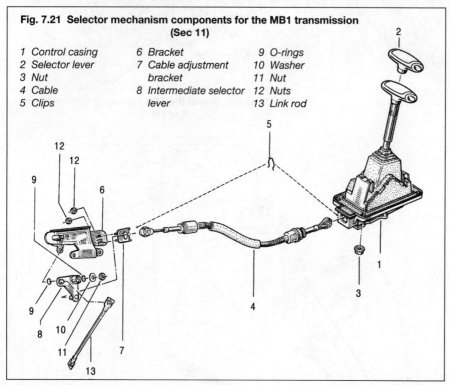

Fig. 7.21 Selector mechanism components for the MB1 transmission (Sec 11)

1 Control casing	6 Bracket	9 O-rings
2 Selector lever	7 Cable adjustment	10 Washer
3 Nut	bracket	11 Nut
4 Cable	8 Intermediate selector	12 Nuts
5 Clips	lever	13 Link rod

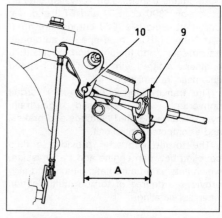

Fig. 7.23 Selector cable adjustment dimension on the MB1 transmission (Sec 11)

A = 131 mm
9 Outer cable location bracket
10 Cable end fitting centre point

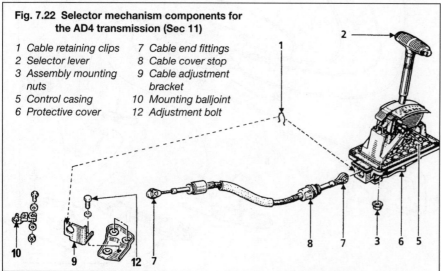

Fig. 7.22 Selector mechanism components for the AD4 transmission (Sec 11)

1 Cable retaining clips	7 Cable end fittings
2 Selector lever	8 Cable cover stop
3 Assembly mounting	9 Cable adjustment
nuts	bracket
5 Control casing	10 Mounting balljoint
6 Protective cover	12 Adjustment bolt

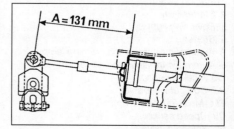

Fig. 7.24 Selector cable adjustment dimension on the AD4 transmission (Sec 11)

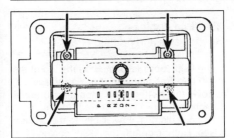

Fig. 7.25 Control casing on the selector assembly (Sec 11)

Adjustment screws arrowed

and the outer cable location bracket is as shown. If this is not the case, loosen the bracket mounting nut(s) and move the bracket as necessary until the dimension is correct. Tighten the nut(s).

6 Inside the car, remove the gaiter from the selector lever, with reference to Chapter 11, Section 25. Loosen the four adjustment screws securing the control casing to the lower assembly. Align the lever mark with the 'D' position, then tighten the screws and refit the gaiter.

7 Loosen the outer cable cover stop by turning it through a quarter-turn. Check that the cable slides freely.

8 Connect the cable to the bottom of the selector lever, and tighten the cover stop by turning it through a quarter-turn.

9 Refit the protective cover to the selector lever assembly.

10 Check that the selector lever moves freely, and that the starter motor will only operate with 'P' or 'N' selected. Also check that the Park function operates correctly. Small adjustments may be made by turning the outer cable cover stop through a quarter-turn, then pulling or pushing the cable as required before tightening the stop again.

11 Lower the car to the ground.

12 Selector mechanism and cable - removal and refitting

1 Apply the handbrake, then jack up the front of the car and support it on axle stands.

2 Inside the car, move the selector lever to position 'D'.

3 Remove the selector lever by pulling it hard upwards.

4 Working under the car, remove the protective cover from under the selector lever.

5 Disconnect the cable end fittings from the bottom of the selector lever and from the intermediate lever on the transmission.

6 Unscrew the four nuts securing the selector lever assembly to the underbody, then lower the assembly. Disconnect the wiring from the selector lever assembly.

7 Extract the clips, and withdraw the cable from the selector lever assembly and from the transmission.

8 Refitting is a reversal of removal, but adjust the cable as described in Section 11.

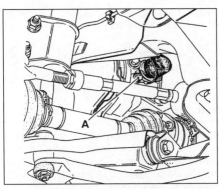

Fig. 7.26 Multi-function switch location on the AD4 transmission (Sec 14)

A Earth wire bolt

13 Speedometer drive - removal and refitting

Removal

MB1 transmission

1 Refer to Section 4.

AD4 transmission

2 Where necessary, remove the air cleaner housing assembly with reference to Chapter 4.

3 Disconnect the speedometer cable from the transmission by unscrewing the collar.

4 Unscrew the retaining sleeve, and withdraw the speedometer drivegear and shaft from the transmission. Recover the O-ring seal.

Refitting

MB1 transmission

5 Refer to Section 4.

AD4 transmission

6 Refitting is a reversal of the removal procedure, using a new O-ring seal.

14 Multi-function switch - removal and refitting

Note: *Before working on the AD4 transmission, read the note given in Section 10.*

Removal

MB1 transmission

1 On the MB1 transmission, the multi-function switch is originally supplied complete with the computer, and for renewal it is necessary to cut the wiring and obtain a new switch, together with a fitting kit. The switch is located on the left-hand end of the transmission.

2 To remove the switch from the transmission, unscrew the mounting bolt and the earth wire bolt, and pull out the switch.

3 If renewing the switch, cut the wiring and connect the new switch, following the instructions given with the kit.

AD4 transmission

4 On the AD4 transmission, the multi-function switch is located on the rear of the transmission, above the left-hand driveshaft.

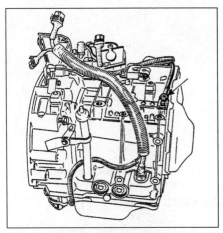

Fig. 7.27 Speed sensor (arrowed) on the AD4 transmission (Sec 16)

5 To remove the switch, first disconnect the switch wiring lead from the computer, on the left-hand side of the engine compartment.

6 Unscrew the mounting bolt and remove the clamp plate. Unscrew the earth wire bolt, then pull the switch out of the transmission.

Refitting

MB1 transmission

7 Check that the O-ring seals are in good condition and renew them if necessary, then insert the switch in the transmission and tighten the mounting bolt. Insert and tighten the earth bolt.

AD4 transmission

8 Refitting is a reversal of the removal procedure, but before using the car on the road, the computer full-throttle position should be validated using Renault test equipment.

15 Kickdown switch (AD4 transmission) - removal and refitting

The kickdown switch is an integral part of the accelerator cable. Refer to Chapter 4.

16 Speed sensor - removal and refitting

Note: *Before working on the AD4 transmission, read the note given in Section 10.*

Removal

MB1 transmission

1 On the MB1 transmission, the speed sensor is originally supplied complete with the computer. It is necessary to cut the wiring and obtain a new switch, together with a fitting kit.

2 The sensor is located on the front of the transmission. To remove the sensor, unscrew the mounting bolt and remove the clamp, then pull out the sensor.

3 If renewing the sensor, cut the wiring and connect the new sensor, following the instructions given with the kit.

7

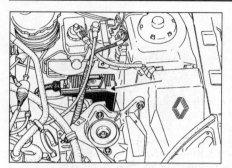

Fig. 7.28 AD4 automatic transmission computer (arrowed) (Sec 18)

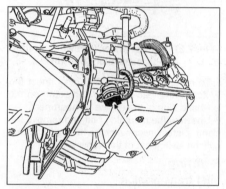

Fig. 7.29 AD4 automatic transmission line pressure sensor (arrowed) (Sec 19)

AD4 transmission

4 On the AD4 transmission, the speed sensor is located on the top left-hand side of the transmission.

5 To remove the sensor, first disconnect the wiring lead from the computer, on the left-hand side of the engine compartment.

6 Remove the battery and battery mounting, with reference to Chapter 12.

7 Unscrew the mounting bolt and remove the clamp plate, then pull the switch out of the transmission.

Refitting

MB1 transmission

8 Check that the O-ring seal is in good condition, and renew if necessary. Insert the sensor in the transmission, and refit the clamp. Insert and tighten the mounting bolt.

AD4 transmission

9 Refitting is a reversal of the removal procedure, but before using the car on the road, the computer full-throttle position should be validated using Renault test equipment.

17 Load potentiometer (AD4 transmission) - removal and refitting

1 The load potentiometer is incorporated in the fuel injection throttle housing. It is preset at the factory, and cannot be repaired or adjusted. No attempt should be made to remove it from the throttle housing.

2 In the event of a fault, the lower half of the throttle housing must be renewed. Refer to Chapter 4 for further information.

18 Computer (AD4 transmission) - removal and refitting

Note: *Before working on the AD4 transmission, read the note given in Section 10.*

1 The computer is mounted on the left-hand side of the engine compartment. If the computer is suspected of having a fault, the car should be taken to a Renault dealer to have the system checked.

2 To remove the computer, disconnect the wiring plugs, noting their positions. Release the strap and withdraw the computer from its bracket. Take care not to allow any dirt or foreign matter to drop into the wiring sockets.

3 Refitting is a reversal of the removal procedure, but before using the car on the road, the computer full-throttle position should be validated using Renault test equipment.

19 Line pressure sensor (AD4 transmission) - removal and refitting

Note: *Before working on the AD4 transmission, read the note given in Section 10.*

1 The line pressure sensor is mounted on the lower front of the transmission.

2 To remove the line pressure sensor, first apply the handbrake, then jack up the front of the car and support on axle stands.

3 Remove the splash guard from under the automatic transmission.

4 Disconnect the appropriate wiring plug from the computer, on the left-hand side of the engine compartment.

5 Unscrew the mounting bolts and withdraw the sensor from the transmission.

6 Refitting is a reversal of the removal procedure, but before using the car on the road, the computer full-throttle position should be validated using Renault test equipment.

20 Differential output oil seals - renewal

Right-hand side differential output oil seal (MB1 transmission)

1 Refer to Section 5 in Part A of this Chapter. The procedure is the same as for the manual gearbox.

Differential output oil seals (AD4 transmission)

2 On the AD4 automatic transmission, access to the differential output oil seals is gained by removing the output flanges. This work should be carried out by a Renault garage, as special tooling is required to overcome the tension of a large spring located behind each of the flanges. Also, it is necessary to validate the full-throttle position using special Renault test equipment.

21 Fluid cooler - removal and refitting

1 The fluid cooler is located on the top of the automatic transmission. For better access, remove the battery and mounting bracket (Chapter 12).

2 Fit hose clamps, if available, to the coolant hoses each side of the fluid cooler. The alternative method is to drain the cooling system completely, with reference to Chapter 1.

3 Loosen the clips and disconnect the hoses from the fluid cooler.

4 Unscrew the special mounting bolts, and remove the fluid cooler from the top of the transmission. There will be some loss of fluid, so some clean rags should be placed around the cooler to absorb it. Make sure that dirt is prevented from entering the hydraulic system.

5 Remove the special bolts, and examine the O-ring seals for damage. If necessary, obtain and fit new O-ring seals.

6 Refitting is a reversal of the removal procedure, but check the transmission fluid level with reference to Chapter 1.

22 Automatic transmission - removal and refitting

Note: *Before working on the AD4 transmission, read the note given in Section 10.*

The automatic transmission must be removed together with the engine, and then separated on the bench. Refer to Chapter 2 for the removal and refitting procedures. On the AD4 transmission, before using the car on the road, the computer full-throttle position should be validated by a Renault dealer.

23 Automatic transmission overhaul - general information

Note: *Check the transmission fluid level and the selector mechanism adjustment before assuming that a fault exists in the transmission itself.*

In the event of a transmission fault occurring, it is first necessary to determine whether it is of an electrical, mechanical or hydraulic nature, and to do this, special test equipment is required. It is therefore essential to have the work carried out by a Renault dealer if a transmission fault is suspected, or if the transmission warning light on the instrument panel illuminates continuously.

If the warning light flashes when the engine is cold and the external temperature is less than -20°C, the automatic transmission fluid temperature is too low. Continue driving until the light goes out. If it flashes under any other circumstances, the fluid temperature is too high; drive at a lower speed until the light goes out.

Do not remove the transmission from the car for possible repair before professional fault diagnosis has been carried out, since most tests require the transmission to be in the vehicle.

Chapter 8 Driveshafts

Contents

Degrees of difficulty

Easy, suitable for novice with little experience	**Fairly easy,** suitable for beginner with some experience	**Fairly difficult,** suitable for competent DIY mechanic 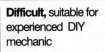	**Difficult,** suitable for experienced DIY mechanic	**Very difficult,** suitable for expert DIY or professional

Specifications

General

Driveshaft type	Equal-length solid steel shafts, splined to inner and outer constant velocity joints, vibration damper fitted on some shafts
Lubricant type/specification	Special grease supplied in sachets with gaiter kits - joints are otherwise pre-packed with grease and sealed

Torque wrench settings

	Nm	lbf ft
Driveshaft retaining nut (renew every time)	250	185
Left-hand driveshaft gaiter retaining plate bolts - manual gearbox and MB1 automatic transmission	25	18
Driveshaft to transmission flange bolts - AD4 automatic transmission .	35	26
Roadwheel bolts	90	66

1 General information

Drive is transmitted from the differential to the front wheels by means of two, equal-length, open driveshafts.

Both driveshafts are fitted with a constant velocity (CV) joint at their outer ends, which may be of the spider-and-yoke type or of the ball-and-cage type. Each joint has an outer member, which is splined at its outer end to accept the wheel hub and is threaded so that it can be fastened to the hub by a large nut. The joint contains either a spring-loaded plunger or six balls within a cage, which engage with the inner member. The complete assembly is protected by a flexible gaiter secured to the driveshaft and joint outer member.

On vehicles equipped with a manual gearbox or with the three-speed (MB1) automatic transmission unit, a different inner constant velocity joint arrangement is fitted to each driveshaft. On the right-hand side, the driveshaft is splined to engage with a tripod

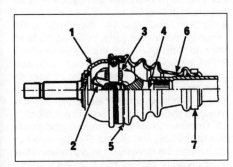

Fig. 8.1 Sectional view of the spider-and-yoke type outer constant velocity joint
(Sec 1)

1 Outer member	5 Outer retaining clip
2 Thrust plunger	6 Gaiter
3 Driveshaft spider	7 Inner retaining clip
4 Driveshaft	

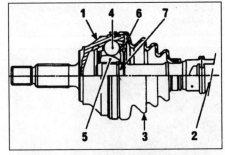

Fig. 8.2 Sectional view of the ball-and-cage type outer constant velocity joint
(Sec 1)

1 Outer member	5 Inner member
2 Driveshaft	6 Ball cage
3 Gaiter	7 Circlip
4 Ball	

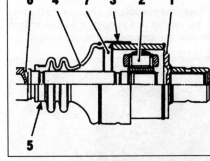

Fig. 8.3 Sectional view of a right-hand inner constant velocity joint - RC490 type
(Sec 1)

1 Outer member	5 Inner retaining clip
2 Tripod joint	6 Driveshaft
3 Metal cover	7 Metal insert
4 Gaiter	

8

2.2 Removing the driveshaft retaining nut and washer

2.4 Using a balljoint separator to release the track rod end balljoint from the swivel hub

2.5 Withdraw the upper bolt, noting which way around it is fitted

joint, containing needle roller bearings and cups. The tripod joint is free to slide within the yoke of the joint outer member, which is splined and retained by a roll pin to the differential sun wheel stub shaft. As on the outer joints, a flexible gaiter secured to the driveshaft and outer member protects the complete assembly. On the left-hand side, the driveshaft also engages with a tripod joint, but the yoke in which the tripod joint is free to slide is an integral part of the differential sun wheel. On this side, the gaiter is secured to the transmission casing with a retaining plate, and to a ball-bearing on the driveshaft with a retaining clip. The bearing allows the driveshaft to turn within the gaiter, which does not revolve.

On vehicles equipped with the four-speed (AD4) automatic transmission, both the left- and right-hand inner joints are the same. Each joint is secured to the transmission drive flange by six retaining bolts. As with the outer joint, the complete assembly is protected by a flexible gaiter which is secured to the driveshaft and outer member.

2 Driveshaft - removal and refitting

Removal

1 Apply the handbrake, then jack up the front of the vehicle and support it on axle stands. Remove the appropriate front roadwheel.

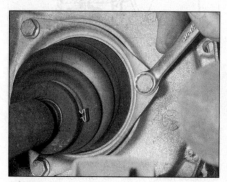

2.8 Remove the left-hand gaiter retaining plate bolts . . .

2 Refit at least two roadwheel bolts to the front hub, and tighten them securely. Have an assistant firmly depress the brake pedal to prevent the front hub from rotating. Using a socket and a long extension bar, slacken and remove the driveshaft retaining nut and washer (photo). This nut is extremely tightBolt the tool to the hub using two wheel bolts, and hold the tool to prevent the hub from rotating as the driveshaft retaining nut is undone. Discard the driveshaft nut; a new one should be used on refitting.

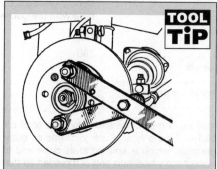

Fig. 8.4 Using a fabricated tool to hold the front hub stationary whilst the driveshaft retaining nut is slackened (Sec 2)

A tool to hold the hub stationary can be fabricated from two lengths of steel strip (one long, one short) and a nut an bolt; the nut and bolt form the pivot of a forked tool.

2.9 . . . and release the tripod joint from the transmission

3 Unscrew the two bolts securing the brake caliper assembly to the swivel hub, and slide the caliper assembly off the disc. Using a piece of wire or string, tie the caliper to the front suspension coil spring, to avoid placing any strain on the hydraulic brake hose.
4 Slacken and remove the nut securing the steering gear track rod end balljoint to the swivel hub. Release the balljoint tapered shank using a universal balljoint separator (photo).
5 Slacken and remove the two nuts and washers from the bolts securing the swivel hub to the suspension strut, noting that the nuts are positioned on the rear side of the strut. Withdraw the upper bolt, but leave the lower bolt in position at this stage (photo). Now proceed as described under the relevant sub-heading.

Left-hand driveshaft - models with manual gearbox or MB1 automatic transmission

6 On models fitted with a manual gearbox, position a suitable container beneath the drain plug, then remove the drain plug and allow the oil to drain from the gearbox. Once the oil has drained, wipe the threads of the drain plug clean, refit it to the gearbox and tighten it securely.
7 On models equipped with automatic transmission, drain the transmission fluid as described in Chapter 1.
8 Slacken and remove the three bolts securing the flexible gaiter retaining plate to the side of the gearbox/transmission (photo).
9 Pull the top of the swivel hub outwards until the driveshaft tripod joint is released from its yoke; be prepared for some oil spillage as the joint is withdrawn (photo). Be careful that the rollers on the end of the tripod do not fall off.
10 Remove the lower bolt securing the swivel hub to the suspension strut. Taking care not to damage the driveshaft gaiters, release the outer constant velocity joint from the hub and remove the driveshaft. Note that locking fluid is applied to the outer constant velocity joint splines during assembly, so it is likely that they will be a tight fit in the hub splines. Use a hammer and a soft metal drift to tap the joint out of the hub, or use an extractor to push the driveshaft out of the hub.

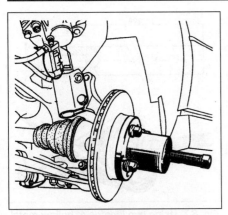

Fig. 8.5 Using an extractor to press the driveshaft out of the front hub (Sec 2)

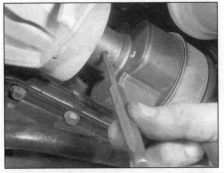

2.11 Drive out the roll pins with a suitable pin punch

2.26 Tighten the driveshaft retaining nut to the specified torque

Right-hand driveshaft - models with manual gearbox or MB1 automatic transmission

11 Rotate the driveshaft until the double roll pin, securing the inner constant velocity joint to the sun wheel shaft, is visible. Using a hammer and a 5 mm diameter pin punch, drive out the double roll pin (photo). New roll pins must be used on refitting.

12 Pull the top of the swivel hub outwards until the inner constant velocity joint splines are released from the sun wheel shaft. Remove the O-ring from the sun wheel shaft splines.

13 Remove the driveshaft as described in paragraph 10.

Both driveshafts - models with AD4 automatic transmission

14 Slacken and remove the six bolts and washers securing the driveshaft inner constant velocity joint to the transmission drive flange, rotating the shaft as necessary to gain access to the bolts.

15 Pull the top of the swivel hub outwards and disengage the inner constant velocity joint from the drive flange.

16 Remove the driveshaft as described in paragraph 10.

Refitting

17 All new driveshafts supplied by Renault are equipped with cardboard or plastic protectors to prevent damage to the gaiters. Even the slightest knock to the gaiter can puncture it, allowing the entry of water or dirt at a later date, which may lead to the premature failure of the joint. If the original driveshaft is being refitted, it is worthwhile making up some cardboard protectors as a precaution. They can be held in position with elastic bands. The protectors should be left on the driveshafts until the end of the refitting procedure.

Left-hand driveshaft - models with manual gearbox or MB1 automatic transmission

18 Wipe clean the side of the gearbox/transmission. Insert the tripod joint into the sun wheel yoke, keeping the driveshaft horizontal as far as possible.

19 Align the gaiter retaining plate with its bolt holes. Refit the retaining bolts, and tighten them to the specified torque. Ensure that the gaiter is not twisted.

20 Check that the splines on the driveshaft outer constant velocity joint and hub are clean and dry. Apply a coat of locking fluid to the splines.

21 Move the top of the swivel hub inwards, at the same time engaging the driveshaft with the hub.

22 Slide the hub fully onto the driveshaft splines, then insert the two suspension strut mounting bolts from the front side of the strut. Refit the washers and nuts to the rear of the bolts, and tighten them to the specified torque (Chapter 10 Specifications).

23 Slide on the washer, then fit the new driveshaft retaining nut, tightening it by hand only at this stage.

24 Reconnect the steering track rod balljoint to the swivel hub, and tighten its retaining nut to the specified torque (Chapter 10 Specifications).

25 Slide the brake caliper assembly into position over the brake disc. Refit the caliper mounting bolts, having first applied a few drops of locking fluid to their threads, and tighten them to the specified torque (Chapter 9 Specifications).

26 Using the method employed during removal to prevent the hub from rotating, tighten a new driveshaft retaining nut to the specified torque (photo). Check that the hub rotates freely, then remove the protectors from the driveshaft, taking great care not to damage the flexible gaiters.

27 Refit the roadwheel. Lower the car to the ground and tighten the roadwheel bolts to the specified torque.

28 Refill the gearbox or automatic transmission with oil or fluid; refer to Chapter 1 for details.

Right-hand driveshaft - models with manual gearbox or MB1 automatic transmission

29 Ensure that the inner constant velocity joint and sun wheel shaft splines are clean and dry. Apply a smear of molybdenum disulphide grease to the splines. Fit a new O-ring over the end of the sun wheel shaft, and slide the O-ring along the shaft until it abuts the transmission oil seal (photo).

30 Engage the driveshaft splines with those of the sun wheel shaft, making sure that the roll pin holes are in alignment (photo). Slide the driveshaft onto the sun wheel shaft until the roll pin holes are aligned.

31 Drive in new roll pins with their slots 90° apart (see Fig. 8.6), then seal the ends of the pins with sealing compound (Renault CAF 4/60 THIXO paste or equivalent) (photo).

2.29 Fit the new O-ring onto the sun wheel shaft . . .

2.30 . . . and engage the driveshaft, ensuring that the roll pin holes (arrowed) are correctly aligned

8

2.31 Seal the ends of the roll pins with a suitable sealing compound

32 Carry out the procedures described in paragraphs 20 to 27.

Both driveshafts - models with AD4 automatic transmission

33 Ensure the inner constant velocity joint and transmission drive flange mating surfaces are clean and dry. Pack the drive flange recess with Molykote BR2 grease (available from your Renault dealer).
34 Engage the driveshaft inner constant velocity joint with the transmission drive flange, then refit the six retaining bolts. Securely tighten the retaining bolts, and wipe off any surplus grease.
35 Carry out the procedures described in paragraphs 20 to 27.

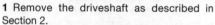

3 Outer constant velocity joint gaiter (models with manual gearbox or MB1 automatic transmission) - renewal

1 Remove the driveshaft as described in Section 2.
2 Cut through the gaiter retaining clip(s) or release the retaining spring and inner collar (as applicable), then slide the gaiter down the shaft to expose the outer constant velocity joint.
3 Scoop out as much grease as possible from the joint, and determine which type of constant velocity joint is fitted. Proceed as described under the relevant sub-heading.

Ball-and-cage type joint

4 Using circlip pliers, expand the joint internal circlip. At the same time, tap the exposed face of the ball hub with a mallet to separate the joint from the driveshaft. Slide off the gaiter and rubber collar.
5 With the constant velocity joint removed from the driveshaft, clean the joint using paraffin, or a suitable solvent, and dry it thoroughly. Carry out a visual inspection of the joint.
6 Move the inner splined driving member from side to side, to expose each ball in turn at the top of its track. Examine the balls for cracks, flat spots or signs of surface pitting.
7 Inspect the ball tracks on the inner and

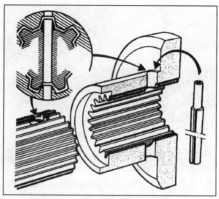

Fig. 8.6 Right-hand driveshaft inner constant velocity joint roll pin arrangement - models with manual gearbox or MB1 automatic transmission (Sec 2)

outer members. If the tracks have widened, the balls will no longer be a tight fit. At the same time, check the ball cage windows for wear or cracking between the windows.
8 If on inspection any of the constant velocity joint components are found to be worn or damaged, it will be necessary to renew the complete driveshaft assembly, since no components are available separately. If the joint is in satisfactory condition, obtain a repair kit from your Renault dealer consisting of a new gaiter, rubber collar, retaining spring, and the correct type and quantity of grease.
9 Tape over the splines on the end of the driveshaft, then slide the rubber collar and gaiter onto the shaft. Locate the inner end of the gaiter on the driveshaft, and secure it in position with the rubber collar.
10 Remove the tape, then slide the constant velocity joint coupling onto the driveshaft until the internal circlip locates in the driveshaft groove.
11 Check that the circlip holds the joint securely on the driveshaft, then pack the joint with the grease supplied. Work the grease well into the ball tracks, and fill the gaiter with any excess.
12 Locate the outer lip of the gaiter in the groove on the joint outer member. With the coupling aligned with the driveshaft, lift the lip

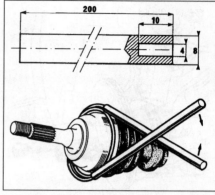

Fig. 8.7 Using two lengths of hollow metal tubing to install the gaiter retaining spring. Tube dimensions in mm (Sec 3)

of the gaiter to equalise the air pressure. Secure the gaiter in position with the large retaining spring, using two lengths of hollow metal tubing to ease the spring into position, as shown in Fig. 8.7.
13 Check that the constant velocity joint moves freely in all directions, then refit the driveshaft to the vehicle as described in Section 2.

Spider-and-yoke type joint

14 Remove the inner constant velocity joint, bearing and gaiter (as applicable), as described in Section 4 or 5 of this Chapter.
15 Where a vibration damper is fitted, clearly mark the position of the damper on the driveshaft, then use a puller or press to remove it from the inner end of the driveshaft, noting which way around it is fitted (photo). Ensure that the legs of the puller or support plate rest only on the damper inner rubber bush, otherwise the damper will distort and break away from the outer metal housing as it is removed.
16 Slide the outer constant velocity joint gaiter off the inner end of the driveshaft.
17 Clean the outer constant velocity joint using paraffin or a suitable solvent, and dry it thoroughly. Carry out a visual inspection of the joint.

3.15 Removing the vibration damper from the driveshaft

3.19 Renault driveshaft gaiter repair kit

3.21 Pack the joint with the grease supplied in the repair kit . . .

3.22 . . . then slide the gaiter into position over the joint

3.23A Fit the large retaining clip . . .

18 Check the driveshaft spider and outer member yoke for signs of wear, pitting or scuffing on their bearing surfaces. Also check that the outer member pivots smoothly and easily, with no traces of roughness.

19 If inspection reveals signs of wear or damage, it will be necessary to renew the driveshaft complete, since no components are available separately. If the joint components are in satisfactory condition, obtain a repair kit consisting of a new gaiter, retaining clips, and the correct type and quantity of grease (photo).

20 Tape over the splines on the inner end of the driveshaft, then carefully slide the outer gaiter onto the shaft.

21 Pack the joint with the grease supplied in the repair kit. Work the grease well into the joint, and fill the gaiter with any excess (photo).

22 Ease the gaiter over the joint, and ensure that the gaiter lips are correctly located in the grooves on the driveshaft and on the joint (photo). With the coupling aligned with the driveshaft, lift the lip of the gaiter to equalise the air pressure.

23 Fit the large metal retaining clip to the gaiter. Remove any slack in the gaiter retaining clip by carefully compressing the raised section of the clip. In the absence of the special tool, a pair of pincers may be used. Secure the small retaining clip using the same procedure (photos). Check that the

constant velocity joint moves freely in all directions before proceeding further.

24 To refit the vibration damper (when applicable), lubricate the driveshaft with a solution of soapy water. Press or drive the vibration damper along the shaft, using a tubular spacer which bears only on the damper inner bush, until it is aligned with the mark made prior to removal.

25 Refit the inner constant velocity joint components as described in Section 4 or 5 (as applicable), then refit the driveshaft to the vehicle as described in Section 2.

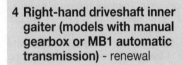

4 Right-hand driveshaft inner gaiter (models with manual gearbox or MB1 automatic transmission) - renewal

1 Remove the driveshaft as described in Section 2.

2 On these models, two different types of inner constant velocity joint are used on the right-hand driveshaft: type GI62 and type RC490. The joints can be identified by the shape of their outer members. The GI62 joint has a smooth, circular outer member; the RC490 joint has a recessed outer member which appears clover-shaped when viewed end-on. Identify the type of joint fitted, then proceed as described under the relevant sub-heading.

GI62-type joint

3 Release the large retaining spring and the inner retaining collar, then slide the gaiter down the shaft to expose the joint.

4 Using pliers, carefully bend up the anti-separation plate tangs at their corners (photo). Slide the outer member off the tripod joint. Be prepared to hold the rollers in place, otherwise they may fall off the tripod ends as the outer member is withdrawn. If necessary, secure the rollers in place using tape after removal of the outer member. The rollers are matched to the tripod joint stems, and it is important that they are not interchanged.

5 Using circlip pliers, extract the circlip securing the tripod joint to the driveshaft (photo). Note that on some models, the joint may be staked in position; if so, relieve the staking using a file. Mark the position of the tripod in relation to the driveshaft, using a dab of paint or a punch.

6 The tripod joint can now be removed (photo). If it is tight, draw the joint off the driveshaft end using a puller. Ensure that the legs of the puller are located behind the joint inner member and do not contact the joint rollers. Alternatively, support the inner member of the tripod joint, and press the shaft out using a hydraulic press, again ensuring that no load is applied to the joint rollers.

7 With the tripod joint removed, slide the gaiter and inner retaining collar off the end of the driveshaft.

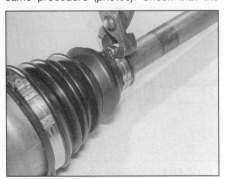

3.23B . . . and the small retaining clip. Note use of pincers to secure the clip

4.4 Bend up the anti-separation tangs with pliers to release the joint outer member (GI62 type)

4.5 Remove the circlip . . .

8

4.6 . . . and withdraw the tripod joint from the driveshaft end

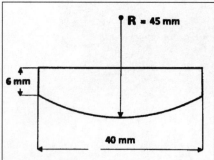

Fig. 8.8 Support plate dimensions for re-forming the anti-separation plate tangs (GI62 type joint) (Sec 4)

8 Wipe clean the joint components, taking care not to remove the alignment marks made on dismantling. **Do not** use paraffin or other solvents to clean this type of joint.

9 Examine the tripod joint, rollers and outer member for any signs of scoring or wear. Check that the rollers move smoothly on the tripod stems. If wear is evident, the tripod joint and roller assembly can be renewed, but it is not possible to obtain a replacement outer member. Obtain a new gaiter, retaining spring/collar and a quantity of the special lubricating grease. These parts are available in the form of a repair kit from your Renault dealer.

10 Tape over the splines on the end of the driveshaft, then carefully slide the inner retaining collar and gaiter onto the shaft.

11 Remove the tape, then, aligning the marks made on dismantling, engage the tripod joint with the driveshaft splines. Use a hammer and soft metal drift to tap the joint onto the shaft, taking great care not to damage the driveshaft splines or joint rollers. Alternatively, support the driveshaft, and press the joint into position using a hydraulic press and suitable tubular spacer which bears only on the joint inner member.

12 Secure the tripod joint in position with the circlip, ensuring that it is correctly located in the driveshaft groove. Where no circlip is fitted, secure the joint in position by staking

the end of the driveshaft in three places, at intervals of 120°, using a hammer and punch.

13 Evenly distribute the grease contained in the repair kit around the tripod joint and inside the outer member. Pack the gaiter with the remainder of the grease.

14 Slide the outer member into position over the tripod joint.

15 Using a piece of 2.5 mm thick steel or similar material, make up a support plate to the dimensions shown in Fig. 8.8.

16 Position the support plate under each anti-separation plate tang in the outer member in turn, and tap the tang down onto the support plate. Remove the plate when all the tangs have been returned to their original shape.

17 Slide the gaiter up the driveshaft. Locate the gaiter in the grooves on the driveshaft and outer member.

18 Slide the inner retaining collar into place over the inner end of the gaiter.

19 Using a blunt rod, carefully lift the outer lip of the gaiter to equalise the air pressure. With the rod in position, compress the joint until the dimension from the inner end of the gaiter to the flat end face of the outer member is as shown in Fig. 8.9. Hold the outer member in this position and withdraw the rod.

20 Slip the new retaining spring into place to secure the outer lip of the gaiter to the outer member. Take care to ensure that the

retaining spring is not overstretched during the fitting process.

21 Check that the constant velocity joint moves freely in all directions, then refit the driveshaft as described in Section 2.

RC490-type joint

22 Using a pair of grips, bend up the metal joint cover at the points where it has been staked into the outer member recesses.

23 Using a pair of snips, cut the gaiter inner retaining clip.

24 Using a soft metal drift, tap the metal joint cover off the outer member. Slide the outer member off the end of the tripod joint. Be prepared to hold the rollers in place, otherwise they may fall off the tripod ends as the outer member is withdrawn. If necessary, secure the rollers in place using tape after removal of the outer member. The rollers are matched to the tripod joint stems, and it is important that they are not interchanged.

25 Remove the tripod joint and gaiter assembly, and examine the joint components for wear, using the information given in paragraphs 6 to 9 of this Section. Obtain a repair kit consisting of a gaiter, retaining clip, metal insert and joint cover, and the correct type and amount of special grease.

26 Fit the metal insert into the inside of the gaiter, then locate the gaiter assembly inside the metal joint cover.

27 Tape over the driveshaft splines, and slide the gaiter and joint cover assembly onto the driveshaft.

28 Refit the tripod joint as described in paragraphs 11 and 12.

29 Evenly distribute the special grease contained in the repair kit around the tripod joint and inside the outer member. Pack the gaiter with the remainder of the grease.

30 Slide the outer member into position over the tripod joint.

31 Slide the metal joint cover onto the outer member until it is flush with the outer member guide panel. Secure the joint cover in position by staking it into the recesses in the outer member, using a hammer and a round-ended punch.

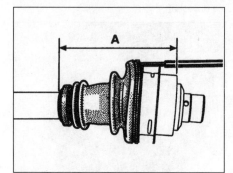

Fig. 8.9 Fitting dimension for the right-hand driveshaft inner constant velocity joint gaiter - GI62 type joint (Sec 4)

A = 153.5 ± 1 mm

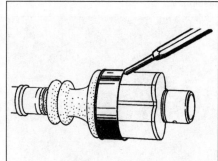

Fig. 8.10 Removing the metal cover from the right-hand driveshaft inner constant velocity joint - RC490 type (Sec 4)

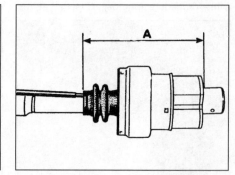

Fig. 8.11 Fitting dimension for the right-hand driveshaft inner constant velocity joint gaiter - RC490 type joint (Sec 4)

A = 156 ± 1 mm

32 Using a blunt rod, carefully lift the inner lip of the gaiter to equalise the air pressure. With the rod in position, compress the joint until the dimension from the inner end of the gaiter to the flat end face of the outer member is as shown in Fig. 8.11. Hold the outer member in this position and withdraw the rod.

33 Fit the small retaining clip to the inner end of the gaiter. Remove any slack in the gaiter retaining clip by carefully compressing the raised section of the clip. In the absence of the special tool, a pair of pincers may be used.

34 Check that the constant velocity joint moves freely in all directions, then refit the driveshaft as described in Section 2.

5 Left-hand driveshaft inner gaiter (models with manual gearbox or MB1 automatic transmission) - renewal

1 Remove the driveshaft as described in Section 2.

2 Using circlip pliers, extract the circlip securing the tripod joint to the driveshaft. Note that on some models, the joint may be staked in position; if so, relieve the stakings using a file. Using a dab of paint or a hammer and punch, mark the position of the tripod joint in relation to the driveshaft, to use as a guide to refitting.

3 The tripod joint can now be removed. If it is tight, draw the joint off the driveshaft end using a puller. Ensure that the legs of the puller are located behind the joint inner member and do not contact the joint rollers. Alternatively, support the inner member of the tripod joint and press the shaft out of the joint, again ensuring that no load is applied to the joint rollers.

4 The gaiter and bearing assembly is removed in the same way, either by drawing the bearing off the driveshaft, or by pressing the driveshaft out of the bearing. Remove the retaining plate, noting which way round it is fitted.

5 Obtain a new gaiter, which is supplied complete with the small bearing.

6 Owing to the lip-type seal used in the bearing, the bearing and gaiter must be pressed into position. If a hammer and tubular drift are used to drive the assembly onto the driveshaft, there is a risk of distorting the seal.

7 Refit the retaining plate to the driveshaft, ensuring that it is fitted the correct way around.

8 Support the driveshaft, and press the gaiter bearing onto the shaft, using a tubular spacer which bears only on the bearing inner race (see Fig. 8.12). Position the bearing so that the distance from the end of the driveshaft to the

inner face of the bearing is as shown in Fig. 8.13.

9 Align the marks made on dismantling, and engage the tripod joint with the driveshaft splines. Use a hammer and soft metal drift to tap the joint onto the shaft, taking care not to damage the driveshaft splines or joint rollers. Alternatively, support the driveshaft, and press the joint into position using a tubular spacer which bears only on the joint inner member.

10 Secure the tripod joint in position with the circlip, ensuring that it is correctly located in the driveshaft groove. Where no circlip is fitted, secure the joint in position by staking the end of the driveshaft in three places, at intervals of 120°, using a hammer and punch.

11 Refit the driveshaft to the vehicle as described in Section 2.

6 Constant velocity joint gaiter renewal (models with AD4 automatic transmission) - general information

1 At the time of writing, no information on driveshaft dismantling was available for these models. If gaiter renewal is necessary, the driveshaft should be removed from the vehicle, as described in Section 2, and taken to a Renault dealer.

7 Driveshaft overhaul - general information

1 If any of the checks described in Chapter 1 reveal wear in a driveshaft joint, first remove the roadwheel trim or centre cap (as appropriate) and check that the driveshaft retaining nut is still correctly tightened; if in doubt, use a torque wrench to check it. Refit the centre cap or trim, and repeat the check on the other driveshaft.

2 Road test the vehicle, and listen for a metallic clicking from the front as the vehicle is driven slowly in a circle on full-lock. If a clicking noise is heard, this indicates wear in the outer constant velocity joint.

3 If vibration, consistent with road speed, is felt through the vehicle when accelerating, there is a possibility of wear in the inner constant velocity joints.

4 Constant velocity joints can be dismantled and inspected for wear as described in Sections 3, 4 and 5.

5 On models with a manual gearbox or MB1 automatic transmission, wear in the outer constant velocity joint can only be rectified by renewing the driveshaft. This is necessary since no outer joint components are available separately. For the inner joint, the tripod joint

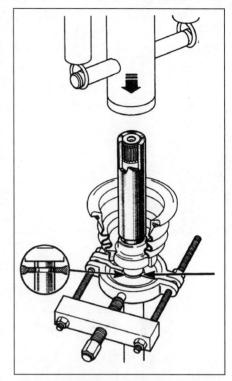

Fig. 8.12 Pressing the inner bearing/gaiter onto the end of the left-hand driveshaft (Sec 5)

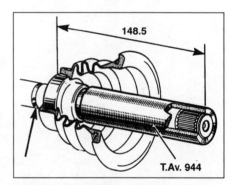

Fig. 8.13 Fitting dimension (in mm) for the left-hand driveshaft inner bearing/gaiter (Sec 5)

and roller assembly is available separately, but wear in any of the other components will also necessitate driveshaft renewal.

6 On models equipped with four-speed (AD4) automatic transmission, wear in either constant velocity joint will necessitate driveshaft renewal; no components for either joint are available separately.

7 On models with ABS, the reluctor ring should be removed from the old driveshaft and fitted to the new one. See Chapter 9, Section 23.

8

Notes

Chapter 9 Braking system

Contents

Degrees of difficulty

Easy, suitable for novice with little experience		**Fairly easy,** suitable for beginner with some experience		**Fairly difficult,** suitable for competent DIY mechanic	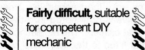	**Difficult,** suitable for experienced DIY mechanic		**Very difficult,** suitable for expert DIY or professional	

Specifications

General

System type	Dual hydraulic circuit, split diagonally, with servo assistance
Front brakes	Disc, with single-piston sliding caliper
Rear brakes	Self-adjusting drum or disc, according to model
Handbrake	Cable-operated, to rear wheels

Front brakes

Disc diameter:

1764 cc (F7P) engine models	259 mm
All other models	238 mm

Disc thickness:

	New	**Minimum**
1108 cc (C1E) engine models	8.0 mm	7.0 mm
1171 cc (E5F and E7F) and 1390 cc (E6J and E7J) engine models . .	12.0 mm	10.5 mm
1721 cc (F2N), 1764 cc (F7P) and 1794 cc (F3P) engine models	20.0 mm	18.0 mm
Disc run-out (all models)	0.07 mm maximum	
Brake pad minimum thickness (friction material and backing plate) . . .	6.0 mm	

Rear drum brakes

Drum diameter:

New	180.25 mm
Maximum diameter after machining	181.25 mm

Brake shoe thickness (friction material and shoe):

New	6.5 mm
Minimum	2.5 mm

Rear disc brakes

Disc diameter	238 mm

Disc thickness:

New	8.0 mm
Minimum	7.0 mm
Disc run-out	0.07 mm maximum
Brake pad minimum thickness (friction material and backing plate) . . .	5.0 mm

9

Anti-lock braking system (ABS)

Wheel sensor-to-reluctor ring clearance:

Front	0.21 to 1.03 mm
Rear	0.40 to 1.50 mm
Sensor electrical resistance	1130 ohms (approx.)

Torque wrench settings

	Nm	lbf ft
Vacuum servo unit mounting nuts	23	17
Flexible brake hose union	13	10
Master cylinder to servo unit nuts	23	17
Master cylinder brake pipe union nuts	13	10
Bendix front brake caliper mounting bolts	100	74
Girling brake caliper:		
Guide pin bolts*	35	26
Mounting bracket-to-swivel hub bolts	100	74
Rear brake caliper:		
Caliper mounting bolts	100	74
Caliper frame mounting bolts	65	48
Rear hub nut*:		
Drum brakes	160	118
Disc brakes	175	129
ABS system components:		
Modulator assembly brake pipe union nuts	12 to 16	9 to 12
Wheel sensor retaining bolts	8 to 10	6 to 7
Roadwheel bolts	90	66

*Renew every time

1 General information

The braking system is of the servo-assisted, dual circuit hydraulic type. The arrangement of the hydraulic system is such that each circuit operates one front and one rear brake from a tandem master cylinder. Under normal circumstances, both circuits operate in unison. However, in the event of hydraulic failure in one circuit, full braking force will still be available at two wheels.

Some larger-engined models are equipped with disc brakes all round as standard, whereas all other models without ABS are fitted with front disc brakes and rear drum brakes. An anti-lock braking system (ABS) is offered as an option on all larger-engined models; on models with ABS, disc brakes are fitted both front and rear. (Refer to Section 22 for further information on ABS operation).

The front disc brakes are actuated by single-piston sliding type calipers, which ensure that equal pressure is applied to each disc pad.

On models with rear drum brakes, the rear brakes incorporate leading and trailing shoes, which are actuated by twin-piston wheel cylinders (one cylinder per drum). The wheel cylinders incorporate integral pressure-regulating valves, which control the hydraulic pressure applied to the rear brakes. The regulating valves help to prevent rear wheel lock-up during emergency braking. As the brake shoe linings wear, footbrake operation automatically operates a self-adjuster mechanism, which effectively lengthens the strut between the shoes and reduces the lining-to-drum clearance.

On models with rear disc brakes, the brakes are actuated by single-piston sliding calipers which incorporate mechanical handbrake mechanisms. A load-sensitive pressure-regulating valve is connected into the brake lines to the rear calipers. The regulating valve is similar to that fitted to the rear wheel cylinders (on rear drum brake models), and helps to prevent rear wheel lock-up during emergency braking. It does this by varying the hydraulic pressure applied to the rear calipers in proportion to the load being carried by the vehicle.

On all models, the handbrake provides an independent mechanical means of rear brake application.

Note: *When servicing any part of the system, work carefully and methodically; also observe scrupulous cleanliness when overhauling any part of the hydraulic system. Always renew components (in axle sets, where applicable) if in doubt about their condition, and use only genuine Renault replacement parts, or at least those of known good quality. Note the warnings given in 'Safety first!' and at relevant points in this Chapter, concerning the dangers of asbestos dust and hydraulic fluid.*

2 Brake pedal - removal and refitting

Removal

1 Remove the clutch pedal as described in Chapter 6.

2 Extract the spring clip, then withdraw the clevis pin securing the servo unit pushrod to the brake pedal. Note the spacer which is located on the inside of the pedal.

3 Fully withdraw the pedal cross-shaft and manoeuvre the brake pedal out of the pedal mounting bracket, noting the fitted positions of the pivot bushes (photo).

4 Inspect the bushes for signs of wear or damage, and renew as necessary.

Refitting

5 Apply a smear of multi-purpose grease to the contact surfaces of the pivot bushes. Fit the bushes to the brake pedal, noting that the thicker bush is fitted on the left-hand side of

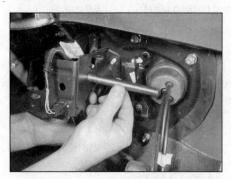

2.3 Withdraw the cross-shaft and manoeuvre the brake pedal out of the mounting bracket

2.5 Ensure that the lugs on the mounting bushes are correctly engaged with the pedal cut-outs (arrowed)

3.7A Remove the spring clip (arrowed) . . .

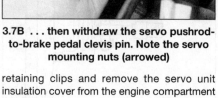

3.7B . . . then withdraw the servo pushrod-to-brake pedal clevis pin. Note the servo mounting nuts (arrowed)

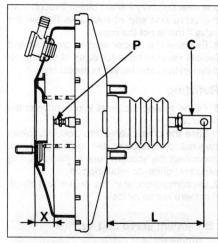

Fig. 9.1 Vacuum servo unit adjustment dimensions (Secs 3 and 8)

C Pushrod clevis   P Pushrod nut
L 110 mm   X 22.3 mm

the pedal. Also ensure that the lug on each bush is correctly located in the pedal cut-out (photo).

6 Slide the pedal and bush assembly into position in the mounting bracket. Ensure that the pedal is correctly engaged with the servo pushrod, then insert the cross-shaft from the right-hand side of the mounting bracket.

7 Position the spacer on the inside of the pedal. Refit the clevis pin, and secure it in position with the spring clip.

8 Refit the clutch pedal as described in Chapter 6.

3 Vacuum servo unit -
testing, removal and refitting

Testing

1 To test the operation of the servo unit, depress the footbrake several times to exhaust the vacuum, then start the engine whilst keeping the pedal firmly depressed. As the engine starts, there should be a noticeable 'give' in the brake pedal as the vacuum builds up. Allow the engine to run for at least two minutes, then switch it off. If the brake pedal is now depressed it should feel normal, but further applications should result in the pedal feeling firmer, with the pedal stroke decreasing with each application.

2 If the servo does not operate as described, inspect the servo unit check valve as described in Section 4. If the check valve is OK, try renewing the servo air filter (Section 5).

3 If the servo still fails to operate satisfactorily, the fault lies within the unit itself. Apart from external components, no spares are available, so a defective servo must be renewed.

Removal

4 Remove the master cylinder as described in Section 8.

5 Slacken the retaining clip, and disconnect the vacuum hose from the check valve on the servo unit. If the original Renault clip is still fitted, cut the clip and discard it; use a standard worm-drive hose clip on refitting.

6 On right-hand drive models, release the

retaining clips and remove the servo unit insulation cover from the engine compartment bulkhead. On left-hand drive models, release the rubber strap and position the expansion tank clear of the servo unit.

7 Working inside the vehicle, extract the spring clip and withdraw the clevis pin securing the servo unit pushrod to the brake pedal. Note the spacer which is located on the inside of the pedal (photos).

8 Remove the four nuts and washers securing the servo unit to the engine compartment bulkhead. Return to the engine compartment and remove the servo, noting the gasket which is fitted to the rear of the unit.

Refitting

9 Prior to refitting, check that the servo unit pushrod is correctly adjusted as follows. With the gasket removed, check the dimensions shown in Fig. 9.1. If adjustment is necessary, dimension 'L' can be altered by slackening the locknut and repositioning the pushrod clevis, and dimension 'X' can be altered by repositioning the nut (P). After adjustment ensure that the clevis locknut is securely tightened.

10 Inspect the check valve sealing grommet for signs of damage or deterioration, and renew if necessary.

11 Fit a new gasket to the rear of the servo unit, and reposition the unit in the engine compartment.

12 Working inside the vehicle, ensure that the servo unit pushrod is correctly engaged with the brake pedal. Refit the servo unit mounting nuts (and washers), and tighten them to the specified torque.

13 Position the spacer on the inside of the pedal. Refit the clevis pin, and secure it in position with the spring clip.

14 On right-hand drive models, refit the servo unit insulation cover to the bulkhead and secure it in position with all the necessary retaining clips. On left-hand drive models, refit the expansion tank to its retaining bracket, and secure it in position with the rubber strap.

15 Reconnect the vacuum hose to the servo unit check valve, and securely tighten its retaining clip.

16 Refit the master cylinder as described in Section 8 of this Chapter.

17 On completion, start the engine and check that there are no air leaks at the servo vacuum hose connection. Check the operation of the servo as described at the beginning of this Section.

4 Vacuum servo unit check valve
- removal, testing and refitting

Removal

1 Slacken the retaining clip, and disconnect the vacuum hose from the check valve on the servo unit (photo). If the original Renault clip is still fitted, cut the clip and discard it; use a standard worm-drive hose clip on refitting.

2 Withdraw the valve from its rubber sealing grommet, using a pulling and twisting motion. Remove the grommet from the servo.

Testing

3 Examine the check valve for signs of damage, and renew if necessary. The valve may be tested by blowing through it in both directions. Air should flow through the valve in

4.1 Vacuum servo unit check valve (arrowed)

9

one direction only - when blown through from the servo unit end of the valve. Renew the valve if this is not the case.

4 Examine the rubber sealing grommet and flexible vacuum hose for signs of damage or deterioration, and renew as necessary.

Refitting

5 Fit the sealing grommet into position in the servo unit.

6 Ease the check valve into position, taking care not to displace or damage the grommet. Reconnect the vacuum hose to the valve, and securely tighten its retaining clip.

7 On completion, start the engine and check that there are no air leaks.

5 Vacuum servo unit air filter - renewal

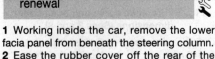

1 Working inside the car, remove the lower facia panel from beneath the steering column.

2 Ease the rubber cover off the rear of the servo unit, and move it up the pushrod.

3 Using a screwdriver or scriber, hook out the old air filter and remove it from the servo.

4 Make a cut in the new filter as shown in Fig. 9.2. Place the filter over the pushrod and into position in the servo end.

5 Refit the rubber cover, then refit the lower facia panel beneath the steering column.

6 Hydraulic system - bleeding

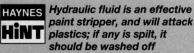

⚠ **Warning: Hydraulic fluid is poisonous; wash off immediately and thoroughly in the case of skin contact, and seek immediate medical advice if any fluid is swallowed or gets into the eyes. Certain types of hydraulic fluid are inflammable, and may ignite when allowed into contact with hot components; when servicing any hydraulic system, it is safest to assume that the fluid is inflammable, and to take precautions against the risk of fire as though it is petrol that is being handled. Finally, it is hygroscopic (it absorbs moisture from the air) - old fluid may be contaminated and unfit for further use. When topping-up or renewing the fluid, always use the recommended type (see Chapter 1), and ensure that it comes from a freshly-opened, previously-sealed container.**

> **HAYNES HiNT** *Hydraulic fluid is an effective paint stripper, and will attack plastics; if any is spilt, it should be washed off immediately using copious quantities of fresh water*

Note: *On models with ABS, if any part of the hydraulic system is disconnected, the system must be fully bled before the vehicle is used. Failure to do so can result in air entering the*

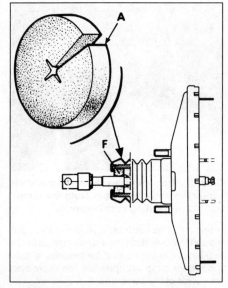

Fig. 9.2 Vacuum servo unit air filter renewal (Sec 5)

A Cut the new filter as shown
F Correct fitted position of filter in servo unit

hydraulic return pump. If the return pump draws in air, it will prove very difficult and even impossible to bleed out.

General

1 The correct operation of any hydraulic system is only possible after removing all air from the components and circuit; this is achieved by bleeding the system.

2 During the bleeding procedure, add only clean, unused hydraulic fluid of the recommended type; never re-use fluid that has already been bled from the system. Ensure that sufficient fluid is available before starting work.

3 If there is any possibility of incorrect fluid being already in the system, the system must be flushed completely with uncontaminated, correct fluid, and new seals should be fitted to the various components.

4 If air has entered the hydraulic system because of a leak, ensure that the fault is cured before proceeding further.

5 Park the vehicle on level ground, switch off the engine and select first or reverse gear (or 'P' on automatic transmission models). Chock the wheels and release the handbrake.

6 Check that all pipes and hoses are secure, unions tight and bleed screws closed. Clean any dirt from around the bleed screws.

7 Unscrew the master cylinder reservoir cap and top the master cylinder reservoir up to the 'MAX' level line; refit the cap loosely. Remember to maintain the fluid level at least above the 'MIN' level line throughout the procedure, or there is a risk of further air entering the system.

8 There are a number of one-man, do-it-yourself brake bleeding kits currently available from motor accessory shops. It is recommended that one of these kits is used

whenever possible, as they greatly simplify the bleeding operation, and also reduce the risk of expelled air and fluid being drawn back into the system. If such a kit is not available, the basic (two-man) method must be used, which is described in detail below.

9 If a kit is to be used, prepare the vehicle as described previously, and follow the kit manufacturer's instructions. The procedure may vary slightly according to the type of kit being used; general procedures are as outlined below in the relevant sub-section.

10 Whichever method is used, the same sequence must be followed (paragraphs 11 and 12) to ensure the removal of all air from the system.

Bleeding sequence

11 If the system has been only partially disconnected, and suitable precautions were taken to minimise fluid loss, it should be necessary only to bleed that part of the system (ie the primary or secondary circuit).

12 If the complete system is to be bled, then it should be done working in the following sequence:

Non-ABS models
(a) Right-hand rear brake.
(b) Left-hand front brake.
(c) Left-hand rear brake.
(d) Right-hand front brake.

ABS models
(a) Left-hand front brake.
(b) Right-hand front brake.
(c) Left-hand rear brake.
(d) Right-hand rear brake.

Bleeding - basic (two-man) method

13 Collect a clean glass jar, a suitable length of plastic or rubber tubing which is a tight fit over the bleed screw, and a ring spanner to fit the screw. The help of an assistant will also be required.

14 Remove the dust cap from the first screw in the sequence. Fit the spanner and tube to the screw, place the other end of the tube in the jar, and pour in sufficient fluid to cover the end of the tube.

15 Ensure that the master cylinder reservoir fluid level is maintained at least above the 'MIN' level line throughout the procedure.

16 Have the assistant fully depress the brake pedal several times to build up pressure, then maintain it on the final stroke.

17 While pedal pressure is maintained, unscrew the bleed screw (approximately one turn) and allow the compressed fluid and air to flow into the jar. The assistant should maintain pedal pressure, following it down to the floor if necessary, and should not release it until instructed to do so. When the flow stops, tighten the bleed screw again. Have the assistant release the pedal slowly.

18 Repeat the steps given in paragraphs 16 and 17 until the fluid emerging from the bleed screw is free from air bubbles. Remember to recheck the fluid level in the master cylinder

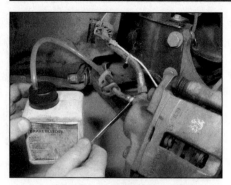

6.21 Bleeding a front brake caliper

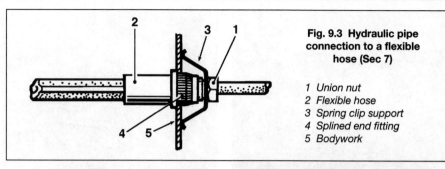

Fig. 9.3 Hydraulic pipe connection to a flexible hose (Sec 7)

1 Union nut
2 Flexible hose
3 Spring clip support
4 Splined end fitting
5 Bodywork

reservoir every five strokes or so. If the master cylinder has been drained and refilled, and air is being bled from the first screw in the sequence, allow approximately five seconds between strokes for the master cylinder passages to refill.

19 When no more air bubbles appear, tighten the bleed screw securely, remove the tube and spanner, and refit the dust cap. Do not overtighten the bleed screw.

20 Repeat the procedure on the remaining screws in the sequence until all air is removed from the system and the brake pedal feels firm.

Bleeding - using a one-way valve kit

21 As their name implies, these kits consist of a length of tubing with a one-way valve fitted, to prevent expelled air and fluid being drawn back into the system; some kits include a translucent container, which can be positioned so that the air bubbles can be more easily seen flowing from the end of the tube (photo).

22 The kit is connected to the bleed screw, which is then opened. The user returns to the driver's seat and depresses the brake pedal with a smooth, steady stroke and slowly releases it; this is repeated until the expelled fluid is clear of air bubbles.

23 Note that these kits simplify work so much that it is easy to forget the master cylinder reservoir fluid level; ensure that this is maintained at least above the 'MIN' level line at all times.

Bleeding - using a pressure-bleeding kit

24 These kits are usually operated by the reservoir of pressurised air contained in the spare tyre, although it may be necessary to reduce the pressure to lower than normal; refer to the instructions supplied with the kit.

25 By connecting a pressurised, fluid-filled container to the master cylinder reservoir, bleeding can be carried out simply by opening each screw in turn (in the specified sequence) and allowing the fluid to flow out until no more air bubbles can be seen in the expelled fluid.

26 This method has the advantage that the large reservoir of fluid provides an additional

safeguard against air being drawn into the system during bleeding.

27 Pressure-bleeding is particularly effective when bleeding 'difficult' systems, or when bleeding the complete system at the time of routine fluid renewal.

All methods

28 When bleeding is complete and firm pedal feel is restored, wash off any spilt fluid, tighten the bleed screws securely and refit their dust caps.

29 Check the hydraulic fluid level, and top-up if necessary (Chapter 1).

30 Discard any hydraulic fluid that has been bled from the system; it will not be fit for re-use.

31 Check the feel of the brake pedal. If it feels at all spongy, air must still be present in the system, and further bleeding is required. Failure to bleed satisfactorily after several repetitions of the bleeding procedure may be due to worn master cylinder seals.

7 Hydraulic pipes and hoses - renewal

Note: *Before starting work, refer to the warning at the beginning of Section 6 concerning the dangers of hydraulic fluid.*

1 If any pipe or hose is to be renewed, minimise fluid loss by removing the master cylinder reservoir cap and then tightening it down onto a piece of polythene (taking care not to damage the sender unit) to obtain an airtight seal. Alternatively, flexible hoses can be sealed, if required, using a proprietary brake hose clamp; metal brake pipe unions can be plugged (if care is taken not to allow dirt into the system) or capped immediately they are disconnected. Place a wad of rag under any union that is to be disconnected, to catch any spilt fluid.

2 If a flexible hose is to be disconnected, unscrew the brake pipe union nut before removing the spring clip which secures the hose to its mounting bracket (photo).

3 To unscrew the union nuts, it is preferable to obtain a brake pipe spanner of the correct size (11 mm/13 mm split ring); these are available from motor accessory shops. Failing this, a close-fitting open-ended spanner will be required, though if the nuts are tight or corroded, their flats may be rounded off if the

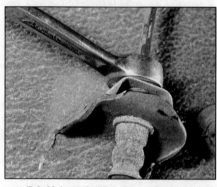

7.2 Using a brake pipe spanner to unscrew a hydraulic union nut

spanner slips. In such a case, a self-locking wrench is often the only way to unscrew a stubborn union, but it follows that the pipe and the damaged nuts must be renewed on reassembly. Always clean a union and surrounding area before disconnecting it. If disconnecting a component with more than one union, make a careful note of the connections before disturbing any of them.

4 If a brake pipe is to be renewed, it can be obtained, cut to length and with the union nuts and end flares in place, from Renault dealers. All that is then necessary is to bend it to shape, following the line of the original, before fitting it to the car. Alternatively, most motor accessory shops can make up brake pipes from kits, but this requires very careful measurement of the original to ensure that the replacement is of the correct length. The safest answer is usually to take the original to the shop as a pattern.

5 On refitting, do not overtighten the union nuts. The specified torque wrench settings (where given) are not high, and it is not necessary to exercise brute force to obtain a sound joint.

6 Ensure that the pipes and hoses are correctly routed with no kinks, and that they are secured in the clips or brackets provided. In the case of flexible hoses, make sure that they cannot contact other components during movement of the steering and/or suspension assemblies.

7 After fitting, remove the polythene from the reservoir (or remove the plugs or clamps, as applicable), and bleed the hydraulic system as described in Section 6. Wash off any spilt fluid, and check carefully for fluid leaks.

9

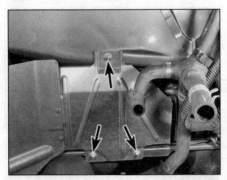

8.1A On right-hand drive models, undo the heat shield retaining nuts (arrowed) . . .

8.1B . . . undo the two retaining nuts . . .

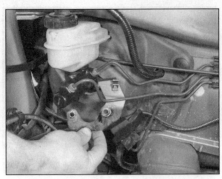

8.1C . . . and remove the cable retaining bracket (engine removed for clarity)

8 Master cylinder -
removal and refitting

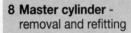

Note: *Before starting work, refer to the warning at the beginning of Section 6 concerning the dangers of hydraulic fluid.*

Removal

1 On right-hand drive models, remove the nuts securing the brake servo heat shield to the engine compartment bulkhead. Manoeuvre the heat shield out of the engine compartment. Undo the two nuts securing the cable retaining bracket to the master cylinder studs, and position the bracket clear of the master cylinder (photos).

2 On all models, remove the master cylinder reservoir cap, having disconnected the sender unit wiring connector, and syphon the hydraulic fluid from the reservoir. **Note:** *Do not syphon the fluid by mouth, as it is poisonous; use a syringe or an old poultry baster.* Alternatively, open any convenient pair of bleed screws in the system (one in each hydraulic circuit) and gently pump the brake pedal to expel the fluid through plastic tubes connected to the screws (see Section 6).

3 Once drained, remove the reservoir from the master cylinder by pulling it upwards.

4 Wipe clean the area around the brake pipe unions on the side of the master cylinder, and place absorbent rags beneath the pipe unions to catch any surplus fluid. Make a note of the correct fitted positions of the unions, then unscrew the union nuts and carefully withdraw the pipes. Plug or tape over the pipe ends and master cylinder orifices, to minimise the loss of brake fluid and to prevent the entry of dirt into the system. Wash off any spilt fluid immediately with cold water.

5 Slacken and remove the two nuts securing the master cylinder to the vacuum servo unit, then withdraw the master cylinder from the engine compartment. Remove the O-ring from the rear of the master cylinder, and discard it.

6 It is not possible to obtain seals or internal components for the master cylinder, therefore if it is faulty, it must be renewed as a complete unit. The reservoir mounting bush seals may be renewed if necessary. The O-ring fitted between the master cylinder and the vacuum servo must be renewed as a matter of course whenever the unit is removed, as a leak at this point will allow atmospheric pressure into the servo unit.

Refitting

7 Before refitting the master cylinder, check that the distance between the tip of the master cylinder end of the pushrod and the front of the servo unit, dimension 'X', is as shown in Fig. 9.1. If necessary, adjust by repositioning the pushrod nut 'P'.

8 Remove all traces of dirt from the master cylinder and servo unit mating surfaces. Fit a new O-ring to the groove on the master cylinder body.

9 Fit the master cylinder to the servo, ensuring that the servo pushrod enters the master cylinder bore centrally. Refit the master cylinder mounting nuts, and tighten them to the specified torque.

10 Wipe clean the brake pipe unions, then refit them to the master cylinder ports. Tighten the union nuts to the specified torque.

11 Carefully align the reservoir with the mounting bush seals. Push the reservoir firmly into position.

12 On right-hand drive models, locate the cable retaining plate on the master cylinder studs, and securely tighten its retaining nuts. Refit the heat shield to the bulkhead and secure it.

13 On all models, refill the master cylinder reservoir with new fluid, and bleed the hydraulic system as described in Section 6.

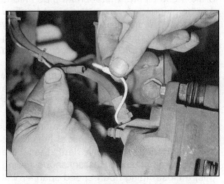

9.2 Disconnecting the pad wear sensor wiring connector (Girling caliper shown)

9 Front brake pads - renewal

> **Warning: Renew both sets of front brake pads at the same time - never renew the pads on only one wheel, as uneven braking may result. Note that the dust created by wear of the pads may contain asbestos, which is a health hazard. Never blow it out with compressed air, and don't inhale any of it. An approved filtering mask should be worn when working on the brakes. DO NOT use petroleum-based solvents to clean brake parts - use brake cleaner or methylated spirit only.**

1 Apply the handbrake, then jack up the front of the vehicle and support it on axle stands. Remove the front roadwheels.

2 Trace the brake pad wear sensor wiring back from the inner pad, and disconnect it at the wiring connector (photo).

3 Push the piston into its bore by pulling the caliper outwards.

Bendix calipers

4 Extract the small spring clip from the pad retaining plate, and then slide the plate out of the caliper (photos).

5 Using pliers if necessary, withdraw the pads from the caliper (photos). Make a note of the correct fitted position of the anti-rattle spring, and remove the spring from each pad.

6 First measure the thickness of each brake

9.4A On Bendix calipers remove the spring clip . . .

9.4B . . . and remove the pad retaining plate

9.5A Withdraw the outer brake pad . . .

9.5B . . . and the inner brake pad

pad (friction material and backing plate). If any pad is worn at any point to the specified minimum thickness or less, all four pads must be renewed. Also, the pads should be renewed if any are fouled with oil or grease; there is no satisfactory way of degreasing friction material once contaminated. If any of the brake pads are worn unevenly or fouled with oil or grease, trace and rectify the cause before reassembly. New brake pads and spring kits are available from Renault dealers.

7 Note that there are two different types of brake pad available for the Bendix caliper. The correct pad type depends on the brake caliper bore/piston diameter. Calipers with a 45 mm diameter bore/piston require pads with symmetrical linings, whereas calipers with a 48 mm diameter bore/piston require pads with offset linings (groove in the friction material not central). The pad types can also be distinguished by the cut-outs in the upper corners of the friction material (see Fig. 9.4); symmetrical linings have two cut-outs, whereas the offset linings have only a single cut-out.

8 If the brake pads are still serviceable, carefully clean them using a clean, fine wire

brush or similar, paying particular attention to the sides and back of the metal backing. Clean out the grooves in the friction material, and pick out any large embedded particles of dirt or debris. Carefully clean the pad locations in the caliper body/mounting bracket.

9 Prior to fitting the pads, check that the guide sleeves are free to slide easily in the caliper body, and check that the rubber guide sleeve gaiters are undamaged. Brush the dust and dirt from the caliper and piston, but *do not inhale it as it is injurious to health*. Inspect the dust seal around the piston for damage, and the piston for evidence of fluid leaks, corrosion or damage. If attention to any of these components is necessary, refer to Section 10. Also inspect the brake disc as described in Section 11.

10 If new brake pads are to be fitted, the caliper piston must be pushed back into the cylinder to make room for them. Either use a G-clamp or similar tool, or use suitable pieces of wood as levers. Provided that the master cylinder reservoir has not been overfilled with hydraulic fluid, there should be no spillage, but keep a careful watch on the fluid level while retracting the piston. If the fluid level rises above the 'MAX' level line at any time, the surplus should be syphoned off (not by mouth - use an old syringe or a poultry baster)

or ejected via a plastic tube connected to the bleed screw (see Section 6).

11 Fit the anti-rattle springs to the pads, so that when the pads are installed in the caliper, the spring end will be located at the opposite end of the pad to the pad retaining plate (see Fig. 9.5).

12 Locate the pads in the caliper, ensuring that the friction material of each pad is against the brake disc, and the pad with the wear sensor is fitted on the inside. Also recheck that the anti-rattle spring ends are at the opposite end of the pad to which the retaining plate is to be inserted. On models with offset linings, looking at the pads from the front of the car, the innermost pad groove must be higher than the outer pad groove (see Fig. 9.6).

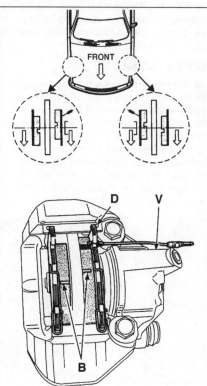

Fig. 9.6 Correct fitting of Bendix offset lining brake pads (Sec 9)

B Grooves
D Pad retaining plate spring clip location
V Bleed screw

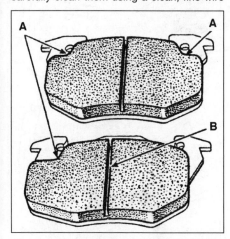

Fig. 9.4 Bendix caliper symmetrical-lining pad (top) and offset-lining pad (bottom). Offset lining has only a single cut-out (A) and the groove in the friction material (B) is not central (Sec 9)

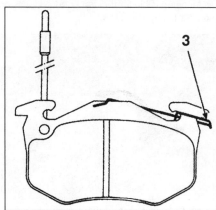

Fig. 9.5 Anti-rattle spring (3) correctly fitted to Bendix inner brake pad (Sec 9)

9

9.12 Correct fitted positions of the Bendix offset lining brake pads

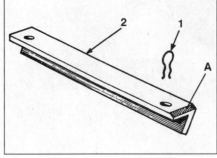

Fig. 9.7 Bendix disc pad retaining plate (2) and spring clip (1) showing filed chamfer (A) (Sec 9)

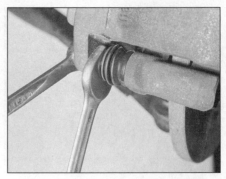

9.19A On Girling calipers, slacken the guide pin bolts whilst holding the guide pins with an open-ended spanner . . .

Make sure that the pads are fitted correctly (photo).

13 Slide the retaining plate into place, and install the small spring clip at its inner end. It may be necessary to file an entry chamfer on the edge of the retaining plate, to enable it to be fitted without difficulty (Fig. 9.7).

14 Reconnect the brake pad wear sensor wiring connector, ensuring that the wire is correctly routed.

15 Depress the brake pedal several times to bring the pads into firm contact with the brake disc.

16 Repeat the above procedure on the other front brake caliper.

17 Refit the roadwheels, then lower the vehicle to the ground and tighten the roadwheel bolts to the specified torque.

18 Check the hydraulic fluid level as described in Chapter 1.

Girling calipers

19 Slacken and remove the caliper upper and lower guide pin bolts, using a slim open-ended spanner to prevent the guide pin itself from rotating (photos). Discard the guide pin bolts; new bolts must be used on refitting. If genuine Renault brake pads are purchased, the bolts will be supplied with the pad set.

20 With the guide pins removed, lift the caliper away from the brake pads and mounting bracket, and tie it to the suspension

strut using a suitable piece of wire. Do not allow the caliper to hang unsupported on the flexible hose.

21 Withdraw the two brake pads from the caliper mounting bracket, and examine them as described above in paragraph 6 and paragraphs 8 to 10, ignoring paragraph 7, which is only applicable to the Bendix caliper.

22 Install the pads in the caliper mounting bracket, ensuring that the friction material of each pad is against the brake disc. The pad with the wear sensor is fitted on the inside (photo).

23 Position the caliper over the pads. Coat the threads of the new lower guide pin bolt with locking fluid, and fit the bolt. Apply locking fluid to the new upper guide pin bolt, press the caliper into position, and fit the bolt. Check that the anti-rattle springs are correctly located (see Fig. 9.8), then tighten the guide pin bolts to the specified torque, starting with the lower bolt.

24 Carry out the operations described above in paragraphs 14 to 18.

All calipers

25 If new pads have been fitted, full braking efficiency will not be obtained until the linings have bedded-in. Be prepared for longer stopping distances, and avoid harsh braking as far as possible for the first hundred miles or so after fitting new pads.

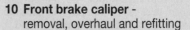

10 Front brake caliper - removal, overhaul and refitting

Note: *Before starting work, refer to the warnings at the beginning of Sections 6 and 9 concerning the dangers of hydraulic fluid and asbestos dust.*

Removal

1 Apply the handbrake, then jack up the front of the vehicle and support it on axle stands. Remove the appropriate roadwheel.

2 Minimise fluid loss, either by removing the master cylinder reservoir cap and then tightening it down onto a piece of polythene to obtain an airtight seal (taking care not to damage the sender unit), or by using a brake hose clamp, a G-clamp or a similar tool with protected jaws to clamp the flexible hose.

Bendix caliper

3 Remove the brake pads as described in Section 9, paragraphs 2 to 5.

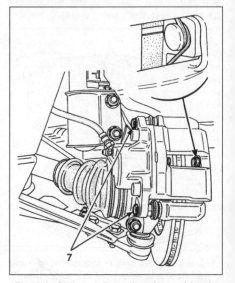

Fig. 9.8 Girling caliper showing guide pin bolts (7) and correct fitted position of anti-rattle spring (Sec 9)

9.19B . . . then withdraw the bolts and lift off the brake caliper

9.22 Ensure that the brake pads are fitted the correct way round, with friction material facing the disc

4 Clean the area around the union, then loosen the brake hose union nut.

5 Slacken the two bolts securing the caliper assembly to the swivel hub, and remove them along with the mounting plate, noting which way around the plate is fitted. Lift the caliper assembly away from the brake disc, and unscrew it from the end of the brake hose.

Girling caliper

6 Clean the area around the hose union, then loosen the brake hose union nut (photo).

7 Slacken and remove the upper and lower caliper guide pin bolts, using a slim open-ended spanner to prevent the guide pin itself from rotating. Discard the guide pin bolts; new bolts must be used on refitting. With the guide pin bolts removed, lift the caliper away from the brake disc, then unscrew the caliper from the end of the brake hose. Note that the brake pads need not be disturbed, and can be left in position in the caliper mounting bracket.

Overhaul

8 With the caliper on the bench, wipe away all traces of dust and dirt, but *avoid inhaling the dust, as it is injurious to health.*

9 On the Girling caliper, using a small flat-bladed screwdriver, carefully prise the dust seal retaining clip out of the caliper bore.

10 On all calipers, withdraw the partially-ejected piston from the caliper body and remove the dust seal. The piston can be withdrawn by hand, or if necessary forced out by applying compressed air to the union bolt hole.

 Caution: The piston may be ejected with some force. Only low pressure should be required, such as is generated by a foot pump.

11 Extract the piston hydraulic seal using a blunt instrument such as a knitting needle or a crochet hook, taking care not to damage the caliper bore.

12 Withdraw the guide sleeves or pins from the caliper body or mounting bracket (as applicable) and remove the rubber gaiters.

13 Thoroughly clean all components using only methylated spirit, isopropyl alcohol or clean hydraulic fluid as a cleaning medium. Never use mineral-based solvents, such as petrol or paraffin, which will attack the hydraulic system rubber components. Dry the components immediately, using compressed air or a clean, lint-free cloth. Use compressed air to blow clear the fluid passages.

14 Check all components and renew any that are worn or damaged. Check particularly the cylinder bore and piston; if they are scratched, worn or corroded in any way, they must be renewed (note that this means the renewal of the complete body assembly). Similarly check the condition of the guide sleeves or pins and their bores; they should be undamaged and (when cleaned) a reasonably tight sliding fit in the body or mounting bracket bores. If there is any doubt about the condition of a component, renew it.

10.6 Slackening the front brake caliper hose union nut (Girling caliper shown)

15 If the assembly is fit for further use, obtain the appropriate repair kit; the components are available from Renault dealers, in various combinations.

16 Renew all rubber seals, dust covers and caps disturbed on dismantling as a matter of course; these should never be re-used.

17 Before commencing reassembly, ensure that all components are absolutely clean and dry.

18 Dip the piston and the new piston (fluid) seal in clean hydraulic fluid. Smear clean fluid on the cylinder bore surface.

19 Fit the new piston (fluid) seal, using only the fingers to manipulate it into the cylinder bore groove. Fit the new dust seal to the piston. Refit the piston to the cylinder bore using a twisting motion, ensuring that the piston enters squarely into the bore. Press the piston fully into the bore, then press the dust seal into the caliper body.

20 On the Girling caliper, install the dust seal retaining clip, ensuring that it is correctly seated in the caliper groove.

21 On all calipers, apply the grease supplied in the repair kit, or a good quality high-temperature brake grease or anti-seize compound to the guide sleeves or pins. Fit the sleeves or pins to the caliper body or mounting bracket. Fit the new rubber gaiters, ensuring that they are correctly located in the grooves on both the sleeve or pin, and body or mounting bracket (as applicable).

Refitting

Bendix caliper

22 Screw the caliper fully onto the flexible hose union nut. Position the caliper over the brake disc, then refit the two caliper mounting bolts and the mounting plate. Note that the mounting plate must be fitted so that its bend curves towards the caliper body (see Fig. 9.9); this is necessary to prevent the plate contacting the driveshaft gaiter when the steering is on full-lock. With the plate correctly positioned, tighten the caliper bolts to the specified torque setting.

23 Tighten the brake hose union nut to the specified torque, then refit the brake pads as described in Section 9.

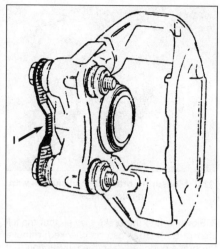

Fig. 9.9 On Bendix caliper, ensure mounting plate (1) is fitted so that its bend curves towards the caliper body (Sec 10)

Girling caliper

24 Screw the caliper body fully onto the flexible hose union nut. Check that the brake pads are still correctly fitted in the caliper mounting bracket.

25 Position the caliper over the pads. Coat the threads of the new lower guide pin bolt with locking fluid, and fit the bolt. Apply locking fluid to the new upper guide pin bolt, press the caliper into position, and fit the bolt. Check that the anti-rattle springs are correctly located (see Fig. 9.8), then tighten the guide pin bolts to the specified torque, starting with the lower bolt.

26 Tighten the brake hose union nut to the specified torque.

All calipers

27 Remove the brake hose clamp or polythene, where fitted, and bleed the hydraulic system as described in Section 6. Providing the precautions described were taken to minimise brake fluid loss, it should only be necessary to bleed the relevant front brake.

28 Refit the roadwheel, then lower the vehicle to the ground and tighten the roadwheel bolts to the specified torque.

---

11 Front brake disc - inspection, removal and refitting

Note: *Before starting work, refer to the warning at the beginning of Section 9 concerning the dangers of asbestos dust. If either disc requires renewal, both should be renewed at the same time, to ensure even and consistent braking. In principle, new pads should be fitted also.*

Inspection

1 Chock the rear wheels, firmly apply the handbrake, jack up the front of the vehicle and support on axle stands. Remove the appropriate front roadwheel.

9

11.3 Measuring brake disc thickness with a micrometer

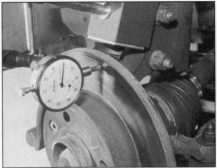

11.4 Measuring brake disc run-out with a dial gauge

11.6A Undo the two bolts (arrowed) securing the caliper to the swivel hub . . .

2 Slowly rotate the brake disc so that the full area of both sides can be checked; remove the brake pads, as described in Section 9, if better access is required to the inboard surface. Light scoring is normal in the area swept by the brake pads, but if heavy scoring is found, the disc must be renewed.

3 It is normal to find a lip of rust and brake dust around the disc's perimeter; this can be scraped off if required. If, however, a lip has formed due to wear of the brake pad swept area, the disc thickness must be measured using a micrometer (photo). Take measurements at several places around the disc at the inside and outside of the pad swept area; if the disc has worn at any point to the specified minimum thickness or less, it must be renewed.

4 If the disc is thought to be warped, it can be checked for run-out, ideally by using a dial gauge mounted on any convenient fixed point, while the disc is slowly rotated (photo). In the absence of a dial gauge, use feeler blades to measure (at several points all around the disc) the clearance between the disc and a fixed point such as the caliper mounting bracket. If the measurements obtained are at the specified maximum or beyond, the disc is excessively warped, and must be renewed; however, it is worth checking first that the hub bearing is in good condition (Chapters 1 and 10). Also try the effect of removing the disc and turning it through 180° to reposition it on the hub; if run-

out is still excessive, the disc must be renewed.

5 Check the disc for cracks (especially around the wheel bolt holes), and for any other wear or damage. Renew the disc if necessary.

Removal

6 Unscrew the two bolts securing the brake caliper to the swivel hub, and slide the caliper assembly off the disc. Using a piece of wire or string, tie the caliper to the front suspension coil spring, to avoid placing any strain on the hydraulic brake hose (photos).

7 If the same disc is to be refitted, use chalk or paint to mark the relationship of the disc to the hub. Remove the two screws securing the brake disc to the hub, and remove the disc. If it is tight, lightly tap its rear face with a hide or plastic mallet.

Refitting

8 Refitting is the reverse of the removal procedure, noting the following points:
(a) Ensure that the mating surfaces of the disc and hub are clean and flat.
(b) If applicable, align the marks made on removal.
(c) Securely tighten the disc retaining screws.
(d) If a new disc has been fitted, use a suitable solvent to wipe any preservative coating from the disc before refitting the caliper.

(e) Apply locking fluid to the threads of the brake caliper mounting bolts, and tighten them to the specified torque.
(f) Refit the roadwheel, then lower the vehicle to the ground and tighten the roadwheel bolts to the specified torque. On completion, depress the brake pedal several times to bring the brake pads into contact with the disc.

12 Load-sensitive pressure-regulating valve (models with rear disc brakes) - general information

1 On models with rear disc brakes, a load-sensitive pressure-regulating valve is incorporated in the hydraulic circuit. The valve regulates the pressure applied to the rear brakes, and reduces the risk of the rear wheels locking under heavy braking. It is mounted underneath the rear of the vehicle, above the rear axle assembly (photo).

2 Removal and refitting of the valve is a straightforward process, but on completion, specialist equipment is required to check and adjust the operation of the valve. Therefore, if the valve is to be removed or if it is thought to be faulty, the work should be entrusted to a Renault dealer, who will have the necessary test equipment.

13 Rear brake drum - removal, inspection and refitting

Note: Before starting work, refer to the warning at the beginning of Section 14 concerning the dangers of asbestos dust. If either drum requires renewal or refinishing, both should be dealt with at the same time, to ensure even and consistent braking. In principle, new shoes should be fitted also.

Removal

1 Chock the front wheels, engage reverse gear (or 'P') and release the handbrake. Jack up the rear of the vehicle and support it on axle stands. Remove the appropriate rear wheel.

11.6B . . . then slide the caliper off the disc, and tie it to the suspension strut spring (Girling caliper shown)

12.1 Load-sensitive pressure-regulating valve - models with rear disc brakes

13.2 Lever out the cap from the centre of the brake drum . . .

13.3 . . . then remove the rear hub nut and thrustwasher

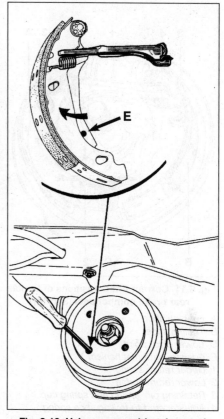

Fig. 9.10 Using a screwdriver inserted through the brake drum to release the handbrake operating lever (Sec 13)

E Handbrake operating lever stop-peg location

2 Using a hammer and suitable large flat-bladed screwdriver, carefully tap and prise the cap out of the centre of the brake drum (photo).

3 Using a socket and long bar, slacken and remove the rear hub nut, and withdraw the thrustwasher (photo). Discard the hub nut; a new nut must used on refitting.

4 It should now be possible to withdraw the brake drum and hub bearing assembly from the stub axle by hand. It may be difficult to remove the drum due to the tightness of the hub bearing on the stub axle, or due to the brake shoes binding on the inner circumference of the drum. If the bearing is tight, tap the periphery of the drum using a hide or plastic mallet, or use a universal puller, secured to the drum with the wheel bolts, to pull it off. If the brake shoes are binding, proceed as follows.

5 First ensure that the handbrake is fully off. From underneath the vehicle, slacken the handbrake cable adjuster locknut, then back off the adjuster nut on the handbrake lever rod. Note that on some models, it will first be necessary to remove the mounting nut(s) and lower the exhaust heat shield to gain access to the adjuster nut.

6 Referring to Fig. 9.10, insert a screwdriver through one of the wheel bolt holes in the brake drum, so that it contacts the handbrake operating lever on the trailing brake shoe (photo). Push the lever until the stop-peg slips behind the brake shoe web, allowing the brake shoes to retract fully. Withdraw the brake drum, and slide the spacer off the stub axle.

Inspection

7 Working carefully, remove all traces of brake dust from the drum, but *avoid inhaling the dust, as it is injurious to health.*

8 Scrub clean the outside of the drum, and check it for obvious signs of wear or damage such as cracks around the roadwheel bolt holes; renew the drum if necessary.

9 Examine carefully the inside of the drum. Light scoring of the friction surface is normal, but if heavy scoring is found, the drum must

be renewed. It is usual to find a lip on the drum's inboard edge which consists of a mixture of rust and brake dust; this should be scraped away to leave a smooth surface which can be polished with fine (120 to 150 grade) emery paper. If the lip is due to the friction surface being recessed by wear, then the drum must be refinished (within the specified limits) or renewed.

10 If the drum is thought to be excessively worn or oval, its internal diameter must be measured at several points using an internal micrometer. Take measurements in pairs, the second at right-angles to the first, and compare the two to check for signs of ovality. Minor ovality can be corrected by machining; otherwise, renew the drum.

Refitting

11 If a new brake drum is to be installed, use a suitable solvent to remove any preservative coating that may have been applied to its interior.

12 Ensure that the handbrake lever stop-peg is correctly repositioned against the edge of the brake shoe web (photo). Apply a smear of gear oil to the stub axle, and slide on the spacer and brake drum, being careful not to get oil onto the brake shoes or the friction surface of the drum. Fit the thrustwasher and a new hub nut; tighten the nut to the specified

torque. Tap the hub cap into place in the centre of the brake drum.

13 Depress the footbrake several times to operate the self-adjusting mechanism.

14 Repeat the above procedure on the remaining rear brake assembly (where necessary), then adjust the handbrake as described in Chapter 1.

15 On completion, refit the roadwheel(s), lower the vehicle to the ground and tighten the wheel bolts to the specified torque.

13.6 Release the handbrake operating lever as described in the text

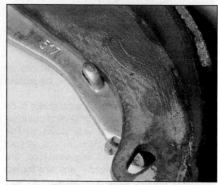

13.12 Check that the handbrake lever stop-peg is correctly positioned against the trailing shoe edge

9

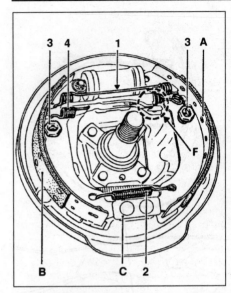

Fig. 9.11 Correct fitted positions of Bendix rear brake components (Sec 14)

A Leading shoe
B Trailing shoe
C Lower pivot point
F Adjuster strut mechanism
1 Upper return spring
2 Lower return spring
3 Retaining pin, spring and spring cup
4 Adjuster strut to trailing shoe spring

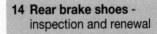

14 Rear brake shoes - inspection and renewal

Warning: Brake shoes must be renewed on both rear wheels at the same time - never renew the shoes on only one wheel, as uneven braking may result. Also, the dust created by wear of the shoes may contain asbestos, which is a health hazard. Never blow it out with compressed air, and don't inhale any of it. An approved filtering mask should be worn when working on the brakes. DO NOT use petroleum-based solvents to clean brake parts - use brake cleaner or methylated spirit only.

Inspection

1 Remove the brake drum as described in Section 13.
2 Working carefully, remove all traces of brake dust from the brake drum, backplate and shoes.
3 Measure the thickness of each brake shoe (friction material and shoe) at several points; if either shoe is worn at any point to the specified minimum thickness or less, all four shoes must be renewed as a set. Also, the shoes should be renewed if any are fouled with oil or grease; there is no satisfactory way of degreasing friction material once contaminated.
4 If any of the brake shoes are worn unevenly,

14.7 On Bendix rear brakes, ease the shoes out of the lower pivot point and disconnect the lower return spring

or fouled with oil or grease, trace and rectify the cause before reassembly.

Renewal

5 The procedure now varies according to which make of brake is fitted.

Bendix brake shoes

6 Using a pair of pliers, remove the shoe retainer spring cups by depressing and turning them through 90° . With the cups removed, lift off the springs and withdraw the retainer pins.
7 Ease the shoes out one at a time from the lower pivot point, to release the tension of the return spring, then disconnect the lower return spring from both shoes (photo).
8 Ease the upper end of both shoes out from their wheel cylinder locations, taking care not to damage the wheel cylinder seals, and disconnect the handbrake cable from the trailing shoe. The brake shoe and adjuster strut assembly can then be manoeuvred out of position and away from the backplate. Do not depress the brake pedal until the brakes are reassembled; wrap a strong elastic band around the wheel cylinder pistons to retain them.
9 With the shoe and adjuster strut assembly on the bench, make a note of the correct fitted positions of the springs and adjuster strut, to use as a guide on reassembly. Release the handbrake lever stop-peg (if not already done), then detach the adjuster strut bolt retaining spring from the leading shoe. Disconnect the upper return spring, then detach the leading shoe and return spring from the trailing shoe and strut assembly. Unhook the spring securing the adjuster strut to the trailing shoe, and separate the two.
10 If genuine Renault brake shoes are being installed, it will be necessary to remove the handbrake lever from the original trailing shoe and fit it to the new shoe. Secure the lever in position with the new retaining clip which is supplied with the brake shoes. All return springs should be renewed, regardless of their apparent condition; spring kits are also available from Renault dealers.
11 Withdraw the adjuster bolt from the strut,

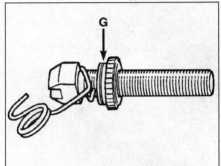

Fig. 9.12 Bendix rear brake left-hand adjuster strut bolt can be identified by groove (G) on adjuster wheel collar (Sec 14)

and carefully examine the assembly for signs of wear or damage, paying particular attention to the threads of the adjuster bolt and the knurled adjuster wheel, and renew if necessary. Note that left-hand and right-hand struts are not interchangeable - they are marked 'G' (gauche/left) and 'D' (droit/right) respectively. Also note that the strut adjuster bolts are not interchangeable; the left-hand strut bolt has a left-hand thread, and the right-hand bolt a right-hand thread. The left-hand bolt can be identified by the groove on its adjuster wheel collar (Fig. 9.12). The right-hand bolt does not have a groove on the adjuster wheel collar, but the bolt itself is painted.
12 Ensure the components on the end of the strut are correctly positioned (photo), then apply a little high melting-point grease to the threads of the adjuster bolt. Screw the adjuster wheel onto the bolt until only a small gap exists between the wheel and the head of the bolt, then install the bolt in the strut.
13 Fit the adjuster strut retaining spring to the trailing shoe, ensuring that the shorter hook of the spring is engaged with the shoe. Attach the adjuster strut to the spring end, then ease the strut into position in its slot in the trailing shoe.
14 Engage the upper return spring with the trailing shoe. Hook the leading shoe onto the other end of the spring, and lever the leading shoe down until the adjuster bolt head is

14.12 Correct fitted position of Bendix adjuster strut spring components

correctly located in its groove. Once the bolt is correctly located, hook its retaining spring into the slot on the leading shoe.

15 Remove the elastic band fitted to the wheel cylinder. Peel back the rubber protective caps, and check the wheel cylinder for fluid leaks or other damage. Also check that both cylinder pistons are free to move easily. Refer to Section 15, if necessary, for information on wheel cylinder renewal.

16 Prior to installation, clean the backplate and apply a thin smear of high-temperature brake grease or anti-seize compound to all those surfaces of the backplate which bear on the shoes, particularly the wheel cylinder pistons and lower pivot point (photo). Do not use too much, and don't allow the lubricant to foul the friction material.

17 Ensure that the handbrake lever stop-peg is correctly located against the edge of the trailing shoe.

18 Manoeuvre the shoe and strut assembly into position on the vehicle. Engage the upper ends of both shoes with the wheel cylinder pistons. Attach the handbrake cable to the trailing shoe lever. Fit the lower return spring to both shoes, and ease the shoes into position on the lower pivot point.

19 Centralise the shoes relative to the backplate by tapping them. Refit the shoe retainer pins and springs, and secure them in position with the spring cups.

20 Using a screwdriver, turn the strut adjuster wheel until the diameter of the shoes is between 179.2 and 179.5 mm. This should allow the brake drum to just pass over the shoes.

21 Slide the drum into position over the linings, but do not refit the hub nut yet.

22 Repeat the above procedure on the remaining rear brake.

23 Once both sets of rear shoes have been renewed, adjust the lining-to-drum clearance by repeatedly depressing the brake pedal. Whilst depressing the pedal, have an assistant listen to the rear drums, to check that the adjuster strut is functioning correctly; if this is so, a clicking sound will be emitted by the strut as the pedal is depressed.

24 Remove both the rear drums, and check that the handbrake lever stop-pegs are still correctly located against the edges of the trailing shoes, and that each lever operates smoothly. If all is well, with the aid of an assistant, adjust the handbrake cable so that the handbrake lever on each rear brake assembly starts to move as the handbrake is moved between the first and second notch (click) of its ratchet mechanism, ie so that the stop-pegs are still in contact with the shoes when the handbrake is on the first notch of the ratchet, but no longer contact the shoes when the handbrake is on the second notch. Once the handbrake adjustment is correct, hold the adjuster nut and securely tighten the locknut. Where necessary, refit the exhaust system heat shield to the vehicle underbody.

25 Refit the brake drums as described in Section 13.

14.16 Apply a little high-melting point grease to shoe contact points on the backplate

26 On completion, check the hydraulic fluid level as described in Chapter 1.

Girling brake shoes

27 Make a note of the correct fitted positions of the springs and adjuster strut, to use as a guide on reassembly.

28 Carefully unhook both the upper and lower return springs, and remove them from the brake shoes.

29 Using a pair of pliers, remove the leading shoe retainer spring cup by depressing it and turning through 90°. With the cup removed, lift off the spring, then withdraw the retainer pin and remove the shoe from the backplate. Unhook the adjusting lever spring, and remove it from the leading shoe.

30 Detach the adjuster strut and remove it from the trailing shoe.

31 Remove the trailing shoe retainer spring cup, spring and pin as described above, then detach the handbrake cable and remove the shoe from the vehicle. Do not depress the brake pedal until the brakes are reassembled; wrap a strong elastic band around the wheel cylinder pistons to retain them.

32 If genuine Renault brake shoes are being installed, it will be necessary to remove the adjusting lever from the original leading shoe and install it on the new shoe. All return springs should be renewed, regardless of their apparent condition; spring kits are also available from Renault dealers.

33 Withdraw the forked end from the adjuster strut. Carefully examine the assembly for signs of wear or damage, paying particular attention to the threads and the knurled adjuster wheel, and renew if necessary. Note that left-hand and right-hand struts are not interchangeable; the left-hand fork has a right-hand thread, and the right-hand fork a left-hand thread. The forks can also be identified by their colour: the left-hand fork is silver, and the right-hand fork is gold.

34 Remove the elastic band fitted to the wheel cylinder. Peel back the rubber protective caps, and check the wheel cylinder for fluid leaks or other damage. Check that both cylinder pistons are free to move easily. Refer to Section 15, if necessary, for information on wheel cylinder renewal.

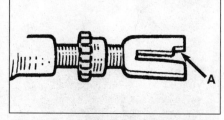

Fig. 9.13 On Girling rear brakes, adjuster strut fork cut-out (A) must engage with leading shoe adjusting lever on refitting (Sec 14)

35 Prior to installation, clean the backplate and apply a thin smear of high-temperature brake grease or anti-seize compound to all those surfaces of the backplate which bear on the shoes, particularly the wheel cylinder pistons and lower pivot point. Do not allow the lubricant to foul the friction material.

36 Ensure that the handbrake lever stop-peg is correctly located against the edge of the trailing shoe.

37 Locate the upper end of the trailing shoe in the wheel cylinder piston, then refit the retainer pin and spring, and secure it in position with the spring cup. Connect the handbrake cable to the lever.

38 Screw in the adjuster wheel until the minimum strut length is obtained, then hook the strut into position on the trailing shoe. Rotate the adjuster strut forked end so that the cut-out of the fork will engage with the leading shoe adjusting lever (Fig. 9.13).

39 Fit the spring to the leading shoe adjusting lever, so that the shorter hook of the spring engages with the lever.

40 Slide the leading shoe assembly into position, ensuring that it is correctly engaged with the adjuster strut fork, and that the fork cut-out is engaged with the adjusting lever. Engage the upper end of the shoe in the wheel cylinder piston, then secure the shoe in position with the retainer pin, spring and spring cup.

41 Install the upper and lower return springs, then tap the shoes to centralise them on the backplate.

42 Using a screwdriver, turn the strut adjuster wheel until the diameter of the shoes is between 178.7 and 179.2 mm. This should just allow the brake drum to pass over the shoes.

43 Slide the drum into position over the linings, but do not refit the hub nut yet.

44 Repeat the above procedure on the remaining rear brake.

45 Carry out the procedures described above in paragraphs 23 to 26.

All brake shoes

46 If new shoes have been fitted, full braking efficiency will not be obtained until the linings have bedded-in. Be prepared for longer stopping distances, and avoid harsh braking as far as possible for the first hundred miles or so after fitting new shoes.

9

15.3 Brake hose clamp fitted to rear brake flexible hose

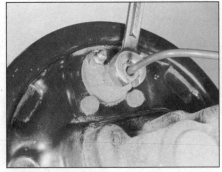

15.4 Unscrewing the union nut from the rear of the wheel cylinder

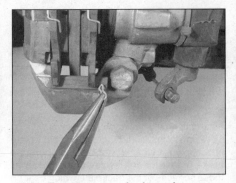

16.2A To remove rear brake pads, remove the spring clip . . .

15 Rear wheel cylinder - removal and refitting

Note: *Before starting work, refer to the warnings at the beginning of Section 6 concerning the dangers of hydraulic fluid, and at the beginning of Section 14 concerning the dangers of asbestos dust.*

Removal

1 Remove the brake drum as described in Section 13.

2 Using pliers, carefully unhook the brake shoe upper return spring and remove it from the brake shoes. Pull the upper ends of the shoes away from the wheel cylinder to disengage them from the pistons.

3 Minimise fluid loss, either by removing the master cylinder reservoir cap and then tightening it down onto a piece of polythene to obtain an airtight seal (taking care not to damage the sender unit), or by using a brake hose clamp, a G-clamp or a similar tool with protected jaws to clamp the flexible hose at the nearest convenient point to the wheel cylinder (photo).

4 Wipe away all traces of dirt around the brake pipe union at the rear of the wheel cylinder, and unscrew the union nut (photo). Carefully ease the pipe out of the wheel cylinder, and plug or tape over its end to prevent dirt entry. Wipe off any spilt fluid immediately.

5 Unscrew the two wheel cylinder retaining bolts from the rear of the backplate. Remove the cylinder, taking care not to allow hydraulic fluid to contaminate the brake shoe linings.

6 It is not possible to overhaul the cylinder, since no components are available separately. If faulty, the complete wheel cylinder assembly must be renewed.

Refitting

7 Ensure the backplate and wheel cylinder mating surfaces are clean, then spread the brake shoes and manoeuvre the wheel cylinder into position.

8 Engage the brake pipe, and screw in the union nut two or three turns to ensure that the thread has started.

9 Insert the two wheel cylinder retaining bolts, and tighten them securely. Now fully tighten the brake pipe union nut.

10 Remove the clamp from the brake hose, or the polythene from the master cylinder reservoir (as applicable).

11 Ensure that the brake shoes are correctly located in the cylinder pistons. Carefully refit the brake shoe upper return spring, using a screwdriver to stretch the spring into position.

12 Refit the brake drum as described in Section 13.

13 Bleed the brake hydraulic system as described in Section 6. Providing suitable precautions were taken to minimise loss of fluid, it should only be necessary to bleed the relevant rear brake.

16 Rear brake pads - inspection and renewal

⚠️ *Warning: Renew both sets of rear brake pads at the same time - never renew the pads on only one wheel, as uneven braking may result. Note that the dust created by wear of the pads may contain asbestos, which is a health hazard. Never blow it out with compressed air, and don't inhale any of it. An approved filtering mask should be worn when working on the brakes. DO NOT use petroleum-based solvents to clean brake parts - use brake cleaner or methylated spirit only.*

Inspection

1 Chock the front wheels, engage reverse gear (or 'P') and release the handbrake. Jack up the rear of the vehicle and support it on axle stands. Remove the rear wheels.

2 Extract the small spring clip from the pad retaining plate. Slide the plate out of the caliper (photos).

3 Withdraw the inner pad from the caliper, using pliers if necessary. Slacken and remove the two outer pad retaining screws, then withdraw the outer pad from the caliper (photos). Make a note of the correct fitted position of the anti-rattle springs, and remove the springs from each pad.

4 First measure the thickness of each brake

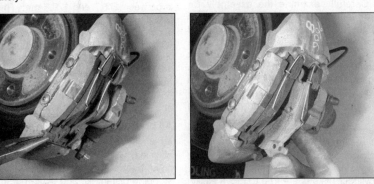

16.2B . . . then withdraw the retaining plate from the caliper

16.3A Slide out the inner brake pad . . .

16.3B . . . undo the two retaining screws . . .

16.3C . . . and withdraw the outer brake pad

16.7 Retract the piston using a square-section bar

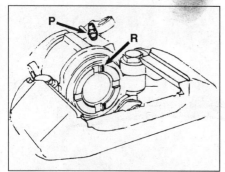

Fig. 9.14 Prior to installing rear brake pads, align groove on caliper piston (R) with bleed screw (P) (Sec 16)

pad (friction material and backing plate). If either pad is worn at any point to the specified minimum thickness or less, all four pads must be renewed. Also, the pads should be renewed if any are fouled with oil or grease; there is no satisfactory way of degreasing friction material once contaminated. If any of the brake pads are worn unevenly, or fouled with oil or grease, trace and rectify the cause before reassembly. New brake pads and spring kits are available from Renault dealers.

5 If the brake pads are still serviceable, carefully clean them using a clean, fine wire brush or similar, paying particular attention to the sides and back of the metal backing. Clean out the grooves in the friction material, and pick out any large embedded particles of dirt or debris. Clean the pad locations in the caliper body/mounting bracket.

6 Prior to fitting the pads, check that the guide sleeves are free to slide easily in the caliper body, and that the guide sleeve rubber gaiters are undamaged. Brush the dust and dirt from the caliper and piston, but *do not inhale it, as it is injurious to health.* Inspect the dust seal around the piston for damage, and the piston for evidence of fluid leaks, corrosion or damage. If attention to any of these components is necessary, refer to Section 17.

Renewal

7 If new brake pads are to be fitted, it will be necessary to retract the piston fully into the

caliper bore by rotating it in a clockwise direction. This can be achieved using a suitable square-section bar, such as the shaft of a suitable screwdriver, which locates snugly in the caliper piston slots (photo). Provided that the master cylinder reservoir has not been overfilled with hydraulic fluid, there should be no spillage, but keep a careful watch on the fluid level while retracting the piston. If the fluid level rises above the 'MAX' level, the surplus should be syphoned off (not by mouth - use an old syringe or a poultry baster), or ejected via a plastic tube connected to the bleed screw (see Section 6).

8 Position the caliper piston so that the small groove scribed across the piston points in the direction of the caliper bleed screw. This is necessary to ensure that the lug on the inner pad will locate with the caliper piston slot on installation (Fig. 9.14).

9 Refit the anti-rattle springs to the pads, so that when the pads are installed in the caliper, the spring end will be located at the opposite end of the pad in relation to the pad retaining plate (photo). The brake pad with the lug on its backing plate is the inner pad.

10 Locate the outer brake pad in the caliper body, ensuring that its friction material is against the brake disc. Insert the retaining screws and tighten them securely.

11 Slide the inner pad into position in the caliper, ensuring that the lug on the pad backing plate is aligned with the slot in the caliper piston. Recheck that the anti-rattle

spring ends on both pads are at the opposite end of the pad to which the retaining plate is to be inserted.

12 Slide the retaining plate into place, and install the small spring clip at its inner end. It may be necessary to file an entry chamfer on the edge of the retaining key, to enable it to be fitted without difficulty.

13 Depress the brake pedal repeatedly until the pads are pressed into firm contact with the brake disc. Check that the inner pad lug is correctly engaged with one of the caliper piston slots (photo).

14 Repeat the procedure on the remaining rear brake caliper.

15 Check the handbrake cable adjustment as described in Chapter 1, then refit the roadwheels and lower the vehicle to the ground. Tighten the roadwheel bolts to the specified torque.

16 Check the hydraulic fluid level as described in Chapter 1.

17 If new pads have been fitted, full braking efficiency will not be obtained until the linings have bedded-in. Be prepared for longer stopping distances, and avoid harsh braking as far as possible for the first hundred miles or so after fitting new pads.

17 Rear brake caliper - removal, overhaul and refitting

Note: *Before starting work, refer to the warnings at the beginning of Section 6 concerning the dangers of hydraulic fluid, and at the beginning of Section 16 concerning the dangers of asbestos dust.*

Removal

1 Chock the front wheels, engage reverse gear (or 'P') and release the handbrake. Jack up the rear of the vehicle and support it on axle stands. Remove the relevant rear wheel.

2 Remove the brake pads as described in paragraphs 2 and 3 of Section 16.

3 Free the handbrake inner cable from the caliper handbrake operating lever, then tap the outer cable out of its bracket on the caliper body.

4 Minimise fluid loss, either by removing the

16.9 Inner brake pad can be identified by its locating lug (arrowed). Note correct fitted positions of anti-rattle springs

16.13 Ensure inner pad locating lug is correctly located in piston slot (arrowed)

9

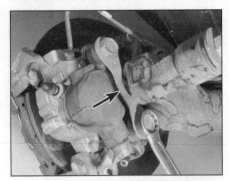

17.6 Remove the caliper mounting bolts, noting which way around the mounting plate is fitted (arrowed)

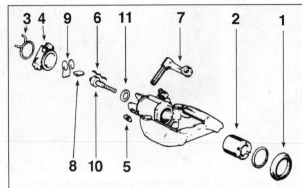

1 Dust seal
2 Piston
3 Retaining clip
4 Handbrake mechanism dust cover
5 Circlip
6 Spring washers
7 Handbrake operating lever
8 Plunger cam
9 Return spring
10 Adjusting screw
11 Thrustwasher

Fig. 9.15 Exploded view of the rear brake caliper (Sec 17)

master cylinder reservoir cap and then tightening it down onto a piece of polythene to obtain an airtight seal (taking care not to damage the sender unit), or by using a brake hose clamp, a G-clamp or a similar tool with protected jaws to clamp the flexible hose at the nearest convenient point to the brake caliper.

5 Wipe away all traces of dirt around the brake pipe union on the caliper, and unscrew the union nut. Carefully ease the pipe out of position, and plug or tape over its end to prevent dirt entry. Wipe off any spilt fluid immediately.

6 Slacken the two bolts securing the caliper assembly to the trailing arm, and remove them along with the mounting plate, noting which way around the plate is fitted (photo). Lift the caliper assembly away from the brake disc.

Overhaul

7 With the caliper on the bench, wipe away all traces of dust and dirt, but *avoid inhaling the dust, as it is injurious to health.*

8 Using a small screwdriver, carefully prise

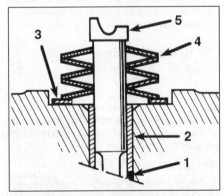

Fig. 9.16 Correct fitted positions of rear brake caliper handbrake mechanism adjuster screw and associated components (Sec 17)

1 O-ring
2 Adjusting screw bush
3 Thrustwasher
4 Correct arrangement of spring washers
5 Adjusting screw

out the dust seal from the caliper bore, taking care not to damage the piston.

9 Remove the piston from the caliper bore by rotating it in an anti-clockwise direction. This can be achieved using a suitable square-section bar, such as the shaft of a suitable screwdriver, which locates snugly in the caliper piston slots. Once the piston turns freely but does not come out any further, the piston can be withdrawn by hand, or if necessary forced out by applying compressed air to the union bolt hole.

 Caution: The piston may be ejected with some force - only low pressure should be required, such as is generated by a foot pump.

10 Using a blunt instrument such as a knitting needle or a crochet hook, extract the piston hydraulic seal, taking care not to damage the caliper bore.

11 Withdraw the guide sleeves from the caliper body, and remove the guide sleeve gaiters.

12 Inspect the caliper components as described in Section 10, paragraphs 13 to 17. Renew as necessary, noting that the inside of the caliper piston must **not** be dismantled. If necessary, the handbrake mechanism can be overhauled as described in the following paragraphs. If it is not wished to overhaul the handbrake mechanism, proceed to paragraph 16.

13 Release the handbrake dust cover retaining clip, and peel the cover away from the rear of the caliper. Make a note of the correct fitted positions of the relative components to use as a guide on reassembly. Remove the circlip from the base of the operating lever shaft, then compress the adjusting screw spring washers, and withdraw the operating lever and dust cover from the caliper body. With the lever withdrawn, remove the return spring, plunger cam, adjusting screw, spring washers and thrustwasher from the rear of the caliper body. Using a suitable pin punch, carefully tap the adjusting screw bush out of the caliper body and remove the O-ring.

14 Clean all the handbrake components in

methylated spirit, and examine them for wear. If there is any sign of wear or damage, the complete handbrake mechanism assembly should be renewed; a kit is available from your Renault dealer.

15 Ensure that all components are clean and dry. Install the O-ring, then press the adjusting screw bush into position until its outer edge is flush with the rear of the caliper body; if necessary, tap the bush into position using a suitable tubular drift. Fit the thrustwasher, then install the adjusting screw and spring washers, ensuring that the washers are positioned as shown in Fig. 9.16. Locate the plunger cam in the end of the adjusting screw, and position the return spring in the caliper housing. Fit the new dust cover to the operating lever, then compress the adjusting screw spring washers and insert the lever shaft through the caliper body, ensuring that it is correctly engaged with the return spring and plunger cam. Secure the operating lever in position with the circlip, then release the spring washers and check the operation of the handbrake mechanism. Apply a smear of high-melting point grease to the operating lever shaft and adjusting screw. Slide the dust cover over the caliper body, and secure it in position with a cable tie.

16 Soak the piston and the new piston (fluid) seal in clean hydraulic fluid. Smear clean fluid on the cylinder bore surface.

17 Fit the new piston (fluid) seal, using only the fingers to manipulate it into the cylinder bore groove, and refit the piston assembly. Turn the piston in a clockwise direction, using the method employed on dismantling, until it is fully retracted into the caliper bore.

18 Fit the dust seal to the caliper, ensuring that it is correctly located in the caliper and also the groove on the piston.

19 Apply the grease supplied in the repair kit, or a good-quality high-temperature brake grease or anti-seize compound to the guide sleeves. Fit the guide sleeves to the caliper body, and fit the new gaiters, ensuring that the gaiters are correctly located in the grooves on both the guide sleeve and caliper body.

17.23 Tap the handbrake outer cable into position using a hammer and punch

18.5 Prising out the rear disc hub cap

18.9A Refit the spacer to the rear of the brake disc . . .

Refitting

20 Position the caliper over the brake disc. Refit the two caliper mounting bolts and the mounting plate, noting that the mounting plate must be fitted so that its bend curves towards the caliper body. With the plate correctly positioned, tighten the caliper bolts to the specified torque.

21 Wipe clean the brake pipe union. Refit the pipe to the caliper, and tighten its union nut securely.

22 Remove the clamp from the brake hose, or the polythene from the master cylinder reservoir (as applicable).

23 Insert the handbrake cable through its bracket on the caliper, and tap the outer cable into position using a hammer and a pin punch (photo). Reconnect the inner cable to the caliper operating lever.

24 Refit the brake pads as described in Section 16.

25 Bleed the hydraulic system as described in Section 6. Note that, providing the precautions described were taken to minimise brake fluid loss, it should only be necessary to bleed the relevant rear brake.

26 Repeatedly apply the brake pedal to bring the pads into contact with the disc. Check and if necessary adjust the handbrake cable as described in Chapter 1.

27 Refit the roadwheel, lower the vehicle to the ground and tighten the wheel bolts to the specified torque. On completion, check the hydraulic fluid level as described in Chapter 1.

---

18 Rear brake disc - inspection, removal and refitting

Note: *Before starting work, refer to the warning at the beginning of Section 16 concerning the dangers of asbestos dust. If either disc requires renewal, both should be renewed at the same time, to ensure even and consistent braking.*

Inspection

1 Chock the front wheels, engage reverse gear (or 'P') and release the handbrake. Jack up the rear of the vehicle and support it on axle stands. Remove the appropriate rear roadwheel.

2 Inspect the disc as described in Section 11, paragraphs 2 to 5.

Removal

3 Remove the brake pads as described in paragraphs 2 and 3 of Section 16.

4 Remove the two caliper frame retaining bolts. Remove the frame from the caliper body.

5 Using a hammer and a large flat-bladed screwdriver, carefully tap and prise the cap out of the centre of the brake disc (photo).

6 Using a socket and long bar, slacken and remove the rear hub nut and withdraw the thrustwasher. Discard the hub nut; a new nut must be used on refitting.

7 It should now be possible to withdraw the brake disc and hub bearing assembly from the stub axle by hand. It may be difficult to

remove the disc, due to the tightness of the hub bearing on the stub axle. If the bearing is tight, tap the periphery of the disc using a hide or plastic mallet, or use a universal puller, secured to the disc with the wheel bolts, to pull it off. Remove the spacer from the rear of the disc, noting which way round it is fitted.

Refitting

8 Prior to refitting the disc, smear the stub axle shaft with gear oil. Be careful not to contaminate the friction surfaces with oil. If a new disc is to be fitted, use a suitable solvent to wipe any preservative coating from its surface.

9 Refit the spacer to the rear of the disc, noting that its slightly bigger protrusion should face the hub bearing. Slide the disc onto the stub axle, and tap it into position using a soft-faced mallet (photos).

10 Slide on the thrustwasher, then fit the new rear hub nut and tighten it to the specified torque (photo). Tap the cap back into position in the centre of the disc.

11 Apply a few drops of locking fluid to the threads of the caliper frame retaining bolts. Offer up the frame and refit the bolts. Tighten both bolts to the specified torque (photos).

12 Refit the brake pads as described in Section 16.

13 Check the handbrake cable adjustment as described in Chapter 1.

14 Refit the roadwheels and lower the vehicle to the ground. Tighten the roadwheel bolts to the specified torque.

18.9B . . . and slide the disc onto the stub axle

18.10 Tighten the rear hub nut to the specified torque

18.11A Apply locking fluid to the retaining bolts, then refit the caliper bracket . . .

9

18.11B . . . and tighten the bolts

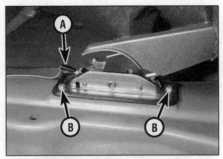

19.5 Handbrake lever warning switch wire (A) and mounting bolts (B) (carpet lifted)

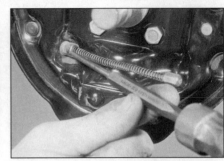

20.5 Drive the handbrake outer cable from the brake backplate (rear drum brake models)

19 Handbrake lever - removal and refitting

1 Chock the front wheels, engage reverse gear (or 'P') and release the handbrake. Jack up the rear of the vehicle and support it on axle stands.
2 Working from underneath the vehicle, undo the nuts securing the exhaust system heat shield to the vehicle underbody. Manoeuvre the heat shield out from under the vehicle.
3 Unscrew the handbrake cable adjuster locknut and the adjuster nut from the end of the handbrake lever linkage rod. Disengage the cable equalizer plate from the end of the linkage rod, then free the rod from its support guide.
4 Working from inside the vehicle, prise off the trim covers and unbolt the seat belt stalk anchorages from both front seats. Make a slit in the carpet, just to the rear of the lever assembly, to provide access to the lever mountings. Alternatively, the seats can be removed and the carpet lifted to gain access to the lever (see Chapter 11).
5 Disconnect the handbrake warning light switch wires from the rear of the lever (photo).
6 Unscrew the two bolts securing the lever to the floor, and remove the assembly from inside the vehicle.
7 Refitting is a reversal of removal. Adjust the handbrake as described in Chapter 1.

20 Handbrake cables - removal and refitting

1 The handbrake cable consists of two sections, a right- and left-hand section, which are linked to the lever assembly by an equalizer plate. Each section can be removed individually as follows.
2 Chock the front wheels, engage reverse gear (or 'P') and release the handbrake. Jack up the rear of the vehicle and support it on axle stands.
3 Working from underneath the vehicle, undo the nut(s) securing the exhaust system heat shield to the vehicle underbody. Lower the rear of the heat shield to gain access to the handbrake cable adjuster nuts.
4 Slacken the cable locknut and adjuster nut

until there is sufficient slack in the inner cable to allow it to be disconnected from the equalizer plate.
5 On models with rear drum brakes, remove the rear brake shoes from the appropriate side as described in Section 14. Using a hammer and pin punch, carefully tap the outer cable from the brake backplate (photo).
6 On models with rear disc brakes, disengage the inner cable from the caliper handbrake lever. Using a hammer and pin punch, tap the outer cable out of its mounting bracket on the caliper (photo).
7 Working along the length of the cable, remove any retaining bolts and screws, and free the cable from the retaining clips and ties. Remove the cable from under the vehicle.
8 Refitting is a reversal of removal. Adjust the handbrake as described in Chapter 1. Note that on models with rear drum brakes, the cable is adjusted before the brake drum is refitted.

21 Stop-light switch - removal, refitting and adjustment

Removal

1 The stop-light switch is located on the pedal bracket beneath the facia.
2 To remove the switch, reach up behind the facia, disconnect the wiring connector and unscrew the switch from the bracket (photos).

Refitting and adjustment

3 Screw the switch back into position in the mounting bracket.

20.6 On rear disc brake models, disconnect the handbrake cable from the brake caliper

4 Connect a continuity tester (ohmmeter or self-powered test light) across the switch terminals. Screw the switch in until an open-circuit is present between the switch terminals (infinite resistance, or light goes out). Gently depress the pedal and check that continuity exists between the switch terminals (zero resistance, or light comes on) after the pedal has travelled approximately 5 mm. If necessary, reposition the switch until it operates as specified.
5 In the absence of a continuity tester, the same adjustment can be made by reconnecting the switch and having an assistant observe the stop-lights (ignition on).
6 Once the stop-light switch is correctly adjusted, remake the original wiring connections, and recheck the operation of the stop-lights.

21.2A Disconnect the wiring connector . . .

21.2B . . . and unscrew the stop-light switch (facia panel removed for clarity)

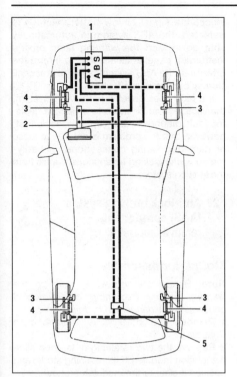

Fig. 9.17 Layout of anti-lock braking system (ABS) components (Sec 22)

1 *Modulator assembly*
2 *Master cylinder and servo unit*
3 *Wheel sensor*
4 *Reluctor ring*
5 *Load-sensitive pressure-regulating valve*

22 Anti-lock braking system (ABS) - general information

ABS is available as an option on the larger-engined Clio models. The purpose of the system is to prevent wheel(s) locking during heavy braking. This is achieved by automatic release of the brake on the relevant wheel, followed by reapplication of the brake.

The main components of the system are four wheel sensors (one per wheel), and a modulator block which contains the ABS computer, the hydraulic solenoid valves and accumulators, and an electrically-driven return pump.

The solenoids are controlled by the computer, which receives signals from the wheel sensors. The sensors detect the speed of rotation of a toothed ring, known as a reluctor ring, attached to the wheel hub. By comparing the speed signals from the four wheels, the computer can determine when a wheel is decelerating at an abnormal rate, and can therefore predict when a wheel is about to lock. During normal operation, the system functions in the same way as a non-ABS braking system does (Fig. 9.18).

If the computer senses that a wheel is about to lock, the ABS system enters the 'pressure-maintain' phase (Fig. 9.19). The

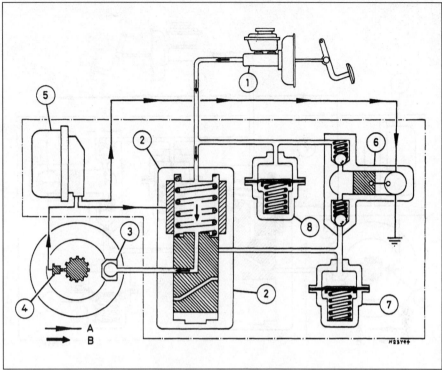

Fig. 9.18 ABS system normal operation (Sec 22)

1 *Master cylinder*
2 *Solenoid valve*
3 *Brake caliper*
4 *Wheel sensor and reluctor ring*
5 *ABS computer*
6 *Return pump*
7 *Accumulator*
8 *Accumulator*
A *Flow of electrical signal*
B *Flow of hydraulic fluid*

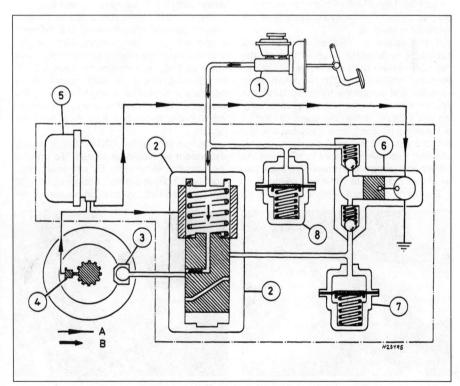

Fig. 9.19 ABS system 'pressure-maintain' phase (Sec 22)

Refer to Fig. 9.18 for key

9

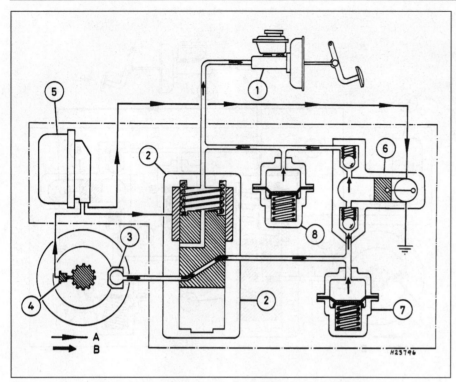

Fig. 9.20 ABS system 'pressure-decrease' phase (Sec 22)

Refer to Fig. 9.18 for key

computer operates the relevant solenoid valve in the modulator block; this isolates the brake caliper on the wheel in question from the master cylinder, effectively sealing-in the hydraulic pressure.

If the speed of rotation of the wheel continues to decrease at an abnormal rate, the ABS system then enters the 'pressure-decrease' phase (Fig. 9.20). The return pump operates and pumps the hydraulic fluid back into the master cylinder, releasing pressure on the brake caliper. When the speed of rotation of the wheel returns to an acceptable rate, the pump stops and the solenoid valve opens, allowing hydraulic pressure to return to the caliper and reapply the brake. This cycle can be carried out at up to 10 times a second.

The action of the solenoid valves and return pump creates pulses in the hydraulic circuit. When the ABS system is functioning, these pulses can be felt through the brake pedal.

The solenoid valves connected to the front calipers operate independently, but the valve connected to the rear calipers, together with the load-sensitive pressure-regulating valve, operates both calipers simultaneously. Since the braking circuit is split diagonally, a separate mechanical plunger valve in the modulator block divides the rear solenoid valve hydraulic outlet into two separate circuits.

The operation of the ABS system is entirely dependent on electrical signals. To prevent the system responding to any inaccurate signals, a built-in safety circuit monitors all signals received by the computer. If an

inaccurate signal or low battery voltage is detected, the ABS system is automatically shut down, and the warning lamp on the instrument panel is illuminated to inform the driver that the ABS system is not operational. Normal braking is unaffected.

If a fault does develop in the ABS system, the vehicle must be taken to a Renault dealer for fault diagnosis and repair. Check first, however, that the problem is not due to loose or damaged wiring connections, or badly-routed wiring picking up spurious signals from the ignition system.

23 Anti-lock braking system (ABS) components - removal and refitting

Modulator assembly

Note: *Before starting work, refer to the warning at the beginning of Section 6 concerning the dangers of hydraulic fluid.*

1 Remove the battery as described in Chapter 12.

2 Remove the two retaining nuts, then slide the ignition module off its retaining studs, and position it clear of the modulator (photo).

3 Release the rubber strap, and position the expansion tank clear of the modulator (photo).

4 On left-hand drive models, remove the brake master cylinder as described in Section 8.

5 Undo the retaining screw, and remove the relay cover from the modulator assembly (photo).

6 Disconnect the large 15-pin connector, the square 4-pin connector and the return pump earth lead from the modulator.

7 On right-hand drive models, remove the master cylinder reservoir filler cap. Place a piece of polythene over the filler neck, and securely refit the cap (taking care not to damage the sender unit). This will minimise brake fluid loss during subsequent operations. On all models, be prepared for some fluid spillage. As a precaution, place absorbent rags beneath the modulator brake pipe unions.

8 Wipe clean the area around the modulator

23.2 Unscrew the retaining nuts, and remove the ignition module from its mounting studs

23.3 Release the expansion tank, and position it clear of the modulator assembly

23.5 Slacken the retaining screw and remove the relay cover

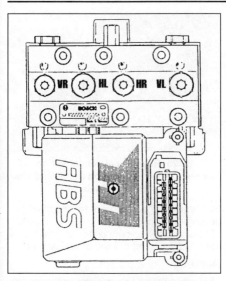

Fig. 9.21 Identification of outlet unions on ABS modulator (Sec 23)

VL Front left (yellow)   HL Rear left (blue)
VR Front right (green)   HR Rear right (red)

brake pipe unions. Make a note of how the pipes are arranged, to use as a reference on refitting; the four outlet pipes are colour-coded, and the modulator outlet unions are marked to aid refitting (Fig. 9.21). Unscrew the union nuts and carefully withdraw the pipes (photo). Plug or tape over the pipe ends and valve orifices, to minimise the loss of brake fluid and to prevent the entry of dirt into the system. Wash off any spilt fluid immediately with cold water.

9 Slacken the mounting nuts, and remove the modulator assembly from the engine compartment. **Note:** *Do not attempt to dismantle the modulator block hydraulic assembly. Overhaul of the unit is a complex job, and should be entrusted to a Renault dealer.*

10 Refitting is the reverse of the removal procedure, noting the following points:
(a) *Tighten the modulator block mounting nuts securely.*
(b) *Refit the brake pipes to the correct unions, and tighten the union nuts to the specified torque.*

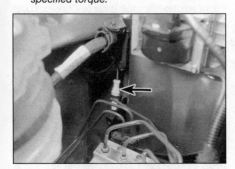

23.15 Front wheel sensor wiring connector (arrowed) is clipped to engine compartment wing valance - right-hand connector shown

23.8 Using a brake pipe spanner to slacken modulator union nuts

23.12B . . . undo the six retaining screws (arrowed) . . .

(c) *Ensure the wiring is correctly routed, and the connectors firmly pressed into position.*
(d) *Before refitting the battery, bleed the complete braking system as described in Section 6. Ensure the system is bled in the correct order, to prevent air entering the return pump.*

ABS computer

11 Disconnect the battery negative terminal. Slacken the retaining screw, and remove the relay cover from the modulator assembly.
12 Disconnect the three wiring connectors from the computer unit. Remove the six Torx retaining screws, and lift the computer away from the modulator assembly (photos).
13 Refitting is a reversal of the removal procedure. Ensure that the computer retaining

23.16 Release the wiring from the suspension strut bracket . . .

23.12A Disconnect the three wiring connectors (arrowed) . . .

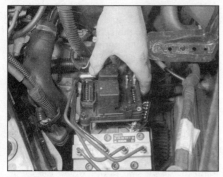

23.12C . . . and remove the ABS computer from the modulator

screws are securely tightened, and that the wiring connectors are firmly reconnected.

Front wheel sensor

14 Chock the rear wheels and apply the handbrake. Jack up the front of the vehicle and support it on axle stands. Remove the appropriate front roadwheel.
15 From inside the engine compartment, release the sensor wiring connector from its clip on the side of the suspension turret. Disconnect the connector, and remove the sensor wiring grommet from the wing valance (photo).
16 Pull the sensor wiring lead through from under the wheel arch, and free the sensor wiring from its bracket on the suspension strut (photo).
17 Remove the bolt securing the sensor to

23.17A . . . undo the retaining screw . . .

9

23.17B ... and withdraw the sensor from the swivel hub

23.20 Checking front wheel sensor-to-reluctor ring clearance

23.25 Rear wheel sensor wiring connectors are located on right-hand side of the vehicle

the swivel hub, and remove the sensor and lead assembly from the vehicle (photos).

18 Prior to refitting, apply a thin coat of multi-purpose grease to the sensor tip.

19 Ensure that the sensor and swivel hub sealing faces are clean. Fit the sensor, using hand pressure only; in particular, do not hit it with a hammer. Fit the sensor retaining bolt, and tighten it to the specified torque.

20 Rotate the hub until one of the reluctor ring teeth is aligned with the sensor tip. Using feeler gauges, measure the air gap between the tooth and the sensor tip (photo). Rotate the hub and repeat the check on several other teeth. If the air gap is not within the range given in the Specifications, then the advice of a Renault dealer must be sought.

21 Ensure the sensor wiring is correctly routed, then clip it back into position in the strut bracket.

22 Feed the wiring through into the engine compartment, and reconnect the wiring connector. Refit the connector to its retaining clip, then refit the sealing grommet to the wing valance.

23 Refit the roadwheel. Lower the vehicle to the ground and tighten the roadwheel bolts to the specified torque.

Rear wheel sensor

24 Chock the front wheels, engage reverse gear (or 'P') and release the handbrake. Jack up the rear of the vehicle and support it on axle stands. Remove the appropriate roadwheel.

25 Trace the wiring back from the sensor to the wiring connector on the right-hand side of the vehicle (photo). Free the connector from

its retaining clip, and disconnect the wiring from the main wiring loom.

26 Work back along the sensor wiring, and free it from any retaining clips.

27 Remove the bolt securing the sensor unit to the trailing arm. Remove the sensor and lead assembly from the vehicle (photo).

28 Prior to refitting, apply a thin coat of multi-purpose grease to the sensor tip.

29 Ensure that the sensor and trailing arm sealing faces are clean. Fit the sensor, using hand pressure only; in particular, do not hit it with a hammer. Tighten the retaining bolt to the specified torque.

30 Rotate the disc until one of the reluctor ring teeth is aligned with the sensor tip. Using feeler gauges, measure the air gap between the tooth and the sensor tip (photo). Rotate the disc and repeat the procedure on several other teeth. If the air gap is not within the range given in the Specifications, then the advice of a Renault dealer must be sought.

31 Ensure that the sensor wiring is correctly routed and retained by all the necessary retaining clips. Reconnect the wiring connector, and clip it into its retaining bracket.

32 Refit the roadwheel. Lower the vehicle to the ground and tighten the roadwheel bolts to the specified torque.

Front reluctor ring
Removal

33 The front reluctor rings are a press fit on the driveshaft outer constant velocity (CV) joints. They should not normally need any attention, but if damage is found such as chipped or missing teeth, they can be renewed as follows.

34 Remove the appropriate driveshaft as described in Chapter 8.

35 Using a heavy-duty puller, carefully draw the reluctor ring off the end of the CV joint.

Refitting

36 Ensure the mating surfaces of the reluctor ring and CV joint are clean and dry. Apply a thin coat of locking fluid to the inside of the ring.

37 Locate the reluctor ring on the CV joint, and tap the ring squarely onto the locating shoulder. Note that, to ensure the correct operation of the ABS system, it is important that the ring is correctly and squarely seated against the CV joint shoulder.

38 Refit the driveshaft to the vehicle as described in Chapter 8, then check the sensor tip air gap as described in paragraph 30 of this Section.

Rear reluctor ring

39 The rear reluctor rings are an integral part of the rear discs, and cannot be renewed separately. Examine the rings for signs of damage such as chipped or missing teeth, and renew as necessary. If renewal is necessary, the brake disc must be renewed as described in Section 18.

Relays

40 Both the solenoid relay and return pump relay are located in the modulator block assembly. To gain access to them, undo the relay cover retaining screw and lift off the cover. Either relay can then be simply pulled out of position (photo). Refer to Chapter 12 for further information on relays.

23.27 Removing a rear wheel sensor retaining bolt

23.30 Checking rear wheel sensor-to-reluctor ring clearance

23.40 Removing an ABS relay from the modulator block

Chapter 10 Suspension and steering

Contents

Degrees of difficulty

Easy, suitable for novice with little experience	Fairly easy, suitable for beginner with some experience	Fairly difficult, suitable for competent DIY mechanic	Difficult, suitable for experienced DIY mechanic	Very difficult, suitable for expert DIY or professional

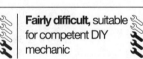

Specifications

Front suspension

Type . . . . Independent by MacPherson struts, with inclined coil springs and integral shock absorbers. Anti-roll bar on larger-engined models

Hub bearing endfloat . . . . 0 to 0.05 mm
Front underbody height (H1 minus H2):
  1390 cc (E6J and E7J) engine models . . . . 82 ± 7.5 mm
  1764 cc (F7P) engine models . . . . 117 ± 7.5 mm
  All other models . . . . 87 ± 7.5 mm

Rear suspension

Type . . . . Trailing arms, with transverse torsion bars (enclosed or open type, according to model) and telescopic shock absorbers. Rear anti-roll bar(s) on all models

Hub bearing endfloat . . . . 0 to 0.03 mm
Rear underbody height checking dimension (H4 minus H5 - see text):
  1108 cc (C1E), 1171 cc (E5F and E7F) and 1390 cc
  (E6J and E7J) engine models . . . . 6 ± 7.5 mm
  1721 cc (F2N) and 1794 cc (F3P) engine models . . . . 4 ± 7.5 mm
  1764 cc (F7P) engine models . . . . 45 ± 7.5 mm

10

Steering

Type . . . . Rack-and-pinion, power-assisted on some models

Wheel alignment and steering angles

Front wheel camber angle at underbody height (H1 minus H2) stated:
  1764 cc (F7P) engine models:
    40 mm . . . . -0° 43' ± 30'
    120 mm . . . . -1° 33' ± 30'
    130 mm . . . . -1° 30' ± 30'
    140 mm . . . . -1° 25' ± 30'
    185 mm . . . . -0° 26' ± 30'
  All other models:
    75 mm . . . . -0° 10' ± 30'
    85 mm . . . . -0° 20' ± 30'
    95 mm . . . . -0° 30' ± 30'
    115 mm . . . . -0° 30' ± 30'
    165 mm . . . . -0° 10' ± 30'
  Maximum difference between left- and right-hand sides . . . . 1°
Castor angle at underbody height (H5 minus H2) stated:
  1764 cc (F7P) engine models:
    41 mm . . . . 3° 05' ± 30'
    60 mm . . . . 2° 35' ± 30'
    79 mm . . . . 2° 05' ± 30'
    98 mm . . . . 1° 35' ± 30'
    117 mm . . . . 1° 05' ± 30'
  All other models:
    41 mm . . . . 2° 55' ± 30'
    60 mm . . . . 2° 25' ± 30'
    79 mm . . . . 1° 55' ± 30'
    98 mm . . . . 1° 25' ± 30'
    117 mm . . . . 0° 55' ± 30'
  Maximum difference between left- and right-hand sides . . . . 1°
Steering axis inclination/kingpin inclination at underbody height (H1 minus H2) stated:
  1764 cc (F7P) engine models:
    40 mm . . . . 8° 38' ± 30'
    120 mm . . . . 10° 47' ± 30'
    130 mm . . . . 10° 56' ± 30'
    140 mm . . . . 11° 03' ± 30'
    185 mm . . . . 11° 08' ± 30'
  All other models:
    75 mm . . . . 9° 50' ± 30'
    85 mm . . . . 10° 00' ± 30'
    95 mm . . . . 10° 20' ± 30'
    115 mm . . . . 10° 40' ± 30'
    165 mm . . . . 11° 10' ± 30'
  Maximum difference between left- and right-hand sides . . . . 1°
Front wheel toe setting (vehicle unladen):
  1764 cc (F7P) engine models . . . . 0° 20' ± 10' (2 ± 1 mm) toe-in
  All other models . . . . 0° ± 10' (1 ± 1 mm) toe-out
Rear wheel camber setting (vehicle unladen) . . . . -0° 50' ± 30'
Rear wheel toe setting (vehicle unladen):
  Open-bar rear axle . . . . 0° 20' ± 10' (2 ± 1 mm) toe-in
  Enclosed-bar rear axle . . . . 0° 15' ± 10' (1.5 ± 1 mm) toe-in

Tyres

Tyre size . . . . 145/70 R 13S, 155/70 R 13S, 165/65 R 13T, 145/80 R 13S, 155/80 R 13T, 165/60 R 14H, 185/60 R 14V or 185/55 R 15V (depending on model)

Pressures - (tyres cold):	Front*	Rear
145/70 R 13S tyres	2.4 bars (35 lbf/in²)	2.4 bars (35 lbf/in²)
155/70 R 13S and 165/65 R 13T tyres	2.1 bars (30 lbf/in²)	2.3 bars (33 lbf/in²)
145/80 R 13S and 155/80 R 13T tyres	2.1 bars (30 lbf/in²)	2.1 bars (30 lbf/in²)
165/60 R 14H tyres	2.3 bars (33 lbf/in²)	2.3 bars (33 lbf/in²)
185/60 R 14V and 185/55 R 15V tyres	2.2 bars (32 lbf/in²)	2.2 bars (32 lbf/in²)

*For models equipped with automatic transmission, add 0.1 bar (1.5 lbf/in²)
Note: Pressures apply only to original equipment tyres, and may vary if any other make or type is fitted; check with the tyre manufacturer or supplier for correct pressures if necessary.

Roadwheels

Type . . . . Pressed-steel or aluminium alloy (depending on model)
Size . . . . 4.5B x 13, 5B x 13, 5.5J x 14, 6J x 14 or 6.5J x 15 (depending on
model)
Maximum run-out at rim . . . . 1.2 mm
Maximum eccentricity on tyre bead locating surface . . . . 0.8 mm

Torque wrench settings

	Nm	lbf ft
Front suspension		
Strut-to-swivel hub bolts	110	81
Strut upper mounting nut	60	44
Anti-roll bar mounting clamp bolts	35	26
Anti-roll bar-to-lower arm bolt	20	15
Lower arm balljoint clamp bolt	55	41
Lower arm balljoint retaining nuts	75	55
Lower arm pivot bolts	110	81
Rear suspension		
Enclosed-bar rear axle:		
Shock absorber upper mounting bolt	80	59
Shock absorber lower mounting bolt	60	44
Anti-roll bar mounting bolts	50	37
Bearing bracket mounting bolts	85	63
Hub nut*	160	118
Open-bar rear axle:		
Shock absorber upper mounting bolt	115	85
Shock absorber lower mounting bolt	85	63
Bearing bracket mounting bolts	100	74
Hub nut*	175	129

Renew this nut every time

	Nm	lbf ft
Steering		
Track rod end balljoint-to-swivel hub retaining nut	35	26
Track rod end balljoint locknut	35	26
Track rod to steering rack	50	37
Steering gear mounting bolts	50	37
Steering wheel nut	40	30
Intermediate shaft universal joint clamp bolt	30	22
Roadwheels		
Wheel bolts	90	66

1 General information

The independent front suspension is of the MacPherson strut type, incorporating coil springs and integral telescopic shock absorbers. The MacPherson struts are located by transverse lower suspension arms, which utilize rubber inner mounting bushes and incorporate a balljoint at the outer ends. The front swivel hubs, which carry the wheel bearings, brake calipers and the hub/disc assemblies, are bolted to the MacPherson struts and connected to the lower arms via the balljoints. A front anti-roll bar is fitted to all larger-engined models. The anti-roll bar is rubber-mounted onto the subframe, and connects both the lower suspension arms.

The rear suspension differs according to which type of rear brake is fitted. Vehicles with rear drum brakes are fitted with an enclosed-bar type rear axle, which consists of two torsion bars within a tubular crossmember which connects both the rear trailing arms. An anti-roll bar is situated just to the rear of the crossmember, and is also connected to both trailing arms.

Vehicles with rear disc brakes are fitted with an open-bar rear axle which consists of two torsion bars, two anti-roll bars and an L-section metal crossmember which is connected to both the trailing arms. The two torsion bars and two anti-roll bars are connected at the centre with a link, and at their outer ends to the trailing arms.

The steering column is connected by a universal joint to an intermediate shaft, which has a second universal joint at its lower end. The lower universal joint is attached to the steering gear pinion by means of a clamp bolt.

The steering gear is mounted onto the front subframe. It is connected by two track rods and balljoints to steering arms projecting rearwards from the swivel hubs. The track rod ends are threaded to enable wheel alignment adjustment.

Power-assisted steering is available on most larger-engined models. On 1.7 litre (F2N engine) and 1.8 litre (F3P engine) models equipped with air conditioning and power steering, an electrically-powered pump is

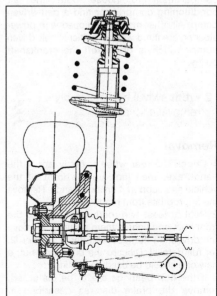

Fig. 10.1 Cross-section of the front suspension (Sec 1)

10

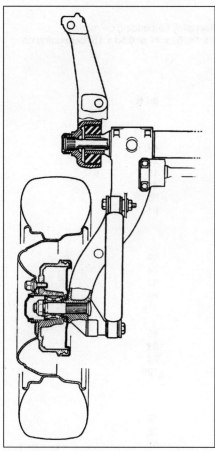

Fig. 10.2 Cross-section of the rear suspension - enclosed-bar axle (open-bar axle similar) (Sec 1)

used to power the steering hydraulic system; this is due to lack of space in the engine compartment to house both the air conditioning compressor and a belt-driven pump. All other models equipped with power steering are fitted with a traditional belt-driven pump, which is driven off the crankshaft pulley.

2 Front swivel hub assembly - removal and refitting

Removal

1 Chock the rear wheels, firmly apply the handbrake, then jack up the front of the vehicle and support it on axle stands. Remove the appropriate front roadwheel.
2 Refit at least two roadwheel bolts to the front hub, and tighten them securely. Remove the driveshaft nut (see Chapter 8, Section 2, for full details). Discard the driveshaft nut; a new one should be used on refitting.
3 If the hub bearings are to be disturbed, remove the brake disc as described in Chapter 9. If not, unscrew the two bolts securing the brake caliper assembly to the swivel hub, and slide the caliper assembly off

2.6 Slackening the lower suspension arm balljoint clamp bolt and nut

the disc. Using a piece of wire or string, tie the caliper to the front suspension coil spring, to avoid straining the brake hose.
4 On models equipped with ABS, undo the retaining bolt and withdraw the front wheel sensor from the swivel hub. Tie the sensor to the suspension strut, so that it does not get damaged during the remainder of the removal procedure.
5 Remove the nut securing the track rod end balljoint to the swivel hub. Release the balljoint tapered shank using a universal balljoint separator.
6 Remove the nut and clamp bolt securing the lower suspension arm to the swivel hub (photo). Carefully lever the balljoint out of the swivel hub, taking care not to damage the balljoint or driveshaft gaiters. Note the plastic protector plate which is fitted to the balljoint shank.
7 Remove the two nuts and washers from the bolts securing the swivel hub to the suspension strut, noting that the nuts are positioned on the rear side of the strut. Withdraw the bolts and support the swivel hub assembly.
8 Release the driveshaft joint from the hub, and remove the swivel hub assembly from the vehicle. Note that locking fluid is applied to the joint splines during assembly, so it is likely that they will be a tight fit in the hub. Use a hammer and soft metal drift to tap the joint out of the hub, or use a puller to draw the swivel hub assembly off the joint splines.

Refitting

9 Ensure that the driveshaft joint and hub splines are clean and dry, then apply a coat of locking fluid to the joint splines.
10 Engage the joint splines with the hub, and slide the hub fully onto the driveshaft. Insert the two swivel hub-to-suspension strut mounting bolts from the front side of the strut, then refit the washers and nuts to the rear of the bolts and tighten them to the specified torque.
11 Slide on the washer and fit the new driveshaft nut, tightening it by hand only at this stage.
12 Ensure that the plastic protector is still fitted to the lower arm balljoint, then locate

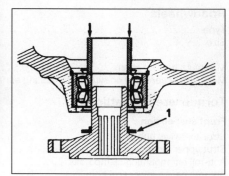

Fig. 10.3 Refitting the hub flange (Sec 3)

1 Thrustwasher

the balljoint shank in the swivel hub. Refit the balljoint clamp bolt, and tighten its retaining nut to the specified torque.
13 Reconnect the track rod end balljoint to the swivel hub, and tighten its retaining nut to the specified torque.
14 On models equipped with ABS, refit the sensor to the hub, and tighten its retaining bolt to the specified torque (see Chapter 9, Section 23).
15 Refit the brake disc (if removed), aligning the marks made on removal, and securely tighten its retaining screws. Slide the brake caliper assembly into position over the brake disc. Apply a few drops of locking fluid to the caliper bolt threads. Refit the bolts and tighten them to the specified torque (Chapter 9).
16 Insert and tighten two wheel bolts. Tighten the driveshaft nut to the specified torque (Chapter 8), using the method employed during removal to prevent the hub from rotating.
17 Check that the hub rotates freely, then refit the roadwheel and lower the vehicle to the ground. Tighten the roadwheel bolts to the specified torque.

3 Front hub bearings - checking, removal and refitting

Note: *The bearing is a sealed, pre-adjusted and pre-lubricated, double-row roller type, and is intended to last the car's entire service life without maintenance or attention. Do not attempt to remove the bearing unless absolutely necessary, as it will be damaged during the removal operation. Never overtighten the driveshaft nut in an attempt to 'adjust' the bearing.*

A press will be required to dismantle and rebuild the assembly; if such a tool is not available, a large bench vice and suitable spacers (such as large sockets) will serve as an adequate substitute. The bearing's inner races are an interference fit on the hub; if the inner race remains on the hub when it is pressed out of the hub carrier, a suitable knife-edged bearing puller will be required to remove it.

4.3 Slacken the suspension strut upper mounting nut, holding the strut piston rod with an Allen key

4.4 Removing the front suspension strut

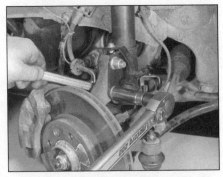

4.6 Tightening a swivel hub-to-strut nut and bolt. Note that the bolt heads face forwards

Checking

1 Wear in the front hub bearings can be checked by measuring the amount of side play present. To do this, a dial gauge should be fixed so that its probe is in contact with the disc face of the hub. The play should be between 0 and 0.05 mm. If it is greater than this, the bearings are worn excessively, and should be renewed.

Removal

2 Remove the swivel hub assembly as described in Section 2, then undo the brake disc shield retaining screws and remove the shield from the hub.

3 Support the swivel hub securely on blocks or in a vice. Using a suitable tubular spacer which bears only on the inner end of the hub flange, press the hub flange out of the bearing. If the bearing outboard inner race remains on the hub, remove it using a suitable bearing puller (see note above), then slide the thrustwasher off the hub flange, noting which way round it is fitted.

4 Extract the bearing retaining circlip from the inner end of the swivel hub assembly.

5 Where necessary, refit the inner race in position over the ball cage, and securely support the inner face of the swivel hub. Using a suitable tubular spacer which bears only on the inner race, press the complete bearing assembly out of the swivel hub.

6 Thoroughly clean the hub and swivel hub, removing all traces of dirt and grease. Polish away any burrs or raised edges which might hinder reassembly. Check for cracks or any other signs of wear or damage, and renew the components if necessary. As noted above, the bearing and its circlip must be renewed whenever they are disturbed. A replacement bearing kit, which consists of the bearing, circlip and thrustwasher, is available from Renault dealers.

Refitting

7 On reassembly, check (if possible) that the new bearing is packed with grease. Apply a light film of oil to the bearing outer race and to the hub flange shaft.

8 Before fitting the new bearing, remove the plastic covers protecting the seals at each end, but leave the inner plastic sleeve in position to hold the inner races together.

9 Securely support the swivel hub, and locate the bearing in the hub. Press the bearing into position, ensuring that it enters the hub squarely, using a suitable tubular spacer which bears only on the outer race.

10 Once the bearing is correctly seated, secure it with the new circlip and remove the plastic sleeve. Apply a smear of grease to the oil seal lips.

11 Slide the thrustwasher onto the hub flange, ensuring that its flat surface is facing the flange. Securely support the outer face of the hub flange.

12 Locate the bearing inner race over the end of the hub flange. Press the bearing onto the hub, using a tubular spacer which bears only on the inner race, until it seats against the thrustwasher. Check that the hub flange rotates freely. Wipe off any excess oil or grease.

13 Refit the brake disc shield to the swivel hub, and tighten its retaining screws.

14 Refit the swivel hub assembly as described in Section 2.

4 Front strut - removal and refitting

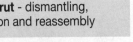

Removal

1 Chock the rear wheels, firmly apply the handbrake, then jack up the front of the vehicle and support it on axle stands. Remove the appropriate roadwheel.

2 Remove the two nuts and washers from the bolts securing the swivel hub to the suspension strut, noting that the nuts are positioned on the rear side of the strut. Withdraw the bolts, and support the swivel hub assembly.

3 From within the engine compartment, remove the strut cover (where fitted). Slacken the suspension strut upper mounting nut, holding the strut piston rod with an Allen key (photo). Remove the nut.

4 Release the strut from the swivel hub, and withdraw it from under the wheel arch, taking

4.7 Refit the upper mounting nut

care to prevent damage to the driveshaft gaiter (photo).

Refitting

5 Manoeuvre the strut assembly into position, taking care not damage the driveshaft gaiter.

6 Insert the two swivel hub-to-suspension strut mounting bolts from the front side of the strut. Refit the washers and nuts to the rear of the bolts, and tighten them to the specified torque (photo).

7 Refit the upper mounting nut (photo) and tighten it to the specified torque. When applicable, refit the cover.

8 Refit the roadwheel, lower the vehicle to the ground and tighten the roadwheel bolts to the specified torque.

5 Front strut - dismantling, inspection and reassembly

> **Warning: Before attempting to dismantle the front suspension strut, a suitable tool to hold the coil spring in compression must be obtained. Adjustable coil spring compressors are readily available, and are recommended for this operation. Any attempt to dismantle the strut without such a tool is likely to result in damage or personal injury.**

Dismantling

1 With the strut removed from the vehicle as described in Section 4, clean away all external dirt. Mount the strut upright in a vice.

5.2 Using spring compressors, compress the suspension strut coil spring . . .

5.3A . . . then prise out the retaining clip . . .

5.3B . . . and remove the retaining collar

2 Remove the upper mounting rubber. Fit the spring compressor, and compress the coil spring until all tension is relieved from the upper mounting plate (photo).

3 Carefully prise the retaining clip out of its groove in the top of the strut piston, then slide off the retaining collar, noting which way around it is fitted (photos). Discard the retaining clip; this must be renewed whenever it is disturbed.

4 Lift off the upper mounting plate, the bearing and the upper spring seat. Remove the coil spring, and slide the rubber dust cover off the strut piston.

Inspection

5 With the strut assembly now completely dismantled, examine all the components for wear, damage or deformation, and check the bearing for smoothness of operation. Renew any of the components as necessary.

6 Examine the strut for signs of fluid leakage. Check the strut piston for signs of pitting along its entire length, and check the strut body for signs of damage. Test the operation of the strut, while holding it in an upright position, by moving the piston through a full stroke and then through short strokes of 50 to 100 mm. In both cases, the resistance felt should be smooth and continuous. If the resistance is jerky or uneven, or if there is any visible sign of wear or damage, renewal is necessary.

7 If any doubt exists about the condition of the coil spring, gradually release the spring compressor, and check the spring for distortion and signs of cracking. Since no minimum free length is specified by Renault, the only way to check the tension of the spring is to compare it to a new component. Renew the spring if it is damaged or distorted, or if there is any doubt as to its condition.

8 Inspect all other components for signs of damage or deterioration, and renew any that are suspect.

Reassembly

9 Ensure that all components are clean and dry. Slide the dust cover into position over the strut piston (photo).

10 Refit the coil spring, followed by the upper spring seat. Ensure that both ends of the spring are correctly located in the spring seats (photos).

11 Fit the strut bearing, with its flat surface at the bottom, then refit the upper mounting plate (photos).

12 Slide the retaining collar onto the strut piston, and secure it in position with the new retaining clip (photo). Ensure that the clip is correctly seated in the strut piston groove. Slowly and carefully release the spring compressor, watching to make sure that both ends of the spring remain correctly located in the spring seats.

13 Fit the upper mounting rubber on the mounting plate (photo).

14 Refit the strut to the vehicle as described in Section 4.

5.9 Refit the rubber dust cover to the strut piston

5.10A Coil spring lower end correctly located in spring seat

5.10B Fitting the upper spring seat

5.11A Fit the strut bearing with its flat surface at the bottom . . .

5.11B . . . then refit the upper mounting plate

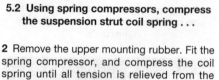

5.12 Refit the retaining collar, and secure it in position with a new retaining clip

5.13 Remove the spring compressors, and refit the mounting rubber to the upper mounting plate

6.4 Removing a front anti-roll bar mounting bolt

6 Front anti-roll bar - removal and refitting

Removal

1 Chock the rear wheels, apply the handbrake, then jack up the front of the vehicle and support it on axle stands. Remove both front roadwheels.
2 Remove the exhaust downpipe as described in Chapter 4.
3 On models fitted with a manual gearbox, disconnect the gearchange mechanism from the gearbox. Refer to Chapter 7.
4 Remove the nuts and bolts securing each end of the anti-roll bar to the lower suspension arms, noting the positions of the rubber mounting spacers (photo).
5 Unbolt the anti-roll bar mounting clamps (photo).
6 Manoeuvre the anti-roll bar out from underneath the vehicle, and remove the rubber mounting bushes.
7 Carefully examine the anti-roll bar components for signs of wear, damage or deterioration, paying particular attention to the mounting bushes. Renew worn components as necessary.

Refitting

8 Lubricate the rubber mounting bushes with a suitable grease (Elf Multi MOS2 or

equivalent). Fit the bushes onto the anti-roll bar.
9 Manoeuvre the anti-roll bar into position underneath the vehicle. Position the bushes so that their splits are against the subframe, then refit the mounting clamps. Do not tighten the mounting clamp nuts or bolts fully yet.
10 Secure the ends of the anti-roll bar to the lower suspension arms, ensuring that the rubber mounting spacers are correctly positioned. Fit the nuts and bolts, but do not tighten them fully yet.
11 On manual gearbox models, reconnect the selector mechanism as described in Chapter 7.
12 Refit the exhaust downpipe as described in Chapter 4.
13 Refit the roadwheels, lower the vehicle to the ground and tighten the wheel bolts to the specified torque.
14 Tighten the anti-roll bar fastenings to the specified torque, starting with the mounting clamps.

7 Front lower arm - removal, overhaul and refitting

Removal

1 Chock the rear wheels, apply the handbrake, jack up the front of the vehicle and support it on axle stands. Remove the appropriate front roadwheel.

2 Disconnect the end of the anti-roll bar from the lower suspension arm, as described in the previous Section.
3 When applicable, remove the bolt securing the support bar to the front wing valance, and the nut securing it to the front pivot bolt. Remove the support bar from the vehicle (photos).
4 Remove the nut and clamp bolt securing the lower suspension arm balljoint to the swivel hub. Carefully lever the balljoint out of the swivel hub, taking care not to damage the balljoint or driveshaft gaiters. Remove the plastic protector plate which is fitted to the balljoint shank (photos).
5 Remove the two pivot nuts and bolts, and remove the lower suspension arm from the vehicle (photo).

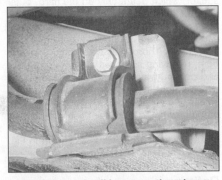

6.5 Front anti-roll bar mounting clamp

7.3A Where fitted, undo the support bar upper mounting bolt . . .

7.3B . . . and remove the bar from the suspension arm pivot bolt

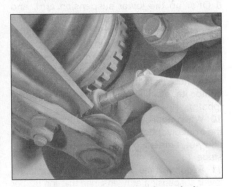

7.4A Remove the balljoint clamp bolt . . .

10

7.4B . . . then free the balljoint from the swivel hub, and remove the plastic protector plate

7.5 Undo the nuts and withdraw the lower suspension arm pivot bolts

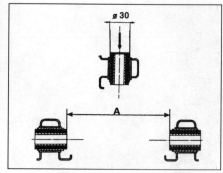

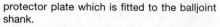

Fig. 10.4 Front suspension lower arm pivot bush fitting dimension (Sec 7)

A = 147 ± 0.5 mm

Overhaul

6 Clean the lower arm and the area around the arm mountings, then check for cracks, distortion or any other signs of damage. On some models, a brake disc cooling shield is clipped to the arm; this should also be removed. Check that the lower arm balljoint moves freely, without any sign of roughness, and that the balljoint gaiter is free from cracks and splits. Examine the shank of the pivot bolts for signs of wear or scoring. Renew worn components as necessary.

7 Inspect the lower arm pivot bushes. If they are worn, cracked, split or perished, they must be renewed. To renew the bushes, support the lower arm, and press the first bush out using a tubular spacer, such as a socket, which bears only on the hard, outer edge of the bush. **Note:** *Remove only one bush at a time from the arm, to ensure that each new bush is correctly positioned on installation.* Thoroughly clean the lower arm bore, removing all traces of dirt and grease, and polish away any burrs or raised edges which might hinder reassembly. Apply a smear of a suitable grease to the outer edge of the new bush. Press the bush into position until the distance 'A' between the inner edges of the lower arm bushes is as shown in Fig. 10.4. Wipe away surplus grease and repeat the procedure on the remaining bush.

Refitting

8 Offer up the lower suspension arm, and insert the two pivot bolts from the rear of the suspension arm. Refit the nuts, but tighten them by hand only at this stage.

9 Refit the plastic protector to the lower arm balljoint, then locate the balljoint shank in the swivel hub. Refit the balljoint clamp bolt, and tighten its retaining nut to the specified torque (photo).

10 When applicable, refit the support bar on the front pivot bolt shank. Fit the retaining bolt and nut, tightening both by hand only at this stage.

11 Reconnect the end of the anti-roll bar to the lower suspension arm, ensuring that the rubber mounting spacers are correctly positioned. Do not fully tighten the fastenings yet.

12 Refit the roadwheel, lower the vehicle and tighten the roadwheel bolts to the specified torque.

13 With the vehicle standing on its wheels, rock the vehicle to settle the lower arm bushes in position. Tighten the lower arm pivot bolts and the anti-roll bar fixing to the specified torque. When the pivot bolts have been tightened, secure the support bar in position (if applicable) by securely tightening its retaining nut and bolt.

8 Front lower arm balljoint - removal and refitting

Removal

1 Chock the rear wheels, apply the handbrake, jack up the front of the vehicle and support it on axle stands. Remove the appropriate front roadwheel.

2 Disconnect the end of the anti-roll bar from the lower suspension arm, as described in Section 6.

3 Slacken the nut securing the support bar (where fitted) to the lower arm front pivot bolt, then slacken both the lower suspension arm pivot bolts.

4 Remove the nut and clamp bolt securing the lower suspension arm to the swivel hub. Carefully lever the balljoint out of the swivel hub, taking care not to damage the balljoint or driveshaft gaiters. Remove the plastic

7.9 Tightening the balljoint clamp bolt

protector plate which is fitted to the balljoint shank.

5 Remove the two nuts and bolts securing the balljoint to the lower suspension arm (photo). Remove the balljoint.

6 Check that the balljoint moves freely, without any sign of roughness or free play. Examine the balljoint gaiter for signs of damage and deterioration such as cracks or splits. Renew the complete balljoint assembly if damaged; it is not possible to renew the balljoint gaiter separately. The balljoint renewal kit obtainable from Renault dealers contains the balljoint, the plastic protector plate and all fixings.

Refitting

7 Fit the balljoint to the suspension arm. Insert the bolts and washers, and tighten the retaining nuts to the specified torque.

8 Fit the plastic protector to the balljoint shank, then locate the shank in the swivel hub. Refit the balljoint clamp bolt, and tighten its retaining nut to the specified torque.

9 Reconnect the end of the anti-roll bar to the lower suspension arm, ensuring that the rubber mounting spacers are correctly positioned. Do not fully tighten the fastenings yet.

10 Refit the roadwheel, lower the vehicle and tighten the roadwheel bolts to the specified torque.

11 With the vehicle standing on its wheels,

8.5 Front lower suspension arm balljoint retaining nuts

9.4 Extract the circlip . . .

9.5 . . . then drive the bearing out of the hub using a suitable tubular drift

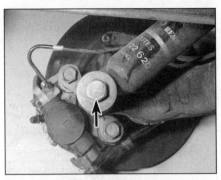

10.2 Rear shock absorber lower mounting bolt (arrowed)

rock the vehicle to settle the lower arm bushes in position. Tighten the lower arm pivot bolts and the anti-roll bar fixing to the specified torque. When the pivot bolts have been tightened, secure the support bar in position (if applicable) by securely tightening its retaining nut and bolt.

9 Rear hub bearings -
checking, removal and refitting

Note: The bearing is a sealed, pre-adjusted and pre-lubricated, double-row tapered-roller type, and is intended to last the car's entire service life without maintenance or attention. Never overtighten the hub nut in an attempt to 'adjust' the bearings.

Checking

1 Chock the front wheels and engage reverse gear (or 'P'). Jack up the rear of the vehicle and support it on axle stands. Remove the appropriate rear roadwheel, and fully release the handbrake.
2 Wear in the rear hub bearings can be checked by measuring the amount of side play present. To do this, a dial test indicator should be fixed so that its probe is in contact with the hub outer face. The play should be between 0 and 0.03 mm. If it is greater than this, the bearings are worn excessively and should be renewed.

Removal

3 Remove the rear brake disc or drum (as applicable), as described in Chapter 9.
4 Using circlip pliers, extract the bearing retaining circlip from the centre of the brake disc or drum (photo).
5 Securely support the disc or drum hub. Press or drive the bearing out of the hub, using a tubular drift of diameter 49 mm (photo).
6 Thoroughly clean the hub, removing all traces of dirt and grease. Polish away any burrs or raised edges which might hinder reassembly. Check the hub for cracks or any other signs of damage, and renew if necessary. The bearing and its circlip must be renewed whenever they are disturbed. A

replacement bearing kit is available from Renault dealers, consisting of the bearing, circlip, spacer, thrustwasher, hub nut and grease cap.

Refitting

7 On reassembly, check (if possible) that the new bearing is packed with grease. Apply a light film of gear oil to the bearing outer race and stub axle.
8 Securely support the hub. Press the bearing into position, ensuring that it enters the hub squarely, using a tube (diameter 51 mm) which presses only on the bearing outer race.
9 Ensure that the bearing is correctly seated against the hub shoulder, and secure it in position with the new circlip. Ensure that the circlip is correctly seated in its groove.
10 Refit the brake disc or drum as described in Chapter 9.

10 Rear shock absorber -
removal, testing and refitting

Removal

1 Chock the front wheels and engage reverse gear (or 'P'). Jack up the rear of the vehicle and support it on axle stands. Remove the appropriate rear roadwheel.
2 Using a jack, raise the trailing arm slightly until the shock absorber is slightly compressed. Remove the lower mounting bolt (photo).
3 Slacken and remove the upper mounting nut and bolt, and remove the shock absorber from the vehicle.

Testing

4 Mount the shock absorber in a vice, and test as described in Section 5 for the front suspension strut. Also check the rubber mounting bushes for damage and deterioration. Renew the shock absorber complete if any damage or wear is evident; the mounting bushes are not available separately. Inspect the mounting bolts for signs of wear or damage, and renew as necessary.

Refitting

5 Prior to refitting the shock absorber, mount it upright in the vice, and operate it fully through several strokes in order to prime it. (This is necessary even if a new unit is being fitted, as it may have been stored horizontally, and so need priming.) Apply a smear of multi-purpose grease to the shock absorber mounting bolts.
6 Offer up the shock absorber, then install its upper mounting bolt and nut, tightening it by hand only at this stage.
7 Refit the lower shock absorber mounting bolt, again tightening it by hand only. Remove the jack from under the trailing arm.
8 Refit the roadwheel, lower the vehicle to the ground and tighten the roadwheel bolts to the specified torque.
9 Rock the vehicle to settle the shock absorber mounting bushes in position, then tighten both the mounting bolts to the specified torque.

11 Rear anti-roll bar (enclosed-bar rear axle) -
removal and refitting

Removal

1 Chock the front wheels and engage reverse gear (or 'P'). Jack up the rear of the vehicle and support it on axle stands.

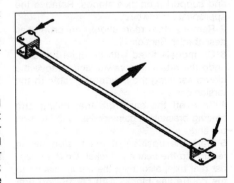

Fig. 10.5 Cutaway ends of rear anti-roll bar brackets (small arrows) must face the front of the vehicle (Sec 11)

10

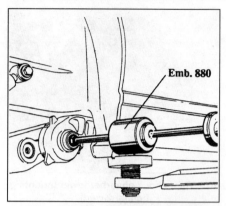

Fig. 10.6 Using a slide hammer (Renault tool Emb.880 shown) to remove a torsion bar (Sec 12)

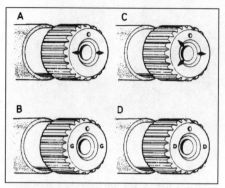

Fig. 10.7 Rear suspension torsion bar identification markings (Sec 12)

A & B - Left-hand   C & D - Right-hand

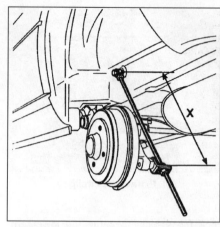

Fig. 10.8 A tool made from a length of threaded bar can be fitted to the shock absorber mounting holes and used to set dimension 'X' - see text (Sec 12)

2 Remove the bolts securing the ends of the anti-roll bar to the rear suspension trailing arms, noting the fitted positions of the handbrake cable retaining clips. Recover the anti-roll bar retaining nut plates from the top of the trailing arms.

3 Withdraw the anti-roll bar from under the vehicle.

Refitting

4 Refit the anti-roll bar to the vehicle, noting that the cutaway ends of the anti-roll bar retaining bolt brackets must face towards the front of the vehicle (see Fig. 10.5).

5 Position the retaining nut plates on the top of the trailing arms. Refit the retaining bolts, ensuring that the handbrake cable retaining clips are correctly positioned. Tighten the anti-roll bar retaining bolts to the specified torque.

6 Lower the vehicle to the ground.

12 Rear torsion bar - removal and refitting

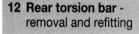

Removal

1 Chock the front wheels and engage reverse gear (or 'P'). Jack up the rear of the vehicle and support it on axle stands. Remove the appropriate roadwheel.

2 Remove the rear shock absorber, as described in Section 10.

3 On models fitted with an open-bar axle, undo the retaining screw(s) and remove the clamp securing the handbrake cable to the torsion bar.

4 Prise off the cap from the trailing arm bearing bracket, to gain access to the torsion bar end.

5 Mark the position of the torsion bar in relation to the bearing bracket. On the open-bar rear axle, also mark the bar in relation to the centre link block. With the trailing arm unsupported, measure the distance between the centres of the shock absorber upper and lower mounting bolt holes. Note the distance;

if the original bar is to be refitted, this will be needed on refitting.

6 The torsion bar can now be withdrawn outwards, using a slide hammer such as Renault tool Emb.880 or a suitable alternative (Fig. 10.6).

> **HAYNES HINT** *It is possible to improvise a torsion bar removal tool by screwing a long bolt with a flat washer into the torsion bar, and placing the jaws of a spanner against the washer. Striking the spanner sharply with a hammer should free the bar.*

7 Once the splines of the torsion bar are free, the bar can be withdrawn completely from its location. Note that the left-hand and right-hand bars are not interchangeable; they can be identified by the markings on their ends (see Fig. 10.7).

Refitting

8 Ensure that the bearing bracket and torsion bar splines are clean and dry. Lubricate the splines with molybdenum disulphide grease, and insert the bar into the bracket.

9 If the original bar is being refitted, position the trailing arm so that the distance between the centres of the upper and lower shock absorber mounting bolt holes is as noted prior to removal.

10 If a new torsion bar is being fitted, the trailing arm must be raised until the distance between the centres of the upper and lower shock absorber mounting bolt holes is 400 mm (372 mm on models with the 1764 cc (F7P) engine). This measurement is known as dimension 'X'. With dimension 'X' correctly set, all the torsional forces will be removed from the bar.

11 Raise the trailing arm to the required position using a jack. Alternatively, a tool similar to that shown in Fig. 10.8 can be fabricated from a length of threaded bar, and used to position the trailing arm.

Original bar (all types)

12 Rotate the torsion bar to align the marks made on removal, then engage the bar with the bearing bracket and bearing/centre link splines. Tap the bar fully into position using a hammer and a soft metal drift.

New bar (enclosed-bar rear axle)

13 Rotate the torsion bar until the position is found where the bar can be freely engaged with the splines on the bearing bracket and bearing. Having found this, tap the bar fully home using a hammer and a soft metal drift.

New bar (open-bar rear axle)

14 Remove the cap from the opposite bearing bracket, and check the position of the drill mark on the end of the torsion bar (photo). Rotate the new bar so that its drill mark is in exactly the same position, then locate its splines with those of the bearing bracket and centre link, and insert the bar. To do this, it may be necessary to rotate the bar slightly, noting that the drill marks on both bars must be no more than two splines away from each other once the splines are engaged. If this proves difficult to achieve, it is permissible to **slightly** lower or raise the trailing arm. If

12.14 Check the position of the circular drill mark (arrowed) on the end of the torsion bar

excessive movement of the arm is required to get the splines to align, it is recommended that the opposite torsion bar is also renewed.

15 Once it is correctly positioned, tap the torsion bar fully into position using a hammer and a soft metal drift.

All types

15 Refit the torsion bar cap(s) to the bearing bracket.

16 On models fitted with an open-bar axle, refit the handbrake cable retaining clamp to the torsion bar, and securely tighten its retaining screw(s).

17 Refit the rear shock absorber, as described in Section 10.

18 Check the rear underbody height, as described in Section 17.

13 Rear trailing arm (enclosed-bar rear axle) - removal and refitting

Removal

1 Chock the front wheels and engage reverse gear (or 'P'). Jack up the rear of the vehicle and support it on axle stands. Remove the appropriate rear roadwheel.

2 Remove the rear anti-roll bar, as described in Section 11.

3 Remove the appropriate torsion bar, as described in Section 12.

4 Remove the brake shoes from the side in question, as described in Chapter 9.

5 Using a hammer and pin punch, tap the handbrake outer cable out of the backplate. Work back along the length of the cable, and release it from any clips or ties securing it to the trailing arm.

6 Minimise brake fluid loss by removing the master cylinder reservoir cap, and then tightening it down onto a piece of polythene to obtain an airtight seal (taking care not to damage the sender unit). Alternatively, use a brake hose clamp, a G-clamp or a similar tool with protected jaws, to clamp the flexible hose at the nearest convenient point to the wheel cylinder.

7 Wipe away all traces of dirt around the brake pipe union at the rear of the wheel cylinder, and unscrew the union nut. Carefully

ease the pipe out of the wheel cylinder, and plug or tape over its end to prevent dirt entry. Wash off any spilt fluid immediately.

8 Support the weight of the trailing arm on a jack.

9 Lift the rear seat cushion to gain access to the trailing arm bearing bracket mounting bolt heads.

10 From underneath the vehicle, unscrew the trailing arm bearing bracket mounting nuts on the side to be removed. Tap the bolts through the bearing bracket with a suitable drift. Loosen the mounting nuts on the opposite bearing bracket.

11 Lower the jack until the trailing arm assembly is clear of the sill, then pull the trailing arm out of the opposite arm, and remove it from the vehicle.

12 Inspect the trailing arm bearings and tracks for signs of wear or damage. Renew them if necessary as described in Section 14. If a new trailing arm is being fitted, undo the four retaining bolts and remove the brake backplate. Fit the backplate to the new trailing arm, apply thread-locking compound to the retaining bolts, and tighten them securely.

Refitting

13 Ensure that the trailing arm bearings are sufficiently greased, then engage the removed trailing arm with the opposite arm. Temporarily position the anti-roll bar over its mounting bolt holes, ensuring that the cutaway ends of the retaining bolt brackets face towards the front of the vehicle. Push the trailing arm into position until the anti-roll bar bracket holes are correctly aligned with the bolt holes in the trailing arms.

14 Once the trailing arms are correctly interlocked, remove the anti-roll bar. Raise the trolley jack, and insert the bearing bracket mounting bolts from inside the vehicle, ensuring that the bolt retaining plate is correctly fitted. Refit the mounting nuts and tighten them to the specified torque. Lower the seat cushion.

15 Refit the brake pipe to the wheel cylinder, and securely tighten its union nut. Remove the clamp from the brake hose or the polythene from the master cylinder reservoir (as applicable).

16 Tap the handbrake cable back into position in the brake backplate, and secure it in position with the ties or clips. Refit the brake shoes as described in Chapter 9.

17 Refit the torsion bar as described in Section 12.

18 Refit the anti-roll bar as described in Section 11.

19 Bleed the brake hydraulic system as described in Chapter 9. Providing the precautions described were taken to minimise brake fluid loss, it should only be necessary to bleed the relevant rear brake.

20 Refit the roadwheel, lower the vehicle to the ground and tighten the wheel bolts to the specified torque.

21 Check the rear underbody height as described in Section 17.

14 Rear trailing arm bearings (enclosed-bar rear axle) - renewal

1 Remove both trailing arms with reference to Section 13.

2 Mount the left-hand trailing arm in a vice.

3 To remove the bearings, it will be necessary to obtain the Renault tool shown in Fig. 10.9, or to fabricate a similar tool as follows. First note the exact position of the bearings.

4 Obtain a threaded rod long enough to reach the inner bearing, a tube of suitable diameter as shown, two thick washers of diameter equal to each bearing, a washer of diameter greater than the tube, and two nuts. Cut two sides off the smaller washer so that just a flat strip with a hole in the centre remains. This will form the swivelling end part shown on the Renault tool. Pass the threaded rod through the hole in the strip, and screw on a nut. Feed the strip and rod through the bearing so that the strip locates behind the bearing. Place the tube over the rod and in contact with the edge of the arm. Place the large washer over the end of the tube, and then screw on the remaining nut. Hold the rod with grips, and tighten the nut to draw out the bearing, then repeat this operation to remove the remaining bearing.

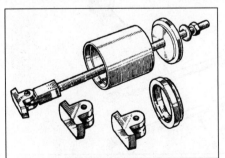

Fig. 10.9 Renault tool for removing the rear suspension trailing arm bearings - enclosed-bar rear axle (Sec 14)

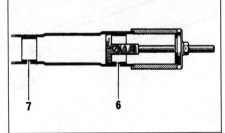

Fig. 10.10 Using the Renault tool to remove trailing arm bearings on enclosed-bar rear axle (Sec 14)

6 Outer bearing   *7 Inner bearing*

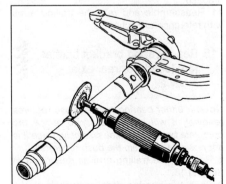

Fig. 10.11 Removing the rear axle bearing inner tracks using a grinder (Sec 14)

10

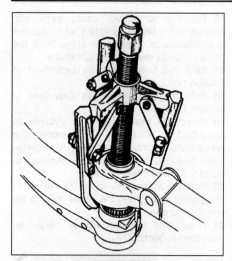

Fig. 10.12 Using a puller to remove the rear suspension bearing bracket from the trailing arm (Sec 15)

5 Thoroughly clean the bore, removing all traces of dirt and grease. Polish away any burrs or raised edges which might hinder reassembly. Apply a smear of multi-purpose grease to the outer edge of the new bearings, then carefully press them squarely into position, using a tubular spacer which presses only on the hard outer edge of the bearing.

6 Mount the right-hand trailing arm in the vice.

7 Mark the exact position of the bearing inner tracks on the trailing arm. Cut or grind almost through both bearing inner tracks, taking care not to damage the tube. Using a cold chisel, split the tracks and remove them from the tube. Also cut and remove the seal from the tube.

8 Clean the arm and fit the new seal.

9 Press on the new inner tracks, making sure that the lead chamfer goes on first. When doing this, if the load is being taken on the axle support assemblies, make sure that the torsion bars are correctly located in the anchor points.

10 It is not necessary to grease the bearing needle races, as they are supplied already greased.

11 Reassemble and refit the trailing arms with reference to Section 13.

15 Rear bearing bracket bushes (enclosed-bar rear axle) - renewal

Note: *If either bearing bracket bush requires renewal, it will be necessary to renew the complete bearing bracket assembly, since it is not possible to obtain the bush separately.*

1 Remove the trailing arm as described in Section 13.

2 Soak the bearing bracket bush in brake fluid for some time, to soften the bush rubber.

3 Wipe off all traces of brake fluid, then mark

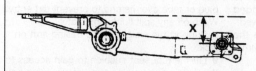

Fig. 10.13 Vertical assembly dimension of the rear suspension bearing bracket and trailing arm (Secs 15 and 16)

$X = 80$ mm

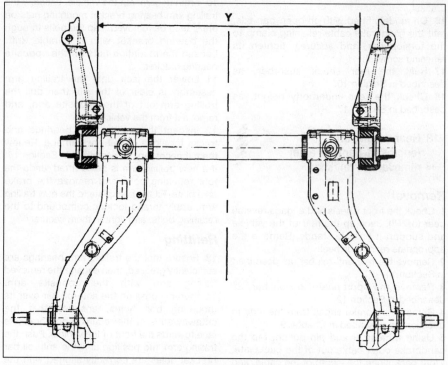

Fig. 10.14 Horizontal assembly dimension of the rear suspension bearing bracket - enclosed-bar axle shown, open-bar axle similar (Secs 15 and 16)

$Y = 1054 \pm 1$ mm

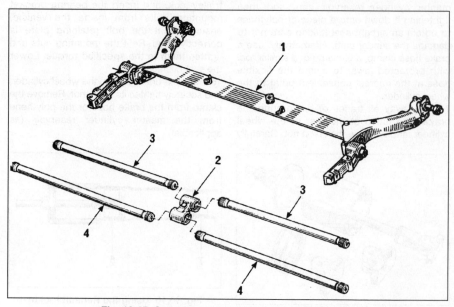

Fig. 10.15 Open-bar rear axle components (Sec 16)

1 Trailing arms and L-shaped crossmember assembly
2 Centre link   3 Anti-roll bars   4 Torsion bars

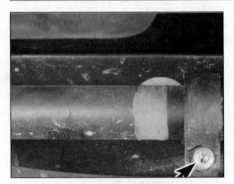

16.7 Handbrake cable retaining clamp screw (arrowed)

the position of the bush inner edge on the trailing arm shaft.

4 Using a two- or three-legged puller, draw the bearing bracket off the trailing arm; the rubber of the bush will tear during this process, leaving the inner part of the bush on the arm.

5 Remove the remaining inner part of the bush by cutting with a hacksaw, taking care not to damage the trailing arm shaft.

6 Thoroughly clean the trailing arm shaft, removing all traces of rubber and grease. Polish away any burrs or raised edges which might hinder reassembly. Apply a smear of suitable grease to the inner edges of the new bush to aid installation.

7 Securely support the trailing arm, then position the new bearing bracket in relation to the trailing arm as shown in Fig. 10.13. Press the bracket onto the trailing arm shaft until the bush reaches the alignment mark made prior to removal.

8 Temporarily refit the trailing arm into position as described in Section 13, paragraph 13. Check that the dimension between the inner bolt hole centres of the left and right-hand bearing brackets is as shown in Fig. 10.14. If this is not the case, remove the arm and adjust the position of the bearing bracket as necessary.

9 Once the bearing bracket is correctly positioned, refit the trailing arm as described in Section 13.

16 Rear axle (open-bar type) - removal, overhaul and refitting

Removal

1 Chock the front wheels and engage reverse gear (or 'P'). Jack up the rear of the vehicle and support it on axle stands. Remove both rear roadwheels.

2 Remove the rear brake calipers as described in Chapter 9.

3 If a new rear axle is to be installed, remove the brake discs as described in Chapter 9.

4 On models equipped with ABS, remove the rear wheel sensors as described in Chapter 9, Section 23.

16.10 Two of the bearing bracket mounting nuts (arrowed) - open-bar rear axle

5 Using a pair of pliers, unhook the brake pressure-regulating valve spring from the top of the axle crossmember.

6 Disconnect the rear brake pipes at the flexible hose unions which are clipped to the top of the axle crossmember. Remove the retaining clips, free the flexible hoses from the crossmember, and plug or tape over the union ends to prevent dirt entry. Wash off any spilt fluid immediately.

7 Remove the screws securing the handbrake cable retaining clamps to the torsion bars (photo).

8 Support the weight of the rear axle on a trolley jack.

9 Lift the rear seat cushion to gain access to the rear axle bearing bracket mounting bolt heads.

10 From underneath the vehicle, unscrew the four rear axle bearing bracket mounting nuts (photo). Tap the bolts up through the bearing bracket with a drift.

11 Withdraw the handbrake cables from the trailing arms. Lower the jack, and pull the assembly out from underneath the vehicle.

Overhaul

12 Remove the two torsion bars with reference to Section 12.

13 Mark the anti-roll bars in relation to the trailing arms and centre link block. Extract them using the same method as that employed when removing the torsion bars.

14 Examine all the components for wear and

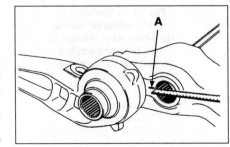

Fig. 10.16 Mark the trailing arms (A) as described in text before refitting the anti-roll bars - open-bar axle (Sec 16)

damage. Check the splines on the torsion bars, anti-roll bars, centre link block and trailing arms. If a trailing arm or the L-shaped crossmember is damaged, it will be necessary to obtain a new rear axle (supplied with the bearing brackets already fitted, but requiring the original torsion bars and anti-roll bars to be fitted). If the bearing bracket bushes require renewal, proceed as described below, noting that it is not possible to obtain either bush separately. If worn, the complete bearing bracket and bush assembly must be renewed. If it is not wished to renew the bushes, proceed to paragraph 18.

15 Obtain a nut of suitable diameter which just fits inside the bearing bracket pivot shaft, and weld it securely to the shaft inner surface. Support the outer surface of the trailing arm. Press out the bearing bracket shaft, using a spacer which bears on the top of the welded nut.

16 Thoroughly clean the trailing arm bore. Polish away any burrs or raised edges which might hinder reassembly. Apply a smear of multi-purpose grease to the outer diameter of the new bracket pivot shaft, to aid installation.

17 Support the inner edge of the trailing arm. Position the new bearing bracket as shown in Fig. 10.13 in relation to the trailing arm. Press the bracket onto the trailing arm shaft until the distance between inner bolt hole centres of the left and right-hand bearing brackets is as shown in Fig. 10.14. Wipe away any excess grease.

18 Commence reassembly of the rear axle by placing it upside-down on blocks of wood positioned under the L-shaped crossmember so that the bearing brackets are free.

19 Use a ruler as shown in Fig. 10.16, and mark each trailing arm on the axis between the centres of the torsion bar and anti-roll bar holes. Clean the anti-roll bar splines, and grease them well with molybdenum disulphide grease. Insert one anti-roll bar so that its drill mark is aligned with the mark made on the trailing arm (Fig. 10.17). Fit the centre link block to the anti-roll bar so that it is parallel with the upper section of the crossmember. Insert the remaining anti-roll bar from the opposite side, also aligning its circular drill mark with the mark made on the trailing arm.

20 It is now necessary to adjust the centre

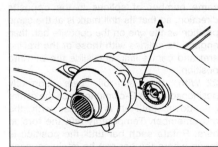

Fig. 10.17 Fit the anti-roll bars so that their drill marks align with the previously-made mark (A) on the trailing arm - open-bar axle (Sec 16)

10

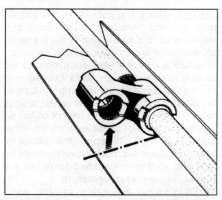

Fig. 10.18 Calculate the amount of adjustment required, then move the centre link away from the crossmember by the specified number of splines - open-bar axle (Sec 16)

link position, in order to avoid any contact with the crossmember during movement of the rear suspension. First measure the movement possible between the centre link and the crossmember. Note this measurement, then refer to the table below to obtain the amount of adjustment, in terms of splines, necessary.

Movement (mm)	Number of splines to compensate
2 to 4	1
5 to 6	2
7 to 8	3
9 to 10	4
11 to 12	5
13 to 14	6
15 to 16	7
17 to 18	8
19 to 20	9

21 Withdraw one of the anti-roll bars from the trailing arm. Slide the centre link off the end of the other anti-roll bar, rotate the link away from the crossmember by the specified number of splines, then refit it to the anti-roll bar end. Disengage the anti-roll bar and centre link assembly from the trailing arm splines. Rotate the bar by the specified number of splines, so that the centre link is parallel to the crossmember again, then relocate the anti-roll bar in the trailing arm splines. Rotate the second anti-roll bar by the same number of splines in the opposite direction, so that its drill mark is at the same position as the one on the opposite bar, then engage its splines with those of the trailing arm and centre link, and slide the bar into position.

22 With the anti-roll bars correctly positioned, use a G-clamp to press the centre link down until it is parallel with the crossmember. Temporarily refit the torsion bars. Rotate each bar until the position is found where the bar can be freely engaged with the splines on the bearing bracket and bearing, then press the bars fully into position.

23 If a new axle is being fitted, remove the brake pipes from the original and fit them to

the new axle. Also transfer the brake caliper mounting brackets, using locking fluid on the threads of the retaining bolts.

Refitting

24 Place the rear axle on a trolley jack, and lift it into position underneath the vehicle. Insert the bearing bracket mounting bolts from inside the vehicle, ensuring that the retaining plate is correctly fitted. Refit the nuts and tighten them to the specified torque.

New axle only

25 Remove the G-clamp from the crossmember, then withdraw the torsion bars again. Position the trailing arms as described in paragraphs 10 and 11 of Section 12.

26 Ensure that the bearing bracket, centre link and torsion bar splines are clean and dry. Lubricate the splines with molybdenum disulphide grease, and insert the bars into the bracket.

27 Rotate the left-hand bar until the position is found where it can be freely engaged with splines on the bearing bracket and centre link. Note the position of the circular drill mark on the left-hand bar. Rotate the right-hand bar so that its drill mark is in the same position, and engage its splines with those of the bearing bracket and centre link; the bar should slide freely into position. If not, it is permissible to rotate the bar slightly, noting that the drill marks on both bars must be no more than two splines away from each other once the splines are engaged. If this proves difficult to achieve, withdraw the left-hand bar and rotate it slightly until it freely

engages with the splines (there will be several positions where this will be the case) before trying to install the right-hand bar. Repeat the procedure until a position is found where both torsion bars slide freely into position and the drill marks are satisfactorily positioned.

28 Tap the torsion bars fully into position, then refit the torsion bar caps to the bearing brackets.

29 Carry out the procedures in the following paragraphs.

New or original axle

30 Refit the rear shock absorbers as described in Section 10.

31 Refit the brake hoses to the brackets on the top of the crossmember, and secure them in position with the retaining clips. Reconnect the brake pipes to the hoses, and securely tighten the union nuts.

32 Hook the brake pressure-regulating valve spring onto its bracket on the top of the crossmember.

33 Feed the handbrake cables through the trailing arms. Refit the handbrake cable retaining clamps to the torsion bars, and tighten their retaining screws.

34 Refit the brake discs, brake calipers and ABS wheel sensors (as applicable); bleed the brake hydraulic system on completion. Refer to the relevant Sections of Chapter 9.

35 Refit the roadwheels, lower the vehicle to the ground and tighten the wheel bolts to the specified torque.

36 Check the rear underbody height as described in Section 17.

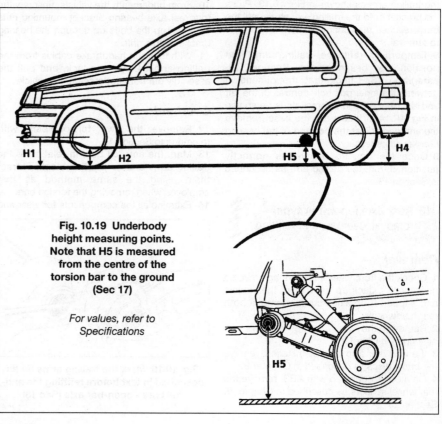

Fig. 10.19 Underbody height measuring points. Note that H5 is measured from the centre of the torsion bar to the ground (Sec 17)

For values, refer to Specifications

17 Underbody height - checking and adjustment

1 Position the unladen vehicle on a level surface, with the tyres correctly inflated and the fuel tank full.

Enclosed-bar rear suspension

2 It is only possible to adjust the rear suspension height in steps of 3 mm. First measure and record the dimensions H4 and H5 on both sides of the vehicle as shown in Fig. 10.19, noting that dimension H5 is measured from the centre of the torsion bar to the ground. Subtract H5 from H4 to find the underbody height checking dimension. Check that this dimension is within the range given in the Specifications at the start of this Chapter. If adjustment is necessary, proceed as follows.

3 Determine by how many splines the torsion bar must be moved, noting that one spline is equal to a 3 mm change of height. For example, if the height needs adjusting by 10 mm, the torsion bar should be moved by three splines.

4 Chock the front wheels and engage reverse gear (or 'P'). Jack up the rear of the vehicle and support it on axle stands. Remove the appropriate rear roadwheel.

5 Position a trolley jack beneath the end of the trailing arm, then remove the torsion bar with reference to Section 12. With the bar removed, raise or lower the trailing arm by the required amount, so that it rotates by the required number of splines in relation to the bearing bracket. Once correctly positioned, adjust the trailing arm position slightly until the torsion bar can be slid freely back into position. Use a hammer and a soft metal drift to tap the bar fully home.

6 Refit the roadwheel, lower the vehicle to the ground and tighten the wheel bolts to the specified torque.

7 Recheck that the underbody height is within the specified range. If not, repeat the adjustment procedure. Note that after making the adjustment, it may be necessary to adjust the headlight beam alignment with reference to Chapter 1.

Open-bar rear suspension

8 Measure and record the dimensions H4 and H5 on both sides of the vehicle as shown in Fig. 10.19, noting that dimension H5 is measured from the centre of the torsion bar to the ground. Subtract H5 from H4 to obtain the underbody height checking dimension. Check that this dimension is within the range given in the Specifications at the start of this Chapter. Note that on these models, the maximum difference between the heights on each side is 5 mm. If necessary, adjust the height(s) using the procedure given under the relevant sub-heading.

Underbody height correct on one side, but difference between sides excessive - new rear axle assembly

9 The difference between the right-hand and left-hand sides is always corrected by altering the position of the anti-roll bar on the lower side, so that the lower side is raised to the same height as the higher side. Calculate the difference in heights between the two sides, then adjust as follows.

10 Mark the relative positions of the two torsion bars and the anti-roll bar to be removed, both on the centre link and on the bearing bracket/trailing arm.

11 Remove both the torsion bars as described in Section 12, then remove the marked anti-roll bar using the same method.

12 Ensure that dimension 'X' is correctly set on both sides, as described in paragraphs 10 and 11 of Section 12. Measure the distance from the centre of the stub axle to the ground, on the side on which the anti-roll bar has been removed. Subtract the side-to-side difference from this height to obtain the correct height at which the stub axle should be, then lower the trailing arm until the stub axle is at the correct height.

13 Refit the anti-roll bar, rotating it until the position is found where it can be freely slid back into position, and tap it fully into position. Using the marks made prior to removal, check that the offset of the anti-roll bar splines in relation to the distance the trailing arm was moved is as shown in the following table.

Trailing arm movement (mm)	Number of splines to compensate
5	2
10	4
15	6
20	8
25	10
30	12
35	14
40	16
45	18
50	20

14 Align the marks made prior to removal, then refit both the torsion bars.

15 Refit the shock absorbers as described in Section 10.

16 Refit the roadwheels, lower the vehicle to the ground and tighten the wheel bolts to the specified torque.

17 Recheck the underbody height. If necessary, repeat the adjustment procedure. Note that after making the adjustment, it will be necessary to adjust the headlight beam alignment with reference to Chapter 1. It is also recommended that the operation of the brake pressure-regulating valve be checked by a Renault dealer at the earliest opportunity.

Underbody heights incorrect on both sides, and difference between sides excessive - new rear axle assembly

18 Calculate the difference in heights between the two sides, and the amount by which the higher side deviates from the specified height.

19 Carry out the procedures described in paragraphs 10 to 13.

20 With both trailing arms at the same height, measure the distance from the centre of the stub axle and the ground on each side. Raise or lower both trailing arms equally so that the distance is increased or decreased (as applicable) by the calculated height deviation of the higher side (paragraph 18).

21 Refit the torsion bars, rotating each one until the position is found where it enters freely, and tap them into position. Using the marks made prior to removal, check that the offset of the bar splines in relation to the distance the trailing arm was moved is as shown in the table in paragraph 13.

22 Carry out the procedures described in paragraphs 15 to 17.

Underbody heights incorrect on both sides, but difference between sides within tolerance - new rear axle assembly

23 In this case, the underbody height is adjusted by repositioning the torsion bars. First calculate the amount by which the actual underbody height dimension deviates from the specified height.

24 Mark the relative positions of the torsion bars in the centre link and bearing bracket, then remove both bars as described in Section 12.

25 Ensure that dimension 'X' is correctly set on both sides, as described in paragraphs 10 and 11 of Section 12. Measure the distance from the centre of the stub axles to the ground. Raise or lower both the trailing arms equally so that the distance is increased or decreased (as applicable) by the calculated deviation (paragraph 23).

26 Refit the torsion bars, rotating each one until the position is found where it enters freely, and tap them into position. Using the marks made prior to removal, check that the offset of the bar splines in relation to the distance the trailing arm was moved is as shown in the table in paragraph 13.

27 Carry out the procedures described in paragraphs 15 to 17.

Underbody height correct on one side, but difference between sides excessive - used rear axle assembly

28 The difference between the right-hand and left-hand sides is always corrected by altering the position of the anti-roll bar on the lower side, so that the lower side is raised to the same height as the higher side. Calculate the difference in heights between the two sides, then proceed as follows.

29 Remove the rear shock absorbers as described in Section 10. Mark the relative positions of the two torsion bars and the anti-roll bar to be removed, both on the centre link and on the bearing bracket/trailing arm.

30 Place a trolley jack under the trailing arm,

10

and raise it until the jack is just supporting the weight. Remove the two torsion bars as described in Section 12. Remove the marked anti-roll bar using the same method.

31 With the bars removed, ensure that the splines are clean and dry. Apply a smear of molybdenum disulphide grease to them.

32 Align the marks made prior to removal, and refit the removed anti-roll bar. If necessary, slightly raise or lower the trailing arm until the position is found where the bar slides freely into position.

33 Once the trailing arm is correctly positioned, so that the bar can freely be removed and refitted without resistance, measure the distance between the centres of the upper and lower shock absorber mounting bolt holes. Call this dimension 'Y'. Remove the anti-roll bar again.

34 Ensuring that dimension 'Y' is still correctly set, measure the distance from the centre of the stub axle to the ground, on the side on which the anti-roll bar has been removed. Subtract the side-to-side height difference from this height, then lower the trailing arm until the stub axle is at the height thus calculated.

35 Refit the anti-roll bar, rotating it until the position is found where it can be freely slid back into position, and tap it into position. Using the marks made prior to removal, check that the offset of the anti-roll bar splines in relation to the distance the trailing arm was moved is as shown in the table in paragraph 13.

36 Carry out the procedures described in paragraphs 14 to 17.

Underbody heights incorrect on both sides, and difference between sides excessive - used rear axle assembly

37 Calculate the difference in heights between the two sides, and the amount by which the higher side deviates from the specified height.

38 Carry out the operations described in paragraphs 29 to 35.

39 With both trailing arms at the same height, measure the distance from the centre of the stub axles to the ground. Raise or lower both trailing arms equally, so that the distance between the stub axles and the ground is

increased or decreased (as applicable) by the calculated height deviation of the higher side (paragraph 37).

40 Refit the torsion bars, rotating each one until the position is found where it can be freely slid back into position, and tap them into position. Using the marks made prior to removal, check that the offset of the bar splines in relation to the distance the trailing arm was moved is as shown in the table in paragraph 13.

41 Carry out the procedures described in paragraphs 15 to 17.

Underbody heights incorrect on both sides, but difference between sides within tolerance - used rear axle assembly

42 In this case, the underbody height is adjusted by repositioning the torsion bars. First calculate the amount by which the actual underbody height deviates from the specified height.

43 Remove the rear shock absorbers as described in Section 10. Mark the relative positions of the torsion bars in the centre link and bearing bracket.

44 Place a trolley jack under the trailing arm, and raise it until the jack is just supporting the weight. Remove the two torsion bars as described in Section 12.

45 With the bars removed, ensure the splines are clean and dry. Apply a smear of molybdenum disulphide grease to them.

46 Align the marks made prior to removal, and refit both torsion bars. If necessary, slightly raise or lower the trailing arm until the position is found where both bars slide freely into position.

47 Once the trailing arms are correctly positioned, so that each bar can freely be removed and refitted without resistance, measure the distance between the centres of the upper and lower shock absorber mounting bolt holes. Call this dimension 'Z'.

48 Ensure that dimension 'Z' is correctly set on both sides, then measure the distance from the centre of the stub axles to the ground. Raise or lower both trailing arms equally, so that the distance between the stub axles and the ground is increased or

decreased (as applicable) by the calculated height deviation (paragraph 42).

49 Refit the torsion bars, rotating each one until the position is found where it can be freely slid back into position, and tap them into position. Using the marks made prior to removal, check that the offset of the bar splines in relation to the distance the trailing arm was moved is as shown in the table in paragraph 13.

50 Carry out the procedures described in paragraphs 15 to 17.

18 Steering wheel - removal and refitting

1 Set the front wheels in the straight-ahead position. Release the steering lock by inserting the ignition key.

2 Carefully ease off the steering wheel centre pad, then remove the steering wheel retaining nut (photo).

3 Mark the steering wheel and steering column shaft in relation to each other, then lift the steering wheel off the column splines. If it is tight, screw the nut back on a few turns. Tap the steering wheel near the centre, using the palm of the hand, or rock it from side to side whilst pulling upwards to release it from the shaft splines. Remove the nut again and lift off the steering wheel.

4 Refitting is a reversal of removal. Align the marks made on removal, and tighten the retaining nut to the specified torque.

19 Steering column - removal, checking and refitting

Removal

1 Disconnect the battery negative lead.

2 Remove the steering wheel as described in Section 18.

3 Pull the bonnet release lever down to expose the release lever assembly securing bolt, then unscrew the bolt. Remove the two retaining screws from the upper edge of the panel, then carefully lower the panel away from the facia to gain the clearance necessary for column removal.

4 On models fitted with a radio/cassette player remote control switch, slide back the cover from the front of the switch assembly, and loosen the switch clamp screw.

5 Working underneath the steering column, undo the three retaining screws, then pull the lower shroud down to release its retaining clips. Withdraw the lower shroud, then remove the two retaining screws from the front of the upper shroud. Remove the upper shroud from the steering column.

6 Slacken the combination switch assembly clamp bolt. Slide the switch assembly off the steering column, disconnecting the wiring connectors as they become accessible. Disconnect the wiring from the ignition switch, and free it from any retaining clips.

18.2 Removing the steering wheel retaining nut

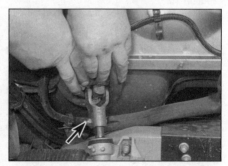

19.9 Fold back the protective cover for access to the universal joint clamp bolt (arrowed) - shown with engine removed for clarity

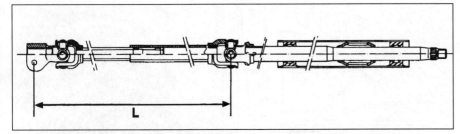

Fig. 10.20 Steering column intermediate shaft checking dimension (L) (Sec 19)

Right-hand drive models with manual steering - 446.0 ± 1 mm
Right-hand drive models with power steering - 396.0 ± 1 mm
Left-hand drive models with manual steering - 396.5 ± 1 mm
Left-hand drive models with power steering - 346.5 ± 1 mm

7 Remove the two bolts securing the facia to the top of the steering column.

8 Apply the handbrake, jack up the front of the vehicle and support it on axle stands.

9 On models with manual steering, working from underneath the vehicle, cut the retaining clip, then fold the rubber protective cover back from the steering gear to gain access to the intermediate shaft universal joint (photo).

10 On all models, mark the exact relationship between the intermediate shaft universal joint and the steering gear drive pinion, using paint marks, or a hammer and punch. Remove the nut and clamp bolt securing the joint to the pinion.

11 From inside the vehicle, slacken and remove the four steering column mounting nuts and bolts. Release the steering column from its mountings and the facia panel. On models equipped with power steering, also release the steering column lower gaiter from the bulkhead.

12 Disengage the universal joint from the steering gear pinion, and remove the steering column assembly from the vehicle.

Checking

13 The intermediate shaft attached to the bottom of the steering column incorporates a telescopic safety feature. In the event of a front-end crash, the shaft collapses and prevents the steering wheel injuring the driver. Before refitting the steering column, the length of the intermediate shaft must be checked. Refer to Fig. 10.20 for the checking dimension. If the length is shorter than specified, the complete steering column must be renewed. Damage to the intermediate shaft is also implied if it is found that the clamp bolt at its base cannot be inserted freely when refitting the column.

14 Check the steering shaft for signs of free play in the column bushes, and check the universal joints for signs of damage or roughness in the joint bearings. If damage or wear is found on the steering shaft universal joints or shaft bushes, the column must be renewed as an assembly.

15 Refer to Chapter 12, Section 15 for information on ignition switch renewal.

Refitting

16 Manoeuvre the steering column assembly into position. Engage the universal joint with the steering gear pinion splines, aligning the marks made prior to removal.

17 Refit and tighten the steering column mounting nuts and bolts. On models equipped with power steering, refit the column gaiter to the bulkhead.

18 From underneath the vehicle, refit the universal joint clamp bolt and nut, and tighten it to the specified torque. On models equipped with manual steering, relocate the protective cover on the steering gear, and secure it in position with a new cable tie. Lower the vehicle to the ground.

19 Refit the two bolts securing the facia to the top of the steering column, and tighten them securely.

20 Ensuring that the wiring is correctly routed, reconnect the wiring connectors to the combination switch assembly. Locate the switch assembly on the top of the steering column, and securely tighten its clamp bolt. Reconnect the wiring to the ignition switch.

21 Refit the upper shroud, tightening its retaining screws securely. Clip the lower shroud onto the steering column, and tighten its retaining screws. Where applicable, refit the radio/cassette remote control switch.

22 Clip the lower facia panel back into position, and secure it in position with its retaining screws, not forgetting the bonnet release lever securing bolt.

23 Refit the steering wheel as described in Section 18, and reconnect the battery negative terminal.

20 Steering gear rubber gaiter - renewal

1 Remove the track rod end balljoint as described in Section 26.

2 Mark the correct fitted position of the gaiter on the track rod. Release the retaining clips, and slide the gaiter off the steering gear housing and track rod end.

3 Thoroughly clean the track rod and the steering gear housing, using fine abrasive paper to polish off any corrosion, burrs or

sharp edges which might damage the sealing lips of the new gaiter on installation.

4 Recover the grease from inside the old gaiter. If it is uncontaminated with dirt or grit, apply it to the track rod inner balljoint. If the old grease is contaminated, or it is suspected that some has been lost, apply some new molybdenum disulphide grease.

5 Grease the inside of the new gaiter. Carefully slide the gaiter onto the track rod, and locate it on the steering gear housing. Align the outer edge of the gaiter with the mark made on the track rod prior to removal, then secure it in position with new retaining clips.

6 Refit the track rod balljoint as described in Section 26.

21 Manual steering gear assembly - removal, overhaul and refitting

Removal

1 Chock the rear wheels, apply the handbrake, jack up the front of the vehicle and support it on axle stands. Remove both front roadwheels.

2 Remove the nuts securing the track rod balljoints to the swivel hubs. Release the balljoint tapered shanks using a universal balljoint separator.

3 Working from underneath the vehicle, cut the retaining clip, then fold the protective cover back from the steering gear to gain access to the intermediate shaft universal joint clamp bolt.

4 Mark the relationship between the intermediate shaft universal joint and the steering gear drive pinion, using a hammer and punch, white paint or similar. Remove the nut and clamp bolt securing the joint to the pinion.

5 Remove the two nuts and bolts securing the steering gear assembly to the rear of the front subframe. Release the steering gear pinion from the universal joint, and manoeuvre the assembly sideways out of position (photos).

21.5A Manual steering gear mounting nut (arrowed) . . .

21.5B ... and bolt (arrowed) - viewed from underneath

Overhaul

General

6 Renewal procedures for the gaiters, the track rod end balljoints and the track rods (complete with inner balljoints) are given in Sections 20, 26 and 27 respectively.

7 Examine the steering gear assembly for signs of wear or damage. Check that the rack moves freely over the full length of its travel, with no signs of roughness or excessive free play between the steering gear pinion and rack. Internal wear or damage can only be cured by renewing the steering gear assembly, but note the points in the following paragraphs.

Thrust plunger adjustment

8 If there is excessive free play of the rack in the steering gear housing, accompanied by a knocking noise, it may be possible to correct this by adjusting the rack thrust plunger. Relieve the staking on the plunger adjusting nut. Using a 10 mm Allen key, tighten the adjusting nut until the free play disappears (but by no more than three flats). Check that the rack still moves freely over its full travel, then secure the adjusting nut by staking it (Fig. 10.21).

Anti-noise bush

9 If a grating noise has been noticed from the steering assembly whilst the steering wheel is being turned, this is probably due to the anti-noise bush being dry. To lubricate this bush, first obtain a sachet of the specified grease (Molykote 33 medium) and a steering gaiter retaining clip, from your Renault dealer. Cut the clip securing the gaiter to the opposite end of the steering gear from the pinion housing (ie the left-hand end on right-hand drive models, right-hand end on left-hand drive models). Peel the gaiter back from the housing. Move the steering rack so that its exposed end is fully extended, then smear the grease over the steering rack surface. Refit the gaiter to the housing, and secure it in position with the new retaining clip.

10 The anti-noise bush can be renewed if necessary. After removing the track rod on the side concerned, prise the old bush out using a

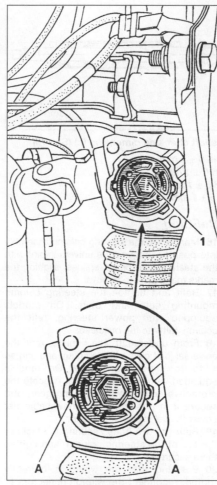

Fig. 10.21 Steering rack thrust plunger adjustment (Sec 21)

1 Adjusting nut   A Staking points

screwdriver. Fit the new bush, making sure that its lugs engage in the slots in the rack housing (Fig. 10.22). Lubricate the bush as just described.

Refitting

11 Manoeuvre the steering gear assembly into position. Engage the universal joint with the steering gear pinion splines, aligning the marks made prior to removal.

12 Insert the steering gear mounting bolts from the rear of the subframe. Refit the nuts and tighten them to the specified torque.

13 Refit the universal joint clamp bolt and nut, and tighten it to the specified torque. Relocate the protective cover on the steering gear, and secure it in position with a new cable tie.

14 Reconnect the track rod balljoints to the swivel hubs, and tighten their retaining nuts to the specified torque.

15 Refit the roadwheels, lower the vehicle to the ground and tighten the wheel bolts to the specified torque.

16 Check the front wheel toe setting as described in Section 28.

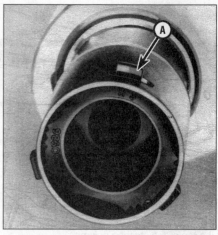

Fig. 10.22 Steering rack anti-noise bush. Lugs (A) must engage in slots (Sec 21)

22 Power steering gear assembly - removal, overhaul and refitting

Removal

1 Remove the battery as described in Chapter 12.

2 Using brake hose clamps, clamp both the supply and return hoses near the power steering fluid reservoir. This will minimise fluid loss during subsequent operations.

3 Chock the rear wheels, firmly apply the handbrake, jack up the front of the vehicle and support on axle stands. Remove the front roadwheels.

4 Remove the nuts securing the track rod balljoints to the swivel hubs. Release the balljoint tapered shanks using a universal balljoint separator.

5 Using a hammer and punch, white paint or similar, mark the relationship between the intermediate shaft universal joint and the steering gear drive pinion. Slacken and remove the nut and clamp bolt securing the joint to the pinion.

6 Remove the nut securing the feed pipe support bracket to the steering gear housing. Mark the pipe unions to ensure they are correctly positioned on reassembly, then unscrew the feed and return pipe union nuts; be prepared for fluid spillage by placing a suitable container beneath the pipes whilst unscrewing the union nuts. The spilt fluid must be disposed of, and new fluid of the specified type (see Chapter 1) used when refilling. Disconnect both pipes. Plug the pipe ends and steering gear orifices, to prevent excessive fluid leakage and the entry of dirt into the hydraulic system.

7 Remove the four mounting nuts and bolts securing the steering gear to the rear of the front subframe. Remove the mounting clamps and brackets from both ends of the steering gear.

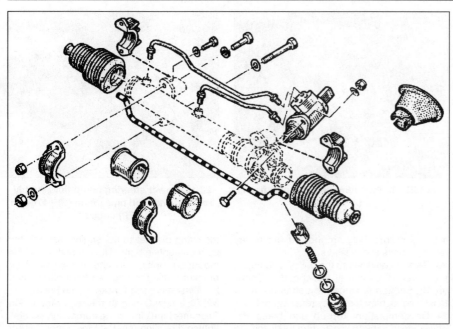

Fig. 10.23 Power-assisted steering gear components (Sec 22)

8 Release the steering gear pinion from the universal joint, and manoeuvre the assembly sideways out of position. Note that on some models, it may be necessary to remove the exhaust system downpipe from the vehicle, as described in Chapter 4, to gain the clearance required.

Overhaul

9 Examine the steering gear for signs of wear or damage. Check that the rack moves freely over the full length of its travel, with no signs of roughness or excessive free play between the pinion and rack. The steering gear must be renewed as an assembly if internal wear or damage is present. Track rod, track rod balljoint and steering gear gaiter renewal procedures are given in Sections 27, 26 and 20 respectively. Note that it is also possible to renew the pinion housing assembly, but it is recommended that this task be entrusted to a Renault dealer.

10 Inspect the steering gear fluid unions for signs of leakage, and check that the union nuts are securely tightened.

11 Examine the steering gear mounting rubbers for signs of damage and deterioration; renew as necessary.

Refitting

12 Manoeuvre the steering gear assembly into position. Aligning the marks made prior to removal, engage the universal joint with the steering gear pinion splines.

13 Locate the mounting brackets and clamps on the steering gear mounting rubbers, then insert the four steering gear mounting bolts from the rear of the subframe. Refit the nuts and tighten the mounting bolts to the specified torque.

14 Refit the exhaust downpipe (if removed) as described in Chapter 4.

15 Refit the universal joint clamp bolt and nut, and tighten it to the specified torque.

16 Wipe clean the feed and return pipe unions, then refit them to their respective positions on the steering gear. Tighten the union nuts. Refit and tighten the nut securing the feed pipe to the steering gear.

17 Reconnect the track rod balljoints to the swivel hubs. Tighten their retaining nuts to the specified torque.

18 Refit the roadwheels, lower the vehicle to the ground and tighten the wheel bolts to the specified torque. Refit the battery.

19 Remove the hose clamps from the reservoir hoses. Top-up the reservoir and bleed the power steering hydraulic system as described in Section 25.

20 On completion, check the front wheel toe setting as described in Section 28.

23 Mechanical power steering pump - removal and refitting

Removal

1390 cc (E6J and E7J) engine models without air conditioning

1 Slacken the three bolts which secure the drivebelt pulley to the pump, then release the drivebelt tension as described in Chapter 12, Section 6. Disengage the drivebelt from the pump pulley.

2 Remove the pulley retaining bolts and washers. Remove the pulley from the pump, noting which way round it is fitted.

3 Mark the pipe unions to ensure they are correctly positioned on reassembly, then

unscrew the feed and return pipe union nuts; be prepared for fluid spillage by placing a container beneath the pipes whilst unscrewing the union nuts. The spilt fluid must be disposed of, and new fluid of the specified type (see Chapter 1) used when refilling. Disconnect both pipes, and plug the pipe ends and pump unions (working quickly to minimise fluid loss), to prevent the entry of dirt into the hydraulic system.

4 Remove the bolts securing the power steering pump and fluid reservoir assembly to its mounting bracket. Remove the pump from the engine compartment.

1390 cc (E6J and E7J) engine models with air conditioning

5 To improve access to the pump, apply the handbrake, then jack up the front of the vehicle and support it on axle stands. To improve access even further, remove the right-hand headlight as described in Chapter 12, Section 18.

6 Slacken the bolt which secures the pulley to the pump. Release the drivebelt tension as described in Chapter 12, Section 6, then disengage the drivebelt from the pump pulley.

7 Unscrew the pulley retaining bolt, and remove the pulley from the pump, noting which way round it is fitted.

8 Clamp the fluid supply hose as near to the power steering pump as possible, using a brake hose clamp. This will minimise fluid loss during subsequent operations.

9 Slacken the retaining clip, and disconnect the fluid supply hose from the rear of the pump. If the original Renault clip is still fitted, cut the clip and discard it; replace it with a standard worm drive hose clip on refitting. Slacken the union nut and disconnect the feed pipe from the pump. Be prepared for some fluid spillage as the pipe and hose are disconnected; plug the hose and pipe ends and the pump unions, to minimise fluid loss and prevent the entry of dirt into the system.

10 Remove the bolts securing the power steering pump to its mounting bracket, noting the fitted positions of its mounting spacer and washers. Remove the pump from the engine compartment.

1721 cc (F2N) and 1794 cc (F3P) engine models without air conditioning

11 Remove the alternator and the right-hand headlight as described in Chapter 12.

12 Disconnect the fluid supply hose and feed pipe as described in paragraphs 8 and 9.

13 Remove the two upper mounting bolts and washers, and slacken the bolt securing the pump to its lower mounting bracket. Remove the two bolts securing the lower mounting bracket to the engine, and remove the pump and bracket assembly from the engine compartment.

1764 cc (F7P) engine models

14 Remove the alternator as described in Chapter 12.

23.15A Undo the pulley retaining screws whilst retaining the pump shaft with an Allen key . . .

23.15B . . . then remove the pulley from the pump

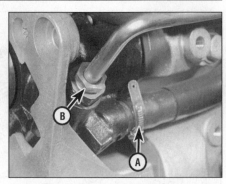

23.16 Power steering pump fluid supply hose (A) and feed pipe union (B) - 1764 cc (F7P) engine

15 Remove the pulley retaining screws, preventing the pulley from rotating by retaining the pump shaft with an Allen key. Remove the pulley, noting which way round it is fitted (photos).

16 Disconnect the fluid supply hose and feed pipe as described in paragraphs 8 and 9 (photo).

17 Undo the two bolts which retain the inlet manifold left-hand support bracket. Remove the bracket from the underside of the manifold (photo).

18 Remove the four large bolts securing the power steering pump rear mounting bracket to the cylinder block, and the small bolt securing the bracket to the rear of the pump (photo). Remove the rear bracket from the engine.

19 Undo the three screws securing the pump to its front mounting bracket, and manoeuvre the pump out from behind the bracket (photo).

Refitting

1390 cc (E6J and E7J) engine models without air conditioning

20 Manoeuvre the pump/reservoir assembly back into position, and tighten the mounting bolts.

21 Renew the fluid pipe union O-rings. Refit the pipes to the pump, using the marks made on removal to ensure that they are correctly reconnected. Tighten the union nuts.

22 Refit the pulley to the pump, making sure

it is the correct way around. Fit the pulley retaining bolts and washers.

23 Refit the drivebelt to the pulley, and adjust the drivebelt tension as described in Chapter 12, Section 6. Once the belt is correctly tensioned, tighten the pulley retaining bolts.

24 On completion, top-up and bleed the power steering hydraulic system as described in Section 25.

1390 cc (E6J and E7J) engine models with air conditioning

25 Manoeuvre the pump into position. Refit the mounting bolts, ensuring that the spacer is correctly positioned, and tighten them.

26 Fit a new O-ring to the feed pipe union. Reconnect the pipe to the pump, and tighten the union nut. Refit the supply pipe to the pump, and tighten its retaining clip. Remove the brake hose clamp used to minimise fluid loss.

27 Refit the pulley and drivebelt as described in paragraphs 22 and 23. Top-up and bleed the power steering hydraulic system as described in Section 25.

28 If removed, refit the headlight as described in Chapter 12, then lower the vehicle to the ground.

1721 cc (F2N) and 1794 cc (F3P) engine models without air conditioning

29 Manoeuvre the pump assembly into position. Refit the two bolts securing the lower

mounting bracket to the engine, and the two upper mounting bolts. Tighten both the upper mounting bolts and the lower mounting bracket bolts, then tighten the bolt securing the lower mounting bracket to the pump.

30 Fit a new O-ring to the feed pipe union. Reconnect the pipe to the pump, and securely tighten the union nut. Refit the supply pipe to the pump, and tighten its retaining clip. Remove the brake hose clamp used to minimise fluid loss.

31 Refit the alternator and the right-hand headlight as described in Chapter 12. Top-up and bleed the power steering hydraulic system as described in Section 25.

1764 cc (F7P) engine models

32 Manoeuvre the pump into position behind the front mounting bracket. Refit the three retaining bolts and tighten them securely.

33 Refit the rear mounting bracket. Tighten the four mounting bolts and the pump rear mounting bolt securely.

34 Refit the inlet manifold support bracket, and tighten the retaining bolts.

35 Fit a new O-ring to the feed pipe union. Reconnect the pipe to the pump and tighten the union nut. Refit the supply pipe to the pump, and tighten the retaining clip. Remove the brake hose clamp used to minimise fluid loss.

36 Refit the pulley to the pump, ensuring it is the correct way round. Fit the washers and the retaining bolts.

23.17 Remove the inlet manifold left-hand support bracket . . .

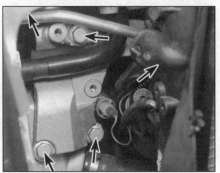

23.18 . . . and the power steering pump rear mounting bracket (mounting bolt locations arrowed - two hidden)

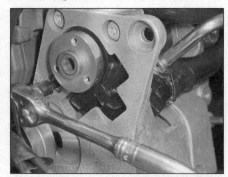

23.19 Undo the three retaining screws and remove the pump from the front mounting bracket - 1764 cc (F7P) engine

37 Refit the alternator as described in Chapter 12. When the belt is correctly tensioned, tighten the pump pulley retaining bolts.
38 On completion, top-up and bleed the power steering hydraulic system as described in Section 25.

24 Electric power steering pump - general information, removal and refitting

General information

1 On 1721 cc (F2N) and 1794 cc (F3P) engine models equipped with air conditioning, an electrically-powered steering pump is used. This is because there is not sufficient space in the engine compartment to mount a conventional belt-driven pump; the air conditioning compressor is situated where the power steering pump would be.
2 The operation of the pump is controlled by a pressure switch in the pump outlet pipe, and by either three or four relays, depending on model. The two power steering pump relays are situated in the engine compartment, mounted on the side of the left-hand suspension turret. On three-relay versions, the system control relay is fixed on the lid of the relay/junction box, which is mounted on the front of the left-hand suspension turret. On four-relay versions, there are two system control relays, both of which are also situated in the relay/junction box; one of them is fixed on the lid of the box, and the other is inside the bottom of the box.
3 Operation of the system is complex. If any fault develops, the vehicle should be taken to a Renault dealer for the fault to be diagnosed. The pump itself can be removed and refitted as follows.

Removal

4 Remove the battery as described in Chapter 12.
5 Apply the handbrake, jack up the front of the vehicle and support it on axle stands.
6 Clamp the feed and return hoses as near to the power steering pump as possible, using brake hose clamps. This will minimise fluid loss during subsequent operations.
7 Mark the pipe unions to ensure they are correctly positioned on reassembly, then unscrew the feed and return pipe union nuts; be prepared for fluid spillage by positioning a container beneath the pipes. The spilt fluid must be disposed of, and new fluid of the specified type (see Chapter 1) used when refilling. Disconnect both pipes, and plug the pipe ends and pump unions (working quickly to minimise fluid loss), to prevent the entry of dirt into the hydraulic system.
8 Remove the two pump upper mounting bracket bolts. From underneath the vehicle, remove the two lower mounting bracket bolts.
9 Lift up the pump until its wiring connectors are accessible, then disconnect them from the

base of the pump. Make a note of the locations of the wires, to ensure they are correctly reconnected on refitting. Remove the pump and reservoir assembly from the engine compartment.

Refitting

10 Manoeuvre the pump/reservoir assembly into position. Reconnect the wiring, using the notes made on removal.
11 Refit and tighten the four mounting bracket retaining bolts.
12 Renew the fluid pipe union O-rings. Refit the pipes to the pump, using the marks made on removal to ensure they are correctly reconnected, and tighten the union nuts.
13 Remove the hose clamps. Refit the battery as described in Chapter 12.
14 Lower the vehicle to the ground. Top-up and bleed the power steering hydraulic system as described in Section 25.

25 Power steering system - bleeding
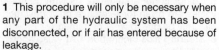

1 This procedure will only be necessary when any part of the hydraulic system has been disconnected, or if air has entered because of leakage.
2 Remove the fluid reservoir filler cap, and top-up the fluid level to the maximum mark, using only the specified fluid. Refer to Chapter 1 'Lubricants, fluids and capacities' for fluid specifications, and to Chapter 1, Section 16 for details of the different types of fluid reservoir markings.
3 With the engine stopped, slowly move the steering from lock-to-lock several times to expel trapped air, then top-up the level in the fluid reservoir. Repeat this procedure until the fluid level in the reservoir does not drop any further.
4 Start the engine. Slowly move the steering from lock-to-lock several times to expel any air remaining in the system. Repeat this procedure until bubbles cease to appear in the fluid reservoir.
5 If, when turning the steering, an abnormal noise is heard from the fluid pipes, it indicates that there is still air in the system. Check this

26.4 Using a universal balljoint separator to release the track rod balljoint from the swivel hub

by turning the wheels to the straight-ahead position and switching off the engine. If the fluid level in the reservoir rises, air is still present in the system, and further bleeding is necessary.
6 Once all traces of air have been removed, stop the engine and allow the system to cool. Once cool, check that the fluid level is up to the maximum mark on the power steering fluid reservoir; top-up if necessary.

26 Track rod end balljoint - removal and refitting

Removal

1 Apply the handbrake, then jack up the front of the vehicle and support it on axle stands. Remove the appropriate front roadwheel.
2 If the balljoint is to be re-used, use a straight-edge and a scriber, or similar, to mark its relationship to the track rod.
3 Holding the balljoint, unscrew its locknut by one quarter of a turn. Do not move the locknut from this position, as it will serve as a reference mark on refitting.
4 Remove the nut securing the track rod balljoint to the swivel hub. Release the balljoint tapered shank using a universal balljoint separator. If the balljoint is to be re-used, protect the threaded end of the shank by screwing the nut back on a few turns before using the separator (photo).
5 Counting the **exact** number of turns necessary to do so, unscrew the balljoint from the track rod end.
6 Count the number of exposed threads between the end of the balljoint and the locknut, and record this figure. If a new balljoint is to be fitted, unscrew the locknut from the old balljoint.
7 Carefully clean the balljoint and the threads. Renew the balljoint if its movement is sloppy or if it is too stiff, if it is excessively worn, or if it is damaged in any way. Carefully check the shank taper and threads. If the balljoint gaiter is damaged, the complete balljoint must be renewed; it is not possible to obtain the gaiter separately.

Refitting

8 If applicable, screw the locknut onto the new balljoint, and position it so that the same number of exposed threads are visible as was noted prior to removal.
9 Screw the balljoint into the track rod by the number of turns noted on removal. This should bring the balljoint locknut to within a quarter of a turn of the end of the track rod, with the alignment marks that were made (if applicable) on removal lined up.
10 Refit the balljoint shank to the swivel hub, and tighten the retaining nut to the specified torque. If difficulty is experienced due to the balljoint shank rotating, jam it by exerting pressure on the underside of the balljoint, using a tyre lever or a jack.

10

11 Refit the roadwheel, lower the vehicle to the ground and tighten the roadwheel bolts to the specified torque.

12 Check the front wheel toe setting as described in Section 28, then tighten the balljoint locknut to the specified torque.

27 Track rod and inner balljoint - removal and refitting

Removal

1 Remove the track rod end balljoint as described in Section 26.

2 Cut the retaining clips, and slide the steering gear gaiter off the track rod.

3 Using a suitable pair of grips, unscrew the track rod inner balljoint from the steering rack end. Prevent the steering rack from turning by holding the balljoint lockwasher with a second pair of grips. Take care not to mark the surfaces of the rack and balljoint.

4 Remove the track rod/inner balljoint assembly and discard the lockwasher; a new one must be used on refitting.

5 Examine the inner balljoint for signs of slackness or tight spots. Check that the track rod itself is straight and free from damage. If necessary, renew the track rod/inner balljoint; the new one will be supplied complete with a new lockwasher and a new end balljoint. It is also recommended that the steering gear gaiter be renewed.

Refitting

6 If a new track rod is being installed, remove the outer balljoint from the track rod end.

7 Locate the new lockwasher assembly on the end of the steering rack, ensuring that its locating tabs are correctly located with the flats on the rack end.

8 Apply a few drops of locking fluid to the inner balljoint threads. Screw the balljoint into the steering rack and tighten it securely. Again, take care not to damage or mark the balljoint or steering rack.

9 Slide the new gaiter onto the track rod end, and locate it on the steering gear housing. Turn the steering from lock-to-lock to check that the gaiter is correctly positioned, then secure it with new retaining clips.

10 Refit the track rod end balljoint as described in Section 26.

28 Wheel alignment and steering angles

General information

1 A car's steering and suspension geometry is defined in four basic settings. For this purpose, all angles are expressed in degrees (toe settings are also expressed as a measurement of length). The steering axis is defined as an imaginary line drawn through the axis of the suspension strut, extended where necessary to contact the ground.

2 **Camber** is the angle between each roadwheel and a vertical line drawn through its centre and tyre contact patch, when viewed from the front or rear of the car. Positive camber is when the roadwheels are tilted outwards from the vertical at the top; negative camber is when they are tilted inwards.

3 Camber is not adjustable. Values are given for reference only. Checking is possible using a camber checking gauge, but if the figure obtained is significantly different from that specified, the vehicle must be taken for careful checking by a professional. Wrong camber settings can only be caused by wear or damage to the body or suspension components.

4 **Castor** is the angle between the steering axis and a vertical line drawn through each roadwheel's centre and tyre contact patch, when viewed from the side of the car. Positive castor is when the steering axis is tilted so that it contacts the ground ahead of the vertical; negative castor is when it contacts the ground behind the vertical.

5 Castor is not adjustable. As with camber, values are given for reference only; deviation can only be due to wear or damage.

6 **Steering axis inclination/SAI** - also known as **kingpin inclination/KPI** - is the angle between the steering axis and a vertical line drawn through each roadwheel's centre and tyre contact patch, when viewed from the front or rear of the car.

7 SAI/KPI is not adjustable, and is given for reference only.

8 **Toe** is the amount by which the roadwheels point outwards or inwards, viewed from above. 'Toe-in' is when the roadwheels point inwards, towards each other at the front, while 'toe-out' is when they splay outwards from each other at the front. The value for toe can be expressed as an angle (taking the centre-line of the car as zero), or as a measurement of length (taking measurements between the inside rims of the wheels at hub height).

9 The front wheel toe setting is adjusted by screwing the balljoints in or out of their track rods to alter the effective length of the track rod assemblies.

10 Rear wheel toe setting is not adjustable, and is given for reference only. While it can be checked, if the figure obtained is significantly different from that specified, the vehicle must be taken for careful checking by a professional, as the fault can only be caused by wear or damage to the body or suspension components.

Checking - general

11 Due to the special measuring equipment necessary to check the wheel alignment, and the skill required to use it properly, the checking and adjustment of these settings is best left to a Renault dealer or similar expert. Most tyre-fitting shops now possess sophisticated checking equipment.

12 For accurate checking, the vehicle must

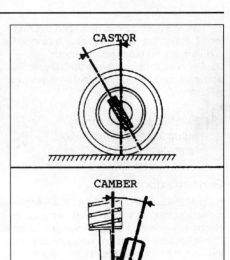

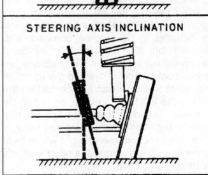

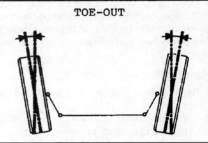

Fig. 10.24 Wheel alignment and steering angles (Sec 28)

be at the kerb weight specified in 'General dimensions and weights'.

13 Before starting work, check first that the tyre sizes and types are as specified, then check tyre pressures and tread wear. Also check roadwheel run-out, the condition of the hub bearings, the steering wheel free play and the condition of the front suspension components (Chapter 1). Correct any faults.

14 Park the vehicle on level ground, with the front roadwheels in the straight-ahead position. Rock both ends to settle the suspension. Release the handbrake and roll the vehicle backwards 1 metre (3 feet), then forwards again, to relieve any stresses in the steering and suspension components.

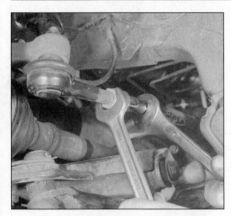

28.22 Adjusting the front wheel toe setting

Toe setting - checking and adjusting

Front wheels - checking

15 Two methods are available to the home mechanic for checking the front wheel toe setting. One method is to use a gauge to measure the distance between the front and rear inside edges of the roadwheels. The other method is to use a scuff plate, in which each front wheel is rolled across a movable plate which records any deviation, or scuff, of the tyre from the straight-ahead position as it moves across the plate. Such gauges are available in relatively-inexpensive form from accessory outlets. It is up to the owner to decide whether the expense is justified, in view of the small amount of use such equipment would normally receive.

16 Prepare the vehicle as described in paragraphs 12 to 14 above.

17 If the measurement procedure is being used, carefully measure the distance between the front edges of the roadwheel rims and the rear edges of the rims. Subtract the rear measurement from the front measurement, and check that the result is within the specified range. If not, adjust the toe setting as described in paragraph 19.

18 If scuff plates are to be used, roll the vehicle backwards, check that the roadwheels are in the straight-ahead position, then roll it across the scuff plates so that each front roadwheel passes squarely over the centre of its respective plate. Note the angle recorded by the scuff plates. To ensure accuracy, repeat the check three times, and take the average of the three readings. If the roadwheels are running parallel, there will of course be no angle recorded; if a deviation value is shown on the scuff plates, compare the reading obtained for each wheel with that specified. If the value recorded is outside the specified tolerance, the toe setting is incorrect, and must be adjusted as follows.

Front wheels - adjusting

19 Apply the handbrake, jack up the front of the vehicle and support it securely on axle stands. Turn the steering wheel onto full-left lock, and record the number of exposed threads on the right-hand track rod end. Now turn the steering onto full-right lock, and record the number of threads on the left-hand side. If there are the same number of threads visible on both sides, then subsequent adjustment should be made equally on both sides. If there are more threads visible on one side than the other, it will be necessary to compensate for this during adjustment. **Note:** *It is important that, after adjustment, the same number of threads be visible on each track rod end.*

20 First clean the track rod threads; if they are corroded, apply penetrating fluid before starting adjustment. Release the rubber gaiter outboard clips, then peel back the gaiters and apply a smear of grease, so that both gaiters are free and will not be twisted or strained as their respective track rods are rotated.

21 Use a straight-edge and a scriber or similar to mark the relationship of each track rod to its balljoint. Holding each track rod in turn, unscrew its locknut fully.

22 Alter the length of the track rods, bearing in mind the note in paragraph 19, by screwing them into or out of the balljoints. Rotate the track rod using an open-ended spanner fitted to the flats provided. Shortening the track rods (screwing them onto their balljoints) will reduce toe-in and increase toe-out. Each complete turn of the track rod effectively adjusts the toe setting by 30' or 3 mm (depending on the method being used) (photo).

23 When the setting is correct, hold the track rods and securely tighten the balljoint locknuts. Check that the balljoints are seated correctly in their sockets, and count the exposed threads. If the number of threads exposed is not the same on both sides, then the adjustment has not been made equally, and problems will be encountered with tyre scrubbing in turns; also, the steering wheel spokes will no longer be horizontal when the wheels are in the straight-ahead position.

24 When the track rod lengths are the same, lower the vehicle to the ground and re-check the toe setting; readjust if necessary. When the setting is correct, tighten the track rod balljoint locknuts to the specified torque. Ensure that the rubber gaiters are seated correctly and are not twisted or strained; secure them in position with new retaining clips.

Rear wheel toe setting

25 The procedure for checking the rear toe setting is same as described for the front in paragraph 17. However, no adjustment is possible.

Notes

Chapter 11 Bodywork and fittings

Contents

Degrees of difficulty

Easy, suitable for novice with little experience
Fairly easy, suitable for beginner with some experience
Fairly difficult, suitable for competent DIY mechanic
Difficult, suitable for experienced DIY mechanic
Very difficult, suitable for expert DIY or professional

Specifications

Torque wrench setting	Nm	lbf ft
Seat belt and seat belt height adjuster mountings*	20	15

*No figure specified by Renault - suggested value only

1 General information and maintenance

General information

The bodyshell and floorpan are manufactured from pressed-steel, and form an integral part of the vehicle's structure (monocoque), without the need for a separate chassis. The Clio is available in 3- and 5-door Hatchback body styles, with a Van version also available in some markets.

Various areas of the structure are strengthened to provide for suspension, steering and engine mounting points, and load distribution.

16-valve models are fitted with front wings manufactured from a polymer compound, which can withstand an impact of up to 10 mph (16 km/h) without sustaining permanent damage.

Corrosion protection is applied to all new vehicles. Various anti-corrosion preparations are used, including galvanising, zinc phosphatisation, and PVC underseal. An 'anti-gravel' undercoat is applied to the front section of the bonnet, to prevent corrosion and paint chipping caused by stones and other road debris hitting the front of the vehicle. Protective wax is injected into the box sections and other hollow cavities.

Extensive use is made of plastic for peripheral components, such as the radiator grille, bumpers and wheel trims, and for much of the interior trim. Plastic wheel arch liners are fitted, to protect the metal body panels against corrosion due to a build-up of road dirt.

Interior fittings are to a high standard on all models, and a wide range of optional equipment is available throughout the range.

Maintenance - bodywork and underframe

Cleaning the vehicle's exterior

The general condition of a vehicle's bodywork is the one thing that significantly affects its value. Maintenance is easy, but needs to be regular. Neglect, particularly after minor damage, can lead quickly to further deterioration and costly repair bills. It is important also to keep watch on those parts of the vehicle not immediately visible, for instance the underbody, inside all the wheel arches and the lower part of the engine compartment.

The basic maintenance routine for the bodywork is washing - preferably with a lot of water, from a hose. This will remove all the loose solids which may have stuck to the vehicle. It is important to flush these off in such a way as to prevent grit from scratching the finish. The wheel arches and underbody need washing in the same way to remove any accumulated mud, which will retain moisture and tend to encourage rust, particularly in Winter, when it is essential that any salt (from that put down on the roads) is washed off. Paradoxically enough, the best time to clean the underbody and wheel arches is in wet weather, when the mud is thoroughly wet and soft. In very wet weather, the underbody is usually cleaned automatically of large accumulations; this is therefore a good time for inspection.

If the vehicle is very dirty, especially underneath or in the engine compartment, it is tempting to use one of the pressure-washers or steam-cleaners available on garage forecourts; while these are quick and effective, especially for the removal of the accumulation of oily grime which sometimes is allowed to become thick in certain areas, their usage does have some disadvantages. If caked-on dirt is simply blasted off the paintwork, its finish soon becomes scratched and dull, and the pressure can allow water to penetrate door and window seals and the lock mechanisms; if the full force of such a jet is

11

directed at the vehicle's underbody, the wax-based protective coating can easily be damaged, and water (with whatever cleaning solvent is used) could be forced into crevices or components that it would not normally reach. Similarly, if such equipment is used to clean the engine compartment, water can be forced into the components of the fuel and electrical systems, and the protective coating can be removed that is applied to many small components during manufacture; this may therefore actually *promote* corrosion (especially inside electrical connectors) and initiate engine problems or other electrical faults. Also, if the jet is pointed directly at any of the oil seals, water can be forced past the seal lips and into the engine or transmission. Great care is required, therefore, if such equipment is used and, in general, regular cleaning by such methods should be avoided.

A much better solution in the long term is just to flush away as much loose dirt as possible using a hose alone, even if this leaves the engine compartment looking 'dirty'. If an oil leak has developed, or if any other accumulation of oil or grease is to be removed, there are one or two excellent grease solvents available which can be brush-applied. The dirt can then be simply hosed off. Take care to replace the wax-based protective coat, if this was affected by the solvent.

Normal washing of the vehicle's bodywork is best carried out using cold or warm water, with a proprietary vehicle shampoo. Tar spots can be removed either by using white spirit, followed by soapy water to remove all traces of spirit, Try to keep water out of the bonnet air intakes, and check afterwards that the heater air inlet box drain tube is clear, so that any water has drained out of the box.

After washing the paintwork, wipe off with a chamois leather to give an unspotted clear finish. A coat of clear protective wax polish will give added protection against chemical pollutants in the air. If the paintwork sheen has dulled or oxidised, use a cleaner/polisher combination to restore the brilliance of the shine. This requires a little effort, but such dulling is usually caused because regular washing has been neglected. Care needs to be taken with metallic paintwork, as special non-abrasive cleaner/polisher is required to avoid damage to the finish.

Brightwork should be treated in the same way as paintwork.

Windscreens and windows can be kept clear of the smeary film which often appears, by the use of proprietary glass cleaner. Never use any form of wax or other body or chromium polish on glass.

Exterior paintwork and body panels check

Once the vehicle has been washed, and all tar spots and other surface blemishes have been cleaned off, check carefully all paintwork, looking closely for chips or scratches; check with particular care vulnerable areas such as the front (bonnet and spoiler) and around the wheel arches. Any damage to the paintwork must be rectified as soon as possible, to comply with the terms of the manufacturer's cosmetic and anti-corrosion warranties; check with a Renault dealer for details.

If a chip or (light) scratch is found that is recent and still free from rust, it can be touched-up using the appropriate touch-up pencil; these can be obtained from Renault dealers. Any more serious damage, or rusted stone chips, can be repaired as described in Section 2, but if damage or corrosion is so severe that a panel must be renewed, seek professional advice as soon as possible.

Always check that any door and ventilator opening drain holes and pipes are completely clear, so that water can drain out.

Underbody sealer check

The wax-based underbody protective coating should be inspected annually, preferably just prior to Winter, when the underbody should be washed down as thoroughly but gently as possible (see the above concerning steam cleaners, etc) and any damage to the coating repaired; if any of the body panels are disturbed for repair or renewed, do not forget to replace the coating and to inject wax into door panels, sills, box sections etc, to maintain the level of protection provided by the vehicle manufacturer.

Maintenance - upholstery and carpets

Mats and carpets should be brushed or vacuum-cleaned regularly, to keep them free of grit. If they are badly stained, remove them from the vehicle for scrubbing or sponging, and make quite sure they are dry before refitting.

Fabric-trimmed seats and interior trim panels can be kept clean by wiping with a damp cloth. If they do become stained (which can be more apparent on light-coloured upholstery) use a little liquid detergent and a soft nail brush to scour the grime out of the grain of the material. Do not forget to keep the headlining clean in the same way as the (fabric) upholstery.

When using liquid cleaners of any sort inside the vehicle, do not over-wet the surfaces being cleaned. Excessive damp could get into the seams and padded interior, causing stains, offensive odours or even rot.

 HAYNES HiNT *If the inside of the vehicle gets wet accidentally, it is worthwhile taking some trouble to dry it out properly, particularly where carpets are involved. Do not leave oil/paraffin-burning or electric heaters inside the vehicle for this purpose.*

2 Minor body damage - repair

Note: *For more detailed information about bodywork repair, Haynes Publishing produce a book by Lindsay Porter called "The Car Bodywork Repair Manual". This incorporates information on such aspects as rust treatment, painting and glass-fibre repairs, as well as details on more ambitious repairs involving welding and panel beating.*

Repairs of minor scratches in bodywork

If the scratch is very superficial, and does not penetrate to the metal of the bodywork, repair is very simple. Lightly rub the area of the scratch with a paintwork renovator, or a very fine cutting paste, to remove loose paint from the scratch, and to clear the surrounding bodywork of wax polish. Rinse the area with clean water.

Apply touch-up paint to the scratch using a fine paint brush; continue to apply fine layers of paint until the surface of the paint in the scratch is level with the surrounding paintwork. Allow the new paint at least two weeks to harden, then blend it into the surrounding paintwork by rubbing the scratch area with a paintwork renovator or a very fine cutting paste. Finally, apply wax polish.

Where the scratch has penetrated right through to the metal of the bodywork, causing the metal to rust, a different repair technique is required. Remove any loose rust from the bottom of the scratch with a penknife, then apply rust-inhibiting paint to prevent the formation of rust in the future. Using a rubber or nylon applicator, fill the scratch with bodystopper paste. If required, this paste can be mixed with cellulose thinners to provide a very thin paste which is ideal for filling narrow scratches. Before the stopper-paste in the scratch hardens, wrap a piece of smooth cotton rag around the top of a finger. Dip the finger in cellulose thinners, and quickly sweep it across the surface of the stopper-paste in the scratch; this will ensure that the surface of the stopper-paste is slightly hollowed. The scratch can now be painted over as described earlier in this Section.

Repairs of dents in bodywork

When deep denting of the vehicle's bodywork has taken place, the first task is to pull the dent out, until the affected bodywork almost attains its original shape. There is little point in trying to restore the original shape completely, as the metal in the damaged area will have stretched on impact, and cannot be reshaped fully to its original contour. It is better to bring the level of the dent up to a point which is about 3 mm below the level of the surrounding bodywork. In cases where the dent is very shallow anyway, it is not worth trying to pull it out at all. If the underside of the dent is accessible, it can be hammered out gently from behind, using a mallet with a wooden or plastic head. Whilst doing this, hold a suitable block of wood firmly against the outside of the panel, to absorb the impact from the hammer blows and thus prevent a large area of the bodywork from being "belled-out".

Should the dent be in a section of the bodywork which has a double skin, or some other factor making it inaccessible from behind, a different technique is called for. Drill several small holes through the metal inside the area - particularly in the deeper section. Then screw long self-tapping screws into the holes, just sufficiently for them to gain a good purchase in the metal. Now the dent can be pulled out by pulling on the protruding heads of the screws with a pair of pliers.

The next stage of the repair is the removal of the paint from the damaged area, and from an inch or so of the surrounding "sound" bodywork. This is accomplished most easily by using a wire brush or abrasive pad on a power drill, although it can be done just as effectively by hand, using sheets of abrasive paper. To complete the preparation for filling, score the surface of the bare metal with a screwdriver or the tang of a file, or alternatively, drill small holes in the affected area. This will provide a really good "key" for the filler paste.

To complete the repair, see the Section on filling and respraying.

Repairs of rust holes or gashes in bodywork

Remove all paint from the affected area, and from an inch or so of the surrounding "sound" bodywork, using an abrasive pad or a wire brush on a power drill. If these are not available, a few sheets of abrasive paper will do the job most effectively. With the paint removed, you will be able to judge the severity of the corrosion, and therefore decide whether to renew the whole panel (if this is possible) or to repair the affected area. New body panels are not as expensive as most people think, and it is often quicker and more satisfactory to fit a new panel than to attempt to repair large areas of corrosion.

Remove all fittings from the affected area, except those which will act as a guide to the original shape of the damaged bodywork (eg headlight shells etc). Then, using tin snips or a hacksaw blade, remove all loose metal and any other metal badly affected by corrosion. Hammer the edges of the hole inwards, in order to create a slight depression for the filler paste.

Wire-brush the affected area to remove the powdery rust from the surface of the remaining metal. Paint the affected area with rust-inhibiting paint, if the back of the rusted area is accessible, treat this also.

Before filling can take place, it will be necessary to block the hole in some way. This can be achieved by the use of aluminium or plastic mesh, or aluminium tape.

Aluminium or plastic mesh, or glass-fibre matting, is probably the best material to use for a large hole. Cut a piece to the approximate size and shape of the hole to be filled, then position it in the hole so that its edges are below the level of the surrounding bodywork. It can be retained in position by several blobs of filler paste around its periphery.

Aluminium tape should be used for small or very narrow holes. Pull a piece off the roll, trim it to the approximate size and shape required, then pull off the backing paper (if used) and stick the tape over the hole; it can be overlapped if the thickness of one piece is insufficient. Burnish down the edges of the tape with the handle of a screwdriver or similar, to ensure that the tape is securely attached to the metal underneath.

Bodywork repairs - filling and respraying

Before using this Section, see the Sections on dent, deep scratch, rust holes and gash repairs.

Many types of bodyfiller are available, but generally speaking, those proprietary kits which contain a tin of filler paste and a tube of resin hardener are best for this type of repair. A wide, flexible plastic or nylon applicator will be found invaluable for imparting a smooth and well-contoured finish to the surface of the filler.

Mix up a little filler on a clean piece of card or board - measure the hardener carefully (follow the maker's instructions on the pack), otherwise the filler will set too rapidly or too slowly. Using the applicator, apply the filler paste to the prepared area; draw the applicator across the surface of the filler to achieve the correct contour and to level the surface. As soon as a contour that approximates to the correct one is achieved, stop working the paste - if you carry on too long, the paste will become sticky and begin to "pick-up" on the applicator. Continue to add thin layers of filler paste at 20-minute intervals, until the level of the filler is just proud of the surrounding bodywork.

Once the filler has hardened, the excess can be removed using a metal plane or file. From then on, progressively-finer grades of abrasive paper should be used, starting with a 40-grade production paper, and finishing with a 400-grade wet-and-dry paper. Always wrap the abrasive paper around a flat rubber, cork, or wooden block - otherwise the surface of the filler will not be completely flat. During the smoothing of the filler surface, the wet-and-dry paper should be periodically rinsed in water. This will ensure that a very smooth finish is imparted to the filler at the final stage.

At this stage, the "dent" should be surrounded by a ring of bare metal, which in turn should be encircled by the finely "feathered" edge of the good paintwork. Rinse the repair area with clean water, until all of the dust produced by the rubbing-down operation has gone.

Spray the whole area with a light coat of primer - this will show up any imperfections in the surface of the filler. Repair these imperfections with fresh filler paste or bodystopper, and once more smooth the surface with abrasive paper. Repeat this spray-and-repair procedure until you are satisfied that the surface of the filler, and the feathered edge of the paintwork, are perfect. Clean the repair area with clean water, and allow to dry fully.

 If bodystopper is used, it can be mixed with cellulose thinners, to form a really thin paste which is ideal for filling small holes.

The repair area is now ready for final spraying. Paint spraying must be carried out in a warm, dry, windless and dust-free atmosphere. This condition can be created artificially if you have access to a large indoor working area, but if you are forced to work in the open, you will have to pick your day very carefully. If you are working indoors, dousing the floor in the work area with water will help to settle the dust which would otherwise be in the atmosphere. If the repair area is confined to one body panel, mask off the surrounding panels; this will help to minimise the effects of a slight mis-match in paint colours. Bodywork fittings (eg chrome strips, door handles etc) will also need to be masked off. Use genuine masking tape, and several thicknesses of newspaper, for the masking operations.

Before commencing to spray, agitate the aerosol can thoroughly, then spray a test area (an old tin, or similar) until the technique is mastered. Cover the repair area with a thick coat of primer; the thickness should be built up using several thin layers of paint, rather than one thick one. Using 400-grade wet-and-dry paper, rub down the surface of the primer until it is really smooth. While doing this, the work area should be thoroughly doused with water, and the wet-and-dry paper periodically rinsed in water. Allow to dry before spraying on more paint.

Spray on the top coat, again building up the thickness by using several thin layers of paint. Start spraying at one edge of the repair area, and then, using a side-to-side motion, work until the whole repair area and about 2 inches of the surrounding original paintwork is covered. Remove all masking material 10 to 15 minutes after spraying on the final coat of paint.

Allow the new paint at least two weeks to harden, then, using a paintwork renovator, or a very fine cutting paste, blend the edges of the paint into the existing paintwork. Finally, apply wax polish.

Plastic components

With the use of more and more plastic body components by the vehicle manufacturers (eg bumpers. spoilers, and in some cases major body panels), rectification of more serious

11

damage to such items has become a matter of either entrusting repair work to a specialist in this field, or renewing complete components. Repair of such damage by the DIY owner is not really feasible, owing to the cost of the equipment and materials required for effecting such repairs. The basic technique involves making a groove along the line of the crack in the plastic, using a rotary burr in a power drill. The damaged part is then welded back together, using a hot-air gun to heat up and fuse a plastic filler rod into the groove. Any excess plastic is then removed, and the area rubbed down to a smooth finish. It is important that a filler rod of the correct plastic is used, as body components can be made of a variety of different types (eg polycarbonate, ABS, polypropylene).

Damage of a less serious nature (abrasions, minor cracks etc) can be repaired by the DIY owner using a two-part epoxy filler repair material. Once mixed in equal proportions, this is used in similar fashion to the bodywork filler used on metal panels. The filler is usually cured in twenty to thirty minutes, ready for sanding and painting.

If the owner is renewing a complete component himself, or if he has repaired it with epoxy filler, he will be left with the problem of finding a suitable paint for finishing which is compatible with the type of plastic used. At one time, the use of a universal paint was not possible, owing to the complex range of plastics encountered in body component applications. Standard paints, generally speaking, will not bond to plastic or rubber satisfactorily. However, it is now possible to obtain a plastic body parts finishing kit which consists of a pre-primer treatment, a primer and coloured top coat. Full instructions are normally supplied with a kit, but basically, the method of use is to first apply the pre-primer to the component concerned, and allow it to dry for up to 30 minutes. Then the primer is applied, and left to dry for about an hour before finally applying the special-coloured top coat. The result is a correctly-coloured component, where the paint will flex with the plastic or rubber, a property that standard paint does not normally posses.

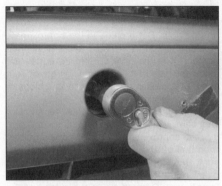

4.4 Unscrewing a front bumper front securing nut

3 Major body damage - repair

Where serious damage has occurred, or large areas need renewal due to neglect, it means that complete new panels will need welding-in; this is best left to professionals. If the damage is due to impact, it will also be necessary to check completely the alignment of the bodyshell; this can only be carried out accurately by a Renault dealer using special jigs. If the body is misaligned, it is dangerous, as the car will not handle properly. In addition, uneven stresses will be imposed on the steering, suspension and possibly transmission, causing abnormal wear or complete failure, particularly of items such as the tyres.

4 Bumpers - removal and refitting

Front bumper

1 On models fitted with front foglights, disconnect the battery negative lead.
2 Remove the radiator grille panel, as described in Section 5.
3 Remove the front number plate. If the number plate is secured by self-tapping screws, take care not to strip the threads in the bumper.

Fig. 11.1 Front wheel arch splash shield lower fixings (A) (Sec 4)

4 Remove the now-exposed bumper securing nut, and recover the washer (photo).
5 To improve access, jack up the front of the car and support it securely on axle stands (see 'Jacking, towing and wheel changing').
6 Working under the right-hand side of the car, remove the securing screw and nut, and detach the wheel arch splash shield from the bumper.
7 Carefully pull the splash shield back from the bumper for access to the bumper side and lower fixings.
8 Unscrew the bumper side securing screw, and recover the washer (photo).
9 Loosen the bumper lower securing nut (there is no need to remove it) (photo).
10 Reach up behind the splash shield, and locate the front foglight wiring connector (photo). Separate the two halves of the connector.
11 Repeat the procedure given in paragraphs 6 to 9 on the left-hand side of the car.
12 Have an assistant support the bumper, then pull the sides of the bumper out from the front wings to release the securing clips.
13 Carefully withdraw the bumper from the car.
14 Refitting is a reversal of removal.

Rear bumper

15 Disconnect the battery negative lead.
16 To improve access, chock the front wheels, then jack up the rear of the car and

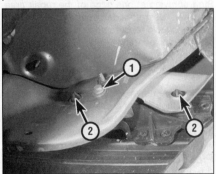

4.8 Front bumper side securing screw (1) and bumper locating clips (2) - wheel arch splash shield removed for clarity

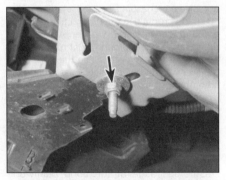

4.9 Front bumper lower securing nut (arrowed)

4.10 Front foglight wiring connector (arrowed) - wheel arch splash shield removed for clarity

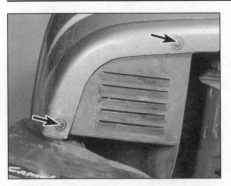

4.19A Left-hand rear wheel arch splash shield securing screws (arrowed)

4.19B Removing the rear wheel arch liner

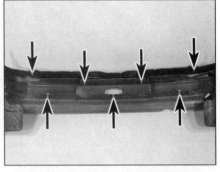

4.21 Rear bumper removed, showing securing nut locations (arrowed)

support it securely on axle stands (see *'Jacking, towing and wheel changing'*).

17 Working on the right-hand side of the car, remove the securing screw and nut, and remove the wheel arch splash shield.

18 Remove the now-exposed bumper side securing screw and washer.

19 Working on the left-hand side of the car, remove the two securing screws and the securing nut, and remove the wheel arch splash shield (photos).

20 Remove the now-exposed bumper side securing screw and washer.

21 Working under the car, remove the seven nuts (four upper and three lower) securing the bumper to the body (photo).

22 Have an assistant support the bumper, then pull the sides of the bumper out from the rear wings to release the locating clips (photo).

23 Carefully lower the bumper from the car. Locate the rear number plate light wiring connector, which is clipped to the body panel below the left-hand rear light cluster, and separate the two halves of the connector (photo).

24 Withdraw the bumper from the car.

25 Refitting is a reversal of removal.

5 Radiator grille panel - removal and refitting

1 Open the bonnet.

2 Carefully prise the covers from the grille panel securing screws, then remove the screws (photos).

3 Using a suitable screwdriver, carefully release the plastic securing clips (two at each end, and one in the centre), and pull the grille panel from the body (photo).

4 Refitting is a reversal of removal.

6 Windscreen cowl panels - removal and refitting

Note: *There are two windscreen cowl panels (one each side). Even if only one panel is to be removed, it is preferable to remove the two panels as an assembly, and then separate the two parts.*

1 Open the bonnet.

2 Remove the windscreen wiper arms, as described in Chapter 12.

4.22 Rear bumper locating clip (arrowed) released from wing

4.23 Separating the two halves of the rear number plate light wiring connector

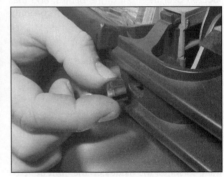

5.2A Prise out the covers . . .

5.2B . . . and remove the front grille panel securing screws

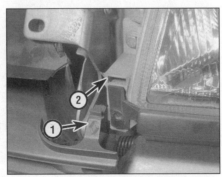

5.3 Front grille panel securing clip (1) released from headlight (2)

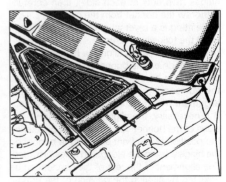

Fig. 11.2 Two of the windscreen cowl panel securing screws (arrowed) (Sec 6)

11

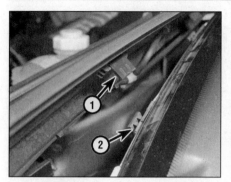

6.4 Windscreen cowl panel securing clip (1) released from body (2)

6.5 Releasing the weatherseal from the windscreen cowl panels

8.3 Bonnet lock assembly securing bolts (arrowed)

3 Remove the five cowl panel securing screws.
4 Carefully release the cowl panel securing clips (located at the base of the windscreen) by pulling the panels upwards (photo).
5 Release the weatherseal from the front edge of the panels, noting how it is located (photo), then withdraw the assembly. Disconnect the washer fluid hose connector as the assembly is withdrawn.
6 If desired, the two halves of the assembly can be unclipped from each other after removal.
7 Refitting is a reversal of removal, bearing in mind the following points.
8 If the two halves of the assembly have been separated, ensure that they are securely clipped together before refitting.
9 Ensure that the weatherseal is correctly located over the front of the panels, and that the panel securing clips are correctly engaged.
10 Refit the windscreen wiper arms with reference to Chapter 12.

7 Bonnet and hinges -
removal and refitting

Bonnet

1 Have an assistant support the bonnet in the open position.
2 If the original bonnet is to be refitted, mark the position of the hinges on the front edge of the bonnet, to aid alignment on refitting.
3 Remove the nuts securing the bonnet to the hinges (two nuts at each side), then carefully withdraw the bonnet from the car.
4 Refitting is a reversal of removal, bearing in mind the following points.
5 Where applicable, align the hinges with the marks made on the bonnet before removal.
6 Close the bonnet (carefully, in case it fouls the surrounding bodywork), and check the alignment with the surrounding body panels.
7 If necessary, the alignment of the front and sides of the bonnet can be adjusted by altering the position of the front edge of the bonnet on the hinges, using the elongated holes provided. The alignment of the rear of the bonnet can be adjusted by altering the position of the bonnet lock assembly on the bulkhead, using the elongated bolt holes provided.

Hinges

8 The bonnet hinges are welded to the body, and cannot easily be removed.

8 Bonnet lock components -
removal and refitting

Bonnet lock assembly

1 Open the bonnet.
2 Mark the position of the lock on the bulkhead, to aid correct alignment when refitting.
3 Unscrew the two securing bolts and recover the washers, then lift the lock assembly from the bulkhead (photo).
4 Remove the securing clip, and release the cable sheath from the lock.
5 Disconnect the end of the bonnet release cable from the lock operating lever, then withdraw the lock assembly.
6 Refitting is a reversal of removal, but align the assembly with the marks made on the bulkhead before removal.
7 On completion, if necessary adjust the alignment of the rear of the bonnet with the surrounding body panels by altering the position of the lock assembly, using the elongated bolt holes provided.

Bonnet lock striker

8 The bonnet lock striker is fixed to the bonnet, and cannot be removed or adjusted.

Bonnet lock release cable/lever assembly

9 Remove the driver's side lower facia panel, as described in Section 26. Note that, where applicable, there is no need to remove the headlight aim adjustment switch and the choke control assembly from the panel.
10 Working in the engine compartment, remove the securing clip, and release the cable sheath from the lock.
11 Disconnect the end of the bonnet release cable from the lock operating lever.
12 Working inside the car, carefully pull the cable through the bulkhead grommet into the car, noting the cable routing.
13 Refitting is a reversal of removal, but ensure that the bulkhead grommet is securely located in the bulkhead, and route the cable as noted during removal.

9 Doors and check straps -
removal, refitting and adjustment

Doors

1 To remove a door, open it fully, and support it under its lower edge on blocks covered with pads of rag.
2 Where applicable, disconnect the battery negative lead, then disconnect the door wiring connector from the door pillar. To disconnect the connector, slide the securing clip up, then pull the connector from its socket (photos).

9.2A Slide the securing clip up . . .

9.2B . . . and disconnect the door wiring connector

9.3 Door check strap securing screw (arrowed)

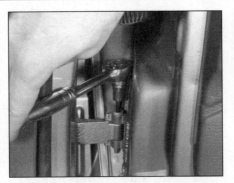

9.4 Loosening a front door hinge pin

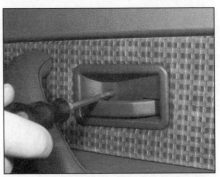

10.1A Remove the interior handle securing screw . . .

3 Using a suitable Torx bit, unscrew the door check strap from the body (photo).

4 Again using a suitable Torx bit, loosen the two hinge pins from the hinges, but do not remove the pins at this stage (photo).

5 Have an assistant support the door, then remove the hinge pins and withdraw the door from the car.

6 Refitting is a reversal of removal.

Adjustment

7 The door hinges are welded onto the door frame and the body pillar, so that there is no provision for adjustment or alignment.

8 Door closure may be adjusted by altering the position of the lock striker on the body pillar, using a suitable Torx bit.

Door check straps

9 Open the door fully.

10.1B . . . then disconnect the operating rod

10 Using a suitable Torx bit, unscrew the door check strap from the body.

11 Remove the door inner trim panel, as described in Section 24.

12 Working at the outer front edge of the door, unscrew the two nuts securing the check strap to the door.

13 Withdraw the check strap through the inside of the door.

14 Refitting is a reversal of removal.

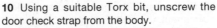

10 Door handle and lock components -
removal and refitting

Door interior handle

1 Remove the single screw securing the door interior handle to the door, then withdraw the handle and disconnect the operating rod from the rear of the handle by releasing the securing clip (photos).

2 Withdraw the handle from the door.

3 Refitting is a reversal of removal.

Front door exterior handle

4 Remove the door inner trim panel as described in Section 24.

5 Working inside the door, release the securing clips, and disconnect the exterior handle operating rod from the lock and the handle.

6 Unscrew the two exterior handle securing nuts from inside the door (photo), then tilt the handle and remove it from outside the door.

7 Refitting is a reversal of removal, but refit the door inner trim panel as described in Section 24.

Rear door exterior handle

8 Remove the door inner trim panel as described in Section 24.

9 Remove the door lock as described later in this Section.

10 Carefully prise the two plastic covers from the door skin for access to the exterior handle securing nuts (photo).

11 Working inside the door, release the securing clip and disconnect the exterior handle operating rod from the handle.

12 Unscrew the two exterior handle securing nuts from inside the door, then tilt the handle and remove it from outside the door.

13 Refitting is a reversal of removal, refitting the door lock as described later in this Section, and the door inner trim panel as described in Section 24.

Front door lock

14 Remove the door inner trim panel as described in Section 24.

15 Working inside the door, release the securing clips and disconnect the exterior handle, lock cylinder and lock button operating rods from the lock. Lift the lock button and operating rod from the door.

16 On models fitted with central locking, disconnect the battery negative lead, if not already done, and disconnect the wiring plug(s) from the lock operating motor (photo).

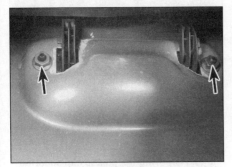

10.6 Front door exterior handle securing nuts (arrowed)

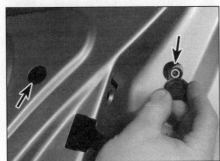

10.10 Rear door exterior handle securing nut locations (arrowed)

10.16 Disconnecting the wiring plugs from the driver's door lock operating motor

11

10.18 Removing a front door lock

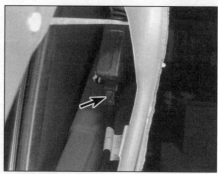

10.22 Disconnect the wiring plug (arrowed) from the lock operating motor

10.23 Unscrewing a rear door lock securing screw

17 Unscrew the three lock securing screws from the rear edge of the door, and recover the washers.

18 Working inside the door, withdraw the lock assembly, complete with the interior handle operating rod, through the door aperture (photo).

19 Refitting is a reversal of removal, refitting the door inner trim panel as described in Section 24.

Rear door lock

20 Remove the door inner trim panel, as described in Section 24.

21 Working inside the door, release the securing clips, and disconnect the exterior handle and lock button operating rods from the lock.

10.24 Withdrawing a rear door lock from the door

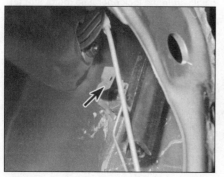

10.28A Remove the metal securing clip (arrowed) . . .

22 On models fitted with central locking, disconnect the battery negative lead, and disconnect the wiring plug from the lock operating motor (photo).

23 Unscrew the three lock securing screws from the rear edge of the door, and recover the washers (photo).

24 Working inside the door, withdraw the lock assembly, complete with the interior handle operating rod, through the door aperture (photo).

25 Refitting is a reversal of removal, refitting the door inner trim panel as described in Section 24.

Front door lock cylinder

26 Remove the door inner trim panel, as described in Section 24.

27 Working inside the door, disconnect the operating rod from the lock cylinder.

28 Slide the metal securing clip from the lock cylinder, then withdraw the lock cylinder from outside the door (photos).

29 Refitting is a reversal of removal, bearing in mind the following points.

30 When refitting the lock cylinder, ensure that the lugs on the assembly engage with the corresponding cut-outs in the door panel.

31 Refit the door inner trim panel as described in Section 24.

Lock striker

32 The lock striker is screwed into the door pillar on the body.

10.28B . . . and withdraw the lock cylinder

33 Before removing the striker, mark its position, so that it can be refitted in exactly the same position.

34 To remove the striker, simply unscrew the securing screw using a suitable Torx bit.

35 Refitting is a reversal of removal, but if necessary, adjust the position of the striker to achieve satisfactory closing of the door.

Central locking components

36 Refer to Section 14.

11 Door window glass and regulators - removal and refitting

Front door window glass

1 Remove the door inner trim panel, as described in Section 24.

2 Fully lower the window. On models with electric windows, temporarily reconnect the battery negative lead, and reconnect the wiring plug to the electric window operating switch, to enable the window to be lowered.

3 Carefully prise the weatherseal from the inner lower edge of the window aperture.

4 Working at the rear of the window glass, release the securing lugs and remove the plastic clip securing the window glass to the guide block (photo).

5 Carefully pull the guide block back to

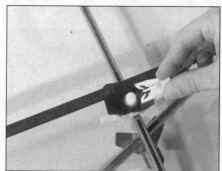

11.4 Removing the plastic clip securing the window glass to the guide block - assembly removed for clarity

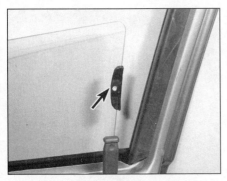

11.6 Guide plate (arrowed) on front door window glass

11.7 Removing the front door window glass

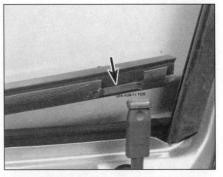

11.10 The cut-out for the lock operating button (arrowed) should be positioned at the rear of the weatherseal

disengage the lug from the hole in the glass, then support the glass.

6 Working inside the rear edge of the door, carefully disengage the guide plate (riveted to the rear upper edge of the glass) from the lower end of the weatherseal/guide channel inside the door (photo).

7 Carefully push the glass upwards, tilt the rear edge forwards, and manipulate the glass out through the inside of the window aperture (photo).

8 Refitting is a reversal of removal, bearing in mind the following points.

9 Ensure that the guide plate at the rear upper edge of the glass is correctly engaged with the weatherseal/guide channel inside the door.

10 When refitting the weatherseal to the lower edge of the window aperture, ensure that the seal is correctly orientated, with the cut-out for the lock operating button at the rear (photo).

11 Before refitting the door inner trim panel, check the operation of the window mechanism, reconnecting the battery and the electric windows operating switch, or refitting the window winder handle, as applicable.

12 Refit the door inner trim panel as described in Section 24.

Rear door fixed window glass

13 Remove the door inner trim panel, as described in Section 24.

14 Fully lower the sliding window glass.

15 Carefully prise the inner and outer weatherseals from the lower edge of the window aperture.

16 Temporarily refit the window winder handle. Raise the window until the screws securing the window glass to the regulator bracket are accessible.

17 If the original sliding glass is to be refitted, mark the positions of the screws securing the window glass to the regulator bracket, then remove the screws and carefully lower the window glass into the door (photo).

18 Remove the upper Torx screw, middle bolt and lower nut securing the window rear guide rail to the door. Support the fixed window glass, and carefully withdraw the

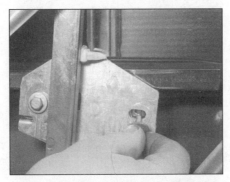

11.17 Removing a rear door window glass-to-regulator bracket securing screw

guide rail through the window aperture (photos).

19 Carefully slide the fixed window glass from its location (photo).

20 Refitting is a reversal of removal, bearing in mind the following points.

21 When refitting the window fixed glass, ensure that the glass engages correctly with the seals around the edges of the window aperture.

22 Where applicable, refit the screws securing the sliding window glass to the regulator bracket in the positions marked before removal.

23 Make sure that the weatherseals are correctly located on the lower edge of the window aperture (photos).

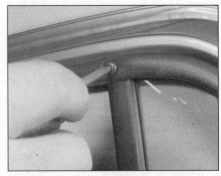

11.18A Remove the upper Torx screw . . .

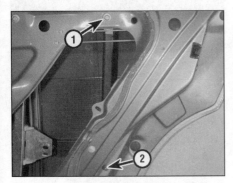

11.18B . . . the middle bolt (1) and the lower nut (2) . . .

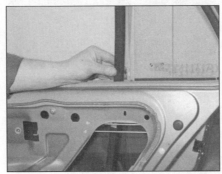

11.18C . . . then withdraw the window guide rail

11.19 Removing the fixed rear window glass

11

11.23A Ensure that the inner . . .

11.23B . . . and outer weatherseals are correctly located

11.27 Lifting the rear door sliding window glass from the door

24 Before refitting the door inner trim panel, temporarily refit the window winder handle, and check the operation of the regulator mechanism. If necessary, the alignment of the glass can be adjusted by altering the position of the screws securing the window glass to the regulator bracket, using the elongated holes provided.

25 Refit the door inner trim panel, as described in Section 24.

Rear door sliding window glass

26 Remove the fixed window glass as described earlier in this Section.

27 Carefully lift the sliding window glass out through the inside of the window aperture (photo).

28 Lower the window glass into the door, then refit the fixed window glass as described earlier in this Section.

Front door window regulator

29 Remove the door inner trim panel, as described in Section 24.

30 Lower the interior handle operating rod clear of the regulator assembly.

31 Fully lower the window. On models with electric windows, temporarily reconnect the battery negative lead, and reconnect the wiring plug to the electric window operating switch, to enable the window to be lowered.

32 Proceed as described in paragraphs 4 and 5.

33 Lift the window glass clear of the regulator assembly, and support it securely (for instance, by using strong adhesive tape to secure the glass to the top edge of the window aperture).

34 On models fitted with electric windows, disconnect the wiring plug from the electric window motor.

35 Remove the five screws securing the regulator assembly to the door (three screws securing the regulator, and the upper and lower screws securing the window guide rail) (photo).

36 Carefully manipulate the regulator assembly out through the rear aperture in the door (photo).

37 Refitting is a reversal of removal, bearing in mind the following points.

38 Before refitting the door inner trim panel, check the operation of the window mechanism, reconnecting the battery and the electric windows operating switch, or refitting the window winder handle, as applicable.

39 Refit the door inner trim panel as described in Section 24.

Rear door window regulator

40 Remove the door inner trim panel, as described in Section 24.

41 Temporarily refit the window winder handle. Lower the window glass until the screws securing the window glass to the regulator bracket are accessible.

42 Mark the positions of the screws securing the window glass to the regulator bracket, then remove the screws and raise the window glass. Support the window glass securely in the fully-raised position (for instance, by using strong adhesive tape to secure the glass to the top edge of the window aperture).

43 Reach up inside the door, and disconnect the lock operating rod from the lock button bellcrank at the front of the door. Note how the operating rod is routed through the regulator mechanism.

44 Remove the four securing screws (two screws securing the regulator, and the upper and lower screws securing the window guide rail), then carefully manipulate the regulator assembly out through the door aperture (photos).

45 Refitting is a reversal of removal, bearing in mind the following points.

46 When reconnecting the lock operating rod to the bellcrank, ensure that the rod is correctly routed around the regulator assembly, as noted before removal.

47 Ensure that the screws securing the sliding window glass to the regulator bracket

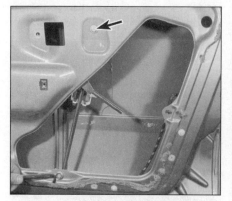

11.35 Removing the lower window guide rail screw from the front door - upper guide rail securing screw arrowed

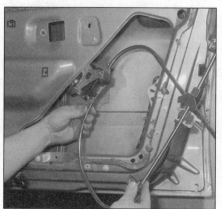

11.36 Removing the front door window regulator assembly

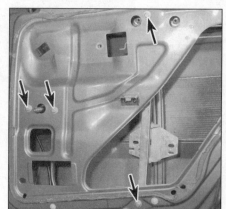

11.44A Rear door window regulator assembly securing screws (arrowed)

11.44B Removing the rear door window regulator mechanism

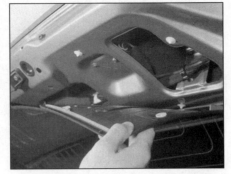

12.3 Removing the tailgate inner trim panel

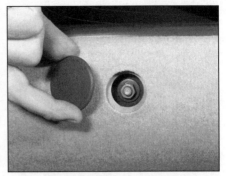

12.7 Removing a tailgate securing nut cover

are refitted in their original positions marked before removal.

48 Before refitting the door inner trim panel, temporarily refit the window winder handle, and check the operation of the regulator mechanism. If necessary, the alignment of the glass can be adjusted by altering the position of the screws securing the window glass to the regulator bracket, using the elongated holes provided.

49 Refit the door inner trim panel as described in Section 24.

Electric window components

50 Refer to Section 15.

12 Tailgate, hinges and support struts - removal and refitting

Tailgate

1 Disconnect the battery negative lead.
2 Open the tailgate.
3 Carefully pull the inner trim panel from the tailgate (photo). If necessary, release the trim clips using a suitable forked tool.
4 Disconnect the wiring from the heated rear window, tailgate wiper motor, and central locking motor, as applicable. Disconnect the washer fluid hose, where applicable; be prepared for fluid spillage.
5 Release the wiring loom and hose grommets from the edge of the tailgate. Tie

string to the wiring loom and hose, and pull them through the edge of the tailgate. Leave the string in position in the tailgate, to aid refitting.
6 Have an assistant support the tailgate. Disconnect the support struts from the tailgate by prising out the retaining clips using a screwdriver.
7 Prise the plastic covers from the headlining for access to the tailgate securing nuts (photo).
8 If the original tailgate is to be refitted, mark the positions of the securing nuts on the hinges.
9 Unscrew the nuts, and carefully withdraw the tailgate from the vehicle.
10 Refitting is a reversal of removal, but where applicable, fit the tailgate securing nuts in their original positions, as noted before removal. Do not fully tighten the tailgate securing nuts until the top of the tailgate is aligned correctly with the roof panel and the rear body pillars.
11 On completion, if adjustment of the tailgate lower edge alignment is required, remove the securing clip and unclip the plastic cover for access to the lock striker on the lower body panel.
12 Alter the position of the lock striker on the body (using screws 'A' in Fig. 11.3 in their elongated holes) to give correct alignment of the lower edge of the tailgate with the surrounding body panels.

13 Once the tailgate is correctly aligned, if necessary adjust the closure using the adjustment screw 'B' in Fig. 11.3.

Hinges

14 The tailgate hinges are welded to the tailgate, and the hinge pins are an interference fit in the hinges. The hinges cannot therefore easily be removed.

Support struts

15 Open the tailgate, and have an assistant support it in the fully-open position.
16 Disconnect the support strut from the tailgate by prising out the securing clip using a suitable screwdriver (photo).
17 Repeat the procedure for the clip securing the strut to the body, and withdraw the strut from the vehicle.
18 Refitting is a reversal of removal.

13 Tailgate lock components - removal and refitting

Lock

1 Open the tailgate, then carefully pull the inner trim panel from the tailgate. If necessary, release the trim clips using a suitable forked tool.
2 Unscrew the two lock securing screws and recover the washers (photo). Withdraw the lock from its location in the tailgate. Turn the

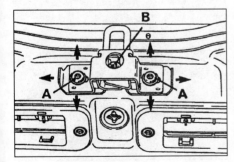

Fig. 11.3 Tailgate lock striker assembly (Sec 12)

A Striker securing screws
B Striker adjustment screw

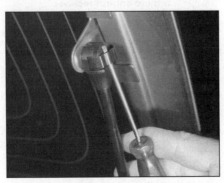

12.16 Prising out a tailgate support strut securing clip

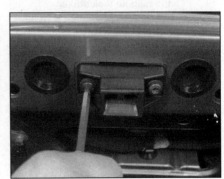

13.2 Unscrewing a tailgate lock securing screw

11

13.7A Remove the securing clip (arrowed) from the rear of the lock barrel . . .

13.7B . . . then withdraw the assembly

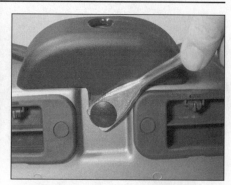

13.10A Remove the securing clip (note the use of a forked tool) . . .

lock as required to disconnect the operating rod.

3 Refitting is a reversal of removal.

Lock barrel assembly

4 Remove the inner trim panel as described in paragraph 1.

5 Reach up through the tailgate aperture, and release the securing clip(s). Disconnect the lock operating rod, and where applicable the central locking motor operating rod, from the lock barrel assembly.

6 Disconnect the water drain hose from the spout at the lower edge of the tailgate.

7 Remove the metal securing clip from the rear of the lock barrel, then withdraw the lock barrel assembly from outside the tailgate (photos).

8 Refitting is a reversal of removal. Ensure that the water drain hose is securely reconnected at both ends.

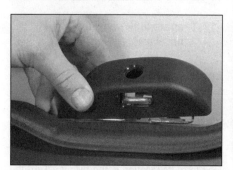

13.10B . . . and withdraw the cover from the tailgate lock striker

14.6 Removing the ashtray housing from the centre console

Lock striker

9 The lock striker is attached to the lower body panel.

10 Remove the securing clip (using a forked tool if necessary), and withdraw the plastic cover from the lock striker assembly (photos).

11 Remove the two securing screws now exposed, recover the washers, and withdraw the lock striker assembly.

12 Refitting is a reversal of removal, but if necessary adjust the position of the striker (by means of the elongated screw holes) to achieve satisfactory alignment of the bottom of the tailgate with the surrounding body panels.

13 On completion, if necessary adjust screw 'B' in Fig. 11.3 to achieve satisfactory closing of the tailgate.

Central locking motor

14 Refer to Section 14.

14.2 Withdrawing the central locking remote control receiver

14.7 Removing the central locking switch from the centre console

14 Central locking components - removal and refitting

Remote control receiver

1 Disconnect the battery negative lead.

2 Carefully grip the front of the receiver unit, and pull it from its location in the roof console (photo).

3 Disconnect the wiring plug and remove the unit.

4 Refitting is a reversal of removal.

Driver's door operating switch

5 The switch is integral with the motor assembly, and cannot be removed independently.

Centre console-mounted switch

6 Pull the ashtray and housing from the centre console (photo). Working through the aperture, push the cigarette lighter/switch mounting panel from the centre console.

7 Disconnect the wiring plugs from the switch, then push the switch out through the top of the panel (photo).

8 Refit by reversing the removal operations.

Front door lock operating motor

9 Remove the door lock, as described in Section 10. Remove the clip securing the lock operating rod to the motor operating rod and the lock lever (photo).

14.9 Removing the clip securing the lock operating rod to the front door central locking motor operating rod and lock lever. Motor securing screw arrowed

14.13 Driving out the roll pin securing the rear door central locking motor to the lock assembly

10 Remove the single screw securing the motor to the lock assembly, then withdraw the motor.

11 Refitting is a reversal of removal. Refit the door lock as described in Section 10.

Rear door lock operating motor

12 Remove the door lock as described in Section 10.

13 Using a suitable pin punch, drive out the roll pin securing the motor to the lock assembly (photo).

14 Disconnect the motor operating arm from the lock, then carefully slide the motor from the locating rail on the lock assembly.

15 Refitting is a reversal of removal. Refit the door lock as described in Section 10.

Tailgate lock operating motor

16 Disconnect the battery negative lead.

17 Open the tailgate. Release the securing clips (preferably using a suitable forked tool) and remove the tailgate interior trim panel.

18 Reach up through the tailgate aperture, and disconnect the motor operating rod from the lock.

19 Remove the single screw securing the motor to the tailgate. Prise off the securing star-clip, taking care not to damage the motor or the tailgate (photo).

20 Withdraw the motor from the tailgate, and disconnect the wiring plug (photo).

21 Refitting is a reversal of removal.

16.2 Removing the mirror trim panel

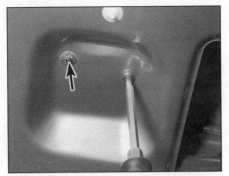

14.19 Removing the tailgate central locking operating motor securing screw - star-clip arrowed

15 Electric window components - removal and refitting

Operating switches

1 Disconnect the battery negative lead.

2 Unscrew the four securing screws, and withdraw the door pocket from the door (photo).

3 Reach behind the door pocket, release the securing clips, and push the switch assembly out through the top of the door pocket (photo).

4 Disconnect the wiring plug(s) and withdraw the switch assembly.

5 Refitting is a reversal of removal.

15.2 Withdrawing the door pocket from the door

16.3A Unscrewing an adjuster housing securing screw - manually-adjustable mirror

14.20 Removing the tailgate central locking operating motor

Operating motors

6 The motors are an integral part of the window regulator assemblies, and cannot be renewed independently.

7 Removal and refitting of the regulator assemblies is described in Section 11.

16 Mirrors - removal, refitting and glass renewal

Door mirror

1 On models fitted with electric mirrors, disconnect the battery negative lead.

2 Carefully prise the mirror trim panel from the front edge of the door (photo).

15.3 Disconnecting an electric window switch wiring plug

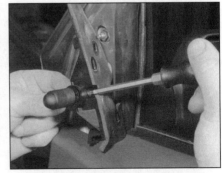

16.3B Unscrewing the adjuster lever clamp screw - manually-adjustable mirror

11

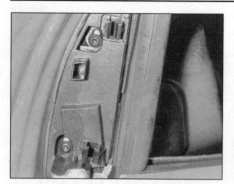

16.4A Mirror cover panel securing screws - electrically-operated mirror

16.4B Removing the mirror cover panel to expose the wiring plug - electrically-operated mirror

16.5 Removing the cover from the mirror lower securing nut

3 On models fitted with manually-adjustable mirrors, unscrew the three adjuster housing securing screws now exposed. Unscrew the clamp screw securing the adjuster lever to the housing, and remove the housing from the door (photos).

4 On models fitted with electrically-operated mirrors, unscrew the two screws securing the mirror cover panel to the door, then unplug the wiring plug(s) from the mirror (photos).

5 Prise the plastic cover from the front edge of the door to expose the mirror front lower securing nut (photo).

6 Unscrew the three mirror securing nuts, and withdraw the mirror from the door. On models with manually-adjustable mirrors, feed the adjuster cable and lever through the aperture

in the door as the mirror is removed (photo).

7 Refitting is a reversal of removal, but on models fitted with manually-operated mirrors, when refitting the adjuster lever to the housing, proceed as follows.

8 Push the adjuster into the housing until the rear lug on the adjuster engages with the slot in the housing, and the front of the adjuster is flush with the end of the housing (photo). Refit and tighten the clamp screw.

Glass renewal - mirror with glass retained by wire clip

9 Using a thin-bladed screwdriver, carefully probe under the edge of the mirror glass, and locate the ends of the wire clip retaining the glass.

10 Support the glass.

11 Push one of the ends of the clip towards the centre of the mirror, then back into the mirror body, to release the glass (photos). Take care not to drop the mirror glass as the clip is released.

12 To refit the glass, re-engage the metal clip with the lugs on the rear of the glass, then push the glass into position in the mirror until the clip engages securely.

Glass renewal - mirror with glass retained by locking ring

13 Using a thin-bladed screwdriver, probe through the hole provided in the bottom of the mirror body, and locate the serrated portion of the locking ring securing the glass.

14 Carefully turn the locking ring, using the screwdriver, until the glass is released.

15 To refit the glass, locate the glass in the mirror using the four guides (C) shown in Fig. 11.4.

16 Insert the screwdriver through the hole in the mirror body, and turn the locking ring in the opposite direction to that used for removal, to lock the glass in position.

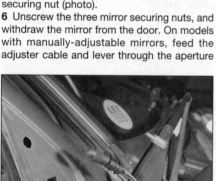

16.6 Withdrawing a manually-adjustable mirror

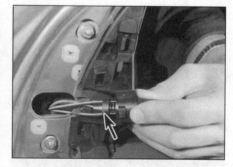

16.8 Ensure that the rear lug (arrowed) on the adjuster engages with the slot in the housing

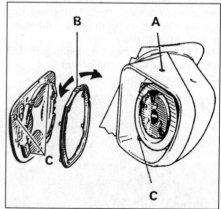

Fig. 11.4 Door mirror glass components - mirror with glass retained by locking ring (Sec 16)

A Access hole (underside of mirror)
B Locking ring serrated portion
C Glass locating guides

16.11A Release the mirror glass retaining clip . . .

16.11B . . . and withdraw the glass

Electric mirror operating switch removal and refitting

17 The switch is located in the driver's door armrest, with the electric window switches.

18 The procedure is as described for the electric window switches in Section 15.

Electric mirror motor removal and refitting

19 The motor is not available separately from the mirror. If it is faulty, the complete mirror assembly must be renewed.

Interior mirror

20 The mirror can be removed by carefully pulling it from the balljoint on the mounting bracket attached to the windscreen.

21 The mounting bracket is fixed to the windscreen using a special adhesive, and should not be disturbed unless absolutely necessary. Note that there is a risk of cracking the windscreen glass if an attempt is made to remove a securely-bonded mounting bracket.

22 Refitting is a reversal of removal. If necessary, the special adhesive required to fix the mounting bracket to the windscreen can be obtained from a Renault dealer.

17 Windscreen and tailgate window glass - general information

Windscreen

1 The windscreen is bonded in position using a special adhesive.

2 Special tools, adhesives and expertise are required for successful removal and refitting of glass fixed by this method. Such work must therefore be entrusted to a Renault dealer, a windscreen specialist, or other competent professional.

Tailgate window glass

Removal

3 The tailgate window glass is secured by the clamping action of the weatherseal.

4 To remove the glass, first remove the tailgate trim panel; pull the panel down to

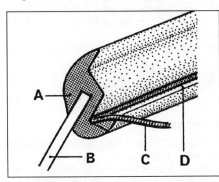

Fig. 11.5 Cross-section of the tailgate window weatherseal, with cord inserted for refitting (Sec 17)

A *Weatherseal* C *Refitting cord*
B *Glass* D *Groove*

release the securing clips, or use a suitable forked tool.

5 Disconnect the battery negative lead, then disconnect the wiring plugs from the heated rear window element.

6 Have an assistant support the window glass from outside the car.

7 Using a blunt flat-bladed screwdriver, carefully prise the weatherseal back from the flange on the tailgate. Withdraw the glass from outside the car.

Refitting

8 Fit the weatherseal to the edge of the glass.

9 Engage a cord in the groove in the weatherseal, as shown in Fig. 11.5, so that the ends of the cord emerge from the groove at the centre of the bottom edge of the glass. Overlap the ends of the cord by approximately 200 mm.

10 Offer the glass up to the tailgate, passing the ends of the cord through inside the tailgate.

11 Slowly pull one of the ends of the cords to seat the weatherseal over the flange on the tailgate.

12 Have an assistant apply gentle pressure to the glass from outside the car, to aid withdrawal of the cord.

13 When the cord reaches the centre of the upper edge of the glass, repeat the operation with the remaining end of the glass.

14 Check that the glass is secure and that the weatherseal is correctly seated, then reconnect the heated rear window wiring, and refit the tailgate trim panel.

18 Rear quarter window components - removal and refitting

Fixed rear quarter window

Removal

1 The window glass is secured by the clamping action of the weatherseal.

2 Remove the B-pillar trim panel, the roof side trim panel, and the rear side trim panel, as described in Section 24.

3 Have an assistant support the window glass from outside the car.

4 Using a blunt flat-bladed screwdriver,

19.2 Removing the cover from a sunroof glass panel-to-hinge securing screw

carefully prise the window weatherseal back from the flange on the body, then withdraw the glass from outside the car.

Refitting

5 The procedure is as described for the tailgate window glass in Section 17, paragraphs 8 to 13 inclusive.

6 Check that the glass is secure and that the weatherseal is correctly seated, then refit the interior trim panels.

Opening rear quarter windows

7 Open the window, then remove the screw securing the window glass to the catch.

8 Have an assistant support the glass. Carefully unscrew the two screws securing the glass to the hinges, and lift the glass from the car.

9 Refitting is a reversal of removal.

Opening rear quarter window catch

10 Open the window, and remove the screw securing the catch to the glass.

11 Remove the two screws securing the catch to the body, then withdraw the catch.

12 Refitting is a reversal of removal.

19 Sunroof components - removal and refitting

Glass panel - models with tilting glass panel

1 Open the sunroof, and remove the screw(s) securing the handle assembly to the glass panel.

2 Hold the panel in the open position, then remove the plastic covers from the screws securing the panel to the hinges (photo).

3 Remove the hinge screws, and lift the panel from the roof.

4 Refitting is a reversal of removal.

Sunroof handle - models with tilting glass panel

5 Open the sunroof, and remove the screw(s) securing the handle assembly to the glass panel.

6 Remove the two securing screws, and withdraw the handle trim panel from the roof (photo).

19.6 Removing a sunroof handle trim panel securing screw

11

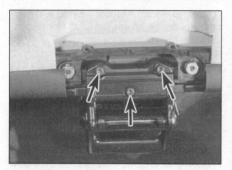

19.7 Sunroof handle-to-roof securing screws (arrowed)

7 Remove the three screws securing the handle assembly to the roof, then withdraw the assembly (photo).
8 Refitting is a reversal of removal.

Tilt/slide sunroof components

9 No information was available for the tilt/slide sunroof components at the time of writing. This type of sunroof is a complex piece of equipment, consisting of a large number of components. It is strongly recommended that the sunroof mechanism is not disturbed unless absolutely necessary. If the sunroof mechanism is faulty, or requires overhaul, consult a Renault dealer for advice.

20 Body exterior fittings - removal and refitting

Engine and wheel arch splash shields

1 Various plastic shields may be fitted to the wheel arches and various engine components to protect against road dirt and moisture.
2 The shields are secured by a combination of plastic clips, screws, nuts and pop-rivets (photo). Removal and refitting should be self-evident. Take particular care not to break plastic clips when removing them. Renew any pop-rivets on refitting where necessary.

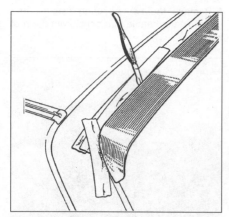

Fig. 11.6 Adhesive tape applied to protect tailgate when removing spoiler - 16-valve models (Sec 20)

20.2 Drilling out a front wheel arch splash shield upper securing rivet

Door rubbing strips

3 Open the door, and remove the plastic cover from the rear edge of the door to expose the rubbing strip securing screw (photo). Remove the screw.
4 Working outside the door, unclip the rubbing strip by pushing it towards the rear of the door.
5 If desired, the rubbing strip securing clips can be removed from the door by turning them one quarter of a turn anti-clockwise with pliers.
6 To refit a rubbing strip, align the holes in the strip with the corresponding clips in the door. Push the strip towards the front of the door to engage the clips.
7 On completion, refit the securing screw and its plastic cover.

Badges

Removal

8 The various badges are secured with adhesives. To remove them, either soften the adhesive using a hot air gun or hairdryer (taking care to avoid damage to the paintwork), or separate the badge from the body by 'sawing' through the adhesive using a length of nylon cord.

Refitting

9 Clean off all traces of adhesive using white spirit, then wash the area with warm soapy water to remove all traces of spirit, and allow to dry. Ensure that the surface to which the new badge is to be fastened is completely clean, and free from grease and dirt.
10 Use the hot air gun to soften the adhesive

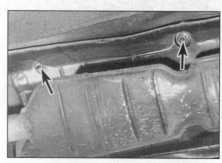

21.3 Front seat inner rail securing nuts (arrowed) - note exhaust system

on the new badge, then press it firmly into position.

Tailgate spoiler (16-valve models)

11 Apply adhesive tape to the tailgate area around the edges of the spoiler, to protect the paint during the following procedure.
12 Using a sharp blade, carefully cut around the edge of the spoiler to release the double-sided adhesive tape securing the spoiler to the tailgate.
13 Remove the two spoiler securing screws (one under each end of the spoiler).
14 Carefully release the spoiler from its retaining clips by sliding it upwards from the tailgate, starting at the ends.
15 Commence refitting by cleaning all remaining adhesive tape from the tailgate, taking care not to damage the paint.
16 If the original spoiler is to be refitted, clean off all old adhesive tape, and apply new tape to the relevant areas.
17 Remove the protective backing from the adhesive tape, then slide the spoiler into position on its retaining clips.
18 Press the spoiler onto the tailgate, ensuring that the adhesive tape holds the spoiler firmly in position.
19 Refit the spoiler securing screws.

Bonnet air duct (16-valve models)

20 Open the bonnet and unscrew the five nuts securing the inner section of the duct.
21 Unscrew the single securing nut, and withdraw the main grille assembly from outside the bonnet.
22 Refitting is a reversal of removal.

21 Seats - removal and refitting

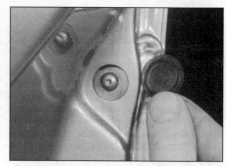

20.3 Removing the cover from a door rubbing strip securing screw

Front seat

1 If desired, remove the seat side trim panels for improved access.
2 Working inside the car, remove the two bolts securing the seat outer rail to the floor.
3 Working underneath the car, unscrew the two nuts securing the seat inner rail to the floor (photo). Depending on which seat is being removed, take care not to burn yourself if the exhaust is still hot.

22.2A Prise off the cover . . .

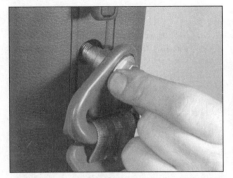

22.2B . . . and remove the upper seat belt mounting bolt

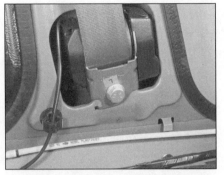

22.3 Footwell side/sill trim panel removed to expose front seat belt inertia reel assembly

4 Carefully lift the seat from the vehicle.
5 Refitting is a reversal of removal.

Rear seat cushion

6 Tilt the seat cushion forwards, then lift it to disengage the securing lugs from the holes in the floor.
7 Refitting is a reversal of removal.

Rear seat back

8 Release the securing catches at the top of the seat backs. Fold the seat backs forwards.
9 Working at the lower edges of the seat backs, release the catches securing the seat backs to the body, then withdraw the assembly from the car.
10 Refitting is a reversal of removal.

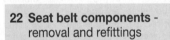

22 Seat belt components -
removal and refittings

 Warning: If the vehicle has been in an accident in which structural damage was sustained, all the seat belt components must be renewed.

Front seat belt

1 Remove the front footwell side/sill trim panel, as described in Section 24.
2 Prise off the cover and unscrew the seat belt upper mounting bolt (photos).

3 Unscrew the mounting bolt, and withdraw the inertia reel assembly from the door pillar (photo). Remove the seat belt assembly from the car.
4 Refitting is a reversal of removal. Tighten the seat belt mounting bolts to the specified torque.

Front seat belt height adjuster

5 Remove the B-pillar trim panel, as described in Section 24.
6 Unscrew the two securing bolts, and withdraw the adjuster from the pillar (photo).
7 Refitting is a reversal of removal. Ensure that the B-pillar trim panel locates correctly over the door aperture weatherseals. Tighten the seat belt mountings to the specified torque.

Rear inertia reel seat belt

8 Prise off the cover, and unscrew the seat belt upper mounting bolt. Note the location of the spacer ring on the bolt (photos).
9 Remove the rear seat cushion, with reference to Section 21, for access to the seat belt lower mounting bolt. Unscrew the bolt.
10 Unclip the seat belt surround from the rear quarter trim panel.
11 Working in the luggage compartment, pull back the side trim to expose the inertia reel mounting bolt (photo).
12 Unscrew the securing bolt and withdraw

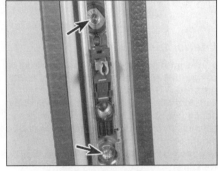

22.6 Front seat belt height adjuster securing bolts (arrowed)

the inertia reel from its location. Feed the seat belt through the hole in the rear quarter trim panel.
13 The seat belt buckle can be unbolted from the floor, after removing the rear seat cushion for access.
14 Refitting is a reversal of removal. Tighten the seat belt mountings to the specified torque.

Rear lap seat belt

15 The seat belt components can be removed, after removing the rear seat cushion for access to the securing bolts.
16 Refitting is a reversal of removal. Tighten the securing bolts to the specified torque.

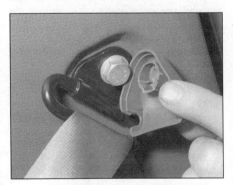

22.8A Prise off the cover . . .

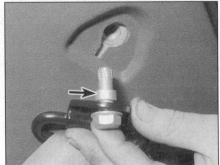

22.8B . . . then unscrew the seat belt upper mounting bolt. Spacer ring arrowed

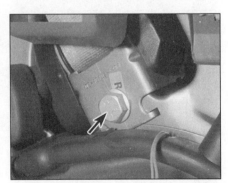

22.11 Luggage compartment side trim pulled back to expose seat belt inertia reel mounting bolt (arrowed)

11

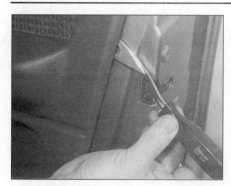

23.4 Suitable forked tool for releasing plastic trim panel securing clips

23 Interior trim -
general information

Interior trim panels

1 The interior trim panels are all secured using either screws or various types of plastic fasteners.

2 Before removing a panel, study it carefully, noting how it is secured. Often, other panels or ancillary components (such as seat belt mountings, grab handles, etc) must be removed before a particular panel can be withdrawn.

3 Once any such components have been removed, check that there are no other panels overlapping the one to be removed. Usually, the sequence to be followed will become obvious on close inspection.

4 Remove all obvious fasteners, such as screws, many of which may have plastic covers fitted. If the panel cannot be freed, it is probably secured by hidden clips or fasteners on the rear of the panel. Such fasteners are usually situated around the edge of the panel, and can be prised up to release them. Note that plastic clips can break quite easily, so it is advisable to have a few replacement clips of the correct type available for refitting. Generally, the best way of releasing such clips is to use a suitable forked tool (photo). If this is not available, an old, broad-bladed screwdriver with the edges rounded-off and

wrapped in insulating tape will serve as a good substitute.

5 The following Section and the accompanying illustrations and photographs describe removal and refitting of all the major trim panels. Note that the type and number of fasteners used often varies during the production run of a particular model, so differences may be noted to the procedures provided for certain vehicles.

6 When removing a panel, **never** use excessive force, or the panel may be damaged. Always check carefully that all fasteners have been removed or released before attempting to withdraw a panel.

7 When refitting, secure the fasteners by pressing them firmly into place. Ensure that all disturbed components are correctly secured, to prevent rattles. If adhesives were found at any point during removal, use white spirit to remove the old adhesive, then wash off the white spirit using soapy water. Use a suitable trim adhesive (a Renault dealer should be able to recommend a proprietary product) on reassembly.

Carpets

8 The passenger compartment floor carpet rests on the floor, and is held in position by the sill trim panels and other surrounding panels and components.

9 Carpet removal and refitting is reasonably straightforward, but very time-consuming, due to the fact that many of the adjoining trim panels must be removed first. It will also be necessary to remove components such as the seats and their mountings, the centre console, etc.

Headlining

10 The headlining is clipped to the roof, and can be withdrawn after all fittings such as the grab handles, sun visors, sunroof trim, door pillar trim panels, rear quarter trim panels, weatherseals, etc, have been removed or prised clear.

11 Note that headlining removal requires considerable skill and patience if it is to be carried out without damage, and is therefore best entrusted to an expert.

24 Interior trim panels -
removal and refitting

Front door inner trim panel

1 On models fitted with central locking, disconnect the battery negative lead.

2 On models fitted with manually-operated windows, note the position of the window winder handle with the window fully open, then release the securing clip and remove the handle. To release the securing clip, make up a forked tool similar to that shown in Fig. 11.7; alternatively, insert a length of wire with a hooked end between the handle and the trim panel, and manipulate it to free the securing clip from the handle. Take care not to damage the trim panel.

3 Remove the single screw securing the door interior handle to the door. Withdraw the handle and disconnect the operating rod from the rear of the handle by releasing the securing clip.

4 Withdraw the handle from the door.

5 Prise the cover from the door grab handle to expose the two securing screws (photo).

6 Remove the securing screws and withdraw the grab handle from the door (photo).

7 Remove the securing screws and withdraw the door pocket from the door. On models fitted with electric windows, reach up behind the door pocket, and disconnect the wiring plugs from the electric window switches. Also disconnect the wiring plug from the electric mirror switch, where applicable.

8 Where applicable, remove the four securing screws and withdraw the loudspeaker cover, loudspeaker and loudspeaker housing from the door. If necessary, remove the mirror trim panel and adjuster housing (see Section 16).

9 Carefully release the securing clips around the edge of the trim panel, preferably using a forked tool. Pull the trim panel from the door, noting that it is secured with sealing compound around its edge (photo).

10 Refitting is a reversal of removal, bearing in mind the following points.

11 When refitting the panel to the door, ensure that the sealing compound provides a

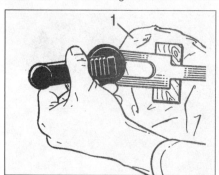

Fig. 11.7 Using a forked tool to release the window regulator handle (Sec 24)

1 Protective cloth

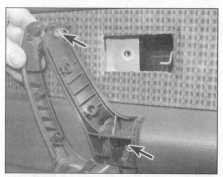

24.5 Prise the cover from the door grab handle to expose the securing screws (arrowed)

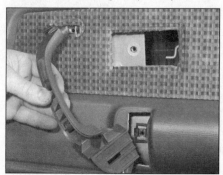

24.6 Withdrawing the front door grab handle

24.9 Pulling the trim panel from the front door

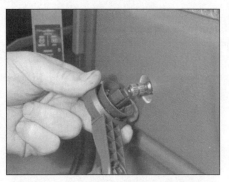

24.13 Removing the rear door window winder handle

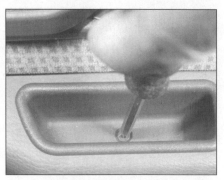

24.14 Removing the tray securing screw from the rear door armrest

24.16 Releasing a rear door ashtray housing securing clip - note the use of tape to protect the trim panel

good seal between the door and the panel. If necessary, apply new sealing compound of a suitable type (available from a Renault dealer).
12 Where applicable, refit the window winder handle in the position noted before removal.

Rear door inner trim panel

Removal

13 Proceed as described in paragraphs 1 to 4 (photo).
14 Remove the securing screw, then withdraw the tray from the armrest (photo).
15 On models fitted with a door bin, unscrew the three securing screws and remove the door bin.
16 On models fitted with an ashtray, pull the ashtray from the housing, then release the housing securing clips (located at the bottom of the housing) using a screwdriver (photo). Pull the ashtray housing from the door.
17 Carefully release the securing clips around the edge of the trim panel, preferably using a forked tool. Pull the trim panel from the door, noting that it is secured with sealing compound around its edge.

Refitting

18 Proceed as described in paragraphs 10 to 12.

Front footwell side/sill trim panel

19 On 3-door models, prise off the plastic

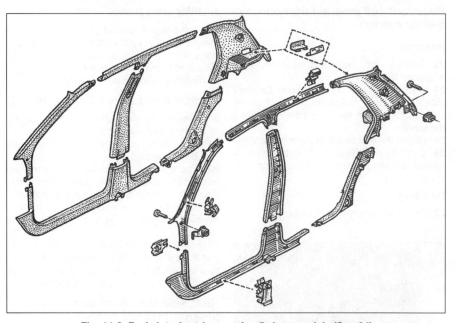

Fig. 11.8 Body interior trim panels - 5-door models (Sec 24)

cover and unscrew the bolt securing the seat belt mounting rail (photo). Note the position of any washers and/or spacers. Carefully manipulate the rail to release the rear end from the trim panel.
20 On 5-door models, unbolt the front seat belt lower mounting (photo).

21 Carefully pull the panel from the body to release the securing clips, and withdraw the panel from the sill (photo).
22 Refitting is a reversal of removal. Ensure that the securing clips are correctly engaged, and tighten the seat belt mounting bolt to the specified torque.

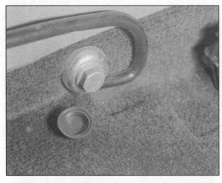

24.19 Cover removed to expose seat belt mounting rail bolt - 3-door model

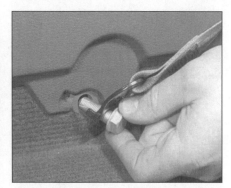

24.20 Removing the front seat belt lower mounting bolt - 5-door model

11

24.21 Removing a front footwell/side trim panel - front seat removed for clarity

24.26 Removing the rear footwell side/sill trim panel - 5-door model

24.32A Prise the clip from the seat belt height adjustment knob . . .

Rear footwell side/sill trim panel - 5-door models

Removal

23 Remove the rear seat cushion, with reference to Section 21.

24 Unclip the rear of the front footwell side/sill trim panel to expose the rear footwell side/sill trim panel securing screw. Remove the screw.

25 Unbolt the rear seat belt lower mounting.

26 Carefully pull the panel from the body to release the securing clips, and withdraw the panel (photo).

Refitting

27 Proceed as described in paragraph 22.

24.32B . . . then pull the knob from the adjuster

A-pillar trim panel

28 Remove the securing screw from the lower edge of the panel, then carefully pull the panel from the pillar to release the securing clips, and withdraw the panel.

29 Refitting is a reversal of removal.

B-pillar trim panel - 5-door models

30 Remove the footwell side/sill trim panels as described earlier in this Section.

31 Prise off the cover, and unbolt the upper seat belt mounting.

32 Where applicable, prise the centre clip from the seat belt height adjustment knob, then pull the knob from the adjuster (photos).

33 Carefully pull the panel from the body to release the securing clips, and withdraw the panel.

34 Refitting is a reversal of removal. Ensure that the panel locates correctly over the door aperture weatherseals, and tighten the seat belt mountings to the specified torque.

Rear side trim/B-pillar trim panel (3-door models)

35 Proceed as described in paragraphs 30 to 32.

36 Remove the rear seat cushion, with reference to Section 21.

37 Unbolt the rear seat belt lower mounting.

38 Carefully release the securing clips around the edge of the trim panel, preferably using a forked tool, then withdraw the panel.

39 Refitting is a reversal of removal. Tighten

the seat belt mountings to the specified torque.

Roof side trim panel

40 Remove the B-pillar trim panel, as described previously in this Section.

41 Carefully pull the roof side trim panel from the body to release the securing clips, and withdraw the panel (photo).

42 Refitting is a reversal of removal. Refit the B-pillar trim panel as described previously in this Section.

Rear quarter trim panel

43 Prise off the plastic cover and unbolt the rear seat belt upper mounting. Note the position of the spacer on the bolt.

44 Lower the rear seat back.

45 Using a suitable Torx bit, unscrew the rear seat back securing lug (photo).

46 Unclip the seat belt surround from the rear quarter trim panel, then feed the seat belt through the hole in the panel into the luggage compartment.

47 Disconnect the battery negative lead, then disconnect the wiring from the loudspeaker, luggage compartment light and light switch, as applicable.

48 Working in the luggage compartment, unscrew the two upper trim panel securing screws from the rear edge of the panel (photo).

49 Pull back the trim from the side of the luggage compartment to expose the two

24.41 Removing the roof side trim panel

24.45 Removing the rear seat back securing lug

24.48 Unscrewing a rear quarter trim panel upper securing screw

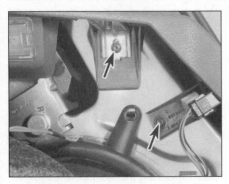

24.49 Rear quarter trim panel lower securing screws (arrowed)

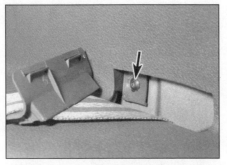

24.50 Rear quarter trim panel securing screw (arrowed) accessible through seat belt aperture

24.51 Removing the rear quarter trim panel

lower securing screws (photo). Note that one of the screws may also secure the luggage compartment light switch. Remove the screws.

50 Unscrew the remaining trim panel securing screw, which is accessible through the hole exposed by removal of the seat belt surround (photo).

51 Carefully pull the panel from the body to release the upper securing clip, and withdraw the panel from the car (photo).

52 Refitting is a reversal of removal. Tighten the seat belt mounting bolt to the specified torque.

Roof console panel

53 Remove the interior light assembly as described in Chapter 12.

54 Remove the two securing screws, then lower the panel from the roof (photo).

55 Where applicable, disconnect the wiring from the remote control central locking receiver, then withdraw the panel (photo).

56 Refitting is a reversal of removal.

Luggage compartment trim panels

57 The luggage compartment trim panels are clipped and glued in place. The edges of the panels can be unclipped for access to most components.

Tailgate trim panel

58 The panel is secured by plastic clips. Carefully pull the panel from the tailgate to release the clips (photo), or release the clips using a suitable forked tool.

Front seat trim panels

59 The seat trim panels are clipped in place, except for the rear outer trim panel, which is

also secured by a single screw through its rear. Removal and refitting are self-evident.

25 Centre console - removal and refitting

1 Disconnect the battery negative lead.

2 Where applicable, unscrew the two securing screws and remove the footrest from the floor next to the centre console (photo).

3 Ensure that the handbrake is fully applied.

4 On models fitted with a manual gearbox, unclip the gear lever gaiter from the centre console (photo).

5 On models fitted with automatic transmission, unclip the selector lever surround from the centre console.

6 Remove the four securing screws (two at the rear, and one each side at the front), and

24.54 Removing a roof console panel securing screw

24.55 Withdrawing the roof console panel

24.58 Removing the tailgate trim panel

25.2 Unscrewing a clutch footrest securing screw

25.4 Unclipping the gear lever gaiter from the centre console

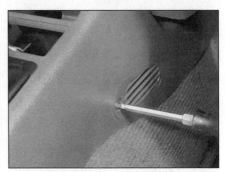

25.6A Remove the front . . .

11

25.6B . . . and rear centre console securing screws

25.8 Centre console partially withdrawn

26.1 Loosening the radio/cassette player remote control switch clamp screw

detach the centre console from the floor (photos). Take care not to strain any wiring.

7 Reach up under the centre console, and disconnect the wiring from the cigarette lighter, cigarette lighter illumination bulb, and any switches mounted in the centre console.

8 Lift the centre console over the gear lever and gaiter (or selector lever surround and selector lever), and withdraw it from the car (photo).

9 Refitting is a reversal of removal.

26 Facia panels and components - removal and refitting

Steering column shrouds

1 On models fitted with a radio/cassette player remote control switch, turn the steering wheel as necessary to gain access to the switch assembly. Slide back the cover from the front of the switch assembly, and loosen the switch clamp screw (photo).

2 Working under the steering column, unscrew the three securing screws, then carefully pull the lower shroud down to release the securing clips. Withdraw the lower shroud (photos).

3 Turn the steering wheel as necessary for access, and remove the two upper shroud securing screws from the front of the shroud. Withdraw the upper shroud (photos).

4 Refitting is a reversal of removal. Where applicable, tighten the radio/cassette player remote control switch clamp screw after refitting the shrouds.

Upper facia trim panel

5 Remove the five securing screws (three along the top edge of the panel, and one at each end of the panel), then lift the panel from the facia (photos).

6 Refitting is a reversal of removal.

Driver's side lower facia panel

7 Pull the bonnet release lever down to expose the release lever assembly securing bolt, then unscrew the bolt.

8 Where applicable, disconnect the choke cable from the carburettor with reference to Chapter 4.

9 Remove the two securing screws from the upper edge of the panel, then carefully lower the panel (photo).

10 On models fitted with a facia-mounted headlight aim adjustment control, carefully pull the knob from the control, using a pair of pliers and a piece of thick paper or card to protect the knob. Remove the two screws

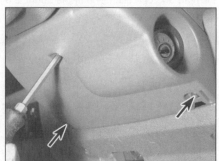

26.2A Unscrew the securing screws (arrowed) . . .

26.2B . . . and withdraw the lower shroud

26.3A Remove the securing screws . . .

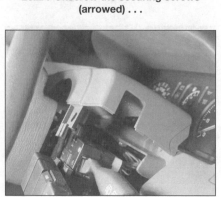

26.3B . . . and withdraw the upper shroud

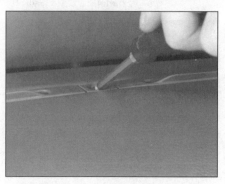

26.5A Unscrewing an upper facia trim panel top securing screw . . .

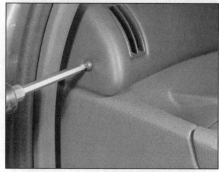

26.5B . . . and side securing screw

26.9 Driver's side lower facia panel securing screw (arrowed)

26.10A Pull the knob from the headlight aim adjustment control . . .

26.10B . . . and remove the two securing screws

securing the control to the facia panel (photos).

11 Take careful note of the routing of any cables and hoses, then unclip the bonnet release lever assembly, the choke control assembly, and the headlight aim adjustment switch, as applicable, from the rear of the panel (photos). Withdraw the panel.

12 Refitting is a reversal of removal, ensuring that any cables and hoses are routed as noted before removal.

13 Where applicable, on completion, check the adjustment of the choke cable as described in Chapter 4.

Radio/cassette player housing

14 Remove the radio/cassette player as described in Chapter 12.

15 Remove the two upper and two lower securing screws. Withdraw the radio/cassette player housing from the facia, feeding the wiring plugs and aerial cable through the rear of the housing as it is withdrawn (photos).

16 Refitting is a reversal of removal. Refit the radio/cassette player with reference to Chapter 12.

Heater ducts and vents

17 Refer to Chapter 3, Section 14.

Glovebox lock

18 The lock striker mounted on the facia also incorporates the glovebox light switch (where applicable). To remove the striker assembly, remove the single securing screw. Withdraw the striker and, where applicable, disconnect the wiring plugs (photo).

19 Two types of lock may be fitted to the glovebox lid. The first type is operated by pushing a button at the top of the lid. This type of lock can be removed after removing the single securing screw.

20 The second type of lock is operated by squeezing two buttons located on the front of the lid. This type of lock is secured by a metal clip on the rear of the lock. Remove the clip and pull the lock out from the front of the glovebox lid.

21 Refitting is a reversal of removal.

Complete facia assembly

Note: *This is an involved procedure, which is likely to take some time. It is advisable to make careful notes as the procedure progresses, to ensure correct refitting of all components, and*

26.11A Releasing the headlight aim adjustment control from the panel

26.11B Bonnet release lever securing clips (arrowed)

26.15A Remove the upper . . .

26.15B . . . and lower securing screws . . .

26.15C . . . and withdraw the radio/ cassette player housing from the facia

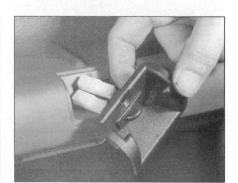

26.18 Removing the glovebox lock striker

11

26.27 Facia wiring harness earth lead securing bolt (arrowed)

26.29A Pull the connector securing clip from the wiring socket . . .

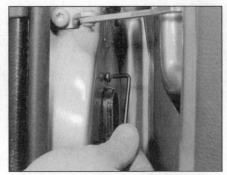

26.29B . . . then unscrew the socket securing screws

correct routing of all wiring, etc. Specific details of fixings and components may vary from model to model, but the following will serve as a guide. Provided plenty of time is allowed, removal of the facia assembly should not present any problems.

22 Disconnect the battery negative lead.

23 Prise the centre cover from the steering wheel, then unscrew the securing nut and remove the steering wheel.

24 Remove the centre console as described in Section 25.

25 Remove both front footwell side/sill trim panels, as described in Section 24.

26 Remove the steering column shrouds, upper facia trim panel, driver's side lower

26.33 Remove the two screws securing the facia assembly to the steering column

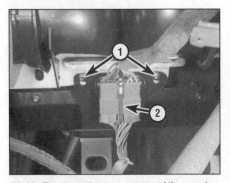

26.36 Remove the two screws (1) securing the facia to the mounting bracket. Note radio/cassette player wiring plug securing clip (2)

facia panel, and radio/cassette player housing, as described earlier in this Section.

27 Working in the footwells, unscrew the securing bolts (one on each side), and disconnect the facia wiring harness earth leads (photo).

28 Where applicable, disconnect the door wiring connectors from the door pillars. To release a connector, slide the securing clip up, then pull the connector from its socket.

29 Pull the connector securing clips from the wiring sockets, then unscrew the two securing screws in each case, and detach the door wiring sockets from the door pillars (photos).

30 On models fitted with a radio/cassette player remote control switch, slide the switch from its bracket on the steering column stalk switch assembly, then withdraw the switch, feeding the wiring through the facia. Note the routing of the wiring as it is withdrawn.

31 Remove the complete steering column stalk switch assembly, as described in Chapter 12, Section 15.

32 Working behind the facia, disconnect the ignition switch wiring connectors.

33 Remove the two screws securing the facia assembly to the steering column (photo).

34 Remove the instrument panel as described in Chapter 12.

35 On 16-valve models, remove the auxiliary instruments, as described in Chapter 12, Section 23.

36 Working through the radio/cassette player housing aperture, remove the two screws

26.39 Heater ducting-to-facia securing screws (arrowed)

securing the facia assembly to the metal mounting bracket (photo).

37 Unclip the radio/cassette player wiring connector from the metal mounting bracket, and separate the two halves of the connector.

38 Remove the two securing screws from the lower edge of the heater control panel. Withdraw the heater control panel from the facia, and disconnect the wiring plug.

39 Working through the radio/cassette player housing aperture, remove the two screws securing the heater ducting to the facia (photo).

40 Locate the rivet securing the facia assembly to the metal bracket on the passenger's side of the heater control panel housing. Using a suitable diameter drill, carefully drill out the rivet.

41 The remaining facia panel wiring harness connectors must now be disconnected.

42 Note that most of the connectors are clipped to the bulkhead or the footwell panels. In certain cases, the two halves of the connector are fastened together by metal or plastic clips. Note the routing of all wiring as the connectors are disconnected (photos).

43 Remove the large plastic securing clip from the top centre of the facia assembly (photo).

44 Working at each corner of the facia assembly, unscrew the four securing nuts (one nut at each corner) (photos).

45 Make a final check to ensure that all wires,

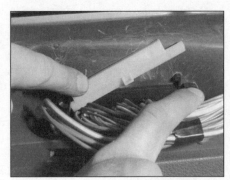

26.42A Removing a plastic clip . . .

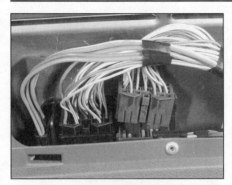

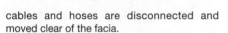

26.42B ... and disconnecting bulkhead wiring connectors

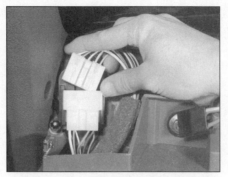

26.42C Disconnecting a bulkhead side wiring connector

26.42D Facia footwell wiring connectors (arrowed)

cables and hoses are disconnected and moved clear of the facia.

46 With the aid of an assistant, carefully withdraw the facia assembly from the bulkhead, and withdraw it from the car through one of the front doors (photo). If the facia is difficult to remove, do not use excessive force - check that all necessary wiring has been disconnected, and that the harnesses are not fouling surrounding components.

47 Refitting is a reversal of removal, bearing in mind the following points.

48 As the facia assembly is offered to the bulkhead, ensure that none of the wiring harnesses, plugs, cables or hoses are trapped.

49 Ensure that all wiring connectors are securely reconnected, and that all connector securing clips are in place, where applicable.

50 Use a new pop rivet, or alternatively a

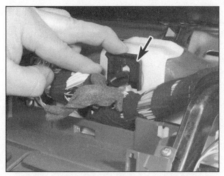

26.43 Removing the plastic securing clip (arrowed) from the top of the facia

suitable nut and bolt, to secure the facia assembly to the metal bracket on the passenger's side of the heater control panel housing (photo).

51 Where applicable, do not fully tighten the

26.44A Facia upper securing nut (arrowed)

radio/cassette player remote control switch clamp screw until the steering column shrouds have been refitted.

52 On completion, check that all switches, lights and instruments function correctly.

26.44B Facia lower securing nut (arrowed)

26.46 Withdrawing the facia assembly

26.50 Nut and bolt (arrowed) used to secure facia to metal bracket

11

Notes

Chapter 12 Electrical system

Contents

Degrees of difficulty

Easy, suitable for novice with little experience	Fairly easy, suitable for beginner with some experience	Fairly difficult, suitable for competent DIY mechanic	Difficult, suitable for experienced DIY mechanic	Very difficult, suitable for expert DIY or professional

Specifications

General

System type	12-volt, negative earth
Battery type	Lead-acid, low-maintenance or 'maintenance-free' (sealed-for-life)
Battery capacity (original equipment)	35, 50 or 60 amp-hours, according to model

Alternator

Make	Valeo
Maximum output - at 13.5 volts and 8000 (alternator) rpm	60, 70, 90 or 110 amps
Regulated voltage	13.5 to 14.8 volts

Starter motor

Make and type	Valeo, Paris-Rhone, Bosch or Mitsubishi, pre-engaged

Fuses in main fusebox

Fuse No	Rating (amps)*	Circuit(s) protected*	Fuse No	Rating (amps)*	Circuit(s) protected*
1	30	Front and rear right-hand electric windows	12	20	Central locking and electric mirror motors
2	15	ABS	13	20	Heated rear window and heated mirrors
3	30	Front and rear left-hand electric windows	14	10	Windscreen wash/wipe timer
4	20	Heated seats	15	-	Positive feed for caravan/trailer wiring
5	15	Courtesy light, radio, alarm	16	10	Direction indicators/hazard warning flashers
6	30	Accessories cut-off (master fuse)	17	10	Rear foglight
7	10	Clock	18	10	Left-hand side and tail lights
8	15	Windscreen wash/wipe motors	19	10	Right-hand side and tail lights
9	15	Cigarette lighter	20	-	Not used
10	10	Stop-lights	21	20	Heater blower
11	10	Radio/cassette player	22	10	Radio, alarm

12

Fuses in auxiliary fusebox

Fuse No	Rating (amps)	Circuit(s) protected*	Fuse No	Rating (amps)	Circuit(s) protected*
1	5	Reversing lights, alarm	8	20	Horn
2	25	Right-hand headlight dipped beam, and headlight wash/wipe	9	15	Left-hand headlight dipped beam (alternative to fuse No 4)
3	15	Left-hand headlight dipped beam	10	25	Engine functions
4	20	Front foglights	11	5	Automatic transmission
5	25	Engine cooling fan	12	15	Right-hand headlight main beam
6	30	Engine functions	13	15	Left-hand headlight main beam
7	25	Right-hand headlight dipped beam, and headlight wash/wipe (alternative to fuse No 2)	14	20	Fuel pump
			15	30	Air conditioning

*Details given are typical. Fuse ratings and circuits protected may vary with territory, model and year. Refer to owner's handbook for further information.

1 General information and precautions

⚠️ **Warning: Before carrying out any work on the electrical system, read through the precautions given in 'Safety first!' at the beginning of this manual.**

The electrical system is of the 12-volt negative earth type, and consists of a 12-volt battery, an alternator with integral voltage regulator, a starter motor, and electrical accessories, components and wiring.

The battery is charged by the alternator, which is belt-driven from a crankshaft-mounted pulley. Most models are fitted with a 'maintenance-free' (sealed-for-life) battery. In certain territories, a low-maintenance battery may be fitted, which will require periodic topping-up of the electrolyte level.

The starter motor is of the pre-engaged type, incorporating an integral solenoid. On starting, the solenoid moves the drive pinion into engagement with the flywheel ring gear before the starter motor is energised. Once the engine has started, a one-way clutch prevents the starter motor being driven by the engine until the pinion disengages from the flywheel.

Further details of the various systems are given in the relevant Sections of this Chapter. While some repair procedures are given, the usual course of action is to renew the component concerned. The owner whose interest extends beyond mere component renewal should obtain a copy of the 'Automobile Electrical & Electronic Systems Manual', available from the publishers of this manual.

Precautions

It is necessary to take extra care when working on the electrical system, to avoid damage to semi-conductor devices (diodes and transistors), and to avoid the risk of personal injury. In addition to the precautions given in 'Safety first!' at the beginning of this manual, observe the following when working on the system:

Always remove rings, watches, etc, before working on the electrical system. Even with the battery disconnected, capacitive discharge could occur if a component's live terminal is earthed through a metal object. This could cause a shock or nasty burn.

Do not reverse the battery connections. Components such as the alternator, fuel injection/ignition system computer, or any component having semi-conductor circuitry, could be irreparably damaged.

If the engine is being started using jump leads and a slave battery, connect the batteries *positive-to-positive* and *negative-to-negative* (see 'Booster battery (jump) starting'). This also applies when connecting a battery charger.

Never disconnect the battery terminals, the alternator, any electrical wiring or any test instruments when the engine is running.

Do not allow the engine to turn the alternator when the alternator is not connected.

Never 'test' for alternator output by 'flashing' the output lead to earth.

Never use an ohmmeter of the type incorporating a hand-cranked generator for circuit or continuity testing.

Always ensure that the battery negative lead is disconnected when working on the electrical system. (The battery terminal connectors fitted as original equipment are particularly easy to disconnect. Just unscrew the wing nut on top of the terminal a few turns; there is no need to remove the connector completely.)

Before using electric-arc welding equipment on the car, disconnect the battery, alternator and components such as the fuel injection/ignition system computer, to protect them.

2 Electrical fault-finding - general information

Note: *Refer to the precautions given in 'Safety first!' and in Section 1 of this Chapter before starting work. The following tests relate to testing of the main electrical circuits, and should not be used to test delicate electronic circuits (such as engine management systems, anti-lock braking systems, etc), particularly where an electronic control module is used.*

General

1 A typical electrical circuit consists of an electrical component, any switches, relays, motors, fuses, fusible links or circuit breakers related to that component, and the wiring and connectors which link the component to both the battery and the chassis. To help to pinpoint a problem in an electrical circuit, wiring diagrams are included at the end of this manual.

2 Before attempting to diagnose an electrical fault, first study the appropriate wiring diagram, to obtain a complete understanding of the components included in the particular circuit concerned. The possible sources of a fault can be narrowed down by noting if other components related to the circuit are operating properly. If several components or circuits fail at one time, the problem is likely to be related to a shared fuse or earth connection.

3 Electrical problems usually stem from simple causes, such as loose or corroded connections, a faulty earth connection, a blown fuse, a melted fusible link, or a faulty relay (refer to Section 13 for details of testing relays). Visually inspect the condition of all fuses, wires and connections in a problem circuit before testing the components. Use the wiring diagrams to determine which terminal connections will need to be checked in order to pinpoint the trouble-spot.

4 The basic tools required for electrical fault-finding include a circuit tester or voltmeter (a 12-volt bulb with a set of test leads can also be used for certain tests); a self-powered test light (sometimes known as a continuity tester); an ohmmeter (to measure resistance); a battery and set of test leads; and a jumper wire, preferably with a circuit breaker or fuse incorporated, which can be used to bypass suspect wires or electrical components. Before attempting to locate a problem with test instruments, use the wiring diagram to determine where to make the connections.

5 To find the source of an intermittent wiring fault (usually due to a poor or dirty connection, or damaged wiring insulation), a 'wiggle' test can be performed on the wiring. This involves wiggling the wiring by hand to see if the fault occurs as the wiring is moved.

It should be possible to narrow down the source of the fault to a particular section of wiring. This method of testing can be used in conjunction with any of the tests described in the following sub-Sections.

6 Apart from problems due to poor connections, two basic types of fault can occur in an electrical circuit - open-circuit, or short-circuit.

7 Open-circuit faults are caused by a break somewhere in the circuit, which prevents current from flowing. An open-circuit fault will prevent a component from working, but will not cause the relevant circuit fuse to blow.

8 Short-circuit faults are caused by a 'short' somewhere in the circuit, which allows the current flowing in the circuit to 'escape' along an alternative route, usually to earth. Short-circuit faults are normally caused by a breakdown in wiring insulation, which allows a feed wire to touch either another wire, or an earthed component such as the bodyshell. A short-circuit fault will normally cause the relevant circuit fuse to blow.

Finding an open-circuit

9 To check for an open-circuit, connect one lead of a circuit tester or voltmeter to the battery negative terminal, or to a known good earth.

10 Connect the other lead to a connector in the circuit being tested, preferably nearest to the battery or fuse.

11 Switch on the circuit, bearing in mind that some circuits are live only when the ignition switch is moved to a particular position.

12 If voltage is present (indicated either by the tester bulb lighting or a voltmeter reading, as applicable), this means that the section of the circuit between the relevant connector and the battery is problem-free.

13 Continue to check the remainder of the circuit in the same fashion.

14 When a point is reached at which no voltage is present, the problem must lie between that point and the previous test point with voltage. Most problems can be traced to a broken, corroded or loose connection.

Finding a short-circuit

15 To check for a short-circuit, first disconnect the load(s) from the circuit (loads are the components which draw current from a circuit, such as bulbs, motors, heating elements, etc).

16 Remove the relevant fuse from the circuit, and connect a circuit tester or voltmeter to the fuse connections.

17 Switch on the circuit, bearing in mind that some circuits are live only when the ignition switch is moved to a particular position.

18 If voltage is present (indicated either by the tester bulb lighting or a voltmeter reading, as applicable), this means that there is a short-circuit.

19 If no voltage is present, but the fuse still blows with the load(s) connected, this indicates an internal fault in the load(s).

Finding an earth fault

20 The battery negative terminal is connected to 'earth' - the metal of the engine/transmission unit and the car body. Most electrical systems are wired so that they only receive a positive feed, the current returning via the metal of the car body. This means that the component mounting and the body form part of that circuit. Loose or corroded mountings can therefore cause a range of electrical faults, ranging from total failure of a circuit, to a puzzling partial fault. In particular, lights may shine dimly (especially when another circuit sharing the same earth point is in operation), motors (eg wiper motors or the radiator cooling fan motor) may run slowly, and the operation of one circuit may have an apparently-unrelated effect on another. Note that on many vehicles, earth straps are used between certain components, such as the engine/transmission and the body, usually where there is no metal-to-metal contact between components due to flexible rubber mountings, etc (photo).

21 To check whether a component is properly earthed, disconnect the battery, and connect one lead of an ohmmeter to a known good earth point. Connect the other lead to the wire or earth connection being tested. The resistance reading should be zero; if not, check the connection as follows.

22 If an earth connection is thought to be faulty, dismantle the connection, and clean back to bare metal both the bodyshell and the wire terminal or the component earth connection mating surface. Be careful to remove all traces of dirt and corrosion, then use a knife to trim away any paint, so that a clean metal-to-metal joint is made. On reassembly, tighten the joint fasteners securely; if a wire terminal is being refitted, use serrated washers between the terminal and the bodyshell to ensure a clean and secure connection. When the connection is remade, prevent the onset of corrosion in the future by applying a coat of petroleum jelly or silicone-based grease, or by spraying on (at regular intervals) a proprietary ignition sealer or a water-dispersant lubricant.

3 Battery - testing and charging

Note: *Refer to the precautions given in 'Safety first!' and in Section 1 of this Chapter.*

Testing

1 Most cars are fitted with a 'maintenance-free' (sealed-for-life) battery, in which case the electrolyte level cannot be checked. In this case, refer to the battery manufacturer's recommendations for maintenance and charging procedures.

2 Where a conventional battery is fitted, the electrolyte level of each cell should be checked, and if necessary topped-up with distilled or de-ionised water, at the intervals

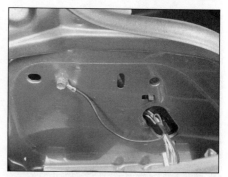

2.20 Rear light cluster earth connection (viewed with light cluster removed)

given in Chapter 1. On some batteries, the case is translucent, and incorporates minimum and maximum level marks.

3 If the car covers a very small annual mileage, it is worthwhile checking the specific gravity of the electrolyte every three months, to determine the state of charge of the battery. Use a hydrometer to make the check, and compare the results with the following table.

	Normal climates	Tropics
Discharged	1.120	1.080
Half-charged	1.200	1.160
Fully-charged	1.280	1.230

4 If the battery condition is suspect, where possible, first check the specific gravity of the electrolyte in each cell. A variation of 0.040 or more between cells indicates loss of electrolyte or deterioration of the internal plates.

5 An accurate test of battery condition can be made by a battery specialist, using a heavy-discharge meter. Alternatively, connect a voltmeter across the battery terminals, and earth the coil HT lead with a suitable wire; operate the starter motor with the headlights, heated rear window and heater blower switched on. If the voltmeter reading remains above 9.6 volts, the battery condition is satisfactory. If the voltmeter reading drops below 9.6 volts, and the battery has already been charged, it is faulty.

Charging

6 In normal use, the battery should not require charging from an external source, unless it is discharged accidentally (for instance by leaving the lights on). Charging can also temporarily revive a failing battery, but if frequent recharging is required (and the alternator output is correct), the battery is worn out.

7 Unless the battery manufacturer advises differently, the charging rate in amps should be no more than one-tenth of the battery capacity in amp-hours (for instance, 6.5 amps for a 65 amp-hour battery). Most domestic battery chargers have an output of 5 amps or so, and these can safely be used overnight. Rapid 'boost' charging is not recommended; if it is not carefully controlled, it can cause serious damage to the battery plates through overheating.

4.4A Unscrew the extended clamp bolt . . .

4.4B . . . and withdraw the battery clamp assembly

8 Both battery terminal leads must be disconnected before connecting the charger leads (disconnect the negative lead first). Connect the charger leads **before** switching on at the mains. When charging is complete, switch off at the mains **before** disconnecting the charger. If this procedure is followed, there is no risk of creating a spark at the battery terminals. Continue to charge the battery until no further rise in specific gravity is noted over a four-hour period, or until vigorous gassing is observed.

9 On completion of charging, check the electrolyte level (if possible) and top-up if necessary, using distilled or de-ionised water.

4 Battery - removal and refitting

Note: *Refer to the precautions given in 'Safety first!' and in Section 1 of this Chapter.*

1 The battery is located at the front left-hand corner of the engine compartment.

2 Disconnect the lead(s) at the negative (earth) terminal by unscrewing the retaining nut and removing the terminal clamp.

3 Disconnect the positive terminal lead(s) in the same way.

4 Unscrew the extended clamp bolt, remove the clamp assembly, then lift the battery from its location (photos). Keep the battery in an upright position, to avoid spilling electrolyte on the bodywork.

5 Refitting is a reversal of removal. Smear petroleum jelly on the terminals when reconnecting the leads. Always reconnect the positive lead first, and the negative lead last.

5 Charging system - testing

Note: *Refer to the warnings given in 'Safety first!' and in Section 1 of this Chapter before starting work.*

1 If the ignition (no-charge) warning light fails to illuminate when the ignition is switched on, first check the security of the alternator wiring connections. If satisfactory, check that the warning light bulb has not blown, and that the bulbholder is secure in its location in the

instrument panel. If the light still fails to illuminate, check the continuity of the warning light feed wire from the alternator to the bulbholder. If all is satisfactory, the alternator is at fault, and should be renewed or taken to an auto-electrician for testing and repair.

2 If the ignition warning light illuminates when the engine is running, stop the engine and check that the drivebelt is correctly tensioned (Section 6) and that the alternator connections are secure. If all is so far satisfactory, check the alternator brushes and slip rings (see Section 8). If the fault persists, the alternator should be renewed, or taken to an auto-electrician for testing and repair.

3 If the alternator output is suspect even though the warning light functions correctly, the regulated voltage may be checked as follows.

4 Connect a voltmeter across the battery terminals, and start the engine.

5 Increase the engine speed until the voltmeter reading remains steady; the reading should be between 13.5 and 14.8 volts.

6 Switch on as many electrical accessories (headlights, heated rear window, heater blower etc) as possible, and check that the alternator maintains the regulated voltage between 13.5 and 14.8 volts. It may be necessary to increase engine speed slightly.

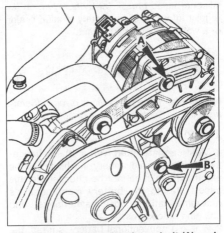

Fig. 12.1 Alternator tensioner bolt (A) and mounting bolt (B) - 1108 cc (C1E) engine (Sec 6)

7 If the regulated voltage is not as stated, the fault may be due to worn brushes, weak brush springs, a faulty voltage regulator, a faulty diode, a severed phase winding, or worn or damaged slip rings. The brushes and slip rings may be checked (see Section 8), but if the fault persists, the alternator should be renewed or taken to an auto-electrician for testing and repair.

6 Auxiliary drivebelts - removal, refitting and tensioning

Note: *Disconnect the battery negative lead before working in the area around the alternator.*

1108 cc (C1E) engine

Removal

1 Working at the top of the alternator, loosen the alternator tensioner nut whilst counterholding the bolt.

2 Similarly, loosen the alternator mounting nut (located at the bottom of the alternator), then pivot the alternator towards the engine to relieve the tension in the drivebelt.

3 Slide the drivebelt from the pulleys.

Refitting

4 Fit the belt around the pulleys, ensuring that the belt is of the correct type if it is being renewed. Take up the slack in the belt by swinging the alternator away from the engine and lightly tightening the mounting and tensioner nuts and bolts.

5 Tension the drivebelt as described in the following paragraphs.

Tensioning

6 Correct tensioning of the drivebelt will ensure that it has a long life. Beware, however, of overtightening, as this can cause excessive wear in the alternator.

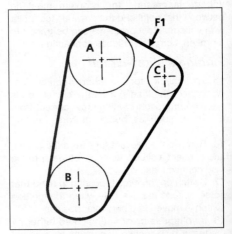

Fig. 12.2 Auxiliary drivebelt tension checking point (F1) - 1108 cc (C1E) engine (Sec 6)

A Water pump pulley *C Alternator pulley*
B Crankshaft pulley *F1 = 4.0 mm*

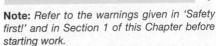

7 The belt should be tensioned so that there is approximately 4 mm of free movement, under firm thumb pressure at the mid-point of the belt run between the water pump pulley and the alternator pulley.

8 With the mounting and tensioner nuts just holding the unit firm, lever the alternator away from the engine until the correct tension is achieved, then tighten the nuts and bolts.

9 When a new belt has been fitted, it will probably stretch slightly when the engine is run. After fitting a new belt, run the engine until the cooling fan cuts in, then stop the engine and re-check the belt tension.

10 With the engine warm, the belt tension should be as described previously in paragraph 7.

1171 cc (E5F and E7F) and 1390 cc (E6J and E7J) engines - alternator drivebelt

Removal

11 Where applicable, remove the power steering pump/air conditioning compressor drivebelt (as applicable), as described later in this Section.

12 Apply the handbrake, then jack up the front of the car and support securely on axle stands (see *'Jacking, towing and wheel changing'*).

13 For improved access, remove the right-hand roadwheel, then remove the wheel arch splash shield, noting that it may be necessary to drill out the securing rivets on certain models.

14 Proceed as described in paragraphs 1 to 3 inclusive.

Refitting

15 Fit the belt around the pulleys, ensuring that the belt is of the correct type if it is being renewed. Take up the slack in the belt by swinging the alternator away from the engine and lightly tightening the mounting and tensioner nuts and bolts.

16 Tension the drivebelt as described in the following paragraphs.

Tensioning

17 Proceed as described in paragraphs 12 and 13, if not already done.

18 Correct tensioning of the drivebelt will ensure that it has a long life. Beware, however, of overtightening, as this can cause excessive wear in the alternator.

19 The belt should be tensioned so that there is between 4.2 and 4.7 mm of free movement, under firm thumb pressure at the mid-point of the belt run between the pulleys.

20 With the mounting and tensioner nuts just holding the unit firm, lever the alternator away from the engine until the correct tension is achieved, then tighten the nuts and bolts.

21 When a new belt has been fitted, it will probably stretch slightly when then engine is run. After fitting a new belt, run the engine until the cooling fan cuts in, then stop the engine and re-check the belt tension.

22 With the engine warm, the belt deflection should be between 6.2 and 7.7 mm, using the method described in paragraph 19.

23 On completion, lower the car to the ground and, where applicable, refit the power steering pump drivebelt, as described later in this Section.

1390 cc (E6J and E7J) engines - power steering pump drivebelt (models without air conditioning)

Removal

24 Working at the top of the pump, loosen the upper pump mounting nut (which secures the pump to the adjuster strut) whilst counterholding the bolt, and loosen the adjuster strut locknut and adjuster nut.

25 Working at the bottom of the pump, loosen the lower pump mounting nut, again counterholding the bolt.

26 Back off the adjuster nut to relieve the tension in the drivebelt, then slip the drivebelt from the pulleys.

Refitting

27 Fit the belt around the pulleys, ensuring that the belt is of the correct type if it is being renewed. Take up the slack in the belt by tightening the adjuster nut.

28 Tension the drivebelt as described in the following paragraphs.

Tensioning

29 Correct tensioning of the drivebelt will ensure that it has a long life. Beware, however, of overtightening, as this can cause excessive wear in the pump.

30 The belt should be tensioned so that there is between 3.2 and 3.5 mm of free movement, under firm thumb pressure at the mid-point of the belt run between the pulleys.

31 With the mounting nuts just holding the pump firm, and the adjuster locknut loosened, turn the adjuster nut until the correct tension is achieved, then tighten the adjuster nut, locknut and pump mounting nuts.

32 When a new belt has been fitted, it will probably stretch slightly when then engine is run. After fitting a new belt, run the engine until the cooling fan cuts in, then stop the engine and re-check the belt tension.

33 With the engine warm, the belt deflection should be between 5.2 and 6.7 mm, using the method described in paragraph 30.

1390 cc (E6J and E7J) engines - air conditioning compressor drivebelt (models without power steering)

Removal

34 Working in the engine compartment, loosen the two belt tensioner pulley bracket securing bolts.

35 Loosen the adjuster strut locknut and the adjuster nut, then back off the adjuster nut to relieve the tension in the drivebelt.

36 Slip the drivebelt from the pulleys.

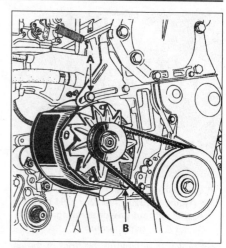

Fig. 12.3 Alternator tensioner bolt (A) and mounting bolt (B) - 1171 cc (E5F and E7F) and 1390 cc (E6J and E7J) engines (Sec 6)

Refitting

37 Fit the belt around the pulleys, ensuring that the belt is of the correct type if it is being renewed. Take up the slack in the belt by tightening the adjuster nut.

38 Tension the drivebelt as described in the following paragraphs.

Tensioning

39 Correct tensioning of the drivebelt will ensure that it has a long life. Beware, however, of overtightening, as this can cause excessive wear in the compressor.

40 At the time of writing, no information was available for the recommended belt tension, but a reasonable approximation would be approximately 4 mm of free movement, under firm thumb pressure at the mid-point of the belt run between the crankshaft and compressor pulleys. Consult a Renault dealer for recommendations if in doubt.

41 With the tensioner pulley bracket securing bolts just holding the bracket firm, turn the adjuster nut until the correct tension is achieved, then tighten the adjuster nut, locknut and pulley bracket securing bolts.

1390 cc (E6J and E7J) engines - air conditioning compressor/power steering pump drivebelt

42 On models fitted with both power steering and air conditioning, the procedure for belt removal, refitting and tensioning is as described previously for models with air conditioning. Consult a Renault dealer for details of the belt tension recommendation.

1721 cc (F2N) and 1794 cc (F3P) engines - models without power steering or air conditioning

Removal

43 Remove the right-hand headlight, as described in Section 18.

44 For improved access, apply the

12

6.46 Alternator adjuster strut-to-engine bolt (1) and adjuster nut (2) - 1794 cc (F3P) model without power steering or air conditioning

6.47 Loosening the alternator upper mounting nut - 1794 cc (F3P) model without power steering or air conditioning

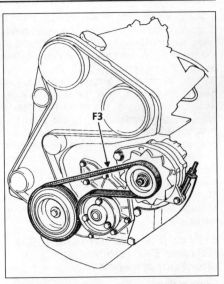

Fig. 12.4 Auxiliary drivebelt tension checking point (F3) - 1721 cc (F2N) and 1794 cc (F3P) models without power steering or air conditioning (Sec 6)

F3 = 2.5 to 3.5 mm cold / 3.5 to 4.5 mm hot

handbrake, then jack up the front of the car and support securely on axle stands (see *'Jacking, towing and wheel changing'*).

45 Remove the right-hand roadwheel, then remove the engine splash shield from the subframe to improve access. Where applicable, unclip the brake pad wear sensor wiring from the splash shield, taking care not to strain the wiring.

46 Working at the bottom of the alternator, loosen the lower mounting nut (which secures the alternator to the adjuster strut) whilst counterholding the bolt. Loosen the adjuster strut locknut and adjuster nut. Also loosen the bolt securing the adjuster strut to the engine (photo).

47 Working at the top of the alternator, loosen the upper alternator mounting nut, again counterholding the bolt (photo).

48 Back off the adjuster nut to relieve the tension in the drivebelt, then slip the drivebelt from the pulleys, taking note of its routing.

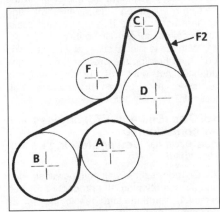

Fig. 12.5 Auxiliary drivebelt tension checking point (F2) - 1721 cc (F2N) and 1794 cc (F3P) models with power steering and without air conditioning (Sec 6)

A Water pump pulley
B Crankshaft pulley
C Alternator pulley
D Power steering pump pulley
F Tensioner pulley
F2 = 3.5 to 4.5 mm cold / 5.5 to 7.0 mm warm

Refitting

49 Fit the belt around the pulleys, ensuring that the belt is of the correct type if it is being renewed. Take up the slack in the belt by tightening the adjuster nut. Ensure that the belt is routed as noted before removal.

50 Tension the drivebelt as described in the following paragraphs.

Tensioning

51 Proceed as described in paragraphs 43 to 47, if not already done.

52 Correct tensioning of the drivebelt will ensure that it has a long life. Beware, however, of overtightening, as this can cause excessive wear in the alternator and/or water pump.

53 The belt should be tensioned so that there is between 2.5 and 3.5 mm of free movement, under firm thumb pressure at the mid-point of the belt run between the crankshaft and alternator pulleys.

54 With the mounting nuts just holding the alternator firm, turn the adjuster nut until the correct tension is achieved, then tighten the adjuster nut, the locknut, the bolt securing the adjuster strut to the engine, and the alternator mounting nuts and bolts.

55 When a new belt has been fitted, it will probably stretch slightly when then engine is run. After fitting a new belt, run the engine until the cooling fan cuts in, then stop the engine and re-check the belt tension.

56 With the engine warm, the belt deflection should be between 3.5 and 4.5 mm, using the method described in paragraph 53.

57 On completion, refit the engine splash shield and the roadwheel. Lower the car to the ground, and refit the headlight as described in Section 18.

1721 cc (F2N) and 1794 cc (F3P) engines - models with power steering and without air conditioning

Removal

58 Proceed as described in paragraphs 43 to 45.

59 Slacken the bolt in the centre of the

tensioner pulley, using a 7 mm Allen key or bit.

60 Using an open-ended spanner on the flats provided, turn the tensioner pulley clockwise to relieve the tension in the drivebelt.

61 Carefully release the drivebelt from the pulleys, noting its routing, and withdraw the belt.

Refitting

62 Fit the belt around the pulleys, ensuring that the belt is of the correct type if it is being renewed. Make sure that the belt is correctly routed round the pulleys, as noted before removal.

63 Tension the drivebelt as described in the following paragraphs.

Tensioning

64 Proceed as described in paragraphs 58 and 59, if not already done.

65 Correct tensioning of the drivebelt will ensure that it has a long life. Beware, however, of overtightening, as this can cause excessive wear in the alternator and/or pumps.

66 The belt should be tensioned so that there is between 3.5 and 4.5 mm of free movement, under firm thumb pressure at the mid-point of the belt run between the alternator and power steering pump pulleys.

67 Turn the tensioner pulley clockwise using a spanner on the flats provided until the correct tension is achieved. Lock the tensioner pulley in position by tightening the Allen bolt whilst counterholding the pulley.

68 Run the engine for approximately 5 minutes, and re-check the drivebelt tension as described in paragraph 66.

69 With the engine warm, the belt deflection should be between 5.5 and 7.0 mm, using the method described in paragraph 66.

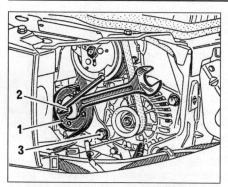

Fig. 12.6 Slackening the auxiliary drivebelt tensioner pulley bolt - 1721 cc (F2N) and 1794 cc (F3P) models with air conditioning (Sec 6)

1 Tensioner pulley
2 Tensioner pulley bolt
3 Alternator lower mounting bolt

70 On completion, refit the engine splash shield and the roadwheel. Lower the car to the ground, and refit the headlight as described in Section 18.

1721 cc (F2N) and 1794 cc (F3P) engines - models with air conditioning

Removal

71 Remove the right-hand headlight as described in Section 18.

72 Slacken the bolt in the centre of the tensioner pulley, using a 7 mm Allen key or bit.

73 Using an open-ended spanner on the flats provided, turn the tensioner pulley clockwise to relieve the tension in the drivebelt.

74 Carefully release the drivebelt from the pulleys, noting its routing, and withdraw the belt.

Refitting

75 Fit the belt around the pulleys, ensuring that the belt is of the correct type if it is being renewed. Make sure that the belt is correctly routed round the pulleys, as noted before removal.

76 Tension the drivebelt as described in the following paragraphs.

Tensioning

77 Proceed as described in paragraphs 71 and 72, if not already done.

78 Correct tensioning of the drivebelt will ensure that it has a long life. Beware, however, of overtightening, as this can cause excessive wear in the alternator and/or compressor.

79 The belt should be tensioned so that there is between 1.5 and 2.5 mm of free movement, under firm thumb pressure at the mid-point of the belt run between the air conditioning compressor and alternator pulleys.

80 Turn the tensioner pulley clockwise using a spanner on the flats provided until the correct tension is achieved. Lock the

6.85 Unscrewing the wiring bracket-to-inner front wing panel screw - 1764 cc (F7P) engine

tensioner pulley in position by tightening the Allen bolt whilst counterholding the pulley.

81 Run the engine for approximately 5 minutes, and re-check the drivebelt tension as described in paragraph 79.

82 Refit the headlight, with reference to Section 18.

1764 cc (F7P) engine

Removal

83 Remove the right-hand headlight, as described in Section 18.

84 On models fitted with a charcoal canister (see Chapter 4, Section 43 or 44), unscrew the two securing bolts, remove the shield from the fuel tank-to-canister hose, then disconnect the hose from the canister.

85 Slacken the screw securing the wiring bracket to the inner front wing panel. Pivot the bracket rearwards to improve access to the belt tensioner pulley bolt (photo).

86 Slacken the bolt in the centre of the tensioner pulley, using a 7 mm Allen key or bit (due to the limited access, if a bit is used, it will have to be turned using a ring spanner).

87 Using an open-ended spanner on the flats provided, turn the tensioner pulley clockwise to relieve the tension in the drivebelt.

88 Apply the handbrake, then jack up the front of the car and support it securely on axle stands (see 'Jacking, towing and wheel changing').

89 Remove the right-hand front roadwheel, then release the securing clip and remove the

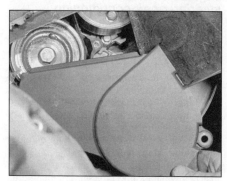

6.90 Removing the timing belt lower cover - 1764 cc (F7P) engine

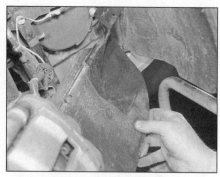

6.89 Removing the right-hand splash shield - 1764 cc (F7P) engine

engine splash shield from the subframe (photo). Note that the brake pad wear sensor wiring must be unclipped from the splash shield before it can be removed. Take care not to strain the wiring.

90 Working under the wheel arch, undo the two securing screws and remove the timing belt lower cover to expose the crankshaft pulley (photo).

91 Working through the headlight aperture, carefully release the drivebelt from the pulleys, noting its routing, and withdraw the belt (photo).

Refitting

92 Fit the belt around the pulleys, ensuring that the belt is of the correct type if it is being renewed. Make sure that the belt is correctly routed round the pulleys, as noted before removal.

93 Tension the drivebelt as described in the following paragraphs.

Tensioning

94 Proceed as described in paragraphs 83 to 86, if not already done.

95 Correct tensioning of the drivebelt will ensure that it has a long life. Beware, however, of overtightening, as this can cause excessive wear in the alternator and/or pump(s).

96 The belt should be tensioned so that there is between 1.5 and 2.5 mm of free movement, under firm thumb pressure at the mid-point of the belt run between the power steering pump

6.91 Releasing the drivebelt from the alternator pulley - 1764 cc (F7P) engine

12

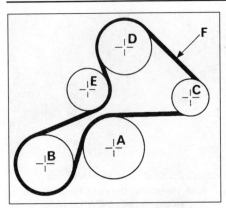

Fig. 12.7 Auxiliary drivebelt tension checking point (F) - 1764 cc (F7P) engine (Sec 6)

A Water pump pulley
B Crankshaft pulley
C Alternator pulley
D Power steering pump or idler pulley
E Tensioner pulley
F = 1.5 to 2.5 mm

or idler pulley (as applicable) and the alternator pulley.

97 Turn the tensioner pulley clockwise using a spanner on the flats provided until the correct tension is achieved. Lock the tensioner pulley in position by tightening the Allen bolt whilst counterholding the pulley (photo).

98 With the drivebelt correctly adjusted, refit and tighten the screw securing the wiring bracket to the inner wing panel.

99 Run the engine for approximately 5 minutes, and re-check the drivebelt tension as described in paragraph 96.

100 Where applicable, reconnect the charcoal canister hose, and refit the hose shield.

101 Refit the headlight, with reference to Section 18.

---

7 Alternator - removal and refitting

1108 cc (C1E) engine

1 Open the bonnet and disconnect the battery leads (negative lead first).

2 Remove the alternator drivebelt, as described in Section 6.

3 Disconnect the wiring plug, and/or disconnect the wires from their terminals on the rear of the alternator, as applicable. Note the locations of the wires for reference when refitting.

4 Unscrew and remove the alternator tensioner nut (located at the top of the alternator). Note the locations of any washers and spacers.

5 Withdraw the tensioner bolt.

6 Similarly, unscrew and remove the alternator mounting nut and bolt (located at

6.97 Counterholding the tensioner pulley while tightening the Allen bolt - 1764 cc (F7P) engine

the bottom of the alternator), then carefully lift the alternator from the engine.

7 Refitting is a reversal of removal, ensuring that any washers and spacers are located as noted before removal. Refit and tension the drivebelt as described in Section 6.

1171 cc (E5F and E7F) and 1390 cc (E6J and E7J) engines

8 Proceed as described in paragraphs 1 to 3 inclusive.

9 Unscrew and remove the alternator tensioner nut (located at the top of the alternator), and recover the washer.

10 Working through the wheel arch, unscrew and remove the alternator mounting nut (located at the bottom of the alternator) whilst counterholding the bolt. Note the locations of any washers and spacers.

11 Withdraw the through-bolt, then lift the alternator, complete with its shield, out through the top of the engine compartment.

12 Refitting is a reversal of removal, bearing in mind the following points.

13 Before refitting the alternator mounting and adjuster bolts, ensure that the alternator shield is in place.

14 Ensure that any washers and spacers are located as noted before removal.

15 Refit and tension the drivebelt as described in Section 6.

1721 cc (F2N) and 1794 cc (F3P) engines - models without power steering or air conditioning

16 Proceed as described in paragraphs 1 to 3 inclusive.

17 Unscrew and remove the bolt securing the alternator to the adjuster strut. Note the locations of any washers and spacers, and recover the nut, where applicable.

18 Unscrew and remove the remaining alternator mounting nut. Note the locations of any washers and spacers.

19 Withdraw the through-bolt, then carefully lift the alternator from the engine.

20 Refitting is a reversal of removal, ensuring that any washers and spacers are located as noted before removal. Refit and tension the drivebelt as described in Section 6.

1721 cc (F2N) and 1794 cc (F3P) engines - models with power steering and without air conditioning

21 Proceed as described in paragraphs 1 to 3 inclusive.

22 Unscrew and remove the nut securing the alternator to the mounting strut, whilst counterholding the bolt. Note the locations of any washers and spacers, and recover the nut, where applicable.

23 Similarly, unscrew and remove the nut securing the alternator to the power steering pump/alternator mounting bracket. Note the locations of any washers and spacers.

24 Withdraw the through-bolt, then carefully lift the alternator from the engine.

25 Refitting is a reversal of removal, ensuring that any washers and spacers are located as noted before removal. Refit and tension the drivebelt as described in Section 6.

1721 cc (F2N) and 1794 cc (F3P) engines - models with air conditioning

26 Disconnect the battery leads (negative lead first).

27 Remove the radiator grille panel and the bonnet as described in Chapter 11.

28 Slacken the alternator drivebelt, and slide it from the alternator pulley as described in Section 6.

29 Disconnect the wiring plug, and/or disconnect the wires from their terminals on the rear of the alternator, as applicable. Note the locations of the wires for reference when refitting.

30 Remove the two nuts and bolts securing the air conditioning compressor to the compressor/alternator mounting bracket, and the single bolt securing the alternator to the bracket, then remove the bracket. Counterhold the bolts as the nuts are unscrewed, and recover any washers and spacers, noting their locations.

31 Unscrew the alternator lower mounting nut, while counterholding the bolt, then remove the bolt, noting the locations of any washers and spacers.

32 Carefully withdraw the alternator through the headlight aperture.

33 Refitting is a reversal of removal. Refit and tension the drivebelt as described in Section 6, and refit the bonnet as described in Chapter 11.

1764 cc (F7P) engine

34 Disconnect the battery leads (negative lead first).

35 Remove the radiator grille panel and the bonnet as described in Chapter 11.

36 Release the securing clips and lift the radiator shroud upper flap for improved access.

37 Slacken the alternator drivebelt and slide

7.38 Alternator main feed terminal (arrowed) viewed through radiator grille panel aperture - 1764 cc (F7P) engine

7.39 Disconnecting the MAP sensor vacuum hose from the inlet manifold - 1764 cc (F7P) engine

7.41 Removing the bolt securing the alternator mounting plate to the power steering pump bracket - 1764 cc (F7P) engine

it from the alternator pulley as described in Section 6.

38 Disconnect the wiring plug, and/or disconnect the wires from their terminals on the rear of the alternator, noting their locations, as applicable (photo).

39 Disconnect the MAP sensor vacuum hose from the inlet manifold (see Chapter 4, Section 36) (photo).

40 Slacken the alternator lower mounting nut.

41 Unscrew the bolt securing the alternator mounting plate to the power steering pump/idler pulley bracket (as applicable). It will be necessary to counterhold the nut at the rear of the bracket as the bolt is unscrewed. Take care not to drop the nut as the bolt is removed (photo).

42 Unscrew the bolt securing the alternator to the mounting plate (photo).

43 Unscrew the two bolts securing the mounting plate to the inlet manifold, and remove the mounting plate (photo).

44 Unscrew the securing bolt and remove the alternator upper mounting bracket (photos).

45 Remove the alternator lower mounting nut, and withdraw the through-bolt from the rear of the alternator. Note the location of the lockplate under the through-bolt head (photos).

46 Carefully manipulate the alternator out through the headlight aperture (photo).

7.42 Removing the bolt securing the alternator to the mounting plate - 1764 cc (F7P) engine

7.43 Removing the alternator mounting plate - 1764 cc (F7P) engine

7.44A Unscrew the securing bolt (arrowed) . . .

7.44B . . . and remove the alternator upper mounting bracket - 1764 cc (F7P) engine

7.45A Remove the alternator lower mounting nut . . .

7.45B . . . then withdraw the through-bolt - 1764 cc (F7P) engine

7.46 Manipulate the alternator out through the headlight aperture - 1764 cc (F7P) engine

12

8.4A Disconnecting the alternator wiring plug - 1794 cc (F3P) engine

8.4B Remove the rubber cover . . .

8.4C . . . then disconnect the alternator main feed wire - 1794 cc (F3P) engine

47 Refitting is a reversal of removal, bearing in mind the following points.

48 Make sure that the lockplate under the head of the alternator mounting through-bolt engages correctly with the rear of the alternator.

49 When refitting the alternator mounting plate, it may be helpful to secure the nut for the mounting plate-to-steering pump (or idler pulley) bracket bolt over its hole in the plate, using tape. This will hold the nut in position as the bolt is refitted. The tape can be removed once the bolt has been fitted and tightened.

50 Refit and tension the drivebelt as described in Section 6.

51 Refit the bonnet as described in Chapter 11.

8 Alternator brushes and regulator - inspection and renewal

1 Open the bonnet and disconnect the battery leads (negative lead first).

2 On 1171 cc (E5F and E7F), 1390 cc (E6J and E7J), and 1764 cc (F7P) engines, remove the alternator as described in Section 7.

3 On 1108 cc (C1E), 1721 cc (F2N) and 1794 cc (F3P) engines, the brush holder/regulator assembly can be removed with the alternator in place - but access is limited.

4 If the alternator is in place, disconnect the wiring plug, and/or disconnect the wires from

their terminals on the rear of the alternator, as applicable, noting their locations (photos).

5 Note the location of the brush holder/regulator assembly wire on the alternator B+ terminal. Unscrew the nut and disconnect the wire (photo).

6 Unclip the plastic cover from the studs on the rear of the alternator, and remove the cover (photo).

7 Unscrew the two screws securing the brush holder/regulator assembly to the rear of the alternator (photo).

8 Disconnect the remaining wire from the terminal on the rear of the alternator, then lift the brush holder/regulator assembly from the alternator (photo).

9 Measure the protrusion of each brush from the brush holder (photo). No minimum dimension is specified by the manufacturers, but as a rough guide, 5 mm should be regarded as a minimum. If either brush is worn below this dimension, the complete brush holder/regulator assembly must be renewed. If the brushes are still serviceable, clean them with a fuel-moistened cloth. Check that the brush spring tension is equal for both brushes and provides a reasonable pressure. The brushes must move freely in their holders.

10 Clean the alternator slip rings with a petrol-moistened cloth. Check for signs of scoring, burning or severe pitting on the surface of the slip rings. It may be possible to have the slip rings renovated by an electrical specialist.

8.5 Disconnecting the brush holder/ regulator assembly wire from the rear of the alternator - 1794 cc (F3P) engine

8.6 Removing the plastic cover from the rear of the alternator - 1794 cc (F3P) engine

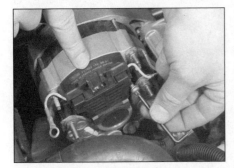

8.7 Unscrewing a brush holder/regulator assembly securing screw - 1794 cc (F3P) engine

8.8 Removing the brush holder/regulator assembly - 1794 cc (F3P) engine

8.9 Measuring the alternator brush protrusion

11 Refit the brush holder/regulator assembly using a reversal of the removal procedure.

12 If the alternator was removed, refit it as described in Section 7.

9 Starting system - testing

Note: *Refer to the precautions given in 'Safety first!' and in Section 1 of this Chapter before starting work.*

1 If the starter motor fails to operate when the ignition key is turned to the appropriate position, the possible causes are as follows.

(a) *The battery is faulty.*

(b) *The electrical connections between the switch, solenoid, battery and starter motor are somewhere failing to pass the necessary current from the battery through the starter to earth.*

(c) *The solenoid is faulty.*

(d) *The starter motor is mechanically or electrically defective.*

2 To check the battery, switch on the headlights. If they dim after a few seconds, this indicates that the battery is discharged - recharge (see Section 3) or renew the battery. If the headlights glow brightly, operate the starter switch and observe the lights. If they dim, then this indicates that current is reaching the starter motor, therefore the fault must lie in the starter motor. If the lights continue to glow brightly (and no clicking sound can be heard from the starter motor solenoid), this indicates that there is a fault in the circuit or solenoid - see the following paragraphs. If the starter motor turns slowly when operated, but the battery is in good condition, then this indicates either that the starter motor is faulty, or there is considerable resistance somewhere in the circuit.

3 If a fault in the circuit is suspected, disconnect the battery leads (including the earth connection to the body), the starter/solenoid wiring and the engine/transmission earth strap. Thoroughly clean the connections, and reconnect the leads and wiring. Use a voltmeter or test light to check that full battery voltage is available at the battery positive lead connection to the solenoid. Smear petroleum jelly around the battery terminals to prevent corrosion - corroded connections are among the most frequent causes of electrical system faults.

4 If the battery and all connections are in good condition, check the circuit by disconnecting the wire from the solenoid blade terminal. Connect a voltmeter or test light between the wire end and a good earth (such as the battery negative terminal), and check that the wire is live when the ignition switch is turned to the 'start' position. If it is, then the circuit is sound - if not, there is a fault in the ignition/starter switch or wiring.

5 The solenoid contacts can be checked by connecting a voltmeter or test light between the battery positive feed connection on the starter side of the solenoid, and earth. When the ignition switch is turned to the 'start' position, there should be a reading or lighted bulb, as applicable. If there is no reading or lighted bulb, the solenoid is faulty and should be renewed.

6 If the circuit and solenoid are proved sound, the fault must lie in the starter motor. Begin checking the starter motor by removing it (see Section 10), and checking the brushes (see Section 11). If the fault does not lie in the brushes, the motor windings must be faulty. In this event, the starter motor must be renewed, unless an auto-electrical specialist can be found who will overhaul the unit at a cost significantly less than that of a new or exchange starter motor.

10 Starter motor - removal and refitting

1108 cc (C1E) engine

1 Open the bonnet and disconnect the battery negative lead.

2 Remove the starter motor heat shield.

3 Disconnect the wiring from the terminals on the starter motor solenoid.

4 Unscrew the bolt securing the starter motor rear support bracket to the cylinder block.

5 Unscrew the three bolts securing the starter motor to the gearbox bellhousing, and withdraw the motor from the engine. Note the locating dowel in the bellhousing bolt hole.

6 Refitting is a reversal of removal. Ensure that the locating dowel is correctly positioned in the bellhousing bolt hole, as noted during removal.

1171 cc (E5F and E7F) and 1390 cc (E6J and E7J) engines

7 Open the bonnet and disconnect the battery negative lead.

8 Unscrew the two bolts and the single nut securing the starter motor to the gearbox/transmission bellhousing.

9 Apply the handbrake, then jack up the front of the car and support it securely on axle stands (see *'Jacking, towing and wheel changing'*).

10 Disconnect the wiring from the terminals on the starter motor solenoid.

11 Unscrew the bolt securing the starter motor rear support bracket to the cylinder block, noting that the bolt also secures a wiring harness support bracket.

12 Carefully withdraw the starter motor sideways from under the right-hand side of the car. Note the locating dowel in the bellhousing bolt hole.

13 Refitting is a reversal of removal. Ensure that the locating dowel is correctly positioned in the bellhousing bolt hole, as noted during removal.

1721 cc (F2N) and 1794 cc (F3P) engines

14 Open the bonnet and disconnect the battery negative lead.

15 Unscrew the clamp screws, and remove the air cleaner trunking to improve access. Note that on certain models, it may be necessary to remove the complete air cleaner assembly.

16 Unscrew the three bolts securing the starter motor to the gearbox bellhousing. Where applicable, note the location of any wiring brackets secured by the bolts.

17 Jack up the front of the car and support it securely on axle stands (see *'Jacking, towing and wheel changing'*).

18 Remove the exhaust downpipe as described in Chapter 4.

19 Unscrew the bolts securing the exhaust heat shield/rear mounting bracket assembly to the cylinder block and the starter motor bracket, and withdraw the assembly (photo).

20 Disconnect the wiring from the terminals on the starter motor solenoid (photo).

21 Working under the car, withdraw the motor from the engine (photo). Note the locating dowel in the bellhousing hole.

22 Refitting is a reversal of removal, bearing in mind the following points.

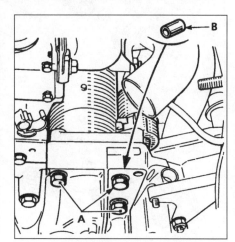

Fig. 12.8 Starter motor securing bolts (A) and locating dowel (B) (Sec 10)

10.19 Removing the exhaust heat shield/starter motor rear mounting bracket assembly - 1794 cc (F3P) engine

12

10.20 Starter motor wiring (arrowed) viewed from underneath the car - 1794 cc (F3P) engine

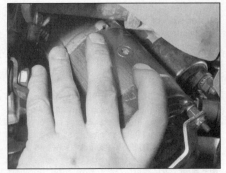

10.21 Withdrawing the starter motor - 1794 cc (F3P) engine

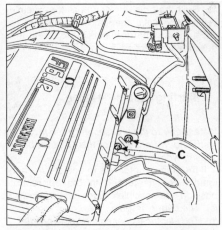

Fig. 12.9 Fuel hose shield securing bolts (C) - 1764 cc (F7P) engine (Sec 10)

23 Ensure that the locating dowel is correctly positioned in the bellhousing bolt hole, as noted during removal.

24 When reconnecting the wiring to the motor, make sure that the terminal on the end of the wiring is positioned clear of the starter motor body and surrounding components.

25 Ensure that the exhaust heat shield is refitted.

26 Refit the exhaust downpipe as described in Chapter 4.

27 Where applicable, refit the any wiring brackets to the starter motor mounting bolts.

1764 cc (F7P) engine

28 Remove the battery as described in Section 4.

29 Unclip the wiring connector which is secured in front of one of the starter motor upper mounting bolts.

30 Unscrew the three bolts securing the starter motor to the gearbox bellhousing. Where applicable, note the location of any wiring brackets secured by the bolts.

31 Remove the two bolts securing the fuel hose shield.

32 Apply the handbrake, then jack up the front of the car and support it securely on axle stands (see 'Jacking, towing and wheel changing').

33 Remove the exhaust downpipe as described in Chapter 4.

34 Unscrew the two securing bolts and

remove the exhaust heat shield.

35 Unscrew the two securing bolts, and remove the exhaust manifold-to-cylinder block bracing bracket.

36 Disconnect the wiring from the terminals on the starter motor solenoid.

37 Working under the car, unscrew the three bolts securing the starter motor rear mounting bracket to the cylinder block, noting that the bolts also secure the fuel hose shield. Withdraw the starter motor from the engine. Note the locating dowel in the bellhousing bolt hole.

38 Refitting is a reversal of removal, bearing in mind the following points.

39 Ensure that the locating dowel is correctly

positioned in the bellhousing bolt hole, as noted during removal.

40 Ensure that the fuel hose shield is in place when refitting the rear starter motor mounting bracket bolts.

41 Make sure that the exhaust heat shield is refitted.

42 Refit the exhaust downpipe as described in Chapter 4.

43 Where applicable, refit any wiring brackets to the starter motor mounting bolts.

44 Refit the battery with reference to Section 4.

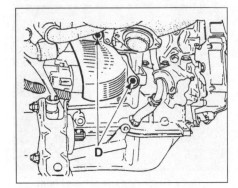

Fig. 12.10 Exhaust heat shield securing bolts (D) - 1764 cc (F7P) engine (Sec 10)

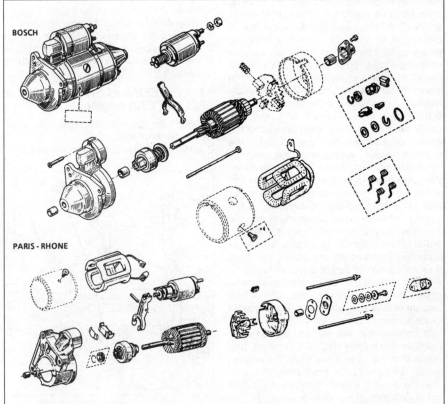

Fig. 12.11 Exploded views of the Bosch and Paris-Rhone starter motors (Sec 11)

11.3 Removing the mounting bracket - Valeo type starter motor

11.4 Removing the armature shaft end cap - Valeo type starter motor

11.5 Unscrewing the bolt from the end of the armature shaft - Valeo type starter motor

11 Starter motor - brush renewal

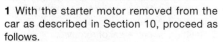

1 With the starter motor removed from the car as described in Section 10, proceed as follows.

2 Make alignment marks on the end cover and the motor body.

3 Where applicable, unscrew the two securing nuts, and remove the mounting bracket from the rear of the starter motor (photo).

4 Where applicable, unscrew the securing screws, and remove the armature shaft end cap (photo).

5 Remove the circlip or the bolt from the end of the armature shaft (photo). Recover the washers and spacers, noting their orientation. On motors with a bolt in the end of the shaft, it will be necessary to counterhold the pinion in order to unscrew the bolt.

6 Unscrew the two through-bolts or through-studs, and withdraw the armature end cover, along with the mounting bracket, when fitted. Release the rubber wiring grommet from the end cover as it is removed. Note that on some types of motor, the brush springs are located in the armature end cover (photos).

7 On motors with brush springs located in the armature end cover, the brushes can now be withdrawn from the brush carrier assembly (photo).

8 On motors with brushes retained by springs

attached to lugs on the brush holder plate, release the springs from the brushes, using a screwdriver. Withdraw the brushes from the brush holder.

9 On motors with brushes mounted in clip-on brush holders, unclip the brush holders from the brush holder plate and recover the springs.

10 Examine the brushes for wear and damage. If they are damaged, or worn to the extent where the springs are unable to exert sufficient pressure to maintain good contact with the commutator, they should be renewed. If renewal is necessary, unsolder the old brushes, and solder new ones into place.

11 Clean the brush holder assemblies, and wipe the commutator with a petrol-moistened cloth. If the commutator is dirty, it may be

cleaned with fine glass paper, then wiped with the cloth.

12 Fit the new brushes using a reversal of the removal procedure, bearing in mind the following points.

13 Ensure that the brushes move freely in their holders.

14 Ensure that the brush springs provide adequate pressure on the brushes, and renew any worn springs.

15 Align the marks made on the end cover and the motor body (paragraph 2).

16 Ensure that the washers and spacers on the end of the armature shaft are fitted as noted before removal.

17 On motors fitted with a bolt in the commutator end of the armature shaft, it will be necessary to proceed as follows in order to

11.6A Unscrewing a starter motor through-bolt - Valeo type starter motor

11.6B Removing the armature end cover. Note brush springs in cover - Valeo type starter motor

11.7 Brush (arrowed) withdrawn from brush carrier assembly - Valeo type starter motor

11.18 Disconnecting the wire from the solenoid - Valeo type starter motor

11.19A Unscrew the two securing screws . . .

12

11.19B . . . and remove the solenoid housing - Valeo type starter motor

11.20 Removing the rubber insert from the edge of the pinion end housing - Valeo type starter motor

11.21 Drive the actuating arm pivot pin from the pinion end housing - Valeo type starter motor

counterhold the armature as the bolt is tightened. This is due to the one-way clutch, which will allow the armature to move independently of the drive pinion as the bolt is tightened.

18 Unscrew the nut securing the wire to the solenoid terminal. Recover the washer (photo).

19 Unscrew the two securing screws, and remove the solenoid housing from the pinion end housing (photos). Recover the solenoid spring.

20 Remove the rubber insert from the edge of the pinion end housing (photo).

21 Using a pin punch, drive out the actuating arm pivot pin from the pinion end housing (photo).

22 Withdraw the pinion end housing, then lift the actuating arm and solenoid plunger from

the armature shaft (photos).

23 Counterhold the end of the armature shaft, taking care not to damage the surface of the shaft, then tighten the bolt in the commutator end of the shaft (photo).

24 Refit the pinion end housing and the solenoid components using a reversal of the removal procedure.

25 Refit the starter motor as described in Section 10.

12 Fusebox - removal and refitting

1 Disconnect the battery negative lead.
2 Reach up under the facia, and unscrew the two securing screws from the rear of the fusebox (photo).

3 Working at the bottom of the fusebox, unscrew the remaining two securing screws.
4 Carefully lower the fusebox from the facia. Disconnect all relevant wiring plugs to enable the fusebox to be withdrawn, noting their locations if necessary.
5 Refitting is a reversal of removal.

13 Fuses and relays - testing and renewal

Fuses

1 Fuses are designed to break a circuit when a predetermined current is reached, in order to protect components and wiring which could be damaged by excessive current flow. Any excessive current flow will be due to a

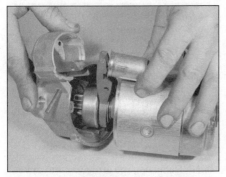

11.22A Withdraw the pinion end housing . . .

11.22B . . . then lift the actuating arm and solenoid plunger from the armature shaft - Valeo type starter motor

11.23 Tightening the armature shaft bolt - Valeo type starter motor

12.2 Fusebox upper securing screws (arrowed) viewed with facia removed

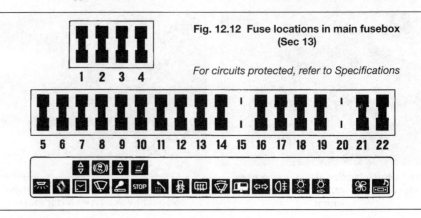

Fig. 12.12 Fuse locations in main fusebox (Sec 13)

For circuits protected, refer to Specifications

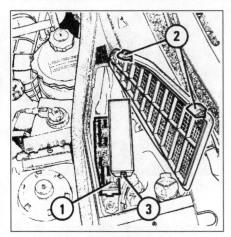

Fig. 12.13 Auxiliary fusebox location in engine compartment (Sec 13)

1 *Fusebox* 3 *Cover securing tab*
2 *Flap securing clips*

fault in the circuit, usually a short-circuit (see Section 2).

2 The main fuses are located in the fusebox, under the facia on the passenger's side.

3 For access to the fuses, press the two plastic securing tabs, then lower the fusebox panel from the facia. The circuits protected by the fuses are marked on a sticker at the bottom of the fusebox panel.

4 Additional fuses are located in an auxiliary fusebox under the bonnet, beneath the hinged flap in the windscreen cowl panel on the left-hand side. For access to these fuses, lift the cowl panel flap, then unclip the cover from the fusebox. The circuits protected by the fuses are marked on the inside of the panel cover. The fuse for the radio/cassette player is mounted on the rear of the unit (photo).

5 A blown fuse can be recognised from its melted or broken wire.

6 To remove a fuse, first ensure that the relevant circuit is switched off.

7 Using the plastic tool provided in the fusebox, pull the fuse from its location.

8 Spare fuses are provided at the lower left-

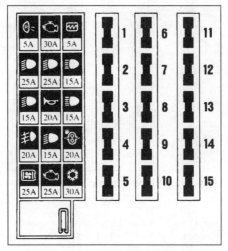

Fig. 12.14 Fuse locations in auxiliary fusebox (Sec 13)

For circuits protected, refer to Specifications

hand side of the main fusebox, and at the left-hand side of the auxiliary fusebox.

9 Before renewing a blown fuse, trace and rectify the cause, and always use a fuse of the correct rating. Never substitute a fuse of a higher rating, or make temporary repairs using wire or metal foil; more serious damage, or even fire, could result.

10 Note that the fuses are colour-coded as follows. Refer to the Specifications for details of the fuse ratings and the circuits protected.

Colour	Rating
Orange	5A
Red	10A
Blue	5A
Yellow	20A
Clear or white	25A
Green	30A

Relays

11 A relay is an electrically-operated switch, which is used for the following reasons.

(a) *A relay can switch a heavy current remotely from the circuit in which the*

13.4 Removing the fuse from the rear of the radio/cassette player

current is flowing, therefore allowing the use of lighter gauge wiring and switch contacts.

(b) *A relay can receive more than one control input, unlike a mechanical switch.*

(c) *A relay can have a 'timer' function - for example, the intermittent wiper relay.*

12 Most of the relays are located in the fusebox. Note that on some models, the tailgate wash/wipe relay is located behind the facia, to the left of the fusebox (photo).

13 Certain models have a relay box located in the engine compartment, on the left-hand side, in front of the suspension turret. The box contains relays for the fuel injection system and/or the power steering pump. Further details can be found in Chapters 4 and 10 respectively.

14 If a circuit controlled by a relay develops a fault, and the relay is suspect, operate the circuit. If the relay is functioning, it should be possible to hear the relay click as it is energised. If this is the case, the fault lies with the components or wiring in the system. If the relay is not being energised, then either the relay is not receiving a switching voltage, or the relay itself is faulty (do not overlook the relay socket terminals when tracing faults). Testing is by the substitution of a known good unit, but be careful; while some relays are identical in appearance and in operation, others look similar, but perform different functions.

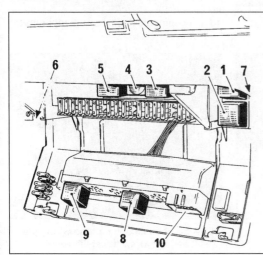

Fig. 12.15 Identification of relays in the main fusebox (Sec 13)

1 *Rear foglight relay*
2 *Heated rear window relay*
3 *'Lights on' warning buzzer*
4 *Direction indicator/hazard flasher relay*
5 *Central locking timer relay*
6 *Windscreen wash/wipe delay relay*
7 *Tailgate wash/wipe delay relay*
8 *Front foglight relay*
9 *External temperature display relay*
10 *Diagnostic socket*

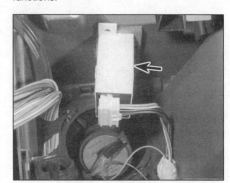

13.12 Tailgate wash/wipe relay (arrowed) viewed with facia removed

12

15.3 Disconnecting a wiring connector from the ignition switch

15.5 Unscrewing the ignition switch grub screw

15.6A Depressing the ignition switch securing clip

14 'Lights on' warning buzzer - removal and refitting

The 'lights on' warning buzzer is located in the fusebox, along with the relays. See Fig. 12.15.

15 Switches - removal and refitting

Ignition switch

1 Disconnect the battery negative lead.
2 Remove the steering column shrouds as described in Chapter 11, Section 26.
3 Follow the switch wiring behind the facia,

15.6B Withdrawing the ignition switch

and disconnect the wiring connectors (photo). If necessary, lower the driver's side facia panel for access, as described in Chapter 11, Section 26. Take note of the routing of the wiring.
4 Insert the ignition key into the switch, and turn it to position 'M' (or '3').
5 Reach up behind the switch, and using a Torx key, remove the grub screw securing the switch assembly to the steering column (photo).
6 Using a screwdriver, depress the securing clip located at the bottom of the switch assembly, then pull the assembly from the steering column using the key. Feed the wiring through the steering column as the switch is withdrawn (photos).
7 To separate the switch from the lock

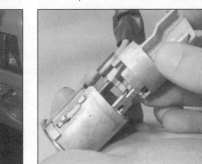

15.7A Removing the ignition switch rear cover . . .

assembly, remove the two securing screws from the rear of the housing, and lift off the rear cover. The switch can now be withdrawn from the lock (photos).
8 Refitting is a reversal of removal, bearing in mind the following points.
9 When refitting the switch to the lock assembly, make sure that the lugs on the lock engage with the cut-outs in the switch. Note that the switch will only fit in one position.
10 Ensure that the switch wiring is routed as noted before removal.

Lighting/direction indicator/horn stalk switch

11 Disconnect the battery negative lead.
12 Remove the steering column shrouds as described in Chapter 11, Section 26.
13 Working under the switch, unscrew the two securing screws and lift the switch from the steering column (photo).
14 Disconnect the wiring plugs, and withdraw the switch (photo).
15 Refitting is a reversal of removal.

Wash/wipe stalk switch

16 Proceed as just described for the lighting/direction indicator/horn stalk switch.

Complete stalk switch assembly

17 Remove the steering wheel as described in Chapter 10.
18 Proceed as described in paragraphs 11 and 12.

15.13 Unscrew the securing screws . . .

15.14 . . . then disconnect the wiring plugs and withdraw the stalk switch

15.7B . . . and withdrawing the switch from the lock

15.21A Loosen the clamp screw . . .

15.21B . . . and withdraw the stalk switch assembly

15.24 Disconnecting the remote control switch wiring from the rear of the radio/cassette player

19 Disconnect the wiring plugs from the switches.

20 Where applicable, slide the radio/cassette player remote control switch from the bracket on the stalk switch assembly.

21 Working at the top of the switch assembly, loosen the clamp screw, then withdraw the assembly from the steering column (photos).

22 Refitting is a reversal of removal. Refit the steering wheel as described in Chapter 10.

Radio/cassette player remote control switch

23 Disconnect the battery negative lead.

24 Remove the radio/cassette player as described in Section 33. Disconnect the remote control switch wiring plug from the rear of the unit (photo).

25 Remove the steering column shrouds as described in Chapter 11, Section 26.

26 Slide the switch from the bracket on the stalk switch assembly (photo).

27 Feed the wiring through from behind the facia, noting its routing, and remove the switch.

28 Refitting is a reversal of removal, bearing in mind the following points.

29 Refit the radio/cassette player with reference to Section 33.

30 Ensure that the wiring is routed as noted during removal.

31 Do not fully tighten the switch clamp screw until the steering column shrouds have been refitted.

Facia-mounted rocker switches

32 Disconnect the battery negative lead.

33 Using two screwdrivers, carefully prise the switch to release it from the facia, taking care not to damage the facia trim (photo).

34 Pull the switch from the facia and disconnect the wiring plug.

35 Refitting is a reversal of removal.

Headlight aim adjustment control

36 The headlight aim adjustment control, hoses and actuators form a complete assembly; no parts are available separately. For further details, see Section 20.

Instrument panel illumination control

37 Disconnect the battery negative lead.

38 Using a screwdriver, carefully prise the

control from the facia, taking care not to damage the facia trim (photo).

39 Pull the control from the facia, and disconnect the wiring plug (photo).

40 Refitting is a reversal of removal.

Central locking operating switches

41 Refer to Chapter 11, Section 14.

Electric window operating switches

42 Refer to Chapter 11, Section 15.

Courtesy light switches

43 The switches are located in the front door pillars.

44 Disconnect the battery negative lead.

45 Pull the rubber cover from the switch (photo).

15.26 Removing the radio/cassette player remote control switch

15.33 Removing a facia-mounted rocker switch

15.38 Prise the instrument panel illumination control from the facia . . .

15.39 . . . then pull the control free and disconnect the wiring plug

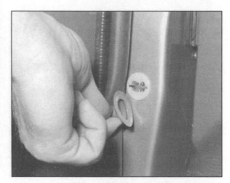

15.45 Pull the rubber cover from the courtesy light switch . . .

12

15.46 . . . then pull the switch from the door pillar

15.50 Disconnecting the wiring from the luggage compartment light switch

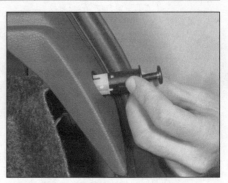

15.51 Removing the luggage compartment light switch

46 Carefully prise the switch from the door pillar, and disconnect the wiring plug (photo). Take care not to allow the wiring to drop down into the door pillar while the switch is removed - tape it to the door pillar if necessary.
47 Refitting is a reversal of removal.

Luggage compartment light switch

48 Disconnect the battery negative lead.
49 Open the tailgate. Pull the luggage compartment right-hand side trim away from the body for access to the rear of the switch.
50 Disconnect the wiring plug from the switch (photo).
51 Push the switch out through the rear quarter trim panel, and remove it (photo).
52 Refitting is a reversal of removal.

16.3 Pulling the wiring plug from the rear of the headlight

Glovebox light switch

53 The glovebox light switch is incorporated in the glovebox lock striker. For details see Chapter 11, Section 26.

Map reading light switch

54 The map reading light switch is integral with the roof console, and cannot be removed separately. Removal and refitting of the roof console panel is described in Chapter 11, Section 24.

16 Bulbs (exterior lights) - renewal

General

1 Whenever a bulb is renewed, note the following points.
(a) *Disconnect the battery negative lead, or at least make sure that the lighting circuit is switched off, before starting work.*
(b) *Remember that if the light has recently been in use, the bulb may be extremely hot.*
(c) *Always check the bulb contacts and/or holder (as applicable). Ensure that there is clean metal-to-metal contact between the bulb contacts and the contacts in the holder, and/or the holder and the wiring plug. Clean off any corrosion or dirt before fitting a new bulb.*
(d) *Ensure that the new bulb is of the correct rating and that it is completely clean*

before fitting; this applies particularly to headlight bulbs.

Headlight

2 Working in the engine compartment, squeeze the retaining clip(s) and remove the plastic cover from the rear of the headlight.
3 Pull the wiring plug from the rear of the headlight bulb (photo).
4 Release the bulb retaining spring clip. Grasp the bulb by its contacts and carefully withdraw it from the headlight unit (photo).
5 When handling the new bulb, use a tissue or clean cloth to avoid touching the glass with the fingers; moisture and grease from the skin can cause blackening and rapid failure of this type of bulb.

> **HAYNES HiNT** *If the headlight glass is accidentally touched, wipe it clean using methylated spirit.*

6 Refitting is a reversal of removal. Note that the headlight cover is refitted with the cut-out for the wiring at the bottom.

Front sidelight

7 Working in the engine compartment, squeeze the retaining clip(s) and remove the plastic cover from the rear of the headlight.
8 Pull the wiring plug from the rear of the sidelight bulbholder.
9 Twist the bulbholder anti-clockwise, and remove it from the rear of the headlight assembly (photo).

16.4 Withdrawing a headlight bulb

16.9 Removing the sidelight bulbholder

16.10 Removing the sidelight bulb from the bulbholder

16.15 Removing a front direction indicator light bulb

16.18 Prise the direction indicator side repeater light from the trim plate . . .

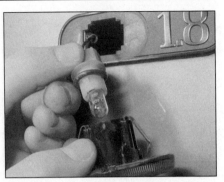
16.19 . . . remove the bulbholder . . .

10 The bulb is a push-fit in the bulbholder (photo).
11 Refitting is a reversal of removal, bearing in mind the following points.
12 Ensure that the rubber seal on the rear of the bulbholder is in good condition and is correctly fitted.
13 Refit the headlight cover with the cut-out for the wiring at the bottom.

Front direction indicator light

14 On models where the bulbholder is accessible from the engine compartment, twist the bulbholder anti-clockwise through a quarter of a turn, then withdraw the bulbholder.
15 On some models, access to the bulbholder is not possible from the engine compartment. Remove the light unit as described in Section 18, then twist the bulbholder anti-clockwise to remove it (photo).
16 The bulb is a bayonet fit in the bulbholder.
17 Refitting is a reversal of removal.

Front direction indicator side repeater light

18 Carefully prise the light unit from the trim plate on the front wing (photo).
19 Twist the bulbholder anti-clockwise, and withdraw it from the rear of the light unit (photo).

20 The bulb is a push-fit in the bulbholder (photo).
21 Refitting is a reversal of removal.

Front foglight

22 Unscrew the two securing screws, and remove the light from the front bumper (photo).
23 Disconnect the wiring plug (photo).
24 Twist the plastic cover anti-clockwise, and withdraw it from the rear of the light.
25 Disconnect the wiring from the rear of the bulb.
26 Release the bulb retaining spring clip, then grasp the bulb by its contacts and carefully withdraw it from the light unit (photo).

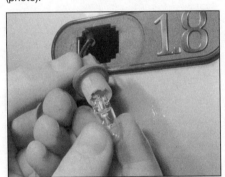
16.20 . . . and withdraw the bulb

27 When handling the new bulb, use a tissue or clean cloth to avoid touching the glass with the fingers; moisture and grease from the skin can cause blackening and rapid failure of this type of bulb. If the glass is accidentally touched, wipe it clean using methylated spirit.
28 Refitting is a reversal of removal.

Rear light cluster

29 Open the tailgate, and unscrew the rear light cluster securing nut (photo).
30 Working outside the car, tilt the light cluster and lift it from the retaining lugs on the body. Disconnect the wiring plug.
31 Squeeze the securing clips towards the centre of the bulbholder, and pull the

16.22 Unscrewing a front foglight securing screw

16.23 Disconnecting a front foglight wiring plug

16.26 Removing a front foglight bulb

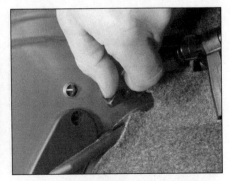
16.29 Unscrewing a rear light cluster securing nut

12

16.31A Squeeze the securing clips . . .

16.31B . . . and remove the rear light cluster bulbholder

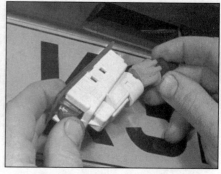

16.34 Removing the rear number plate light and wiring plug

bulbholder from the rear of the light cluster (photos).

32 The bulbs are a bayonet fit in their holders.

33 Refitting is a reversal of removal.

Rear number plate light

34 Using a screwdriver, carefully prise the number plate light assembly from its location in the bumper. Disconnect the wiring plug (photo).

35 Release the securing tabs and unclip the lens from the light assembly (photo).

36 Pull the bulb from its contacts.

37 Refitting is a reversal of removal. Ensure that the rubber seal is correctly located when refitting the lens.

17 Bulbs (interior lights) - renewal

General

1 Refer to Section 16, paragraph 1.

Courtesy light

2 Unclip the cover from the roof console for access to the bulb (photo).

3 The bulb is a push-fit in its contacts (photo).

4 Refitting is a reversal of removal.

Map reading light

5 Proceed as described previously for the courtesy light.

Glovebox light

6 Prise the light assembly from its location in the glovebox (photo).

7 The bulb is a push-fit in its contacts.

8 Refitting is a reversal of removal.

Luggage compartment light

9 The procedure is as described previously for the glovebox light.

Instrument panel illumination and warning light bulbs

10 Refer to Section 22.

Auxiliary gauge illumination bulbs

11 Refer to Section 23.

Cigarette lighter illumination bulb

12 Pull the ashtray and housing from the centre console. Working through the aperture, push the cigarette lighter mounting panel from the centre console (photo).

13 Disconnect the wiring from the cigarette lighter, and from any switches mounted in the panel (photo).

14 Working at the rear of the cigarette lighter, pull the plastic cover from the bulb (photo).

15 Using a small screwdriver inserted through the rear of the bulbholder, push the bulb from its contacts.

16 Push the new bulb into the bulbholder, then reassemble and refit the cigarette lighter using a reversal of the removal procedure.

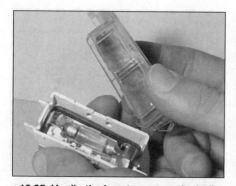

16.35 Unclip the lens to expose the bulb

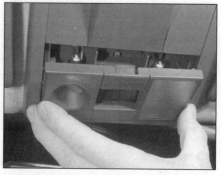

17.2 Remove the cover from the courtesy light . . .

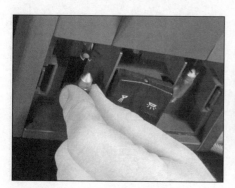

17.3 . . . for access to the bulb

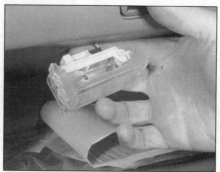

17.6 Removing the glovebox light

17.12 Push the cigarette lighter mounting panel from the centre console . . .

17.13 . . . and disconnect the wiring plugs

17.14 Pull off the plastic cover to expose the cigarette lighter bulb

18.6A Unscrewing the headlight front securing screw. Upper securing screws arrowed

Ashtray illumination bulb

17 The ashtray shares the same illumination bulb as the cigarette lighter. Refer to paragraphs 12 to 16 for details.

Clock illumination bulb

18 Remove the clock, as described in Section 24.
19 Twist the bulbholder anti-clockwise, and pull it from the rear of the clock.
20 The bulb is integral with the bulbholder.
21 Refitting is a reversal of removal.

18 Exterior light units - removal and refitting

Headlight unit

1 Remove the direction indicator light unit, as described later in this Section.
2 Remove the radiator grille panel, as described in Chapter 11.
3 Working at the rear of the headlight, squeeze the retaining clip(s) and remove the plastic cover from the rear of the headlight.
4 Disconnect the wiring plugs from the rear of the headlight and sidelight bulbs.
5 On models fitted with a remote headlight aim adjustment control, free the actuator from the rear of the headlight, by twisting the actuator anti-clockwise and pulling it sharply to release the balljoint.
6 Unscrew the three upper securing screws,

and the single front securing screw, then withdraw the headlight (photos).
7 Refitting is a reversal of removal, but where applicable, take care not to damage the headlight aim adjustment actuator balljoint when reconnecting it to the headlight.

Beam alignment

8 On completion, the headlight beam alignment should be checked, ideally using optical setting equipment. This check can be carried out by a Renault dealer or a suitably-equipped garage. The beam alignment is adjusted using the two screws provided on the light unit (photos).

Front direction indicator light unit

9 Open the bonnet and disconnect the battery negative lead.

18.6B Withdrawing a headlight

10 Working through the hole in the top of the engine compartment front panel, depress the direction indicator light unit retaining tab with a screwdriver (photo).
11 Pull the light unit forwards, and disconnect the wiring plug from the bulbholder (photo).
12 Refitting is a reversal of removal. Ensure that the weatherseal is correctly located between the light unit and the body.

Front direction indicator side repeater light

13 Disconnect the battery negative lead.
14 Carefully prise the light unit from the trim plate on the front wing (photo). Disconnect the wiring plug from the rear of the bulbholder.
15 Refitting is a reversal of removal.

18.8A Headlight beam vertical adjustment screw . . .

18.8B . . . and horizontal adjustment screw

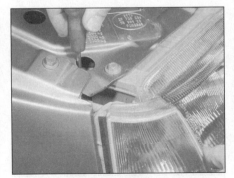

18.10 Depressing the direction indicator light unit retaining tab

18.11 Disconnecting the direction indicator light unit wiring plug

12

18.14 Removing the front direction indicator side repeater light

Front foglight

16 Open the bonnet and disconnect the battery negative lead.
17 Unscrew the two securing screws from the front of the foglight, then pull the light from the bumper, and disconnect the wiring plug.
18 Refitting is a reversal of removal. If necessary, the vertical alignment of the beam can be adjusted by turning screw 'B' shown in Fig. 12.16.

Rear light cluster

19 Open the bonnet and disconnect the battery negative lead.
20 Open the tailgate and unscrew the plastic nut from the rear of the light cluster.
21 Withdraw the light cluster from outside the car, taking care not to damage the retaining lugs at the bottom of the light, then disconnect the wiring plugs (photo).
22 Where applicable, feed the wiring through the grommet in the bottom of the light cluster. Withdraw the assembly from the car.
23 Refitting is a reversal of removal, bearing in mind the following points.
24 Where applicable, make sure that the wiring grommet is correctly positioned in the bottom of the light cluster.
25 Ensure that the retaining lugs at the base of the light cluster are correctly engaged with the corresponding holes in the body.
26 Make sure that the weatherseal is correctly located between the light cluster and the body.

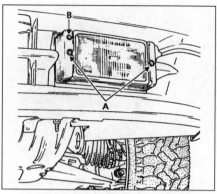

Fig. 12.16 Front foglight securing screws (A), and beam adjustment screw (B) (Sec 18)

Rear number plate light

27 Using a screwdriver, carefully prise the light unit from its location in the bumper.
28 Disconnect the wiring and withdraw the unit from the car.
29 Refitting is a reversal of removal.

19 Interior light units - removal and refitting

Courtesy light

1 Disconnect the battery negative lead.
2 On models with a separate courtesy light and no map reading light, carefully prise the light assembly from its location in the roof panel, and disconnect the wiring plugs.
3 On models with a map reading light, carefully prise the cover from the light assembly. Using a screwdriver, release the securing lugs and pull the light assembly (complete with map reading light and switch) from the roof console panel. Disconnect the wiring plug and withdraw the light unit (photos).
4 Refitting is a reversal of removal.

Map reading light

5 The map reading light is integral with the courtesy light assembly. Removal and refitting are as described previously in this Section.

18.21 Disconnecting a wiring plug from a rear light unit

Glovebox light

6 Disconnect the battery negative lead.
7 Prise the light from its location, and disconnect the wiring plug.
8 Refitting is a reversal of removal.

Luggage compartment light

9 Proceed as described previously for the glovebox light.

20 Headlight aim adjustment components - removal and refitting

Note: *The headlight aim adjustment control, hoses and actuators form a complete assembly, and no parts are available separately.*
1 Remove the driver's side lower facia panel, as described in Chapter 11, Section 26.
2 Working in the engine compartment, release the actuator from each headlight in turn by twisting anti-clockwise and pulling sharply to release the balljoint (photo).
3 Pull the adjuster hose grommet from the bulkhead into the car interior, then manipulate the actuators and the hoses through the hole in the bulkhead into the inside of the car. Take note of the routing of the hoses.
4 Withdraw the assembly from the car.
5 Refitting is a reversal of removal, bearing in mind the following points.
6 Route the hoses as noted during removal.
7 Take care not to damage the actuator balljoints when reconnecting them to the headlights.

19.3A Releasing a courtesy light assembly securing lug

19.3B Disconnecting the courtesy light wiring plug

20.2 Headlight aim adjustment actuator (arrowed) removed from headlight

8 Make sure that the bulkhead grommet is securely located.
9 Refit the driver's side lower facia panel with reference to Chapter 11, Section 26.

21 Instrument panel - removal and refitting

1 Disconnect the battery negative lead.
2 Remove the steering column shrouds and the facia upper trim panel, as described in Chapter 11, Section 26.
3 Unscrew the four securing screws (two upper and two lower) (photos).
4 Pull the instrument panel forwards from the facia, at the same time disconnecting the wiring plugs and the speedometer cable from the rear of the assembly (photos). Where applicable, disconnect the wiring plug from the speed sensor on the end of the speedometer cable.
5 Withdraw the panel from the facia.
6 Refitting is a reversal of removal, ensuring that all wiring plugs and the speedometer cable are securely reconnected.
7 Where applicable, do not fully tighten the radio/cassette player remote control switch clamp screw until the steering column shrouds have been refitted.

22 Instrument panel components - removal and refitting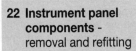

1 With the instrument panel removed as described in Section 21, proceed as follows.

Panel illumination and warning light bulbs

2 Twist the appropriate bulbholder anti-clockwise, and withdraw it from the printed circuit board on the rear of the instrument panel (photo).
3 The bulbs cannot be separated from the bulbholders.
4 Refitting is a reversal of removal.

Instruments

5 Pull the trip meter reset button from the front of the instrument panel.
6 Working at the rear of the instrument panel,

21.3A Unscrew the instrument panel upper . . .

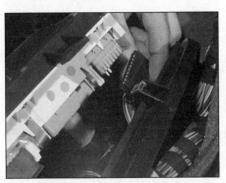

21.4A Disconnect the wiring plugs . . .

remove the securing screws, or release the securing clips, as applicable. Withdraw the lens/cover assembly from the front of the instrument panel.
7 Remove the appropriate securing screws from the rear of the panel, then withdraw the instrument from the front of the panel (photo).
8 Refitting is a reversal of removal.

Printed circuit board

Note: *No attempt should be made to remove the printed circuit board unless it is known to be faulty, and is to be renewed.*

9 Remove the illumination and warning light bulbs, and the instruments, as described previously in this Section.
10 The circuit board can now be unclipped

21.3B . . . and lower securing screws

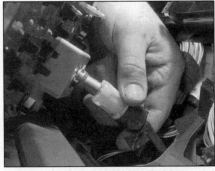

21.4B . . . and the speedometer cable from the rear of the instrument panel

from the rear of the instrument panel, noting how it is located.
11 Refitting is a reversal of removal, remembering that the circuit board is fragile. Ensure that it is located as noted before removal, and that it fits correctly over the wiring connectors at the top of the panel.

23 Auxiliary gauges (16-valve models) - removal and refitting

1 Disconnect the battery negative lead.
2 Remove the facia upper trim panel, as described in Chapter 11, Section 26.
3 Remove the two securing screws from the top of the auxiliary gauge panel, then pull panel forwards from the facia (photo).

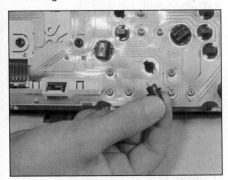

22.2 Withdrawing an instrument panel bulb

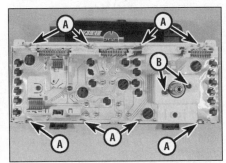

22.7 Instrument lens/cover assembly securing screws (A), and speedometer securing screws (B)

23.3 Auxiliary gauge panel securing screws (arrowed) - 16-valve model

12

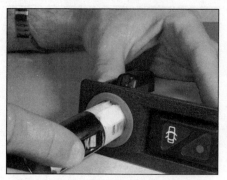

26.5 Withdrawing the cigarette lighter metal housing . . .

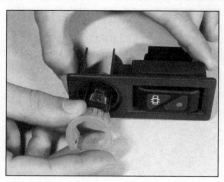

26.6 . . . and the plastic surround

27.1 Horn location (arrowed) viewed from underneath with front wheel arch splash shield removed

4 Reach up behind the panel and disconnect the wiring plugs, then withdraw the panel from the facia.

5 To remove the components from the instrument panel, proceed as described in Section 22 for the main instruments.

6 Refitting is a reversal of removal.

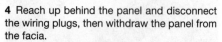

24 Clock - removal and refitting

Lower facia-mounted clock

1 Disconnect the battery negative lead.

2 Remove the radio/cassette player, as described in Section 33.

3 Remove the radio/cassette player housing, as described in Chapter 11, Section 26.

4 Reach up behind the clock, and push it out through the front of the facia.

5 Disconnect the wiring plug and withdraw the clock.

6 Refitting is a reversal of removal.

Upper facia-mounted clock

7 No information was available regarding the removal and refitting of the upper facia-mounted clock at the time of writing.

25 Outside air temperature display components - removal and refitting

Instrument panel-mounted display

1 Proceed as described in Section 22 for the instruments.

Upper facia-mounted display

2 No information was available regarding the removal and refitting of the upper facia-mounted display at the time of writing.

Sensor

3 The sensor is mounted in the bottom of the left-hand exterior mirror.

4 Disconnect the battery negative lead.

5 Open the door, and carefully prise the

mirror trim panel from the front edge of the door.

6 Unscrew the two screws securing the mirror cover panel to the door, then disconnect the sensor wiring plug.

7 Remove the mirror glass, as described in Chapter 11, Section 16.

8 Remove the single securing screw, and withdraw the sensor from the mirror, feeding the wiring through the mirror body.

9 Refitting is a reversal of removal.

26 Cigarette lighter - removal and refitting

1 Disconnect the battery negative lead.

2 Pull the ashtray and housing from the centre console. Working through the aperture, push the cigarette lighter mounting panel from the centre console.

3 Disconnect the wiring plugs from the cigarette lighter, and from any switches mounted in the panel.

4 Remove the cigarette lighter from its housing.

5 Working through the cigarette lighter aperture, depress the retaining lugs, and prise the metal housing from the plastic surround. Push the housing out through the front of the assembly (photo).

6 Release the securing lugs and push the plastic surround out through the front of the mounting panel (photo).

7 Refitting is a reversal of removal.

27 Horn - removal and refitting

1 The horn is located behind the right-hand side of the front bumper (photo).

2 Disconnect the battery negative lead.

3 To gain access to the horn, remove the front wheel arch splash shield. Refer to Chapter 11, Section 20 if necessary.

4 Disconnect the wiring from the horn.

5 Unscrew the securing nut, and withdraw the horn from its mounting bracket.

6 Refitting is a reversal of removal.

28 Speedometer cable - removal and refitting

1 Withdraw the instrument panel from the facia, and disconnect the speedometer cable from the rear of the panel, as described in Section 21.

2 Working in the engine compartment, unclip or unscrew the end of the speedometer cable from the gearbox or transmission casing. Where applicable, note how the securing clip locates on the gearbox/transmission, as several different types of clip are used.

3 Working behind the left-hand side of the facia, locate the speedometer cable grommet in the bulkhead.

4 Pull the grommet and the cable through the bulkhead into the passenger compartment, noting the routing of the cable.

5 Take note of the routing of the cable behind the facia, then feed the cable through the facia and withdraw it from the car. Where applicable, take care not to damage the speed sensor unit on the speedometer end of the cable.

6 Refitting is a reversal of removal, bearing in mind the following points.

7 Make sure that the cable is routed as noted before removal, and be careful not to kink or twist it between the bulkhead and the rear of the instrument panel.

8 Ensure that the bulkhead grommet is correctly located.

29 Wiper arms - removal and refitting

1 The wiper motor should be in the parked position before removing the wiper arm. Mark the position of the blade on the glass with adhesive tape, as a guide to refitting.

2 If both windscreen wiper arms are to be removed, identify them so that they can be refitted in their original positions (the arms are of different lengths).

3 Lift the hinged cover, and remove the nut securing the arm to the spindle (photo).

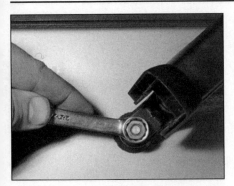

29.3 Unscrewing the tailgate wiper arm securing nut

4 Pull or prise the arm from the spindle, using a screwdriver if necessary. Take care not to damage the trim or paintwork.
5 Refitting is a reversal of removal. Position the arms so that the blades align with the tape applied to the glass before removal.

30 Windscreen wiper motor and linkage - removal and refitting

1 Make sure that the wipers are in the parked position.
2 Disconnect the battery negative lead.
3 Remove the windscreen wiper arms, as described in Section 29.
4 Remove the windscreen cowl panels, as described in Chapter 11.
5 Release the securing clip (pull upwards), and disconnect the wiring plug from the motor (photo).
6 Unscrew the securing bolt, and remove the car jack from its location on the right-hand side of the scuttle.
7 To provide improved clearance for removal, carefully turn the motor and the spindles by hand to position the drive link (connecting the motor to the drive arms) approximately 110° from its parked position - see Fig. 12.17.
8 Unscrew the motor/linkage assembly securing bolts (one at each end, and one or two lower bolts, depending on model), and recover the washers (photo).

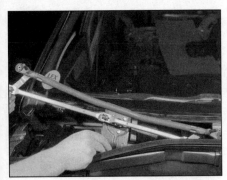

30.11 Withdrawing the windscreen wiper motor/linkage assembly

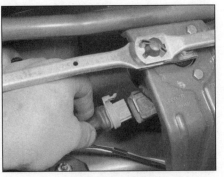

30.5 Disconnecting the windscreen wiper motor wiring plug

9 Note how the assembly is mounted on the body brackets. (On right-hand drive models, the right-hand end of the assembly is mounted under the bracket on the body, and the left-hand end is mounted on top of the bracket.)
10 If necessary, carefully remove the windscreen cowl panel securing clips from the body to provide additional clearance.
11 Manipulate the assembly out through the right-hand side of the scuttle, taking care not to damage surrounding components (photo).
12 Refitting is a reversal of removal, bearing in mind the following points.
13 Ensure that the assembly is mounted on the body brackets as noted before removal.
14 Before refitting the windscreen cowl panels, turn the drive link and the spindles by hand to the parked position.
15 Refit the windscreen wiper arms with reference to Section 29.

31 Tailgate wiper motor and linkage - removal and refitting

1 Disconnect the battery negative lead.
2 Remove the tailgate wiper arm, as described in Section 29.
3 Open the tailgate, then remove the tailgate interior trim panel.
4 Disconnect the motor wiring plug.
5 Unscrew the three screws securing the

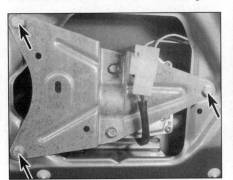

31.5A Tailgate wiper motor securing screws (arrowed)

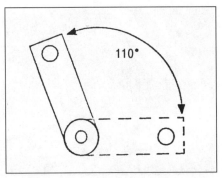

Fig. 12.17 Drive link positioned for removal of windscreen wiper motor and linkage assembly (Sec 30)

30.8 Unscrewing a windscreen wiper motor lower securing bolt

motor mounting bracket to the tailgate. Withdraw the motor and bracket (photos).
6 Refitting is a reversal of removal.

32 Windscreen/tailgate/headlight washer system components - removal and refitting

Fluid reservoir

1 Disconnect the battery negative lead.
2 Remove the windscreen wiper arms as described in Section 29.
3 Remove the windscreen cowl panels as described in Chapter 11.

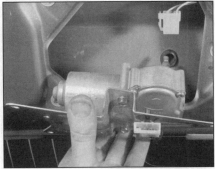

31.5B Withdrawing the tailgate wiper motor assembly

12

32.4 Unscrewing the washer fluid reservoir securing bolt

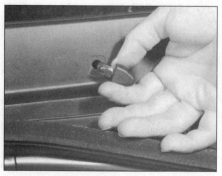

32.11A Removing a windscreen washer nozzle

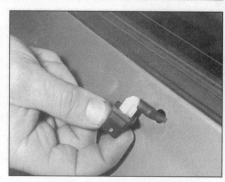

32.11B Removing the tailgate washer nozzle

4 Unscrew the fluid reservoir securing bolt (photo), then lift the reservoir sufficiently to disconnect the hoses and wiring from the pumps. Be prepared for fluid spillage.
5 Manipulate the reservoir out from the scuttle.
6 Refitting is a reversal of removal. Refit the windscreen wiper arms with reference to Section 29.

Fluid pumps

7 Disconnect the battery negative lead.
8 Open the access hatch in the windscreen cowl panel.
9 Carefully pull the appropriate pump from the reservoir, then disconnect the wiring plug

33.4 Using DIN standard removal clips to remove the radio/cassette player

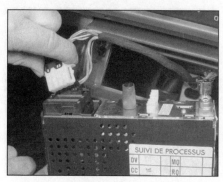

33.5 Disconnecting the radio/cassette player wiring plugs

and the hose. Be prepared for fluid spillage.
10 Refitting is a reversal of removal.

Washer nozzles

11 Carefully prise the nozzle from its mounting, then disconnect the fluid hose (photos).
12 Refitting is a reversal of removal. The nozzle can be adjusted by inserting a pin into the jet, and swivelling it to the required position.

33 Radio/cassette player -
removal and refitting

Removal

1 All the radio/cassette players fitted to the Clio range have DIN standard fixings. A pair of removal clips, obtainable from in-car entertainment specialists, will be required for removal.
2 Disconnect the battery negative lead.
3 Where applicable, prise the plastic covers from the sides of the radio/cassette player.
4 Insert the clips into the holes at the sides of the unit until they snap into place. Pull the clips rearwards (away from the facia) to release the unit (photo).
5 Withdraw the radio/cassette unit from the facia. Disconnect the wiring plugs and the aerial cable from the rear of the unit (photo).

Refitting

6 Reconnect the wiring plugs and the aerial cable to the rear of the unit.
7 Push the unit into its housing in the facia until the retaining lugs snap into place.
8 Where applicable, refit the covers to the sides of the unit, then reconnect the battery negative lead.

34 Loudspeakers -
removal and refitting

Front door-mounted loudspeakers

1 Disconnect the battery negative lead.
2 Remove the four securing screws, then withdraw the loudspeaker cover and the loudspeaker from the housing in the door. Note that the loudspeaker housing is also secured by the four screws (photos).
3 Disconnect the wiring and remove the loudspeaker.
4 Refitting is a reversal of removal. Ensure that the plastic housing is correctly positioned in the door before refitting the loudspeaker.

Front facia-mounted loudspeakers

5 Disconnect the battery negative lead.
6 Remove the three securing screws (two lower screws, and one screw at the side nearest the door), and withdraw the facia side vent.

34.2A Withdrawing a front door-mounted loudspeaker

34.2B Removing the front door-mounted loudspeaker housing

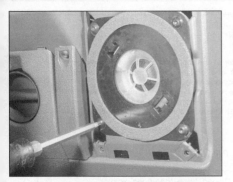

34.7 Unscrewing a facia-mounted loudspeaker securing screw

34.8 Removing a facia-mounted loudspeaker

34.11A Rear loudspeaker securing screws (arrowed)

7 Unscrew the four loudspeaker securing screws (photo).
8 Withdraw the loudspeaker from the facia, and disconnect the wiring plugs (photo).
9 Refitting is a reversal of removal.

Rear loudspeakers

10 Disconnect the battery negative lead.
11 Open the tailgate. Working under the rear quarter trim panel, unscrew the four loudspeaker securing screws. Lower the loudspeaker and its plastic shield, then disconnect the wiring plugs (photos).
12 Refitting is a reversal of removal.

35 Radio aerial - removal and refitting

Aerial assembly

1 Remove the courtesy light assembly, as described in Section 19, to expose the base of the aerial.
2 Unscrew the securing sleeve, and disconnect the aerial lead from the base of the aerial.
3 Unscrew the securing nut and recover the washers, then lift the aerial from the roof (photo).
4 Refitting is a reversal of removal.

Aerial lead

5 With the lead disconnected from the aerial

as described previously in this Section, tie a length of string to the end of the lead.
6 Observe the routing of the lead. Remove the A-pillar trim panel for access to the appropriate side, as described in Chapter 11, Section 24.
7 Remove the radio/cassette player and disconnect the aerial lead from the rear of the unit, as described in Section 33.
8 Pull the lower end of the lead. Feed the lead down the A-pillar, behind the facia, and out through the radio/cassette player aperture.
9 Untie the string from the lead, and leave it in position to aid refitting.
10 Refitting is a reversal of removal, using the string to pull the lead into position. Take care not to damage the lead or surrounding

34.11B Withdrawing the rear loudspeaker and its plastic shield

components when feeding it through behind the facia.
11 Refit the radio/cassette player with reference to Section 33.

36 Anti-theft alarm system - general information

Certain models are fitted with an anti-theft alarm system, which uses various sensing systems and warning sirens, depending on model.

No information was available for the alarm systems at the time of writing. Any faults should be referred to a Renault dealer for diagnosis.

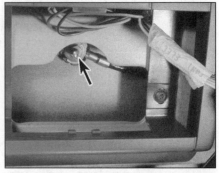

35.3 Radio aerial securing nut (arrowed)

NOTES:

1. All diagrams are divided into numbered circuits depending on function e.g. Diagram 2: Exterior lighting.
2. Items are arranged in relation to a plan view of the vehicle.
3. Wires may interconnect between diagrams and are located by using a grid reference e.g. 2/A1 denotes a position on diagram 2 grid location A1.
4. Complex items appear on the diagrams as blocks and are expanded on the internal connections page.
5. Brackets show how the circuit may be connected in more than one way.
6. Not all items are fitted to all models.

INTERNAL CONNECTION DETAILS

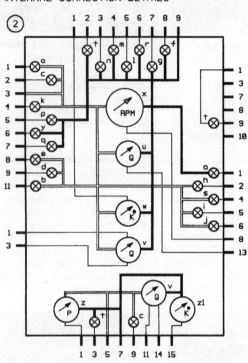

ENGINE FUSEBOX

FUSE	RATING	CIRCUIT
1	5A	Reversing lamps, alarm
2	25A	RH Dipped beam, headlamp washer
3	15A	LH Dipped beam
4	20A	Foglamps front
5	25A	Cooling fan
6	30A	Engine functions
7		Spare
8	20A	Horn
9		Spare
10	25A	Engine functions
11	5A	Automatic transmission
12	15A	RH Main beam
13	15A	LH Main beam
14	20A	Fuel pump
15	30A	Air conditioning

PASSENGER FUSEBOX

FUSE	RATING	CIRCUIT
1	30A	RH Electric window
2	15A	Anti-lock braking system
3	30A	LH Electric window
4	20A	Heated Seats
5	15A	Interior lamp, radio, alarm
6	30A	Accessories cut-off
7	10A	Clock
8	15A	Wash/wipe front
9	15A	Cigar lighter
10	10A	Stop lamps
11	10A	Radio
12	20A	Central locking, electric mirrors
13	20A	Heated rear window and mirrors
14	10A	Wiper timer front
15		+ ve before ignition for caravan option
16	10A	Direction indicator flasher unit
17	10A	Foglamp rear
18	10A	LH Side/tail lamps
19	10A	RH Side/tail lamps
20		Spare
21	20A	Heater blower
22	10A	Radio, alarm

KEY TO INSTRUMENT CLUSTER (ITEM 2)

a = Hazard Warning Lamp
b = No Charge Warning Lamp
c = Oil Pressure Warning Lamp
d = Brake System Warning Lamp
e = High Temp. Warning Lamp
f = Rear Foglamp Warning Lamp
g = Heated Rear Window Warning Lamp
h = Brake Pad Wear Warning Lamp
i = Choke Warning Lamp
j = Auto. Trans./Injection Warning Lamp
k = Pre Heater Warning Lamp
l = Sidelamp Warning Lamp
m = Dipped Beam Warning Lamp
n = Main Beam Warning Lamp
o = Low Fuel Warning Lamp
p = Direction Indicator LH
q = Direction Indicator RH
r = Front Foglamp Warning Lamp
s = ABS Warning Lamp
t = Instrument Illumination
u = Fuel Gauge
v = Oil Level Gauge
w = Coolant Temp. Gauge
x = Tachometer
y = Direction Indicator Repeater
z = Oil Pressure Gauge
z1 = Oil Temp. Gauge

KEY TO SYMBOLS

PLUG-IN CONNECTOR

EARTH

BULB

DIODE

LINE CONNECTORS   B3   A107

FUSE/
FUSIBLE LINK

WIRE COLOURS

Ba	White	No	Black
Be	Blue	Or	Orange
Bj	Beige	Rg	Red
Cy	Clear	Sa	Pink
Gr	Grey	Ve	Green
Ja	Yellow	Vi	Mauve
Ma	Brown		

H24400
T.M.MARKE

Notes, internal connection details, key to symbols and wire colours

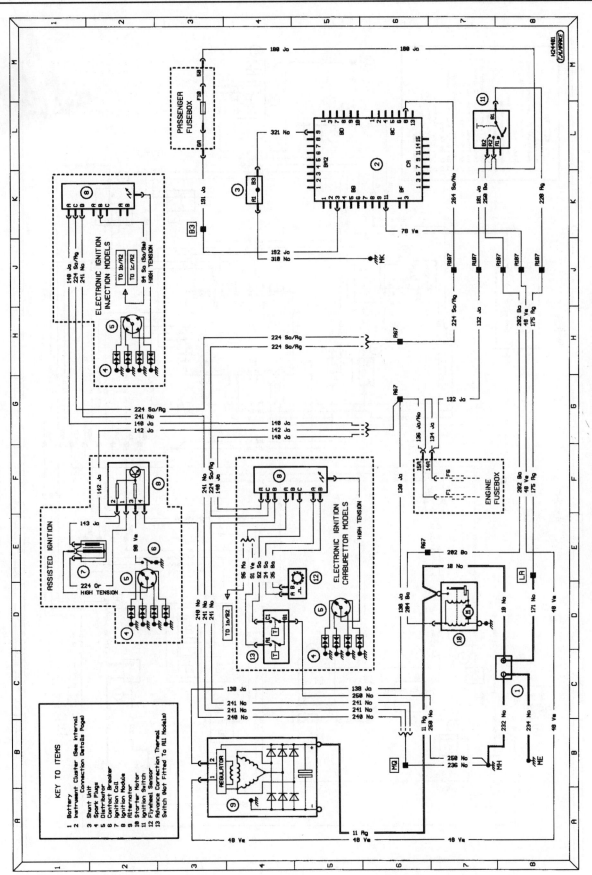

Diagram 1: Starting (manual transmission), charging and ignition

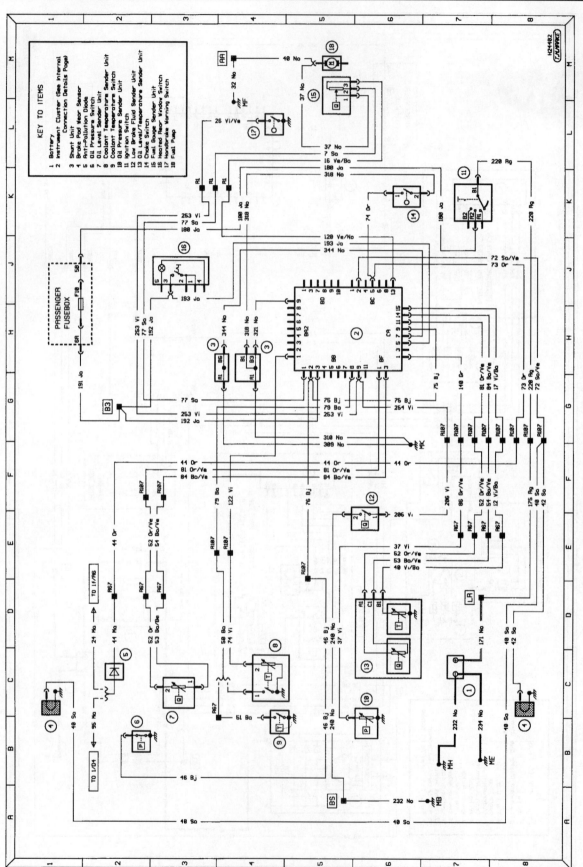

KEY TO ITEMS

1 Battery
2 Instrument Cluster (See Internal Connection Details Page)
3 Shunt Unit
4 Brake Pad Wear Sensor
5 Anti-Pollution Diode
6 Oil Pressure Switch
7 Oil Level Sender Unit
8 Coolant Temperature Sender Unit
9 Coolant Temperature Switch
10 Oil Pressure Sender Unit
11 Ignition Switch
12 Low Brake Fluid Sender Unit
13 Oil Level/Temperature Sender Unit
14 Choke Switch
15 Fuel Gauge Sender Unit
16 Heated Rear Window Switch
17 Handbrake Warning Switch
18 Fuel Pump

PASSENGER FUSEBOX

TO 1f/MB

TO 1/D4

Diagram 1a: Warning lamps and gauges

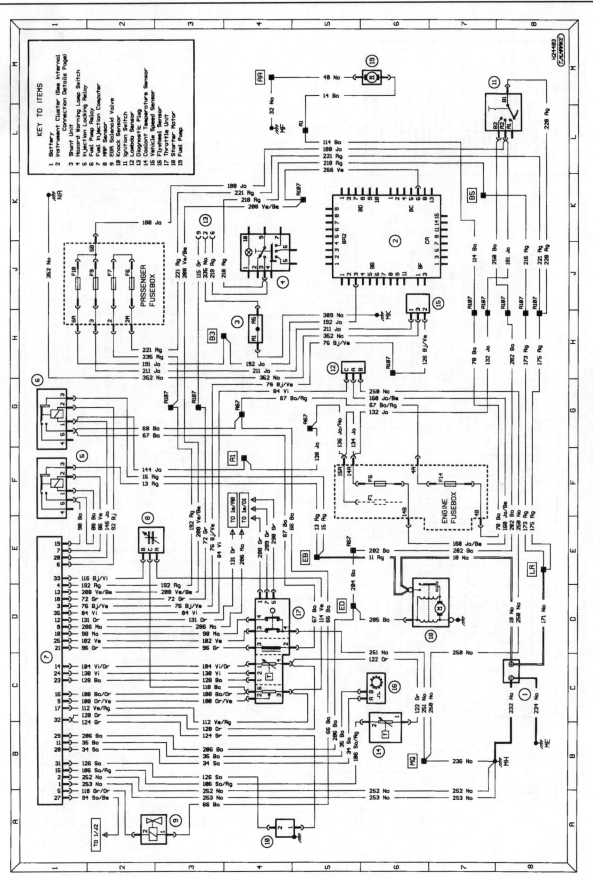

Diagram 1b: Single-point fuel injection

12

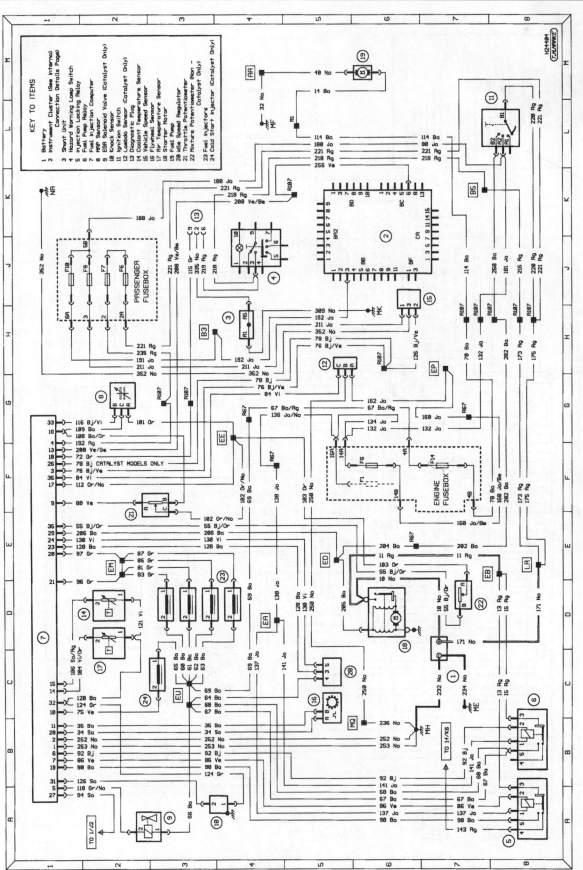

Diagram 1c: Multi-point fuel injection

KEY TO ITEMS

1 Battery
2 Instrument Cluster (See Internal Connection Details Page)
3 Shunt Unit
4 Hazard Warning Loop Switch
5 Injection Locking Relay
6 Fuel Pump Relay
7 Fuel Injection Computer
8 MAP Sensor
9 EGR Solenoid Valve (Catalyst Only)
10 Knock Sensor
11 Ignition Switch
12 Lambda Sensor (Catalyst Only)
13 Diagnostic Plug
14 Coolant Temperature Sensor
15 Vehicle Speed Sensor
16 Air Temperature Sensor
17 Flywheel Sensor
18 Starter Motor
19 Fuel Pump
20 Idle Speed Regulator
21 Throttle Potentiometer
22 Mixture Potentiometer (Non – Catalyst Only)
23 Fuel Injectors
24 Cold Start Injector (Catalyst Only)

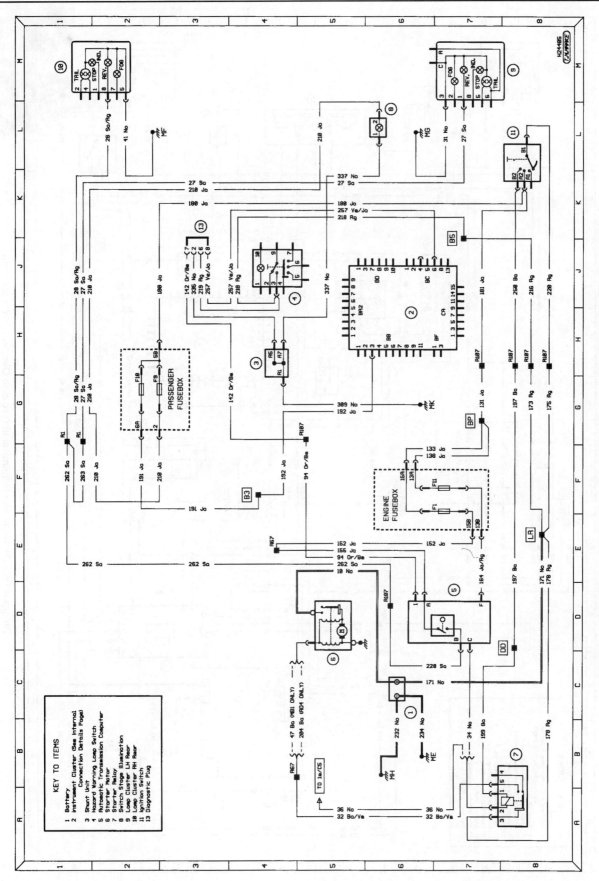

Diagram 1d: MB1 automatic transmission (including starting for AD4 automatic transmission)

KEY TO ITEMS

1 Battery
2 Instrument Cluster (See Internal Connection Details Page)
3 Shunt Unit
4 Hazard Warning Loop Switch
5 Automatic Transmission Computer
6 Starter Motor
7 Starter Relay
8 Switch Stage Illumination
9 Loop Cluster LH Rear
10 Loop Cluster RH Rear
11 Ignition Switch
13 Diagnostic Plug

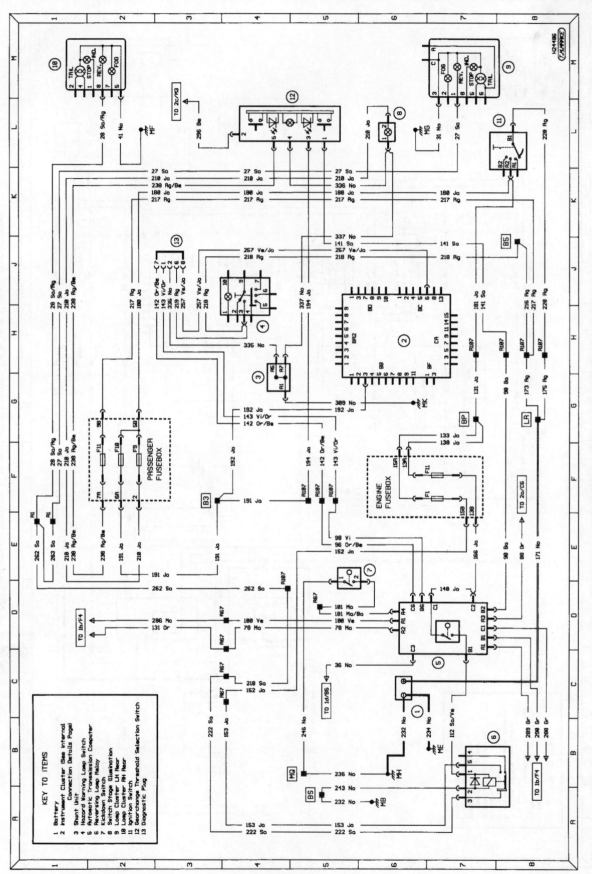

Diagram 1e: AD4 automatic transmission (see also Diagram 1d for starting)

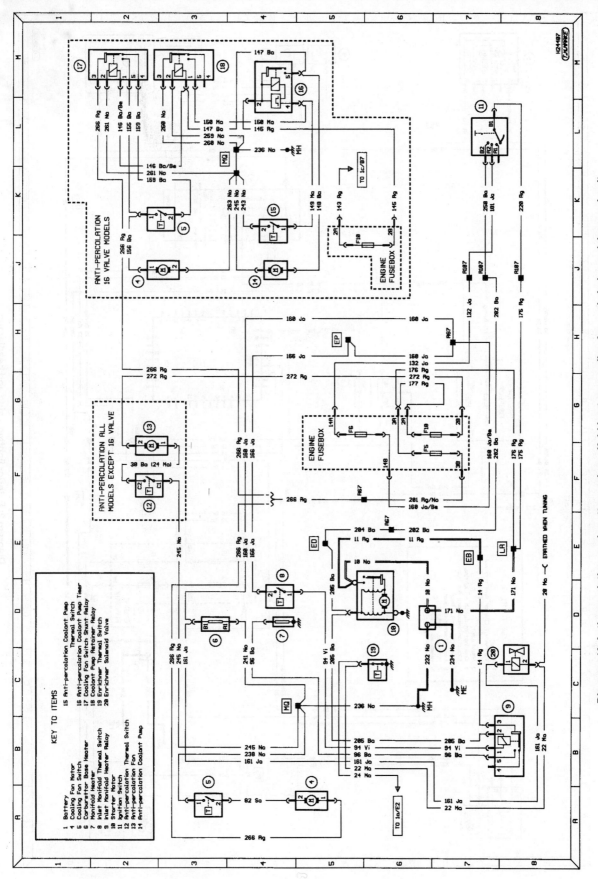

Diagram 1f: Anti-percolation, carburettor and manifold heaters, cooling fan and choke enrichener circuit

KEY TO ITEMS

1 Battery
4 Cooling Fan Motor
5 Cooling Fan Switch
6 Carburettor Base Heater
7 Manifold Heater
8 Inlet Manifold Thermal Switch
9 Inlet Manifold Heater Relay
10 Starter Motor
11 Ignition Switch
12 Anti-percolation Thermal Switch
13 Anti-percolation Fan
14 Anti-percolation Coolant Pump

15 Anti-percolation Coolant Pump
    Thermal Switch
16 Anti-percolation Coolant Pump Timer
17 Cooling Fan Switch Shunt Relay
18 Coolant Pump Retainer Relay
19 Enrichener Thermal Switch
20 Enrichener Solenoid Valve

12

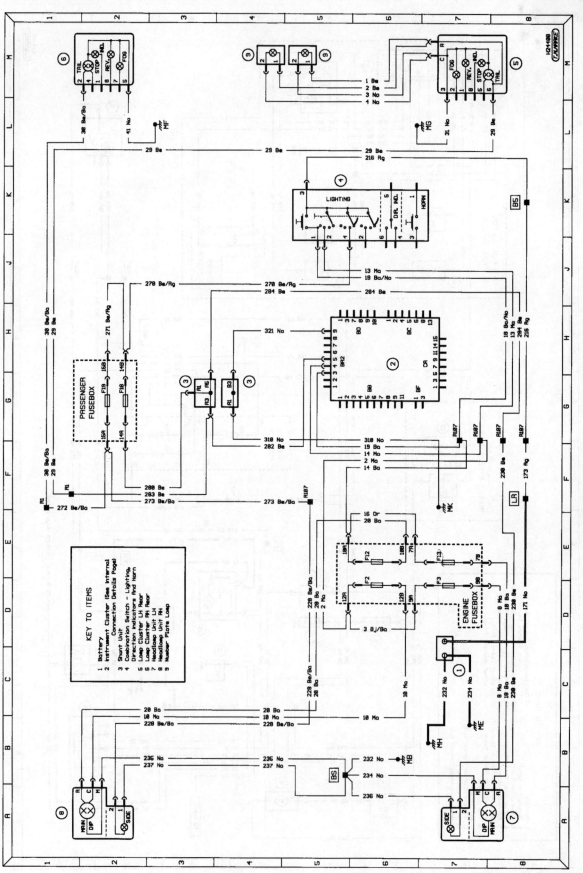

Diagram 2: Exterior lighting - side and headlamps

KEY TO ITEMS

1 Battery
2 Instrument Cluster (See Internal Connection Details Page)
3 Shunt Unit
4 Combination Switch - Lighting, Direction Indicators And Horn
5 Loop Cluster LH Rear
6 Loop Cluster RH Rear
7 Headlamp Unit LH
8 Headlamp Unit RH
9 Number Plate Loop

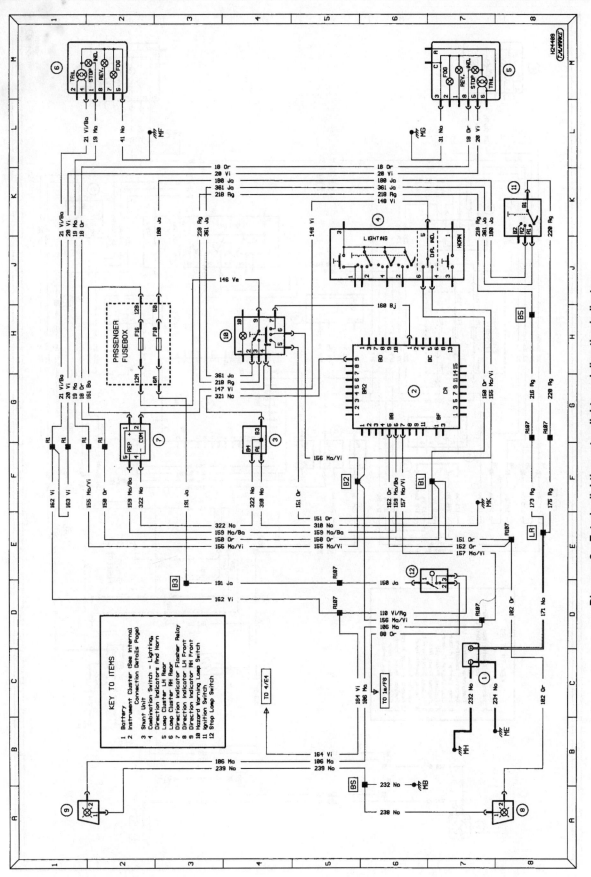

Diagram 2a: Exterior lighting - stop-lights and direction indicators

KEY TO ITEMS

1 Battery
2 Instrument Cluster (See internal Connection Details Page)
3 Shunt Unit
4 Combination Switch - Lighting Indicators And Horn
5 Direction Indicators And Horn
6 Lamp Cluster LH Rear
7 Lamp Cluster RH Rear
8 Direction Indicator LH Front
9 Direction Indicator RH Front
10 Hazard Warning Lamp Switch
11 Ignition Switch
12 Stop Lamp Switch

12

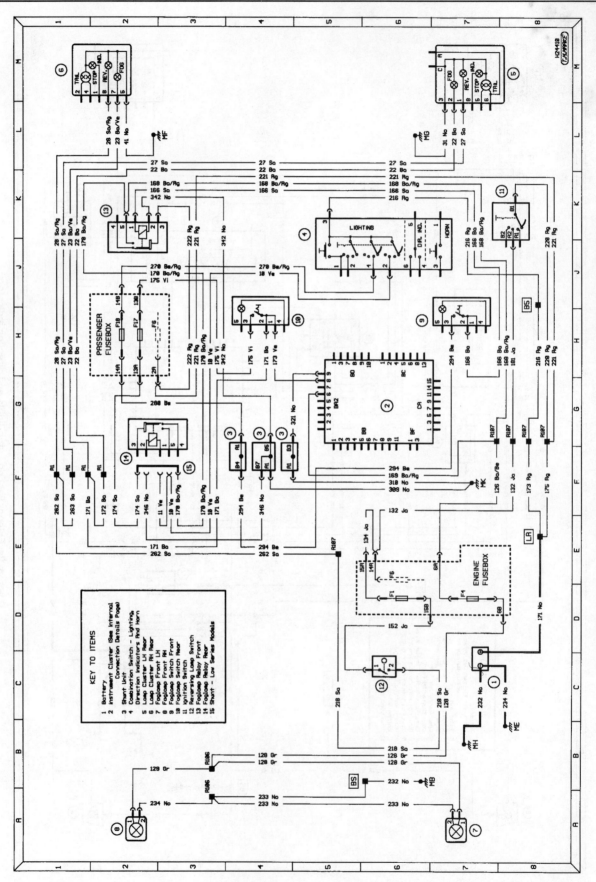

Diagram 2b: Exterior lighting - reversing lights (manual gearbox models only) and foglights

KEY TO ITEMS

1 Battery
2 Instrument Cluster (See internal
   Connection Details Page)
3 Shunt Unit
4 Combination Switch - Lighting
   Direction Indicators And Horn
5 Lamp Cluster LH Rear
6 Lamp Cluster RH Rear
7 Fogloamp Front LH
8 Fogloamp Front RH
9 Fogloamp Switch Front
10 Fogloamp Switch Rear
11 Ignition Switch
12 Reversing Lamp Switch
13 Fogloamp Relay Front
14 Fogloamp Relay Rear
15 Shunt - Low Series Models

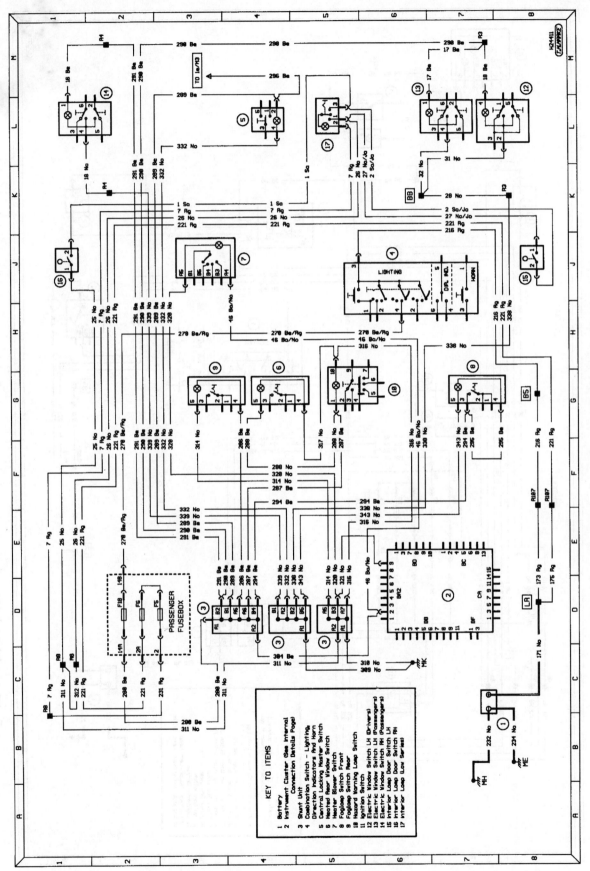

Diagram 2c: Interior lighting - switch illumination and low-series interior light

KEY TO ITEMS

1 Battery
2 Instrument Cluster (See Internal Connection Details on Page)
3 Shunt Unit
4 Combination Switch - Lighting Direction Indicators And Horn
5 Central Locking Master Switch
6 Heated Rear Window Switch
7 Heater Blower Switch
8 Foglamp Switch Front
9 Foglamp Switch Rear
10 Hazard Warning Lamp Switch
11 Ignition Switch
12 Electric Window Switch LH (Drivers)
13 Electric Window Switch LH (Passenger)
14 Electric Window Switch RH (Passenger)
15 Interior Lamp Door Switch LH
16 Interior Lamp Door Switch RH
17 Interior Lamp (Low Series)

12

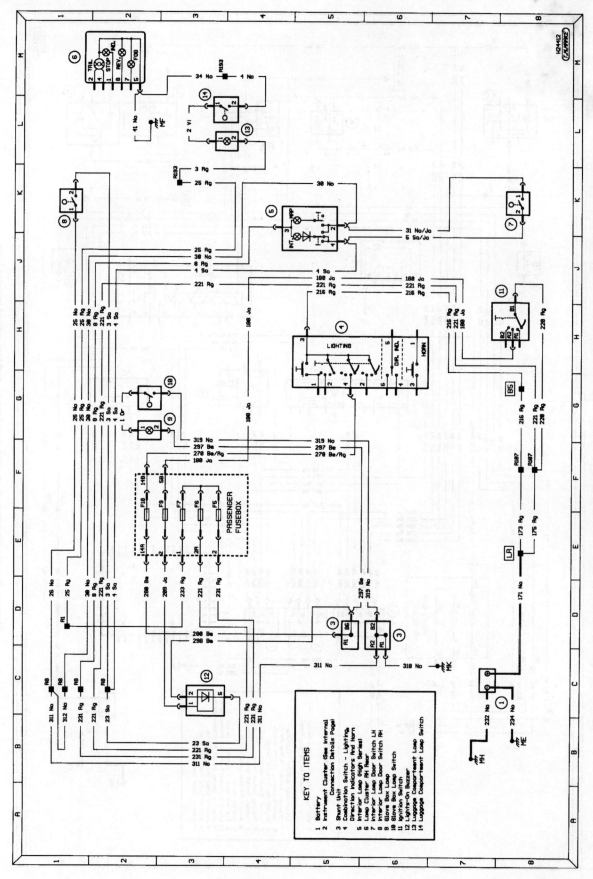

Diagram 2d: Interior lighting - lights-on buzzer, high-series interior light, luggage compartment and glovebox lights

KEY TO ITEMS

1 Battery
2 Instrument Cluster (See internal
    Connection Details Page)
3 Shunt Unit
4 Combination Switch - Lighting
    Direction Indicators And Horn
5 Interior Loop (High Series)
6 Loop Cluster RH Rear
7 Interior Loop Door Switch LH
8 Interior Loop Door Switch RH
9 Glove Box Loop
10 Glove Box Loop Switch
11 Ignition Switch
12 Lights-On Buzzer
13 Luggage Compartment Loop
14 Luggage Compartment Loop Switch

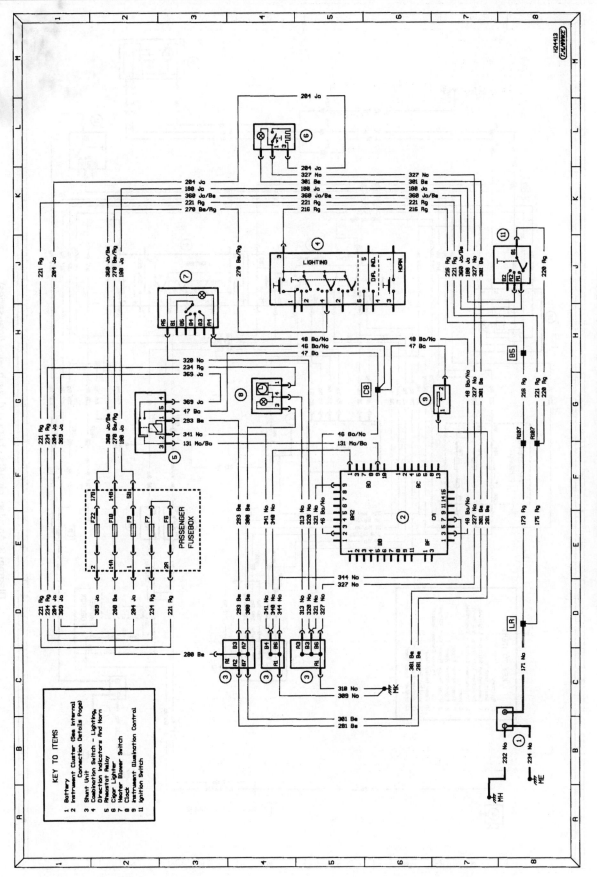

Diagram 2e: Interior lighting - dimmer circuit, clock and cigar lighter

KEY TO ITEMS
1 Battery
2 Instrument Cluster (See Internal
   Connection Details Page)
3 Shunt Unit
4 Combination Switch - Lighting,
   Direction Indicators And Horn
5 Rheostat Relay
6 Cigar Lighter
7 Heater Blower Switch
9 Clock
9 Instrument Illumination Control
11 Ignition Switch

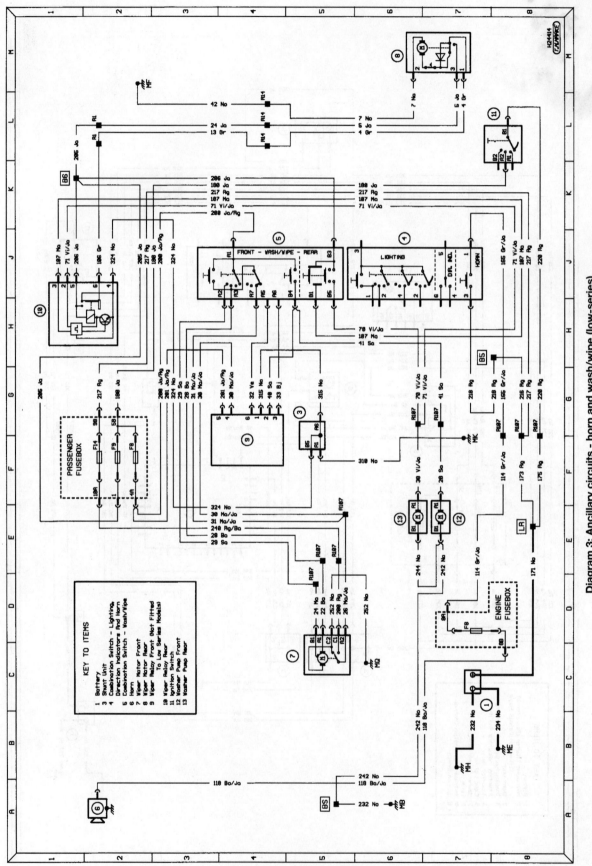

Diagram 3: Ancillary circuits - horn and wash/wipe (low-series)

KEY TO ITEMS

1 Battery
3 Shunt Unit
4 Combination Switch - Lighting and Horn
5 Combination Switch - Wash/Wipe
6 Horn
7 Viper Motor Front
8 Viper Motor Rear
9 Viper Relay Front (Not Fitted To Low Series Models)
10 Viper Relay Rear
11 Ignition Switch
12 Washer Pump Front
13 Washer Pump Rear

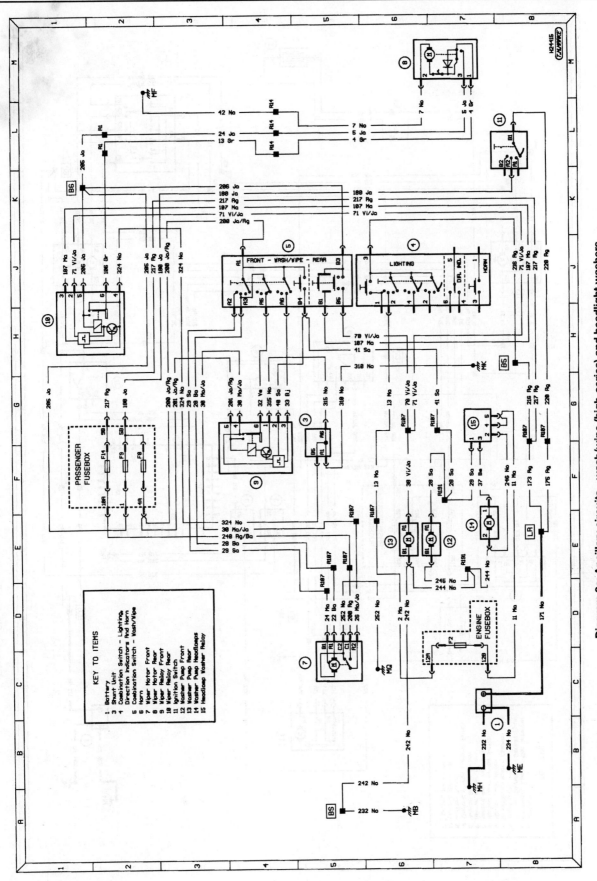

Diagram 3a: Ancillary circuits - wash/wipe (high-series) and headlight washers

KEY TO ITEMS

1 Battery
3 Shunt Unit
4 Combination Switch - Lighting
      Direction Indicators And Horn
5 Combination Switch - Wash/Wipe
6 Horn
7 Wiper Motor Front
8 Wiper Motor Rear
9 Wiper Relay Front
10 Wiper Relay Rear
11 Ignition Switch
12 Washer Pump Front
13 Washer Pump Rear
14 Washer Pump Headlamps
15 Headlamp Washer Relay

PASSENGER FUSEBOX

ENGINE FUSEBOX

FRONT - WASH/WIPE - REAR

LIGHTING

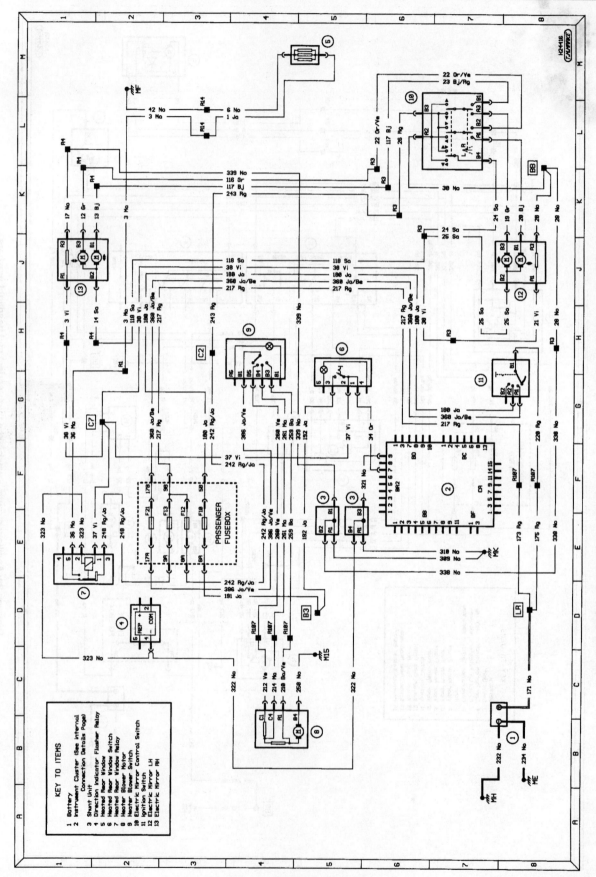

Diagram 3b: Ancillary circuits - heated rear window, heater blower and electric door mirrors

KEY TO ITEMS
1 Battery
2 Instrument Cluster (See internal Connection Details Page)
3 Shunt Unit
4 Direction Indicator Flasher Relay
5 Heated Rear Window
6 Heated Rear Window Switch
7 Heated Rear Window Relay
8 Heater Blower Motor
9 Heater Blower Switch
10 Electric Mirror Control Switch
11 Ignition Switch
12 Electric Mirror LH
13 Electric Mirror RH

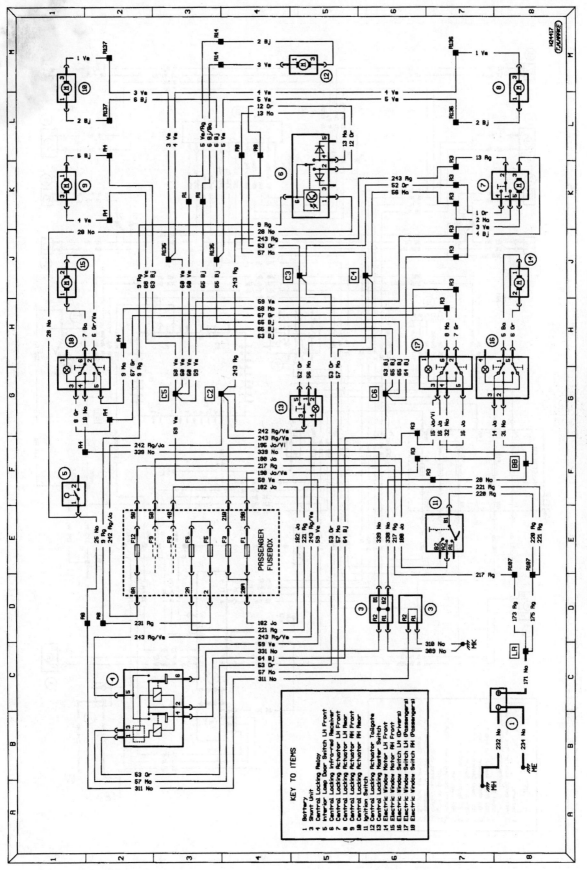

Diagram 3c: Ancillary circuits - electric windows and central locking

KEY TO ITEMS

1 Battery
3 Shunt Unit
4 Central Locking Relay
5 Interior Lamp Door Switch Front
6 Central Locking Infra-red Receiver
7 Central Locking Actuator LH Front
8 Central Locking Actuator LH Rear
9 Central Locking Actuator RH Front
10 Central Locking Actuator RH Rear
11 Ignition Switch
12 Central Locking Actuator Tailgate
13 Central Locking Master Switch
14 Electric Window Motor LH Front
15 Electric Window Motor RH Front
16 Electric Window Switch LH (Drivers)
17 Electric Window Switch LH (Passengers)
18 Electric Window Switch RH (Passengers)

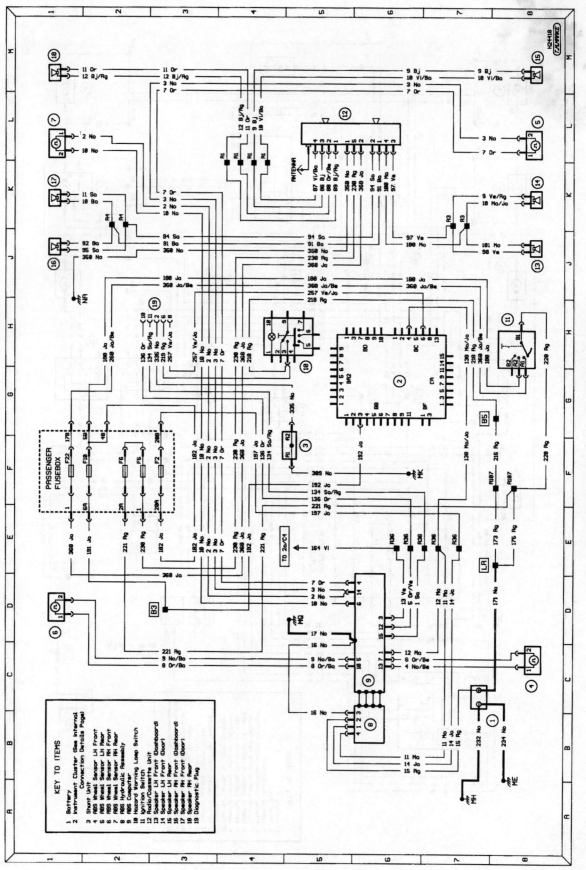

Diagram 4: Anti-lock braking system and radio/cassette

KEY TO ITEMS

1 Battery
2 Instrument Cluster (See internal Connection Details Page)
3 Shunt Unit
4 ABS Wheel Sensor LH Front
5 ABS Wheel Sensor LH Rear
6 ABS Wheel Sensor RH Front
7 ABS Wheel Sensor RH Rear
8 ABS Computer
9 ABS Hydraulic Assembly
10 Hazard Warning Loop Switch
11 Ignition Switch
12 Radio/Cassette Unit
13 Speaker LH Front (Dashboard)
14 Speaker LH Front (Door)
15 Speaker LH Rear
16 Speaker RH Front (Dashboard)
17 Speaker RH Front (Door)
18 Speaker RH Rear
19 Diagnostic Plug

Tools and Working Facilities

Introduction

A selection of good tools is a fundamental requirement for anyone contemplating the maintenance and repair of a motor vehicle. For the owner who does not possess any, their purchase will prove a considerable expense, offsetting some of the savings made by doing-it-yourself. However, provided that the tools purchased meet the relevant national safety standards and are of good quality, they will last for many years and prove an extremely worthwhile investment.

To help the average owner to decide which tools are needed to carry out the various tasks detailed in this manual, we have compiled three lists of tools under the following headings: *Maintenance and minor repair, Repair and overhaul*, and *Special*. Newcomers to practical mechanics should start off with the *Maintenance and minor repair* tool kit, and confine themselves to the simpler jobs around the vehicle. Then, as confidence and experience grow, more difficult tasks can be undertaken, with extra tools being purchased as, and when, they are needed. In this way, a *Maintenance and minor repair* tool kit can be built up into a *Repair and overhaul* tool kit over a considerable period of time, without any major cash outlays. The experienced do-it-yourselfer will have a tool kit good enough for most repair and overhaul procedures, and will add tools from the *Special* category when it is felt that the expense is justified by the amount of use to which these tools will be put.

Maintenance and minor repair tool kit

The tools given in this list should be considered as a minimum requirement if routine maintenance, servicing and minor repair operations are to be undertaken. We recommend the purchase of combination spanners (ring one end, open-ended the other); although more expensive than open-ended ones, they do give the advantages of both types of spanner.

☐ *Combination spanners:*
    Metric - 8, 9, 10, 11, 12, 13, 14, 15, 16, 17, 19, 21, 22, 24 & 26 mm
☐ *Adjustable spanner - 35 mm jaw (approx)*
☐ *Engine sump drain plug key (8 mm square)*
☐ *Set of feeler gauges*
☐ *Spark plug spanner (with rubber insert)*
☐ *Spark plug gap adjustment tool*
☐ *Brake bleed nipple spanner*
☐ *Screwdrivers:*
    Flat blade - approx 100 mm long x 6 mm dia
    Cross blade - approx 100 mm long x 6 mm dia
☐ *Combination pliers*
☐ *Hacksaw (junior)*
☐ *Tyre pump*
☐ *Tyre pressure gauge*
☐ *Oil can*
☐ *Oil filter removal tool*
☐ *Fine emery cloth*
☐ *Wire brush (small)*
☐ *Funnel (medium size)*

Repair and overhaul tool kit

These tools are virtually essential for anyone undertaking any major repairs to a motor vehicle, and are additional to those given in the *Maintenance and minor repair* list. Included in this list is a comprehensive set of sockets. Although these are expensive, they will be found invaluable as they are so versatile - particularly if various drives are included in the set. We recommend the half-inch square-drive type, as this can be used with most proprietary torque wrenches. If you cannot afford a socket set, even bought piecemeal, then inexpensive tubular box spanners are a useful alternative.

The tools in this list will occasionally need to be supplemented by tools from the *Special* list:

☐ *Sockets (or box spanners) to cover range in previous list*
☐ *Reversible ratchet drive (for use with sockets) (see illustration)*
☐ *Extension piece, 250 mm (for use with sockets)*
☐ *Universal joint (for use with sockets)*
☐ *Torque wrench (for use with sockets)*
☐ *Self-locking grips*
☐ *Ball pein hammer*
☐ *Soft-faced mallet (plastic/aluminium or rubber)*
☐ *Screwdrivers:*
    Flat blade - long & sturdy, short (chubby), and narrow (electrician's) types
    Cross blade - Long & sturdy, and short (chubby) types
☐ *Pliers:*
    Long-nosed
    Side cutters (electrician's)
    Circlip (internal and external)
☐ *Cold chisel - 25 mm*
☐ *Scriber*
☐ *Scraper*
☐ *Centre-punch*
☐ *Pin punch*
☐ *Hacksaw*
☐ *Brake hose clamp*
☐ *Brake bleeding kit*
☐ *Selection of twist drills*

☐ *Steel rule/straight-edge*
☐ *Allen keys (inc. splined/Torx type) (see illustrations)*
☐ *Selection of files*
☐ *Wire brush*
☐ *Axle stands*
☐ *Jack (strong trolley or hydraulic type)*
☐ *Light with extension lead*

Special tools

The tools in this list are those which are not used regularly, are expensive to buy, or which need to be used in accordance with their manufacturers' instructions. Unless relatively difficult mechanical jobs are undertaken frequently, it will not be economic to buy many of these tools. Where this is the case, you could consider clubbing together with friends (or joining a motorists' club) to make a joint purchase, or borrowing the tools against a deposit from a local garage or tool hire specialist. It is worth noting that many of the larger DIY superstores now carry a large range of special tools for hire at modest rates.

The following list contains only those tools and instruments freely available to the public, and not those special tools produced by the vehicle manufacturer specifically for its dealer network. You will find occasional references to these manufacturers' special tools in the text of this manual. Generally, an alternative method of doing the job without the vehicle manufacturers' special tool is given. However, sometimes there is no alternative to using them. Where this is the case and the relevant tool cannot be bought or borrowed, you will have to entrust the work to a franchised garage.

☐ *Valve spring compressor (see illustration)*
☐ *Valve grinding tool*
☐ *Piston ring compressor (see illustration)*
☐ *Piston ring removal/installation tool (see illustration)*
☐ *Cylinder bore hone (see illustration)*
☐ *Balljoint separator*
☐ *Coil spring compressors (where applicable)*
☐ *Two/three-legged hub and bearing puller (see illustration)*

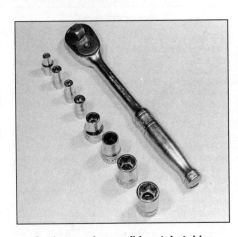

Sockets and reversible ratchet drive

Spline bit set

Tools and Working Facilities

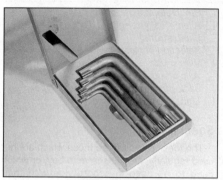

Spline key set

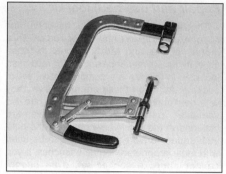

Valve spring compressor

Piston ring compressor

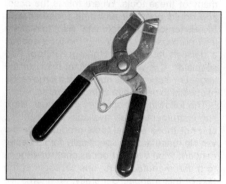

Piston ring removal/installation tool

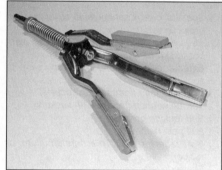

Cylinder bore hone

Three-legged hub and bearing puller

Micrometer set

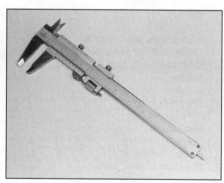

Vernier calipers

Dial test indicator and magnetic stand

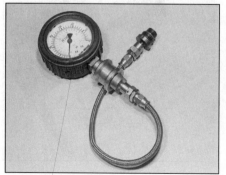

Compression testing gauge

Clutch plate alignment set

Brake shoe steady spring cup removal tool

- ☐ Impact screwdriver
- ☐ Micrometer and/or vernier calipers (**see illustrations**)
- ☐ Dial gauge (**see illustration**)
- ☐ Universal electrical multi-meter
- ☐ Cylinder compression gauge (**see illustration**)
- ☐ Clutch plate alignment set (**see illustration**)
- ☐ Brake shoe steady spring cup removal tool (**see illustration**)
- ☐ Bush and bearing removal/installation set (**see illustration**)
- ☐ Stud extractors (**see illustration**)
- ☐ Tap and die set (**see illustration**)
- ☐ Lifting tackle
- ☐ Trolley jack

Buying tools

For practically all tools, a tool factor is the best source, since he will have a very comprehensive range compared with the average garage or accessory shop. Having said that, accessory shops often offer excellent quality tools at discount prices, so it pays to shop around.

Remember, you don't have to buy the most expensive items on the shelf, but it is always advisable to steer clear of the very cheap tools. There are plenty of good tools around at reasonable prices, but always aim to purchase items which meet the relevant national safety standards. If in doubt, ask the proprietor or manager of the shop for advice before making a purchase.

Care and maintenance of tools

Having purchased a reasonable tool kit, it is necessary to keep the tools in a clean and serviceable condition. After use, always wipe off any dirt, grease and metal particles using a clean, dry cloth, before putting the tools away. Never leave them lying around after they have been used. A simple tool rack on the garage or workshop wall for items such as screwdrivers and pliers is a good idea. Store all normal spanners and sockets in a metal box. Any measuring instruments, gauges, meters, etc, must be carefully stored where they cannot be damaged or become rusty.

Take a little care when tools are used. Hammer heads inevitably become marked, and screwdrivers lose the keen edge on their blades from time to time. A little timely attention with emery cloth or a file will soon restore items like this to a good serviceable finish.

Working facilities

Not to be forgotten when discussing tools is the workshop itself. If anything more than routine maintenance is to be carried out, some form of suitable working area becomes essential.

It is appreciated that many an owner-mechanic is forced by circumstances to remove an engine or similar item without the benefit of a garage or workshop. Having done this, any repairs should always be done under the cover of a roof.

Wherever possible, any dismantling should be done on a clean, flat workbench or table at a suitable working height.

Any workbench needs a vice; one with a jaw opening of 100 mm is suitable for most jobs. As mentioned previously, some clean dry storage space is also required for tools, as well as for any lubricants, cleaning fluids, touch-up paints and so on, which become necessary.

Another item which may be required, and which has a much more general usage, is an electric drill with a chuck capacity of at least 8 mm. This, together with a good range of twist drills, is virtually essential for fitting accessories.

Last, but not least, always keep a supply of old newspapers and clean, lint-free rags available, and try to keep any working area as clean as possible.

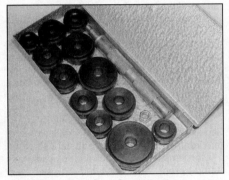

Bush and bearing removal/installation set

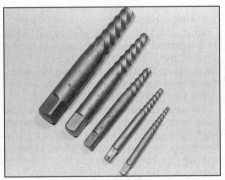

Stud extractor set

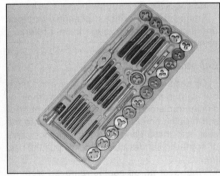

Tap and die set

Whenever servicing, repair or overhaul work is carried out on the car or its components, it is necessary to observe the following procedures and instructions. This will assist in carrying out the operation efficiently and to a professional standard of workmanship.

Joint mating faces and gaskets

When separating components at their mating faces, never insert screwdrivers or similar implements into the joint between the faces in order to prise them apart. This can cause severe damage which results in oil leaks, coolant leaks, etc upon reassembly. Separation is usually achieved by tapping along the joint with a soft-faced hammer in order to break the seal. However, note that this method may not be suitable where dowels are used for component location.

Where a gasket is used between the mating faces of two components, ensure that it is renewed on reassembly, and fit it dry unless otherwise stated in the repair procedure. Make sure that the mating faces are clean and dry, with all traces of old gasket removed. When cleaning a joint face, use a tool which is not likely to score or damage the face, and remove any burrs or nicks with an oilstone or fine file.

Make sure that tapped holes are cleaned with a pipe cleaner, and keep them free of jointing compound, if this is being used, unless specifically instructed otherwise.

Ensure that all orifices, channels or pipes are clear, and blow through them, preferably using compressed air.

Oil seals

Oil seals can be removed by levering them out with a wide flat-bladed screwdriver or similar implement. Alternatively, a number of self-tapping screws may be screwed into the seal, and these used as a purchase for pliers or some similar device in order to pull the seal free.

Whenever an oil seal is removed from its working location, either individually or as part of an assembly, it should be renewed.

The very fine sealing lip of the seal is easily damaged, and will not seal if the surface it contacts is not completely clean and free from scratches, nicks or grooves. Protect the lips of the seal from any surface which may damage them in the course of fitting. Use tape or a conical sleeve where possible. Lubricate the seal lips with oil before fitting and, on dual-lipped seals, fill the space between the lips with grease.

Unless otherwise stated, oil seals must be fitted with their sealing lips toward the lubricant to be sealed.

Use a tubular drift or block of wood of the appropriate size to install the seal and, if the seal housing is shouldered, drive the seal down to the shoulder. If the seal housing is unshouldered, the seal should be fitted with its face flush with the housing top face (unless otherwise instructed).

Screw threads and fastenings

Seized nuts, bolts and screws are quite a common occurrence where corrosion has set in, and the use of penetrating oil or releasing fluid will often overcome this problem if the offending item is soaked for a while before attempting to release it. The use of an impact driver may also provide a means of releasing such stubborn fastening devices, when used in conjunction with the appropriate screwdriver bit or socket. If none of these methods works, it may be necessary to resort to the careful application of heat, or the use of a hacksaw or nut splitter device.

Studs are usually removed by locking two nuts together on the threaded part, and then using a spanner on the lower nut to unscrew the stud. Studs or bolts which have broken off below the surface of the component in which they are mounted can sometimes be removed using a proprietary stud extractor. Always ensure that a blind tapped hole is completely free from oil, grease, water or other fluid before installing the bolt or stud. Failure to do this could cause the housing to crack due to the hydraulic action of the bolt or stud as it is screwed in.

When tightening a castellated nut to accept a split pin, tighten the nut to the specified torque, where applicable, and then tighten further to the next split pin hole. Never slacken the nut to align the split pin hole, unless stated in the repair procedure.

When checking or retightening a nut or bolt to a specified torque setting, slacken the nut or bolt by a quarter of a turn, and then retighten to the specified setting. However, this should not be attempted where angular tightening has been used.

For some screw fastenings, notably cylinder head bolts or nuts, torque wrench settings are no longer specified for the latter stages of tightening, "angle-tightening" being called up instead. Typically, a fairly low torque wrench setting will be applied to the bolts/nuts in the correct sequence, followed by one or more stages of tightening through specified angles.

Locknuts, locktabs and washers

Any fastening which will rotate against a component or housing in the course of tightening should always have a washer between it and the relevant component or housing.

Spring or split washers should always be renewed when they are used to lock a critical component such as a big-end bearing retaining bolt or nut. Locktabs which are folded over to retain a nut or bolt should always be renewed.

Self-locking nuts can be re-used in non-critical areas, providing resistance can be felt when the locking portion passes over the bolt or stud thread. However, it should be noted that self-locking stiffnuts tend to lose their effectiveness after long periods of use, and in such cases should be renewed as a matter of course.

Split pins must always be replaced with new ones of the correct size for the hole.

When thread-locking compound is found on the threads of a fastener which is to be re-used, it should be cleaned off with a wire brush and solvent, and fresh compound applied on reassembly.

Special tools

Some repair procedures in this manual entail the use of special tools such as a press, two or three-legged pullers, spring compressors, etc. Wherever possible, suitable readily-available alternatives to the manufacturer's special tools are described, and are shown in use. Unless you are highly-skilled and have a thorough understanding of the procedures described, never attempt to bypass the use of any special tool when the procedure described specifies its use. Not only is there a very great risk of personal injury, but expensive damage could be caused to the components involved.

Environmental considerations

When disposing of used engine oil, brake fluid, antifreeze, etc, give due consideration to any detrimental environmental effects. Do not, for instance, pour any of the above liquids down drains into the general sewage system, or onto the ground to soak away. Many local council refuse tips provide a facility for waste oil disposal, as do some garages. If none of these facilities are available, consult your local Environmental Health Department for further advice.

With the universal tightening-up of legislation regarding the emission of environmentally-harmful substances from motor vehicles, most current vehicles have tamperproof devices fitted to the main adjustment points of the fuel system. These devices are primarily designed to prevent unqualified persons from adjusting the fuel/air mixture, with the chance of a consequent increase in toxic emissions. If such devices are encountered during servicing or overhaul, they should, wherever possible, be renewed or refitted in accordance with the vehicle manufacturer's requirements or current legislation.

OIL CARE
FOLLOW THE CODE

OIL BANK LINE
0800 66 33 66

Note: It is antisocial and illegal to dump oil down the drain. To find the location of your local oil recycling bank, call this number free.

Buying spare parts

Spare parts are available from many sources; for example, Renault garages, other garages and accessory shops, and motor factors. Our advice regarding spare part sources is as follows.

Officially-appointed Renault garages - This is the best source for parts which are peculiar to your car, and which are not generally available (eg complete cylinder heads, gearbox components, badges, interior trim etc). It is also the only place at which you should buy parts if the vehicle is still under warranty. To be sure of obtaining the correct parts, it will be necessary to give the storeman your car's vehicle identification number, and if

VIN plate location (A)

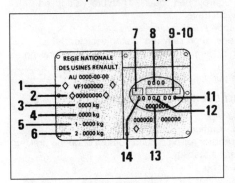

VIN plate details

1 Vehicle type
2 Chassis number
3 Gross vehicle weight
4 Gross train weight
5 Maximum permitted total weight on front axle
6 Maximum permitted total weight on rear axle
7 Vehicle special features
8 Vehicle type
9 Equipment number
10 Options
11 Vehicle equipment level
12 Original paint reference
13 Fabrication number
14 Upholstery/trim code number

possible, take the old parts along for positive identification. Many parts are available under a factory exchange scheme - any parts returned should always be clean. It obviously makes good sense to go straight to the specialists on your car for this type of part, as they are best equipped to supply you.

Other garages and accessory shops - These are often very good places to buy materials and components needed for the maintenance of your car (eg oil filters, spark plugs, bulbs, drivebelts, oils and greases, touch-up paint, filler paste, etc). They also sell general accessories, usually have convenient opening hours, charge lower prices and can often be found not far from home.

Motor factors - Good factors will stock all the more important components which wear out comparatively quickly (eg exhaust systems, brake pads, seals and hydraulic parts, clutch components, bearing shells, pistons, valves etc). Motor factors will often provide new or reconditioned components on a part exchange basis - this can save a considerable amount of money.

Vehicle identification numbers

Modifications are a continuing and unpublicised process in vehicle manufacture, quite apart from major model changes. Spare parts manuals and lists are compiled upon a numerical basis, the individual vehicle identification numbers being essential to correct identification of the component concerned.

When ordering spare parts, always give as much information as possible. Quote the car model, year of manufacture, body and engine numbers as appropriate.

The *Vehicle Identification Number (VIN) plate* is riveted to the top of the body front panel, and can be viewed once the bonnet is open. The plate carries the VIN number, chassis number, vehicle weight information and various other information, depending on territory.

The *chassis number* is given on the VIN plate.

The *engine number* is located on a metal plate attached to the engine. The location of the plate varies according to engine type, and the various locations are shown in the accompanying illustrations.

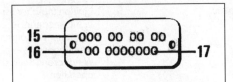

Engine number plate details

15 Engine type
16 Engine suffix
17 Engine number

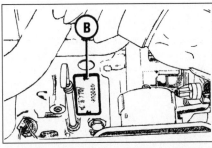

Engine number plate location (B) - 1721 cc (F2N) and 1794 cc (F3P) engines

Engine number plate location (B) - 1764 cc (F7P) engine

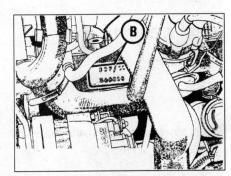

Engine number plate location (B) - 1108 cc (C1E) engine

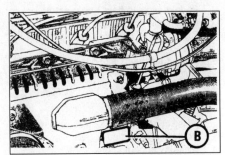

Engine number plate location (B) - 1171 cc (E5F and E7F) and 1390 cc (E6J and E7J) engines

Engine   1

- [ ] Engine fails to rotate when attempting to start
- [ ] Engine rotates but will not start
- [ ] Engine difficult to start when cold
- [ ] Engine difficult to start when hot
- [ ] Starter motor noisy or excessively rough in engagement
- [ ] Engine starts but stops immediately
- [ ] Engine idles erratically
- [ ] Engine misfires at idle speed
- [ ] Engine misfires throughout the driving speed range
- [ ] Engine hesitates on acceleration
- [ ] Engine stalls
- [ ] Engine lacks power
- [ ] Engine backfires
- [ ] Oil pressure warning light illuminated with engine running
- [ ] Engine runs-on after switching off
- [ ] Engine noises

Cooling system   2

- [ ] Overheating
- [ ] Overcooling
- [ ] External coolant leakage
- [ ] Internal coolant leakage
- [ ] Corrosion

Fuel and exhaust systems   3

- [ ] Excessive fuel consumption
- [ ] Fuel leakage and/or fuel odour
- [ ] Excessive noise or fumes from exhaust system

Clutch   4

- [ ] Pedal travels to floor - no pressure or very little resistance
- [ ] Clutch fails to disengage (unable to select gears)
- [ ] Clutch slips (engine speed increases with no increase in vehicle speed)
- [ ] Judder as clutch is engaged
- [ ] Noise when depressing or releasing clutch pedal

Manual gearbox   5

- [ ] Noisy in neutral with engine running
- [ ] Noisy in one particular gear
- [ ] Difficulty engaging gears
- [ ] Jumps out of gear
- [ ] Vibration
- [ ] Lubricant leaks

Automatic transmission   6

- [ ] Fluid leakage
- [ ] Transmission fluid brown, or has burned smell
- [ ] General gear selection problems
- [ ] Transmission will not downshift (kickdown) with accelerator fully depressed
- [ ] Engine will not start in any gear, or starts in gears other than Park or Neutral
- [ ] Transmission slips, shifts roughly, is noisy or has no drive in forward or reverse gears

Driveshafts   7

- [ ] Clicking or knocking noise on turns (at slow speed on full-lock)
- [ ] Vibration when accelerating or decelerating

Braking system   8

- [ ] Vehicle pulls to one side under braking
- [ ] Noise (grinding or high-pitched squeal) when brakes applied
- [ ] Excessive brake pedal travel
- [ ] Brake pedal feels spongy when depressed
- [ ] Excessive brake pedal effort required to stop vehicle
- [ ] Judder felt through brake pedal or steering wheel when braking
- [ ] Brakes binding
- [ ] Rear wheels locking under normal braking

Suspension and steering systems   9

- [ ] Vehicle pulls to one side
- [ ] Wheel wobble and vibration
- [ ] Excessive pitching and/or rolling around corners or during braking
- [ ] Wandering or general instability
- [ ] Excessively-stiff steering
- [ ] Excessive play in steering
- [ ] Lack of power assistance
- [ ] Tyre wear excessive

Electrical system   10

- [ ] Battery will not hold a charge for more than a few days
- [ ] Ignition (no-charge) warning light remains illuminated with engine running
- [ ] Ignition (no-charge) warning light fails to come on
- [ ] Lights inoperative
- [ ] Instrument readings inaccurate or erratic
- [ ] Horn inoperative, or unsatisfactory in operation
- [ ] Windscreen/tailgate wipers inoperative, or unsatisfactory in operation
- [ ] Windscreen/tailgate washers inoperative, or unsatisfactory in operation
- [ ] Electric windows inoperative, or unsatisfactory in operation
- [ ] Central locking system inoperative, or unsatisfactory in operation

Introduction

The vehicle owner who does his or her own maintenance according to the recommended service schedules should not have to use this section of the manual very often. Modern component reliability is such that, provided those items subject to wear or deterioration are inspected or renewed at the specified intervals, sudden failure is comparatively rare. Faults do not usually just happen as a result of sudden failure, but develop over a period of time. Major mechanical failures in particular are usually preceded by characteristic symptoms over hundreds or even thousands of miles. Those components which do occasionally fail without warning are often small and easily carried in the vehicle.

With any fault finding, the first step is to decide where to begin investigations. Sometimes this is obvious, but on other occasions a little detective work will be necessary. The owner who makes half a dozen haphazard adjustments or replacements may be successful in curing a fault (or its symptoms), but will be none the wiser if the fault recurs, and ultimately may have spent more time and money than was necessary. A calm and logical approach will be found to be more satisfactory in the long run. Always take into account any warning signs or abnormalities that may have been noticed in the period preceding the fault - power loss, high or low gauge readings, unusual smells, etc - and remember that failure of components such as fuses or spark plugs may only be pointers to some underlying fault.

The pages which follow provide an easy-reference guide to the more common problems which may occur during the operation of the vehicle. These problems and their possible causes are grouped under headings denoting various components or systems, such as Engine, Cooling system, etc. The Chapter and/or Section which deals with the problem is also shown in brackets. Whatever the fault, certain basic principles apply. These are as follows:

Verify the fault. This is simply a matter of being sure that you know what the symptoms are before starting work. This is particularly important if you are investigating a fault for someone else who may not have described it very accurately.

Don't overlook the obvious. For example, if the vehicle won't start, is there petrol in the tank? (Don't take anyone else's word on this particular point, and don't trust the fuel gauge either!) If an electrical fault is indicated, look for loose or broken wires before digging out the test gear.

Cure the disease, not the symptom. Substituting a flat battery with a fully-charged one will get you off the hard shoulder, but if the underlying cause is not attended to, the new battery will go the same way. Similarly, changing oil-fouled spark plugs for a new set will get you moving again, but remember that the reason for the fouling (if it wasn't simply an incorrect grade of plug) will have to be established and corrected.

Don't take anything for granted. Particularly, don't forget that a 'new' component may itself be defective (especially if it's been rattling around in the boot for months), and don't leave components out of a fault diagnosis sequence just because they are new or recently fitted. When you do finally diagnose a difficult fault, you'll probably realise that all the evidence was there from the start.

1 Engine

Engine fails to rotate when attempting to start

☐ Battery terminal connections loose or corroded (Chapter 1).
☐ Battery discharged or faulty (Chapter 12).
☐ Broken, loose or disconnected wiring in the starting circuit (Chapter 12).
☐ Defective starter solenoid or switch (Chapter 12).
☐ Defective starter motor (Chapter 12).
☐ Flywheel ring gear or starter pinion teeth loose or broken (Chapter 1 or 12).
☐ Engine earth strap broken or disconnected (Chapter 12).
☐ Automatic transmission not in Park/Neutral position, or starter inhibitor switch faulty (Chapter 7, Part B).

Engine rotates but will not start

☐ Battery discharged - engine rotates slowly (Chapter 12).
☐ Battery terminal connections loose or corroded (Chapter 1).
☐ Ignition components damp or damaged (Chapters 1 and 5).
☐ Broken, loose or disconnected wiring in the ignition circuit (Chapters 1 and 5).
☐ Contact breaker points dirty or incorrectly gapped - 1108 cc (C1E) engine only (Chapter 1).
☐ Worn, faulty or incorrectly-gapped spark plugs (Chapter 1).
☐ Choke mechanism sticking, incorrectly adjusted, or faulty - carburettor engines only (Chapter 4, Part A).
☐ Fuel injection system fault - where applicable (Chapter 4, Part B).
☐ Major mechanical failure - eg camshaft drive (Chapter 2).

Engine difficult to start when cold

☐ Battery discharged (Chapter 12).
☐ Battery terminal connections loose or corroded (Chapter 1).
☐ Contact breaker points dirty or incorrectly gapped - 1108 cc (C1E) engine only (Chapter 1).
☐ Worn, faulty or incorrectly-gapped spark plugs (Chapter 1).
☐ Choke mechanism sticking, incorrectly adjusted, or faulty - carburettor engines only (Chapter 4, Part A).
☐ Fuel injection system fault - where applicable (Chapter 4, Part B)
☐ Other ignition system fault (Chapters 1 and 5).
☐ Low cylinder compressions (Chapter 2).

Engine difficult to start when hot

☐ Air cleaner element dirty or clogged (Chapter 1).
☐ Choke mechanism sticking, incorrectly adjusted, or faulty - carburettor engines only (Chapter 4, Part A).
☐ Carburettor float chamber flooding (Chapter 4, Part A).
☐ Fuel injection system fault (Chapter 4, Section B).
☐ Low cylinder compressions (Chapter 2).

Starter motor noisy or excessively rough in engagement

☐ Starter pinion or flywheel ring gear teeth loose or broken (Chapter 12).
☐ Starter motor mounting bolts loose or missing (Chapter 12).
☐ Starter motor internal components worn or damaged (Chapter 12).

Engine starts but stops immediately

☐ Insufficient fuel reaching carburettor/fuel injector(s) (Chapter 4).
☐ Loose or faulty electrical connections in the ignition circuit (Chapters 1 and 5).
☐ Vacuum leak at the carburettor/throttle body or inlet manifold (Chapter 4).
☐ Blocked carburettor jet(s) or internal passages - where applicable (Chapter 4, Part A).
☐ Fuel injection system fault (Chapter 4, Part B).

Engine idles erratically

☐ Incorrectly-adjusted idle speed and/or mixture settings (Chapters 1 and 4).
☐ Air cleaner element clogged (Chapter 1).
☐ Vacuum leak at the carburettor/throttle body, inlet manifold or associated hoses (Chapter 4).
☐ Fuel injection system fault (Chapter 4, Part B).
☐ Contact breaker points dirty or incorrectly-gapped - 1108 cc (C1E) engine only (Chapter 1).
☐ Worn, faulty or incorrectly-gapped spark plugs (Chapter 1).
☐ Incorrectly-adjusted valve clearances - where applicable (Chapter 2).
☐ Uneven or low cylinder compressions (Chapter 2).
☐ Camshaft lobes worn (Chapter 2).
☐ Timing belt/chain incorrectly-tensioned (Chapter 2).

Fault Finding

Engine misfires at idle speed

☐ Contact breaker points dirty or incorrectly gapped - 1108 cc (C1E) engine only (Chapter 1).
☐ Worn, faulty or incorrectly-gapped spark plugs (Chapter 1).
☐ Faulty spark plug HT leads (Chapter 1).
☐ Incorrectly-adjusted idle mixture settings (Chapters 1 and 4).
☐ Incorrect ignition timing (Chapters 1 and 5).
☐ Vacuum leak at the carburettor, inlet manifold or associated hoses (Chapter 4).
☐ Fuel injection system fault (Chapter 4, Part B).
☐ Distributor cap cracked or tracking internally (Chapter 1).
☐ Incorrectly-adjusted valve clearances - where applicable (Chapter 2).
☐ Uneven or low cylinder compressions (Chapter 2).
☐ Disconnected, leaking or perished crankcase ventilation hoses (Chapters 1 and 4).

Engine misfires throughout the driving speed range

☐ Blocked carburettor jet(s) or internal passages - where applicable (Chapter 4).
☐ Carburettor worn or incorrectly adjusted - where applicable (Chapters 1 and 4).
☐ Fuel injection system fault (Chapter 4, Part B).
☐ Fuel filter choked (Chapter 1).
☐ Fuel pump faulty or delivery pressure low (Chapter 4).
☐ Fuel tank vent blocked or fuel pipes restricted (Chapter 4).
☐ Vacuum leak at the carburettor, inlet manifold or associated hoses (Chapter 4).
☐ Contact breaker points dirty or incorrectly gapped - 1108 cc (C1E) engine only (Chapter 1).
☐ Worn, faulty or incorrectly-gapped spark plugs (Chapter 1).
☐ Faulty spark plug HT leads (Chapter 1).
☐ Distributor cap cracked or tracking internally (Chapter 1).
☐ Faulty ignition coil (Chapter 5).
☐ Uneven or low cylinder compressions (Chapter 2).

Engine hesitates on acceleration

☐ Contact breaker points dirty or incorrectly gapped - 1108 cc (C1E) engine only (Chapter 1).
☐ Worn, faulty or incorrectly-gapped spark plugs (Chapter 1).
☐ Carburettor accelerator pump faulty - where applicable (Chapter 4).
☐ Blocked carburettor jets or internal passages - where applicable (Chapter 4).
☐ Vacuum leak at the carburettor, inlet manifold or associated hoses (Chapter 4).
☐ Carburettor worn or incorrectly adjusted (Chapters 1 and 4).
☐ Fuel injection system fault (Chapter 4, Part B).

Engine stalls

☐ Incorrectly-adjusted idle speed and/or mixture settings (Chapters 1 and 4).
☐ Blocked carburettor jet(s) or internal passages - where applicable (Chapter 4).
☐ Vacuum leak at the carburettor, inlet manifold or associated hoses (Chapter 4).
☐ Fuel filter choked (Chapter 1).
☐ Fuel pump faulty or delivery pressure low (Chapter 4).
☐ Fuel tank vent blocked or fuel pipes restricted (Chapter 4).
☐ Fuel injection system fault (Chapter 4, Part B).

Engine lacks power

☐ Incorrect ignition timing (Chapters 1 and 5).
☐ Carburettor worn or incorrectly adjusted - where applicable (Chapters 1 and 4).
☐ Fuel injection system fault (Chapter 4, Part B).
☐ Timing belt/chain incorrectly fitted or tensioned (Chapter 2).
☐ Fuel filter choked (Chapter 1).
☐ Fuel pump faulty or delivery pressure low (Chapter 4).
☐ Uneven or low cylinder compressions (Chapter 2).

☐ Contact breaker points dirty or incorrectly gapped - 1108 cc (C1E) engine only (Chapter 1).
☐ Worn, faulty or incorrectly-gapped spark plugs (Chapter 1).
☐ Vacuum leak at the carburettor, inlet manifold or associated hoses (Chapter 4).
☐ Brakes binding (Chapters 1 and 9).
☐ Clutch slipping (Chapter 6).
☐ Automatic transmission fluid level incorrect (Chapters 1 and 7).

Engine backfires

☐ Ignition timing incorrect (Chapters 1 and 5).
☐ Timing belt/chain incorrectly fitted or tensioned (Chapter 2).
☐ Carburettor worn or incorrectly adjusted - where applicable (Chapters 1 and 4).
☐ Vacuum leak at the carburettor, inlet manifold or associated hoses (Chapter 4).

Oil pressure warning light illuminated with engine running

☐ Low oil level or incorrect grade (Chapter 1).
☐ Faulty oil pressure transmitter (sender) unit (Chapter 2).
☐ Worn engine bearings and/or oil pump (Chapter 2).
☐ Overheating (see below).
☐ Oil pressure relief valve defective (Chapter 2).
☐ Oil pick-up strainer clogged (Chapter 2).

Engine runs-on after switching off

☐ Idle speed excessively high (Chapters 1 and 4).
☐ Faulty anti-run-on solenoid - where applicable (Chapter 4).
☐ Excessive carbon build-up in engine (Chapter 2).
☐ High engine operating temperature (Chapter 3).

Engine noises

Pre-ignition (pinking) or knocking during acceleration or under load

☐ Ignition timing incorrect (Chapters 1 and 5).
☐ Incorrect grade of fuel (Chapter 4).
☐ Vacuum leak at the carburettor, inlet manifold or associated hoses (Chapter 4).
☐ Excessive carbon build-up in engine (Chapter 2).
☐ Worn or damaged distributor or other ignition system component (Chapter 5).
☐ Carburettor worn or incorrectly adjusted - where applicable (Chapters 1 and 4).
☐ Fuel injection system fault (Chapter 4, Part B).

Whistling or wheezing noises

☐ Leaking inlet manifold or carburettor gasket (Chapter 4).
☐ Leaking exhaust manifold gasket or pipe-to-manifold joint (Chapters 1 and 4).
☐ Leaking vacuum hose (Chapters 4, 5 and 9).
☐ Blowing cylinder head gasket (Chapter 2).

Tapping or rattling noises

☐ Incorrect valve clearances - where applicable (Chapter 2).
☐ Worn valve gear or camshaft (Chapter 2).
☐ Worn timing chain or tensioner - where applicable (Chapter 2).
☐ Ancillary component fault - water pump, alternator etc (Chapters 3 and 12).

Knocking or thumping noises

☐ Worn big-end bearings - regular heavy knocking, perhaps less under load (Chapter 2).
☐ Worn main bearings - rumbling and knocking, perhaps worsening under load (Chapter 2).
☐ Piston slap - most noticeable when cold (Chapter 2).
☐ Ancillary component fault - alternator, water pump etc (Chapters 3 and 12).

2 Cooling system

Overheating

- [ ] Insufficient coolant in system (Chapters 1 and 3).
- [ ] Thermostat faulty (Chapter 3).
- [ ] Radiator core blocked or grille restricted (Chapter 3).
- [ ] Electric cooling fan or thermoswitch faulty (Chapter 3).
- [ ] Pressure cap faulty (Chapter 3).
- [ ] Water pump drivebelt worn, or incorrectly adjusted - where applicable (Chapters 1, 3 and 12).
- [ ] Ignition timing incorrect (Chapters 1 and 5).
- [ ] Inaccurate temperature gauge sender unit (Chapter 3).
- [ ] Air-lock in cooling system (Chapter 1).

Overcooling

- [ ] Thermostat faulty (Chapter 3).
- [ ] Inaccurate temperature gauge sender unit (Chapter 3).

External coolant leakage

- [ ] Deteriorated or damaged hoses or hose clips (Chapter 1).
- [ ] Radiator core or heater radiator leaking (Chapter 3).
- [ ] Pressure cap faulty (Chapter 3).
- [ ] Water pump seal leaking (Chapter 3).
- [ ] Boiling due to overheating (Chapter 3).
- [ ] Core plug leaking (Chapter 2).

Internal coolant leakage

- [ ] Leaking cylinder head gasket (Chapter 2).
- [ ] Cracked cylinder head or cylinder bore (Chapter 2).
- [ ] Cylinder liner base seals leaking - when applicable (Chapter 2).

Corrosion

- [ ] Infrequent draining and flushing (Chapter 1).
- [ ] Incorrect antifreeze mixture or inappropriate type (Chapter 1).

3 Fuel and exhaust system

Excessive fuel consumption

- [ ] Unsympathetic driving style or unfavourable conditions.
- [ ] Air cleaner element dirty or clogged (Chapter 1).
- [ ] Carburettor worn or incorrectly adjusted - where applicable (Chapters 1 and 4).
- [ ] Choke cable incorrectly adjusted or choke sticking - where applicable (Chapter 4).
- [ ] Fuel injection system fault (Chapter 4, Part B).
- [ ] Ignition timing incorrect (Chapters 1 and 5).
- [ ] Tyres under-inflated (Chapter 1).

Fuel leakage and/or fuel odour

- [ ] Damaged or corroded fuel tank, pipes or connections (Chapter 1).
- [ ] Carburettor float chamber flooding - where applicable (Chapter 4).

Excessive noise or fumes from exhaust system

- [ ] Leaking exhaust system or manifold joints (Chapters 1 and 4).
- [ ] Leaking, corroded or damaged silencers or pipe (Chapters 1 and 4).
- [ ] Broken mountings causing body or suspension contact (Chapters 1 and 4).

4 Clutch

Judder as clutch is engaged

- [ ] Clutch disc linings contaminated with oil or grease (Chapter 6).
- [ ] Clutch disc linings excessively worn (Chapter 6).
- [ ] Clutch cable sticking or frayed (Chapter 6).
- [ ] Faulty or distorted pressure plate or diaphragm spring (Chapter 6).
- [ ] Worn or loose engine or gearbox mountings (Chapters 2 and 7).
- [ ] Clutch disc hub or gearbox input shaft splines worn (Chapter 6).

Clutch fails to disengage (unable to select gears)

- [ ] Faulty clutch pedal self-adjust mechanism (Chapter 6).
- [ ] Clutch disc sticking on gearbox input shaft splines (Chapter 6).
- [ ] Clutch disc sticking to flywheel or pressure plate (Chapter 6).
- [ ] Faulty pressure plate assembly (Chapter 6).
- [ ] Gearbox input shaft seized in crankshaft spigot bearing - where applicable (Chapter 2).
- [ ] Clutch release mechanism worn or incorrectly assembled (Chapter 6).

Pedal travels to floor - no pressure or very little resistance

- [ ] Broken clutch cable (Chapter 6).
- [ ] Faulty clutch pedal self-adjust mechanism (Chapter 6).
- [ ] Broken clutch release bearing or fork (Chapter 6).
- [ ] Broken diaphragm spring in clutch pressure plate (Chapter 6).

Clutch slips

- [ ] Faulty clutch pedal self-adjust mechanism (Chapter 6).
- [ ] Clutch disc linings excessively worn (Chapter 6).
- [ ] Clutch disc linings contaminated with oil or grease (Chapter 6).
- [ ] Faulty pressure plate or weak diaphragm spring (Chapter 6).

Noise when depressing or releasing clutch pedal

- [ ] Worn clutch release bearing (Chapter 6).
- [ ] Worn or dry clutch pedal bushes (Chapter 6).
- [ ] Faulty pressure plate assembly (Chapter 6).
- [ ] Pressure plate diaphragm spring broken (Chapter 6).
- [ ] Broken clutch disc cushioning springs (Chapter 6).

5 Manual gearbox

Noisy in one particular gear

- [ ] Worn, damaged or chipped gear teeth (Chapter 7).*

Vibration

- [ ] Lack of oil (Chapter 1).
- [ ] Worn bearings (Chapter 7).*

Noisy in neutral with engine running

- [ ] Input shaft bearings worn - noise apparent with clutch pedal released but not when depressed (Chapter 7).*
- [ ] Clutch release bearing worn - noise apparent with clutch pedal depressed, possibly less when released (Chapter 6).

continued over

Lubricant leaks

☐ Leaking differential output oil seal (Chapter 7).
☐ Leaking housing joint (Chapter 7).*
☐ Leaking input shaft oil seal (Chapter 7).*

Difficulty engaging gears

☐ Clutch fault (Chapter 6).
☐ Worn or damaged gear linkage (Chapter 7).
☐ Incorrectly-adjusted gear linkage (Chapter 7).
☐ Worn synchroniser units (Chapter 7).*

Jumps out of gear

☐ Worn or damaged gear linkage (Chapter 7).
☐ Incorrectly-adjusted gear linkage (Chapter 7).
☐ Worn synchroniser units (Chapter 7).*
☐ Worn selector forks (Chapter 7).*

Although the corrective action necessary to remedy the symptoms described is beyond the scope of the home mechanic, the above information should be helpful in isolating the cause of the condition so that the owner can communicate clearly with a professional mechanic.

6 Automatic transmission

Note: *Due to the complexity of the automatic transmission, it is difficult for the home mechanic to properly diagnose and service this unit. For problems other than the following, the vehicle should be taken to a dealer service department or automatic transmission specialist.*

Fluid leakage

☐ Automatic transmission fluid is usually deep red in colour. Fluid leaks should not be confused with engine oil which can easily be blown onto the transmission by air flow.
☐ To determine the source of a leak, first remove all built-up dirt and grime from the transmission housing and surrounding areas using a degreasing agent or by steam cleaning. Drive the vehicle at low speed so air flow will not blow the leak far from its source. Raise and support the vehicle and determine where the leak is coming from. The following are common areas of leakage.
   (a) Oil pan (Chapters 1 and 7).
   (b) Dipstick tube (Chapters 1 and 7).
   (c) Transmission to oil cooler fluid pipes/unions (Chapter 7).

General gear selection problems

☐ Chapter 7, Part B, deals with checking and adjusting the selector linkage on automatic transmissions. The following are common problems which may be caused by a poorly-adjusted linkage.
   (a) Engine starting in gears other than Park or Neutral.
   (b) Indicator on gear selector lever pointing to a gear other than the one actually being used.
   (c) Vehicle moves when in Park or Neutral.
   (d) Poor gear shift quality, or erratic gear changes.
☐ Refer to Chapter 7, Part B for the selector linkage adjustment procedure.

Transmission fluid brown, or has burned smell

☐ Transmission fluid level low or fluid in need of renewal (Chapter 1).

Transmission will not downshift (kickdown) with accelerator pedal fully depressed

☐ Low transmission fluid level (Chapter 1).
☐ Incorrect selector mechanism adjustment (Chapter 7, Part B).

Engine will not start in any gear, or starts in gears other than Park or Neutral

☐ Incorrect starter/inhibitor switch adjustment (Chapter 7, Part B).
☐ Incorrect selector mechanism adjustment (Chapter 7, Part B).

Transmission slips, shifts roughly, is noisy or has no drive in forward or reverse gears

☐ There are many possible causes for the above problems, but the home mechanic should be concerned with only one - fluid level. Before taking the vehicle to a dealer or transmission specialist, check the fluid level and condition of the fluid as described in Chapter 1. Correct the fluid level as necessary, or change the fluid and filter if needed. If the problem persists, professional help will be necessary.

7 Driveshafts

Clicking or knocking noise on turns (at slow speed on full- lock)

☐ Lack of constant velocity joint lubricant (Chapter 8).
☐ Worn outer constant velocity joint (Chapter 8).

Vibration when accelerating or decelerating

☐ Worn inner constant velocity joint (Chapter 8).
☐ Bent or distorted driveshaft (Chapter 8).

8 Braking system

Note: *Before assuming that a brake problem exists, make sure that the tyres are in good condition and correctly inflated, that the front wheel alignment is correct and that the vehicle is not loaded with weight in an unequal manner. Apart from checking the condition of all pipe and hose connections, any faults occurring on the anti-lock braking system should be referred to a Renault dealer for diagnosis.*

Vehicle pulls to one side under braking

☐ Worn, defective, damaged or contaminated front or rear brake pads/shoes on one side (Chapters 1 and 9).
☐ Seized or partially-seized front or rear brake caliper/wheel cylinder piston (Chapter 9).
☐ A mixture of pad/shoe lining materials fitted (Chapters 1 and 9).
☐ Brake caliper mounting bolts loose (Chapter 9).
☐ Rear brake backplate mounting bolts loose (Chapter 9).
☐ Worn or damaged steering or suspension components (Chapter 10).

Noise (grinding or high-pitched squeal) when brakes applied

☐ Brake pad or shoe friction lining material worn down to metal backing (Chapters 1 and 9).
☐ Excessive corrosion of brake disc or drum. (May be apparent after the vehicle has been standing for some time (Chapters 1 and 9).
☐ Foreign object (stone chipping etc) trapped between brake disc and splash shield (Chapters 1 and 9).

Excessive brake pedal travel

- ☐ Inoperative rear brake self-adjust mechanism (Chapters 1 and 9).
- ☐ Faulty master cylinder (Chapter 9).
- ☐ Air in hydraulic system (Chapter 9).
- ☐ Faulty vacuum servo unit (Chapter 9).

Brake pedal feels spongy when depressed

- ☐ Air in hydraulic system (Chapter 9).
- ☐ Deteriorated flexible rubber brake hoses (Chapters 1 and 9).
- ☐ Master cylinder mounting nuts loose (Chapter 9).
- ☐ Faulty master cylinder (Chapter 9).

High brake pedal effort required to stop vehicle

- ☐ Faulty vacuum servo unit (Chapter 9).
- ☐ Disconnected, damaged or insecure brake servo vacuum hose (Chapter 9).
- ☐ Primary or secondary hydraulic circuit failure (Chapter 9).
- ☐ Seized brake caliper or wheel cylinder piston(s) (Chapter 9).
- ☐ Brake pads or brake shoes incorrectly fitted (Chapters 1 and 9).
- ☐ Incorrect grade of brake pads shoes fitted (Chapters 1 and 9).
- ☐ Brake pads or brake shoe linings contaminated (Chapters 1 and 9).

Judder felt through brake pedal or steering wheel when braking

- ☐ Excessive run-out or distortion of front discs or rear drums (Chapter 9).
- ☐ Brake pad or brake shoe linings worn (Chapters 1 and 9).
- ☐ Brake caliper or rear brake backplate mounting bolts loose (Chapter 9).
- ☐ Wear in suspension or steering components or mountings (Chapter 10).

Brakes binding

- ☐ Seized brake caliper or wheel cylinder piston(s) (Chapter 9).
- ☐ Incorrectly-adjusted handbrake mechanism or linkage (Chapters 1 and 9).
- ☐ Faulty master cylinder (Chapter 9).

Rear wheels locking under normal braking

- ☐ Rear brake shoe linings contaminated (Chapters 1 and 9).
- ☐ Faulty brake pressure regulator (Chapter 9).

9 Suspension and steering systems

Note: *Before diagnosing suspension or steering faults, be sure that the trouble is not due to incorrect tyre pressures, mixtures of tyre types or binding brakes.*

Vehicle pulls to one side

- ☐ Defective tyre (Chapter 1).
- ☐ Excessive wear in suspension or steering components (Chapter 10).
- ☐ Incorrect front wheel alignment (Chapter 10).
- ☐ Accident damage to steering or suspension components (Chapter 10).

Wheel wobble and vibration

- ☐ Front roadwheels out of balance - vibration felt mainly through the steering wheel (Chapter 10).
- ☐ Rear roadwheels out of balance - vibration felt throughout the vehicle (Chapter 10).
- ☐ Roadwheels damaged or distorted (Chapter 10).
- ☐ Faulty or damaged tyre (Chapter 1).
- ☐ Worn steering or suspension joints, bushes or components (Chapter 10).
- ☐ Wheel bolts loose (Chapter 10).

Excessive pitching and/or rolling around corners or during braking

- ☐ Defective shock absorbers (Chapter 10).
- ☐ Broken or weak coil spring and/or suspension component (Chapter 10).
- ☐ Worn or damaged anti-roll bar or mountings (Chapter 10).

Wandering or general instability

- ☐ Incorrect front wheel alignment (Chapter 10).
- ☐ Worn steering or suspension joints, bushes or components (Chapter 10).
- ☐ Roadwheels out of balance (Chapter 10).
- ☐ Faulty or damaged tyre (Chapter 1).
- ☐ Wheel bolts loose (Chapter 10).
- ☐ Defective shock absorbers (Chapter 10).

Excessively-stiff steering

- ☐ Lack of steering gear lubricant (Chapter 10).
- ☐ Seized tie-rod end balljoint or suspension balljoint (Chapter 10).
- ☐ Broken or incorrectly-adjusted power steering pump drivebelt (Chapter 12).
- ☐ Incorrect front wheel alignment (Chapter 10).
- ☐ Steering rack or column bent or damaged (Chapter 10).

Excessive play in steering

- ☐ Worn steering column universal joint(s) or intermediate coupling (Chapter 10).
- ☐ Worn steering tie-rod end balljoints (Chapter 10).
- ☐ Worn rack-and-pinion steering gear (Chapter 10).
- ☐ Worn steering or suspension joints, bushes or components (Chapter 10).

Lack of power assistance

- ☐ Broken or incorrectly-adjusted power steering pump drivebelt (Chapter 12).
- ☐ Incorrect power steering fluid level (Chapter 1).
- ☐ Restriction in power steering fluid hoses (Chapter 10).
- ☐ Faulty power steering pump (Chapter 10).
- ☐ Faulty rack-and-pinion steering gear (Chapter 10).

Tyre wear excessive

Tyres worn on inside or outside edges

- ☐ Tyres under-inflated - wear on both edges (Chapter 1).
- ☐ Incorrect camber or castor angles - wear on one edge only (Chapter 10).
- ☐ Worn steering or suspension joints, bushes or components (Chapter 10).
- ☐ Excessively-hard cornering.
- ☐ Accident damage.

Tyre treads exhibit feathered edges

- ☐ Incorrect toe setting (Chapter 10).

Tyres worn in centre of tread

- ☐ Tyres over-inflated (Chapter 1).

Tyres worn on inside and outside edges

- ☐ Tyres under-inflated (Chapter 1).

Tyres worn unevenly

- ☐ Tyres out of balance (Chapter 1).
- ☐ Excessive wheel or tyre run-out (Chapter 1).
- ☐ Worn shock absorbers (Chapter 10).
- ☐ Faulty tyre (Chapter 1).

10 Electrical system

Note: *For problems associated with the starting system, refer to the faults listed under 'Engine' earlier in this Section.*

Battery will not hold a charge for more than a few days

- ☐ Battery defective internally (Chapter 12).
- ☐ Battery electrolyte level low (Chapter 1).
- ☐ Battery terminal connections loose or corroded (Chapter 1).
- ☐ Alternator drivebelt worn or incorrectly adjusted (Chapter 12).
- ☐ Alternator not charging at correct output (Chapter 12).
- ☐ Alternator or voltage regulator faulty (Chapter 12).
- ☐ Short-circuit causing continual battery drain (Chapter 12).

Ignition (no-charge) warning light remains illuminated with engine running

- ☐ Alternator drivebelt broken, worn, or incorrectly adjusted (Chapter 12).
- ☐ Alternator brushes worn, sticking, or dirty (Chapter 12).
- ☐ Alternator brush springs weak or broken (Chapter 12).
- ☐ Internal fault in alternator or voltage regulator (Chapter 12).
- ☐ Broken, disconnected, or loose wiring in charging circuit (Chapter 12).

Ignition (no-charge) warning light fails to come on

- ☐ Warning light bulb blown (Chapter 12).
- ☐ Broken, disconnected, or loose wiring in warning light circuit (Chapter 12).
- ☐ Alternator faulty (Chapter 12).

Lights inoperative

- ☐ Bulb blown (Chapter 12).
- ☐ Corrosion of bulb or bulbholder contacts (Chapter 12).
- ☐ Blown fuse (Chapter 12).
- ☐ Faulty relay (Chapter 12).
- ☐ Broken, loose, or disconnected wiring (Chapter 12).
- ☐ Faulty switch (Chapter 12).

Instrument readings inaccurate or erratic

Instrument readings increase with engine speed

- ☐ Faulty voltage regulator (Chapter 12).

Fuel or temperature gauges give no reading

- ☐ Faulty gauge sender unit (Chapters 3 or 4).
- ☐ Wiring open-circuit (Chapter 12).
- ☐ Faulty gauge (Chapter 12).

Fuel or temperature gauges give continuous maximum reading

- ☐ Faulty gauge sender unit (Chapters 3 or 4).
- ☐ Wiring short-circuit (Chapter 12).
- ☐ Faulty gauge (Chapter 12).

Horn inoperative, or unsatisfactory in operation

Horn operates all the time

- ☐ Horn push either earthed or stuck down (Chapter 12).
- ☐ Horn cable to horn push earthed (Chapter 12).

Horn fails to operate

- ☐ Blown fuse (Chapter 12).
- ☐ Cable or cable connections loose, broken or disconnected (Chapter 12).
- ☐ Faulty horn (Chapter 12).

Horn emits intermittent or unsatisfactory sound

- ☐ Cable connections loose (Chapter 12).
- ☐ Horn mountings loose (Chapter 12).
- ☐ Faulty horn (Chapter 12).

Wipers inoperative, or unsatisfactory in operation

Wipers fail to operate or operate very slowly

- ☐ Wiper blades stuck to screen, or linkage seized or binding (Chapter 12).
- ☐ Blown fuse (Chapter 12).
- ☐ Cable or cable connections loose or disconnected (Chapter 12).
- ☐ Faulty relay (Chapter 12).
- ☐ Faulty wiper motor (Chapter 12).

Wipers sweep over too large or too small an area of the glass

- ☐ Wiper arms incorrectly positioned on spindles (Chapter 12).
- ☐ Excessive wear of wiper linkage (Chapter 12).
- ☐ Wiper motor or linkage mountings loose or insecure (Chapter 12).

Wiper blades fail to clean the glass effectively

- ☐ Wiper blade rubbers worn or perished (Chapter 1).
- ☐ Wiper arm springs broken or arm pivots seized (Chapter 12).
- ☐ Insufficient windscreen washer additive to adequately remove road film (Chapter 1).

Washers inoperative, or unsatisfactory in operation

One or more washer jets inoperative

- ☐ Blocked washer jet (Chapter 12).
- ☐ Disconnected, kinked or restricted fluid hose (Chapter 12).
- ☐ Insufficient fluid in washer reservoir (Chapter 1).

Washer pump fails to operate

- ☐ Broken or disconnected wiring or connections (Chapter 12).
- ☐ Blown fuse (Chapter 12).
- ☐ Faulty washer switch (Chapter 12).
- ☐ Faulty washer pump (Chapter 12).
- ☐ Fluid in washer reservoir frozen.

Electric windows unsatisfactory in operation

Window glass will only move in one direction

- ☐ Faulty switch (Chapter 11).

Window glass slow to move

- ☐ Incorrectly-adjusted door glass guide channels (Chapter 11).
- ☐ Regulator damaged, or in need of lubrication (Chapter 11).
- ☐ Door internal components or trim fouling regulator (Chapter 11).
- ☐ Faulty motor (Chapter 11).

Window glass fails to move

- ☐ Incorrectly-adjusted door glass guide channels (Chapter 11).
- ☐ Blown fuse (Chapter 12).
- ☐ Faulty relay (Chapter 12).
- ☐ Broken or disconnected wiring or connections (Chapter 12).
- ☐ Faulty motor (Chapter 11).

Central locking unsatisfactory in operation

Complete system failure

- ☐ Blown fuse (Chapter 12).
- ☐ Faulty relay (Chapter 12).
- ☐ Broken or disconnected wiring or connections (Chapter 12).

Latch locks but will not unlock, or unlocks but will not lock

- ☐ Faulty master switch (Chapter 11).
- ☐ Disconnected latch operating rods or levers (Chapter 11).
- ☐ Faulty relay (Chapter 12).

One solenoid/motor fails to operate

- ☐ Broken or disconnected wiring or connections (Chapter 12).
- ☐ Faulty motor (Chapter 11).
- ☐ Binding or disconnected latch operating rods or levers (Chapter 11).
- ☐ Fault in door latch (Chapter 11).

A

ABS (Anti-lock brake system) A system, usually electronically controlled, that senses incipient wheel lockup during braking and relieves hydraulic pressure at wheels that are about to skid.

Air bag An inflatable bag hidden in the steering wheel (driver's side) or the dash or glovebox (passenger side). In a head-on collision, the bags inflate, preventing the driver and front passenger from being thrown forward into the steering wheel or windscreen.

Air cleaner A metal or plastic housing, containing a filter element, which removes dust and dirt from the air being drawn into the engine.

Air filter element The actual filter in an air cleaner system, usually manufactured from pleated paper and requiring renewal at regular intervals.

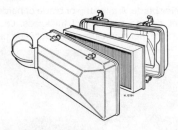

Air filter

Allen key A hexagonal wrench which fits into a recessed hexagonal hole.

Alligator clip A long-nosed spring-loaded metal clip with meshing teeth. Used to make temporary electrical connections.

Alternator A component in the electrical system which converts mechanical energy from a drivebelt into electrical energy to charge the battery and to operate the starting system, ignition system and electrical accessories.

Ampere (amp) A unit of measurement for the flow of electric current. One amp is the amount of current produced by one volt acting through a resistance of one ohm.

Anaerobic sealer A substance used to prevent bolts and screws from loosening. Anaerobic means that it does not require oxygen for activation. The Loctite brand is widely used.

Antifreeze A substance (usually ethylene glycol) mixed with water, and added to a vehicle's cooling system, to prevent freezing of the coolant in winter. Antifreeze also contains chemicals to inhibit corrosion and the formation of rust and other deposits that would tend to clog the radiator and coolant passages and reduce cooling efficiency.

Anti-seize compound A coating that reduces the risk of seizing on fasteners that are subjected to high temperatures, such as exhaust manifold bolts and nuts.

Asbestos A natural fibrous mineral with great heat resistance, commonly used in the composition of brake friction materials.

Asbestos is a health hazard and the dust created by brake systems should never be inhaled or ingested.

Axle A shaft on which a wheel revolves, or which revolves with a wheel. Also, a solid beam that connects the two wheels at one end of the vehicle. An axle which also transmits power to the wheels is known as a live axle.

Axleshaft A single rotating shaft, on either side of the differential, which delivers power from the final drive assembly to the drive wheels. Also called a driveshaft or a halfshaft.

B

Ball bearing An anti-friction bearing consisting of a hardened inner and outer race with hardened steel balls between two races.

Bearing The curved surface on a shaft or in a bore, or the part assembled into either, that permits relative motion between them with minimum wear and friction.

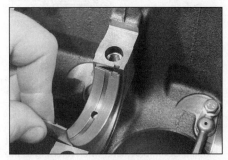

Bearing

Big-end bearing The bearing in the end of the connecting rod that's attached to the crankshaft.

Bleed nipple A valve on a brake wheel cylinder, caliper or other hydraulic component that is opened to purge the hydraulic system of air. Also called a bleed screw.

Brake bleeding Procedure for removing air from lines of a hydraulic brake system.

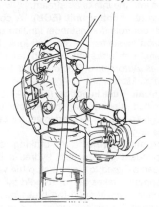

Brake bleeding

Brake disc The component of a disc brake that rotates with the wheels.

Brake drum The component of a drum brake that rotates with the wheels.

Brake linings The friction material which contacts the brake disc or drum to retard the vehicle's speed. The linings are bonded or riveted to the brake pads or shoes.

Brake pads The replaceable friction pads that pinch the brake disc when the brakes are applied. Brake pads consist of a friction material bonded or riveted to a rigid backing plate.

Brake shoe The crescent-shaped carrier to which the brake linings are mounted and which forces the lining against the rotating drum during braking.

Braking systems For more information on braking systems, consult the *Haynes Automotive Brake Manual*.

Breaker bar A long socket wrench handle providing greater leverage.

Bulkhead The insulated partition between the engine and the passenger compartment.

C

Caliper The non-rotating part of a disc-brake assembly that straddles the disc and carries the brake pads. The caliper also contains the hydraulic components that cause the pads to pinch the disc when the brakes are applied. A caliper is also a measuring tool that can be set to measure inside or outside dimensions of an object.

Camshaft A rotating shaft on which a series of cam lobes operate the valve mechanisms. The camshaft may be driven by gears, by sprockets and chain or by sprockets and a belt.

Canister A container in an evaporative emission control system; contains activated charcoal granules to trap vapours from the fuel system.

Canister

Carburettor A device which mixes fuel with air in the proper proportions to provide a desired power output from a spark ignition internal combustion engine.

Castellated Resembling the parapets along the top of a castle wall. For example, a castellated balljoint stud nut.

Castor In wheel alignment, the backward or forward tilt of the steering axis. Castor is positive when the steering axis is inclined rearward at the top.

Catalytic converter A silencer-like device in the exhaust system which converts certain pollutants in the exhaust gases into less harmful substances.

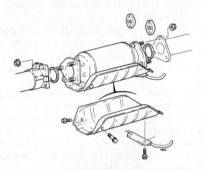

Catalytic converter

Circlip A ring-shaped clip used to prevent endwise movement of cylindrical parts and shafts. An internal circlip is installed in a groove in a housing; an external circlip fits into a groove on the outside of a cylindrical piece such as a shaft.

Clearance The amount of space between two parts. For example, between a piston and a cylinder, between a bearing and a journal, etc.

Coil spring A spiral of elastic steel found in various sizes throughout a vehicle, for example as a springing medium in the suspension and in the valve train.

Compression Reduction in volume, and increase in pressure and temperature, of a gas, caused by squeezing it into a smaller space.

Compression ratio The relationship between cylinder volume when the piston is at top dead centre and cylinder volume when the piston is at bottom dead centre.

Constant velocity (CV) joint A type of universal joint that cancels out vibrations caused by driving power being transmitted through an angle.

Core plug A disc or cup-shaped metal device inserted in a hole in a casting through which core was removed when the casting was formed. Also known as a freeze plug or expansion plug.

Crankcase The lower part of the engine block in which the crankshaft rotates.

Crankshaft The main rotating member, or shaft, running the length of the crankcase, with offset "throws" to which the connecting rods are attached.

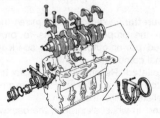

Crankshaft assembly

Crocodile clip See Alligator clip

D

Diagnostic code Code numbers obtained by accessing the diagnostic mode of an engine management computer. This code can be used to determine the area in the system where a malfunction may be located.

Disc brake A brake design incorporating a rotating disc onto which brake pads are squeezed. The resulting friction converts the energy of a moving vehicle into heat.

Double-overhead cam (DOHC) An engine that uses two overhead camshafts, usually one for the intake valves and one for the exhaust valves.

Drivebelt(s) The belt(s) used to drive accessories such as the alternator, water pump, power steering pump, air conditioning compressor, etc. off the crankshaft pulley.

Accessory drivebelts

Driveshaft Any shaft used to transmit motion. Commonly used when referring to the axleshafts on a front wheel drive vehicle.

Drum brake A type of brake using a drum-shaped metal cylinder attached to the inner surface of the wheel. When the brake pedal is pressed, curved brake shoes with friction linings press against the inside of the drum to slow or stop the vehicle.

E

EGR valve A valve used to introduce exhaust gases into the intake air stream.

Electronic control unit (ECU) A computer which controls (for instance) ignition and fuel injection systems, or an anti-lock braking system. For more information refer to the *Haynes Automotive Electrical and Electronic Systems Manual*.

Electronic Fuel Injection (EFI) A computer controlled fuel system that distributes fuel through an injector located in each intake port of the engine.

Emergency brake A braking system, independent of the main hydraulic system, that can be used to slow or stop the vehicle if the primary brakes fail, or to hold the vehicle stationary even though the brake pedal isn't depressed. It usually consists of a hand lever that actuates either front or rear brakes mechanically through a series of cables and linkages. Also known as a handbrake or parking brake.

Endfloat The amount of lengthwise movement between two parts. As applied to a crankshaft, the distance that the crankshaft can move forward and back in the cylinder block.

Engine management system (EMS) A computer controlled system which manages the fuel injection and the ignition systems in an integrated fashion.

Exhaust manifold A part with several passages through which exhaust gases leave the engine combustion chambers and enter the exhaust pipe.

F

Fan clutch A viscous (fluid) drive coupling device which permits variable engine fan speeds in relation to engine speeds.

Feeler blade A thin strip or blade of hardened steel, ground to an exact thickness, used to check or measure clearances between parts.

Feeler blade

Firing order The order in which the engine cylinders fire, or deliver their power strokes, beginning with the number one cylinder.

Flywheel A heavy spinning wheel in which energy is absorbed and stored by means of momentum. On cars, the flywheel is attached to the crankshaft to smooth out firing impulses.

Free play The amount of travel before any action takes place. The "looseness" in a linkage, or an assembly of parts, between the initial application of force and actual movement. For example, the distance the brake pedal moves before the pistons in the master cylinder are actuated.

Fuse An electrical device which protects a circuit against accidental overload. The typical fuse contains a soft piece of metal which is calibrated to melt at a predetermined current flow (expressed as amps) and break the circuit.

Fusible link A circuit protection device consisting of a conductor surrounded by heat-resistant insulation. The conductor is smaller than the wire it protects, so it acts as the weakest link in the circuit. Unlike a blown fuse, a failed fusible link must frequently be cut from the wire for replacement.

G

Gap The distance the spark must travel in jumping from the centre electrode to the side electrode in a spark plug. Also refers to the spacing between the points in a contact breaker assembly in a conventional points-type ignition, or to the distance between the reluctor or rotor and the pickup coil in an electronic ignition.

Adjusting spark plug gap

Gasket Any thin, soft material - usually cork, cardboard, asbestos or soft metal - installed between two metal surfaces to ensure a good seal. For instance, the cylinder head gasket seals the joint between the block and the cylinder head.

Gasket

Gauge An instrument panel display used to monitor engine conditions. A gauge with a movable pointer on a dial or a fixed scale is an analogue gauge. A gauge with a numerical readout is called a digital gauge.

H

Halfshaft A rotating shaft that transmits power from the final drive unit to a drive wheel, usually when referring to a live rear axle.

Harmonic balancer A device designed to reduce torsion or twisting vibration in the crankshaft. May be incorporated in the crankshaft pulley. Also known as a vibration damper.

Hone An abrasive tool for correcting small irregularities or differences in diameter in an engine cylinder, brake cylinder, etc.

Hydraulic tappet A tappet that utilises hydraulic pressure from the engine's lubrication system to maintain zero clearance (constant contact with both camshaft and valve stem). Automatically adjusts to variation in valve stem length. Hydraulic tappets also reduce valve noise.

I

Ignition timing The moment at which the spark plug fires, usually expressed in the number of crankshaft degrees before the piston reaches the top of its stroke.

Inlet manifold A tube or housing with passages through which flows the air-fuel mixture (carburettor vehicles and vehicles with throttle body injection) or air only (port fuel-injected vehicles) to the port openings in the cylinder head.

J

Jump start Starting the engine of a vehicle with a discharged or weak battery by attaching jump leads from the weak battery to a charged or helper battery.

L

Load Sensing Proportioning Valve (LSPV) A brake hydraulic system control valve that works like a proportioning valve, but also takes into consideration the amount of weight carried by the rear axle.

Locknut A nut used to lock an adjustment nut, or other threaded component, in place. For example, a locknut is employed to keep the adjusting nut on the rocker arm in position.

Lockwasher A form of washer designed to prevent an attaching nut from working loose.

M

MacPherson strut A type of front suspension system devised by Earle MacPherson at Ford of England. In its original form, a simple lateral link with the anti-roll bar creates the lower control arm. A long strut - an integral coil spring and shock absorber - is mounted between the body and the steering knuckle. Many modern so-called MacPherson strut systems use a conventional lower A-arm and don't rely on the anti-roll bar for location.

Multimeter An electrical test instrument with the capability to measure voltage, current and resistance.

N

NOx Oxides of Nitrogen. A common toxic pollutant emitted by petrol and diesel engines at higher temperatures.

O

Ohm The unit of electrical resistance. One volt applied to a resistance of one ohm will produce a current of one amp.

Ohmmeter An instrument for measuring electrical resistance.

O-ring A type of sealing ring made of a special rubber-like material; in use, the O-ring is compressed into a groove to provide the sealing action.

Overhead cam (ohc) engine An engine with the camshaft(s) located on top of the cylinder head(s).

Overhead valve (ohv) engine An engine with the valves located in the cylinder head, but with the camshaft located in the engine block.

Oxygen sensor A device installed in the engine exhaust manifold, which senses the oxygen content in the exhaust and converts this information into an electric current. Also called a Lambda sensor.

P

Phillips screw A type of screw head having a cross instead of a slot for a corresponding type of screwdriver.

Plastigage A thin strip of plastic thread, available in different sizes, used for measuring clearances. For example, a strip of Plastigage is laid across a bearing journal. The parts are assembled and dismantled; the width of the crushed strip indicates the clearance between journal and bearing.

Plastigage

Propeller shaft The long hollow tube with universal joints at both ends that carries power from the transmission to the differential on front-engined rear wheel drive vehicles.

Proportioning valve A hydraulic control valve which limits the amount of pressure to the rear brakes during panic stops to prevent wheel lock-up.

R

Rack-and-pinion steering A steering system with a pinion gear on the end of the steering shaft that mates with a rack (think of a geared wheel opened up and laid flat). When the steering wheel is turned, the pinion turns, moving the rack to the left or right. This movement is transmitted through the track rods to the steering arms at the wheels.

Radiator A liquid-to-air heat transfer device designed to reduce the temperature of the coolant in an internal combustion engine cooling system.

Refrigerant Any substance used as a heat transfer agent in an air-conditioning system. R-12 has been the principle refrigerant for many years; recently, however, manufacturers have begun using R-134a, a non-CFC substance that is considered less harmful to the ozone in the upper atmosphere.

Rocker arm A lever arm that rocks on a shaft or pivots on a stud. In an overhead valve engine, the rocker arm converts the upward movement of the pushrod into a downward movement to open a valve.

Glossary of Technical Terms

Rotor In a distributor, the rotating device inside the cap that connects the centre electrode and the outer terminals as it turns, distributing the high voltage from the coil secondary winding to the proper spark plug. Also, that part of an alternator which rotates inside the stator. Also, the rotating assembly of a turbocharger, including the compressor wheel, shaft and turbine wheel.

Runout The amount of wobble (in-and-out movement) of a gear or wheel as it's rotated. The amount a shaft rotates "out-of-true." The out-of-round condition of a rotating part.

S

Sealant A liquid or paste used to prevent leakage at a joint. Sometimes used in conjunction with a gasket.

Sealed beam lamp An older headlight design which integrates the reflector, lens and filaments into a hermetically-sealed one-piece unit. When a filament burns out or the lens cracks, the entire unit is simply replaced.

Serpentine drivebelt A single, long, wide accessory drivebelt that's used on some newer vehicles to drive all the accessories, instead of a series of smaller, shorter belts. Serpentine drivebelts are usually tensioned by an automatic tensioner.

Serpentine drivebelt

Shim Thin spacer, commonly used to adjust the clearance or relative positions between two parts. For example, shims inserted into or under bucket tappets control valve clearances. Clearance is adjusted by changing the thickness of the shim.

Slide hammer A special puller that screws into or hooks onto a component such as a shaft or bearing; a heavy sliding handle on the shaft bottoms against the end of the shaft to knock the component free.

Sprocket A tooth or projection on the periphery of a wheel, shaped to engage with a chain or drivebelt. Commonly used to refer to the sprocket wheel itself.

Starter inhibitor switch On vehicles with an automatic transmission, a switch that prevents starting if the vehicle is not in Neutral or Park.

Strut See MacPherson strut.

T

Tappet A cylindrical component which transmits motion from the cam to the valve stem, either directly or via a pushrod and rocker arm. Also called a cam follower.

Thermostat A heat-controlled valve that regulates the flow of coolant between the cylinder block and the radiator, so maintaining optimum engine operating temperature. A thermostat is also used in some air cleaners in which the temperature is regulated.

Thrust bearing The bearing in the clutch assembly that is moved in to the release levers by clutch pedal action to disengage the clutch. Also referred to as a release bearing.

Timing belt A toothed belt which drives the camshaft. Serious engine damage may result if it breaks in service.

Timing chain A chain which drives the camshaft.

Toe-in The amount the front wheels are closer together at the front than at the rear. On rear wheel drive vehicles, a slight amount of toe-in is usually specified to keep the front wheels running parallel on the road by offsetting other forces that tend to spread the wheels apart.

Toe-out The amount the front wheels are closer together at the rear than at the front. On front wheel drive vehicles, a slight amount of toe-out is usually specified.

Tools For full information on choosing and using tools, refer to the *Haynes Automotive Tools Manual*.

Tracer A stripe of a second colour applied to a wire insulator to distinguish that wire from another one with the same colour insulator.

Tune-up A process of accurate and careful adjustments and parts replacement to obtain the best possible engine performance.

Turbocharger A centrifugal device, driven by exhaust gases, that pressurises the intake air. Normally used to increase the power output from a given engine displacement, but can also be used primarily to reduce exhaust emissions (as on VW's "Umwelt" Diesel engine).

U

Universal joint or U-joint A double-pivoted connection for transmitting power from a driving to a driven shaft through an angle. A U-joint consists of two Y-shaped yokes and a cross-shaped member called the spider.

V

Valve A device through which the flow of liquid, gas, vacuum, or loose material in bulk may be started, stopped, or regulated by a movable part that opens, shuts, or partially obstructs one or more ports or passageways. A valve is also the movable part of such a device.

Valve clearance The clearance between the valve tip (the end of the valve stem) and the rocker arm or tappet. The valve clearance is measured when the valve is closed.

Vernier caliper A precision measuring instrument that measures inside and outside dimensions. Not quite as accurate as a micrometer, but more convenient.

Viscosity The thickness of a liquid or its resistance to flow.

Volt A unit for expressing electrical "pressure" in a circuit. One volt that will produce a current of one ampere through a resistance of one ohm.

W

Welding Various processes used to join metal items by heating the areas to be joined to a molten state and fusing them together. For more information refer to the *Haynes Automotive Welding Manual*.

Wiring diagram A drawing portraying the components and wires in a vehicle's electrical system, using standardised symbols. For more information refer to the *Haynes Automotive Electrical and Electronic Systems Manual*.

Note: *References throughout this index relate to Chapter•page number*

Preserving Our Motoring Heritage

< *The Model J Duesenberg Derham Tourster. Only eight of these magnificent cars were ever built – this is the only example to be found outside the United States of America*

Almost every car you've ever loved, loathed or desired is gathered under one roof at the Haynes Motor Museum. Over 300 immaculately presented cars and motorbikes represent every aspect of our motoring heritage, from elegant reminders of bygone days, such as the superb Model J Duesenberg to curiosities like the bug-eyed BMW Isetta. There are also many old friends and flames. Perhaps you remember the 1959 Ford Popular that you did your courting in? The magnificent 'Red Collection' is a spectacle of classic sports cars including AC, Alfa Romeo, Austin Healey, Ferrari, Lamborghini, Maserati, MG, Riley, Porsche and Triumph.

A Perfect Day Out

Each and every vehicle at the Haynes Motor Museum has played its part in the history and culture of Motoring. Today, they make a wonderful spectacle and a great day out for all the family. Bring the kids, bring Mum and Dad, but above all bring your camera to capture those golden memories for ever. You will also find an impressive array of motoring memorabilia, a comfortable 70 seat video cinema and one of the most extensive transport book shops in Britain. The Pit Stop Cafe serves everything from a cup of tea to wholesome, home-made meals or, if you prefer, you can enjoy the large picnic area nestled in the beautiful rural surroundings of Somerset.

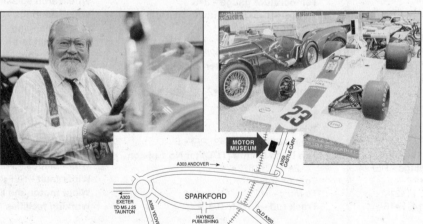

> *John Haynes O.B.E., Founder and Chairman of the museum at the wheel of a Haynes Light 12.*

< *Graham Hill's Lola Cosworth Formula 1 car next to a 1934 Riley Sports.*

The Museum is situated on the A359 Yeovil to Frome road at Sparkford, just off the A303 in Somerset. It is about 40 miles south of Bristol, and 25 minutes drive from the M5 intersection at Taunton.
Open 9.30am - 5.30pm (10.00am - 4.00pm Winter) 7 days a week, *except Christmas Day, Boxing Day and New Years Day*
Special rates available for schools, coach parties and outings Charitable Trust No. 292048